McGraw-Hill Circuit Encyclopedia and Troubleshooting Guide

McGraw-Hill Circuit Encyclopedia and Troubleshooting Guide

Volume 3

John D. Lenk

McGraw-Hill

New York San Francisco Washington, D.C. Auckland Bogotá
Caracas Lisbon London Madrid Mexico City Milan
Montreal New Delhi San Juan Singapore
Sydney Tokyo Toronto

McGraw-Hill

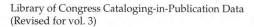

A Division of The McGraw·Hill Companies

Library of Congress Cataloging-in-Publication Data
(Revised for vol. 3)

Lenk, John D.
    McGraw-Hill circuit encyclopedia and troubleshooting
guide.

    Cover title: McGraw-Hill circuit encyclopedia &
troubleshooting guide.
    Includes indexes.
    1. Electronic circuits. 2. Electronic circuits—Test-
ing. I. McGraw-Hill circuit encyclopedia & troubleshoot-
ing guide. II. Title.
TK7867.L463 1993   621.3815   92-33275
ISBN 0-07-037603-4 (v. 1)
ISBN 0-07-038076-7 (pbk.)
ISBN 0-07-037610-7 (v. 2)
ISBN 0-07-037611-5 (pbk.)
ISBN 0-07-037716-2 (v. 3)
ISBN 0-07-037717-0 (pbk.)

1 2 3 4 5 6 7 8 9   DOC/DOC   9 0 0 9 8 7 6

ISBN 0-07-037717-0

*The sponsoring editor of this book was Steve Chapman, the editor was B.J. Peterson, the super-
vising editor was Andrew Yoder and the executive editor was Lori Flaherty. The production
supervisor was Katherine G. Brown. This book was set in Palatino. It was composed in Blue
Ridge Summit, Pa.*

Printed and bound by R.R. Donnelley and Sons, Crawsfordsville, Indiana.

McGraw-Hill books are available at special quantity discounts to use as premiums
and sales promotions, or for use in corporate training programs. For more
information, please write to the Director of Special Sales, McGraw-Hill, 11 West 19th
Street, New York, NY 10011. Or contact your local bookstore.

Product or brand names used in this book may be trade names or trademarks. Where
we believe that there may be proprietary claims to such trade names or trademarks, the
name has been used with an initial capital or it has been capitalized in the style used by
the name claimant. Regardless of the capitalization used, all such names have been
used in an editorial manner without any intent to convey endorsement of or other
affiliation with the name claimant. Neither the author nor the publisher intends to
express any judgment as to the validity or legal status of any such proprietary claims.

Dedication

Greetings from the Villa Buttercup!
To my wonderful wife, Irene,
thank you for being by my side all these years!
To my lovely family, Karen, Tom, Brandon, and Justin.
And to our Lambie and Suzzie, be happy wherever you are!
To my special readers: May good fortune find your doorway, bringing you good
health and happy things. Thank you for buying my books!
And special thanks to Steve Chapman, Stephen Fitzgerald, Leslie Wenger, Patrick
Hansard, Kate Hertzog, Roland Phelps, George Corey, Florence Trimble, Janet
Gomolson, Gemma Velten, Fran Minerva, Jane Stark, Ann Wilson, Andrew Yoder
(best-selling author), and Robert McGraw of McGraw-Hill for making me an
international bestseller again!
This is book number 85.
Abundance!

Acknowledgments

Many professionals have contributed to this book. I gratefully acknowledge the tremendous effort needed to produce this book. Such a comprehensive work is impossible for one person, and I thank all who contributed, both directly and indirectly.

I give special thanks to the following: Alan Haun of Analog Devices, Syd Coppersmith of Dallas Semiconductor, Rosie Hinojosa of EXAR Corporation, Jeff Salter of GEC Plessey, John Allen, Helen Cox, and Linda da Costa of Harris Semiconductors, Ron Denchfield and Bob Scott of Linear Technology Corporation, David Fullagar and William Levin of Maxim Integrated Products, Fred Swymer of Microsemi Corporation, Linda Capcara of Motorola, Inc., Andrew Jenkins and Shantha Natarajan of National Semiconductor, Antonio Ortiz of Optical Electronics Incorporated, Lawrence Fogel of Philips Semiconductors, John Marlow of Raytheon Electronics Semiconductor Division, Anthony Armstrong of Semtech Corporation, Ed Oxner and Robert Decker of Siliconix Incorporated, Amy Sullivan of Texas Instruments, and Alan Campbell of Unitrode Corporation.

I also wish to thank Joseph A. Labok of Los Angeles Valley College for help and encouragement throughout the years.

And a very special thanks to Steve Chapman, Stephen Fitzgerald, Leslie Wenger, Patrick Hansard, Kate Hertzog, Roland Phelps, George Corey, Florence Trimble, Janet Gomolson, Gemma Velten, Fran Minerva, Jane Stark, Melanie Holscher, Fred Perkins, Robert McGraw, Barbara McCann, Kimberly Martin, Jane Schmidt, Tracy Baer, Judith Reiss, Charles Love, Betty Crawford, Jeanne Myers, Peggy Lamb, Thomas Kowalczyk, Suzanne Barbeuf, Jaclyn Boone, Katherine Brown, Kathy Green, Jen Priest, Donna Namorato, Susan Kagey, Bob Ostrander, Lori Flaherty, Stacey Spurlock, Midge Haramis, B.J. Peterson, and Andrew Yoder (best-selling author) of the McGraw-Hill Professional Publishing organization for having that much confidence in me.

And to Irene, my wife and Super Agent, I extend my thanks. Without her help, this book could not have been written.

Contents

Introduction

When you have finished reading this encyclopedia, you should be able to recognize more than 700 circuits that are commonly used in all phases of electronics. You also will understand how the circuits operate and where they fit into electronic equipment and systems. This information alone makes the book an excellent one-stop source or reference for anyone (student, experimenter, technician, or designer) who is involved with electronic circuits.

However, this book is much more than a collection of circuits and descriptions. First, all circuits are grouped by function. With each functional group, the book has a practical guide for testing and troubleshooting that type of circuit. Thus, if you are building a circuit, and the circuit fails to perform as outlined, you are given a specific troubleshooting approach to locate problems.

Second, you can put the circuit to work as it is, because actual circuits with proven component values are given in full detail. On those circuits where performance (such as frequency range, power outputs, etc.) depend on circuit values, you are given information as to how circuit values can be selected to meet a certain performance goal. This information is particularly useful to the student or experimenter, but it also can be an effective time-saver for the designer.

Circuit sources
and addresses

The source for each circuit is given at the end of the circuit description. The source information includes the publication title and date, as well as the page number or numbers. This information makes it possible for you to contact the original source for further information on the circuit or circuit component. To this end, the complete mailing address and telephone numbers for each source are included in this section. When writing, give complete information, including publication date and page number of the original source. Notice that all circuit diagrams have been reproduced directly from the original source, without redrawing, by permission of the original publisher in each case.

AIE Magnetics
701 Murfreeboro Road
Nashville, TN 37210
(615) 244-9024

Analog Devices
One Technology Way
PO Box 9106
Norwood, MA 02062-9106
(617) 329-4700
Fax (617) 326-8703

Dallas Semiconductor
4401 S. Beltwood Parkway
Dallas, TX 75244-3292
(214) 450-0400

EXAR Corporation
2222 Qume Drive
PO Box 49007
San Jose, CA 95161-9007
(408) 434-6400
Fax (408) 943-8245

GEC Plessey Semiconductors
Cheney Manor
Swindon, Wiltshire
United Kingdom SN2 2QW
0793 51800
Fax 0793 518411

Harris Semiconductor
PO Box 883
Melbourne, FL 32902-0883
(407) 724-7000
Fax (407) 724-3937
1-800-442-7747

Linear Technology Corporation
1630 McCarthy Boulevard
Milpitas, CA 95036-7487
(408) 432-1900
Fax (408) 434-0507
1-800-637-5545

Magnetics Division of Spang and
  Company
900 East Butler
PO Box 391
Butler, PA 16003
(412) 282-8282

Maxim Integrated Products
120 San Gabriel Drive
Sunnyvale, CA 94086
(408) 737-7600
Fax (408) 737-7194
1-800-998-8800

Motorola, Inc.
Semiconductor Products Sector
Public Relations Department
5102 N. 56th Street
Phoenix, AZ 85018
(602) 952-3000

National Semiconductor Corporation
2900 Semiconductor Drive
PO Box 58090
Santa Clara, CA 95052-8090
(408) 721-5000
1-800-272-9959

Optical Electronics, Inc.
PO Box 11140
Tucson, AZ 85734
(602) 889-8811

Philips Semiconductors
811 E. Arques Avenue
PO Box 3409
Sunnyvale, CA 94088-3409
(408) 991-2000

Raytheon Company Semiconductor
  Division
350 Ellis Street
PO Box 7106
Mountain View, CA 94039-7016
(415) 968-9211
Fax (415) 966-7742
1-800-722-7074

Semtech Corporation
652 Mitchell Road
Newbury Park, CA 91320
(805) 498-2111

Siliconix Incorporated
2201 Laurelwood Road
Santa Clara, CA 95054
(408) 988-8000

Unitrode Corporation
8 Suburban Park Drive
Billerica, MA 01821
(508) 670-9086

Substitutions and cross-reference tables

Substitutions can often be made for the semiconductor and IC types that are specified on the circuit diagrams. Newer components, not available when the original source was published, might actually improve the performance of the circuit. Electrical characteristics, terminal connections, and such critical ratings as voltage, current, frequency, and duty cycle must, of course, be taken into account if experimenting without referring to substitution guides.

Semiconductor and IC substitution guides can usually be purchased at electronic parts-supply stores. In the absence of any substitution guides, the following cross-reference tables will help in locating possible substitute ICs.

General Cross References

INDUSTRY TYPE	RAYTHEON DIRECT REPLACEMENT	RAYTHEON FUNCTIONAL REPLACEMENT	INDUSTRY TYPE	RAYTHEON DIRECT REPLACEMENT	RAYTHEON FUNCTIONAL REPLACEMENT
ADVFC32		RC4153	ICL7660		RC4391
ADOP07	OP-07		ICL7680		RC4190
ADOP27	OP-27		ICL8013		RC4200
ADOP37	OP-37		LF155	LF155	
ADREF01	REF-01		LF156	LF156	
ADREF02	REF-02		LF157	LF157	
AD101	LM101		LH2101	LH2101	
AD558		DAC-4888	LH2108	LH2108	
AD565	DAC-8565		LH2111	LH2111	
AD581		REF-01	LM101	LM101	
AD586		REF-02	LM111	LM111	
AD647		RC4207	LM108	LM108	
AD654		RC4152	LM124	LM124	
AD707		RC4077	LM148	LM148	
AD708		RC4277	LM324	LM324	
AD741	RC741		LM331		RC4152
AD767		DAC-4881	LM348	LM348	
AM686		RC4805	LM368-5.0		REF-02
AM6012	DAC-6012		LM368-10		REF-01
CA124	LM124		LM369		REF-01
CA324	LM324		LM607		RC4077
CA139	LM139		LM741	RC741	
CA339	LM339		LM833	RC5532	
CA741	RC741		LM1458		RC4558
CS3842		RC4190	LM1851	LM1851	
CMP-04		LM139	LM1851		RC4145
CMP-05		RC4805	LM2900	LM2900	
DAC-08	DAC-08		LM2901		LM339
DAC-10	DAC-10		LM2902		LM324
DAC-80		DAC-4881	LM3900	LM3900	
DAC-100		DAC-10	LP165	LP165	
DAC-312	DAC-6012		LP365	LP365	
DAC0800	DAC-08		LT-1001	LT-1001	
DAC0801	DAC-08		LT-1012	LT-1012	
DAC0830		DAC-4888	LT-1012		RC4097
DAC-888		DAC-4888	LT-1019		REF-01
DAC1208		DAC-4881	LT-1019		REF-02
DAC1218		DAC-6012	LT-1024		RC4207
DAC1219		DAC-6012	LT-1028		OP-37
DAC1230		DAC-4881	LT-1054		RC4391
DAC8222		DAC-4881	LT-1070		RC4190
HA-OP27	OP-07		LT-1084		RC4292
HA-OP27	OP-27		MAX400		RC4077
HA-OP37	OP-37		MAX630	RC4193	
HA-3182	RC3182		MAX630		RC4190
HA-4741	RC4741		MAX634	RC4391	
HA-5147		OP-47	MC1741	RC741	
HSOP07	OP-07		MC1747	RC747	
HSOP27	OP-27		MC3403	RC3403	
HSOP37	OP-37		MC4558	RC4558	

RAYTHEON

General Cross References (Continued)

INDUSTRY TYPE	RAYTHEON DIRECT REPLACEMENT	RAYTHEON FUNCTIONAL REPLACEMENT	INDUSTRY TYPE	RAYTHEON DIRECT REPLACEMENT	RAYTHEON FUNCTIONAL REPLACEMENT
MC4741	RC4741		SG741	RC741	
MPREF01	REF-01		SI-9100		RC4292
MPREF02	REF-02		SSM-2134		RC5534
MPOP07	OP-07		TA7504	RC741	
MPOP27	OP-27		TA75339	LM339	
MPOP37	OP-37		TL494		RC4190
MP108	LM108		TL496		RC4190
MP155	LM155		TL497		RC4190
MP156	LM156		TL510		RC4805
MP157	LM157		TSC9400		RC4151
NE5532	RC5532		TSC9401		RC4151
NE5534	RC5534		TSC9402		RC4151
OPA156		LM156	UC1842		RC4292
OPA27		OP-27	VFC-32		RC4153
OPA37		OP-37	XR-2207	XR-2207	
OP-02		RC741	XR-2208		RC4200
OP-04		RC747	XR-2211	XR-2211	
OP-07	OP-07		XR-3403	RC3403	
OP-14		RC4558	XR-4136	RC4136	
OP-16		LF156	XR-4194	RC4194	
OP-27	OP-27		XR-4195	RC4195	
OP-37	OP-37		XR-5532	RC5532	
OP-77	OP-77		XR-5534	RC5534	
OP-97		RC4097	µA101	LM101	
OP-200		RC4207, RC4277	µA108	LM108	
			µA111	LM111	
OP-207		RC4207	µA124	LM124	
OP-227		RC4227	µA139	LM139	
OP-270		RC4227	µA148	LM148	
PM-108	LM108		µA324	LM324	
PM-139	LM139		µA339	LM339	
PM-148	LM148		µA348	LM348	
PM-155	LM155		µA741	RC741	
PM-156	LM156		µA747	RC747	
PM-157	LM157				
PM-339	LM339				
PM-348	LM348				
PM-741	RC741				
PM-747	RC747				
RC4136	RC4136				
RC4151	RC4151				
RC4152	RC4152				
RC4558	RC4558				
RC4559	RC4559				
REF-01	REF-01				
REF-02	REF-02				
REF-05		REF-02			
REF-10		REF-01			
SE5534		RC5534			
SG101	LM101				
SG124	LM124				

RAYTHEON

Precision Operational Amplifier Cross Reference

ANALOG DEV.	RAYTHEON	PACKAGE	ANALOG DEV.	RAYTHEON	PACKAGE
AD OP-07AH	*OP-07AT	TO-99	AD OP-37AH/883	OP-37AT/883B	TO-99
AD OP-07AH/883	*OP-07AT/883B	TO-99	AD OP-37AQ	OP-37AD	CERAMIC
AD OP-07CN	*OP-07CN	PLASTIC	AD OP-37AQ/883	OP-37AD/883B	CERAMIC
AD OP-07CR	*OP-07CM	SO-8	AD OP-37BH	OP-37BT	TO-99
AD OP-07Q/883	*OP-07D/883B	CERAMIC	AD OP-37BH/883	OP-37BT/883B	TO-99
AD OP-07DN	*OP-07DN	PLASTIC	AD OP-37BQ	OP-37BD	CERAMIC
AD OP-07EN	*OP-07EN	PLASTIC	AD OP-37BQ/883	OP-37BD/883B	CERAMIC
AD OP-07H	*OP-07T	TO-99	AD OP-37CH	OP-37CT	TO-99
AD OP-07H/883	*OP-07T/883B	TO-99	AD OP-37CH/883	OP-37CT/883B	TO-99
AD OP-07Q	*OP-07D	CERAMIC	AD OP-37CQ	OP-37CD	CERAMIC
AD OP-07AQ	*OP-07AD	CERAMIC	AD OP-37CQ/883	OP-37CD/883B	CERAMIC
AD OP-07AQ/883B	*OP-07AD/883B	CERAMIC	AD OP-37EN	OP-37EN	PLASTIC
			AD OP-37FN	OP-37FN	PLASTIC
AD OP-27AH	OP-27AT	TO-99	AD OP-37GN	OP-37GN	PLASTIC
AD OP-27AH/883	OP-27AT/883B	TO-99			
AD OP-27AQ	OP-27AD	CERAMIC	AD707AQ	*RC4077FD	CERAMIC
AD OP-27AQ/883	OP-27AD/883B	CERAMIC	AD707CH	*RM4077AT	TO-99
AD OP-27BH	OP-27BT	TO-99	AD707CH/883	*RM4077AT/883B	TO-99
AD OP-27BH/883	OP-27BT/883B	TO-99	AD707CQ	*RM4077AD	CERAMIC
AD OP-27BQ	OP-27BD	CERAMIC	AD707CQ/883	*RM4077AD/883B	CERAMIC
AD OP-27BQ/883	OP-27BD/883B	CERAMIC	AD707JN	*RC4077FN	PLASTIC
AD OP-27CH	OP-27CT	TO-99	AD707JR	*RC4077FM	SO-8
AD OP-27CH/883	OP-27CT/883B	TO-99	AD707KN	*RC4077EN	PLASTIC
AD OP-27CQ	OP-27CD	CERAMIC	AD707KR	*RC4077EM	SO-8
AD OP-27CQ/883	OP-27CD/883B	CERAMIC	AD707SH	*RC4077AT	TO-99
AD OP-27EN	OP-27EN	PLASTIC	AD707SH/883B	*RC4077AT/883B	TO-99
AD OP-27FN	OP-27FN	PLASTIC	AD707SQ	*RC4077AD	CERAMIC
AD OP-27GN	OP-27GN	PLASTIC	AD707SQ/883	*RC4077AD/883B	CERAMIC
			AD707TH	*RC4077AT	TO-99
AD OP-37AE	OP-37AL	LCC	AD707TH/883B	*RC4077AT/883B	TO-99
AD OP-37AE/883	OP-37AL/883B	LCC	AD707TQ	*RC4077AD	CERAMIC
AD OP-37AH	OP-37AT	TO-99	AD707TQ/883	*RC4077AD/883B	CERAMIC

BURR BROWN	RAYTHEON	PACKAGE	BURR BROWN	RAYTHEON	PACKAGE
OPA27AJ/883	*OP-27AT/883B	TO-99	OPA37AJ	*OP-37AT	TO-99
OPA27BJ/883	*OP-27BT/883B	TO-99	OPA37AJ/883	*OP-37AT/883B	TO-99
OPA27CJ	*OP-27CT/883B	TO-99	OPA37AZ	*OP-37AD	CERAMIC
OPA27AJ	*OP-27AT	TO-99	OPA37AZ/883	*OP-37AD/883B	CERAMIC
OPA27AZ	*OP-27AD	CERAMIC	OPA37BJ	*OP-37BT	TO-99
OPA27BJ	*OP-27BT	TO-99	OPA37BJ/883	*OP-37BT/883B	TO-99
OPA27BZ	*OP-27BD	CERAMIC	OPA37BZ	*OP-37BD	CERAMIC
OPA27CJ	*OP-27CT	TO-99	OPA37BZ/883	*OP37-BD/883B	CERAMIC
OPA27CZ	*OP-27CD	CERAMIC	OPA37CJ	*OP-37CT	TO-99
OPA27EP	*OP-27EN	PLASTIC	OPA37CJ/883	*OP-37CT/883B	TO-99
OPA27FP	*OP-27FN	PLASTIC	OPA37CJ/883	*OP-37CD/883B	CERAMIC
OPA27GP	*OP-27GN	PLASTIC	OPA37CZ	*OP-37CD	CERAMIC
OPA27GU	*OP-27GM	SO-8	OPA37EP	*OP-37EN	PLASTIC
OPA27GZ	*OP-27GD	CERAMIC	OPA37FP	*OP-37FN	PLASTIC
OPA27AZ/883	*OP-27AD/883B	CERAMIC	OPA37GP	*OP-37GN	PLASTIC
OPA27BZ/883	*OP-27BD/883B	CERAMIC	OPA37GU	*OP-27GM	SO-8
OPA27CZ/883	*OP-27CD/883B	CERAMIC			

* Denotes functionally equivalent types.

RAYTHEON

Precision Operational Amplifier Cross Reference (Continued)

LTC	RAYTHEON	PACKAGE	LTC	RAYTHEON	PACKAGE
OP-07AH	OP-07AT	TO-99	LM108AH	LM108AT	TO-99
OP-07AH/883B	OP-07AT/883B	TO-99	LM108AH/883B	LM108AT/883B	TO-99
OP-07AJ8	OP-07AD	CERAMIC	LM108AJ8/883B	LM108AD/883B	CERAMIC
OP-07AJ8/883B	OP-07AD/883B	CERAMIC	LM108H	LM108T	TO-99
OP-07CN8	OP-07CN	PLASTIC	LM108H/883B	LM108T/883B	TO-99
OP-07CS8	OP-07CM	SO-8	LM108J8/883B	LM108D/883B	CERAMIC
OP-07EN8	OP-07EN	PLASTIC			
OP-07H	OP-07T	TO-99	LT1001ACH	LT-1001ACT	TO-99
OP-07H/883B	OP-07T/883B	TO-99	LT1001ACN8	LT-1001ACN	PLASTIC
OP-07J8	OP-07D	CERAMIC	LT1001AMH/883B	LT-1001AMT/883B	TO-99
OP-07J8/883B	OP-07D/883B	CERAMIC	LT1001AMJ8	LT-1001AMD	CERAMIC
			LT1001AMJ8/883	LT-1001AMD/883B	CERAMIC
OP-27AH	OP-27AT	TO-99	LT1001CH	LT-1001CT	TO-99
OP-27AH/883B	OP-27AT/883B	TO-99	LT1001CN8	LT-1001CN	PLASTIC
OP-27AJ8	OP-27AD	CERAMIC	LT1001CS8	LT-1001CM	SO-8
OP-27AJ8/883B	OP-27AD/883B	CERAMIC	LT1001MH	LT-1001MT	TO-99
OP-27CH	OP-27CT	TO-99	LT1001MH/883B	LT-1001MT/883B	TO-99
OP-27CH/883B	OP-27CT/883B	TO-99	LT1001MJ8	LT-1001MD	CERAMIC
OP-27CJ8	OP-27CD	CERAMIC	LT1001MJ8/883B	LT-1001MD/883B	CERAMIC
OP-27CJ8/883B	OP-27CD/883B	CERAMIC			
OP-27EN8	OP-27EN	PLASTIC	OP-227EN	*RC4227FN	PLASTIC
OP-27GN8	OP-27GN	PLASTIC	OP-227GN	*RC4227GN	PLASTIC
			OP-227AJ	*RM4227BD	CERAMIC
OP-37AH	OP-37AT	TO-99	OP-227AJ/883B	*RM4227BD/883B	CERAMIC
OP-37AH/883B	OP-37AT/883B	TO-99			
OP-37AJ8	OP-37AD	CERAMIC			
OP-37AJ8/883B	OP-37AD/883B	CERAMIC			
OP-37CH	OP-37CT	TO-99			
OP-37CH/883B	OP-37CT/883B	TO-99			
OP-37CJ8	OP-37CD	CERAMIC			
OP-37CJ8/883B	OP-37CD/883B	CERAMIC			
OP-37EN8	OP-37EN	PLASTIC			
OP-37GN8	OP-37GN	PLASTIC			

*Denotes functionally equivalent types.
NOTE: LTC OP-227 contains two die in a 14-pin package.
Raytheon's 4227 is a monolithic IC in an 8-pin package.

RAYTHEON

Precision Operational Amplifier Cross Reference (Continued)

PMI	RAYTHEON	PACKAGE	PMI	RAYTHEON	PACKAGE
OP07AJ	OP-07AT	TO-99	OP77AJ	OP-77AT	TO-99
OP07AJ/883	OP-07AT/883B	TO-99	OP77AJ/883	OP-77AT/883B	TO-99
OP07AZ	OP-07AD	CERAMIC	OP77AZ	OP-77AD	CERAMIC
OP07AZ/883	OP-07AD/883B	CERAMIC	OP77AZ/883	OP-77AD/883B	CERAMIC
OP07CP	OP-07CN	PLASTIC	OP77BJ	OP-77BT	TO-99
OP07CS	OP-07CM	SO-8	OP77BJ/883	OP-77BT/883B	TO-99
OP07DP	OP-07DN	PLASTIC	OP77BRC/883	OP-77BL/883B	LCC
OP07DS	OP-07DM	SO-8	OP77BZ	OP-77BD	CERAMIC
OP07EP	OP-07EN	PLASTIC	OP77BZ/883	OP-77BD/883B	CERAMIC
OP07J	OP-07T	TO-99	OP77EP	OP-77EN	PLASTIC
OP07J/883	OP-07T/883B	TO-99	OP77FP	OP-77FN	PLASTIC
OP07RC/883	OP-07L/883B	LCC	OP77FS	OP-77FM	SO-8
OP07Z	OP-07D	CERAMIC	OP77GP	OP-77GN	PLASTIC
OP07Z/883	OP-07D/883B	CERAMIC	OP77GS	OP-77GM	SO-8
OP27AJ	OP-27AT	TO-99	PM108AZ	LM108AD	CERAMIC
OP27AJ/883	OP-27AT/883B	TO-99	PM108AZ/883	LM108AD/883B	CERAMIC
OP27AZ	OP-27AD	CERAMIC	PM108AJ	LM108AT	TO-99
OP27AZ/883	OP-27AD/883B	CERAMIC	PM108AJ/883	LM108AT/883B	TO-99
OP27BJ	OP-27BT	TO-99	PM108ARC	LM108AL	LCC
OP27BJ/883	OP-27BT/883B	TO-99	PM108ARC/883	LM108AL/883B	LCC
OP27BRC/883	OP-27BL/883B	LCC	PM108DZ	LM108D	CERAMIC
OP27BZ	OP-27BD	CERAMIC	PM108DZ/883	LM108D/883B	CERAMIC
OP27BZ/883	OP-27BD/883B	CERAMIC	PM108J	LM108T	TO-99
OP27CJ	OP-27CT	TO-99	PM108J/883	LM108T/883B	TO-99
OP27CJ/883	OP-27CT/883B	TO-99			
OP27CZ	OP-27CD	CERAMIC	PM2108AQ	LH2108AD	CERAMIC
OP27CZ/883	OP-27CD/883B	CERAMIC	PM2108AQ/883	LH2108AD/883B	CERAMIC
OP27EP	OP-27EN	PLASTIC	PM2108Q	LH2108D	CERAMIC
OP27FP	OP-27FN	PLASTIC	PM2108Q/883	LH2108D/883B	CERAMIC
OP27FS	OP-27FM	SO-8			
OP27GS	OP-27GM	SO-8	OP207AY/883	*RM4207BD/883B	CERAMIC
OP27GP	OP-27GN	PLASTIC	OP207AY	*RM4207BD	CERAMIC
OP37AJ	OP-37AT	TO-99	OP227AY	*RM4227BD	CERAMIC
OP37AJ/883	OP-37AT/883B	TO-99	OP227AY/883	*RM4227BD/883B	CERAMIC
OP37AZ	OP-37AD	CERAMIC	OP227BY/883	*RM4227BD/883B	CERAMIC
OP37AZ/883	OP-37AD/883B	CERAMIC	OP227GY	*RC4227GN	PLASTIC
OP37BJ	OP-37BT	TO-99			
OP37BJ/883	OP-37BT/883B	TO-99			
OP37BRC/883	OP-37BL/883B	LCC			
OP37BZ	OP-37BD	CERAMIC			
OP37BZ/883	OP-37BD/883B	CERAMIC			
OP37CJ	OP-37CT	TO-99			
OP37CJ/883	OP-37CT/883B	TO-99			
OP37CZ	OP-37CD	CERAMIC			
OP37CZ/883	OP-37CD/883B	CERAMIC			
OP37EP	OP-37EN	PLASTIC			
OP37FP	OP-37FN	PLASTIC			

* Denotes functionally equivalent types.
NOTE: PMI's OP207/227 contains two die in a 14-pin package.
Raytheon's 4207/4227 is a monolithic IC in an 8-pin package.

RAYTHEON

General Purpose Operational Amplifier Cross Reference

Raytheon	PMI	FSC	AMD	Motorola	National	RCA	Signetics	T.I.
LH2101A			LH2101A		LH2101A		LH2101A	
LH2111			LH2111		LH2111			
LM101A		μA101A	LM101A	LM101A	LM101A	CA101A	LM101A	
LM111		μA111	LM111	LM111	LM111	CA111	LM111	
LM124		μA124	LM124	LM124	LM124	CA124	LM124	LM124
LM139	PM139	μA139	LM139	LM139	LM139	CA139	LM139	LM139
LM148	PM148	μA148	LM148		LM148		LM148	
LM301A		μA301A	LM301A	LM301A	LM301A	CA301A	LM301A	LM301A
LM324		μA324	LM324	LM324	LM324	CA324	LM324	LM324
LM339	PM339	μA339	LM339	LM339	LM339	CA339	LM339	LM339
LM348		μA348	LM348		LM348		LM348	LM348
LM2900		μA2900			LM2900			
LM3900		μA3900			LM3900			LM3900
RC3403A		μA3403		MC3403				MC3403
RC4136	OP-09	μA4136						RC4136
RC4156		μA148*		MC4741	LM348*			LM348*
RC4157		μA148/348*		MC4741*	LM348*			LM348*
RC4558		μA4558		MC4558				RC4558
RC4559		μA4558*		MC4558*				RC4559
RC4741N				MC3-4741-5				
RM4741D				MC1-4741-2				
RC5532							NE5532	NE5532
RC5532A							NE5532A	NE5532A
RC5534							NE5534	NE5534
RC5534A							NE5534A	NE5534A
RC741	OP-02	μA741		MC1741	LM741	CA741	CA741	
RC747	OP-04	μA747		MC1747	LM747	CA747	CA747	
RC747S	OP-04	μA747			LM747			

*Functional Equivalent

RAYTHEON

Data Conversion Cross Reference

Raythen	PMI	AMD	Motorola	NSC	Devices	Analog Power	Micro-Datel
DAC-08AD	DAC-08AQ	AMDAC-08AQ	MC1408L8	DAC-08AQ	AD-1508-9D	MP-7523*	DAC-IC8BC*
DAC-08D	DAC-08Q	AMDAC-08Q		DAC-08Q	AD-1508-9D	MP-7523*	DAC-IC8BC*
DAC-08ED	DAC-08EQ	AMDAC-08EQ		DAC-08EQ	AD-1408-8D	MP-7523*	DAC-IC8UP*
DAC-08EN	DAC-08EP	AMDAC-08EN		DAC-08EP			DAC-IC8UP*
DAC-08CN	DAC-08CP	AMDAC-08CN	MC1408P6	DAC-08CP			DAC-IC8UP*
DAC-10BD	DAC-10BX			DAC-1020 LD*	AD7520/30/33*	MP-7520/30/33*	DAC- HF10BMM*
DAC-10CD	DAC-10CX			DAC-1021/22LD8*	AD7520/30/33*	MP-7520/30/33*	DAC- HF10BMM*
DAC-10FD	DAC-10FX			DAC-1020 LCN*	AD7520/30/33*	MP-7520 30/33*	DAC- HF10BMC*
DAC-10GD	DAC-10GX			DAC-1021/22LCN*	AD7520/30/33*	MP-7520/30/33*	DAC-HF10BMC*
DAC- 6012AMD		AM6012ADM		DAC-1220 LD*	AD6012ADM	MP-7531/41*	DAC-HF12BMM*
DAC- 6012MD	DAC-312 BR*	AM6012DM		DAC-1221/22LD*	AD6012DM	MP-7531/41*	DAC- HF12BMM*
DAC- 6012ACN		AM6012ADC		DAC-1220 LCN*	AD6012ADC	MP-7531/41*	DAC- HF12BMC*
DAC- 6012CN	DAC-312FR*	AM6012DC		DAC-1221/22LCN*	AD6012DC	MP-7531/41*	DAC- HF12BMC*
DAC-8565DS*				MC3412L	DAC-1208 AD-I*	AD565JD/BIN	
DAC-8565JS*				MC3412L	DAC-1280 HCD-I*	AD565JD/BIN	
DAC-8565SS*					DAC-1280 HCD-I*	AD565SD/BIN	

*Functional Equivalent

RAYTHEON

Special Functions Cross Reference

Raytheon	Teledyne	Analog Devices	EXAR	Motorola	Datel	Burr Brown
RC4151	4780*	AD451*	XR4151		VFQ-1C*	VFC-32KF*
RC4152	4781*	AD452*	XR4151*		VFQ-2C*	VFC-42BP*
RC4153	4782*	AD537*			VFQ-3C*	VFC-52BP*
RC4200/A		AD539*		MC1494*		4202K* & 4205K*
XR2207			XR2207			
XR2211			XR2211			
RC4444				MC3416		

*Functional Equivalent

Voltage Regulator and Voltage Reference Cross Reference

Raytheon	EXAR	Maxim	T.I.	Analog Devices	Motorola	NSC
REF-01	REF-01		MP-5501	AD581*	MC1504AU10*	LH0070-0*
REF-01A	REF-01A		MP-5501A	AD581*		LH0070-1*
REF-01C	REF-01C		MP-5501C	AD581*	MC1404U10*	LH0070-2*
REF-01D	REF-01D		MP-5501D	AD581*	MC1404U10*	
REF-01E	REF-01E		MP-5501E	AD581*		
REF-01H	REF-01H		MP-5501H	AD581*	MC1404AU10*	
REF-02	REF-02		MP-5502		MC1504AU5*	LM136-5.0*
REF-02A	REF-02A		MP-5502A			LM136A-5.0*
REF-02C	REF-02C		MP-5502C		MC1404U5*	LM336-5.0*
REF-02D	REF-02D		MP-5502D		MC1404U5*	LM336-5.0*
REF-02E	REF-02E		MP-5502E			LM336A-5.0*
REF-02H	REF-02H		MP-5502H		MC1404AU5*	
RC4190		MAX630*				
RC4193		MAX630*				
RC4391		MAX634*				
RC4194	XR4194CN					
RC4195	XR4195CP				MC1468/ MC1568*	LM325/326*

*Functional Equivalent

RAYTHEON

Typical IC packages and pin connections

Not all circuits give power connections and pin locations for ICs (integrated circuits), but you can get this information from manufacturers' data sheets. Also, looking through other circuits might turn up another diagram on which the desired connections are shown for the same IC.

The diagrams show a few typical pin connections. Notice that the functions shown in the following diagrams apply only to that specific IC, and are included to show the normal pin-numbering sequence only. As shown, numbering normally starts with 1 (beginning at the top) for the first pin counterclockwise from the notched (or otherwise marked) end and continues in sequence. The highest number is next to the notch (or mark) on the other side of the IC.

Notice that these guides show only the most common pin-connection configurations, including metal can, DIP (dual-in-line package), SO DIP (small-outline DIP) LCC (leadless chip carrier), multipin DIP, and surface mount.

Connection Information

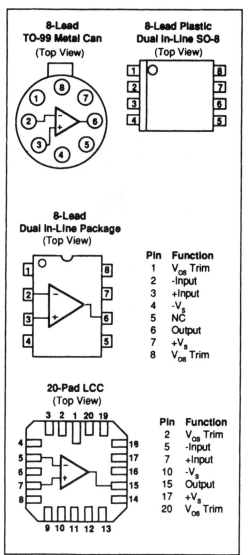

8-Lead TO-99 Metal Can
(Top View)

8-Lead Plastic Dual In-Line SO-8
(Top View)

8-Lead Dual In-Line Package
(Top View)

Pin	Function
1	V_{OS} Trim
2	-Input
3	+Input
4	$-V_s$
5	NC
6	Output
7	$+V_s$
8	V_{OS} Trim

20-Pad LCC
(Top View)

Pin	Function
2	V_{OS} Trim
5	-Input
7	+Input
10	$-V_s$
15	Output
17	$+V_s$
20	V_{OS} Trim

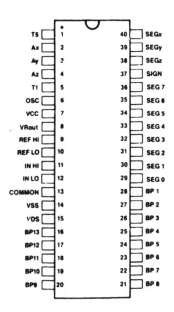

T5	1	40	SEGx
Ax	2	39	SEGy
Ay	3	38	SEGz
Az	4	37	SIGN
T1	5	36	SEG 7
OSC	6	35	SEG 6
VCC	7	34	SEG 5
VRout	8	33	SEG 4
REF HI	9	32	SEG 3
REF LO	10	31	SEG 2
IN HI	11	30	SEG 1
IN LO	12	29	SEG 0
COMMON	13	28	BP 1
VSS	14	27	BP 2
VDS	15	26	BP 3
BP13	16	25	BP 4
BP12	17	24	BP 5
BP11	18	23	BP 6
BP10	19	22	BP 7
BP9	20	21	BP 8

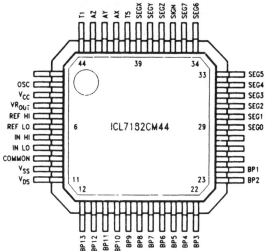

ICL7132CM44

Abbreviations and reference symbols

Most electronics manufacturers and publications outside the United States use some different symbols for electronics components, as well as a different abbreviation system for values or units of measure.

For example, in the illustration, notice the resistor symbol (a rectangular box) at pin 2 of the SL6442. Also notice the capacitor at the same pin. One half of the symbol is solid and the other half is open. The symbol is used wherever polarity must be observed when connecting the capacitor into the circuit. When polarity is of no particular concern, both halves of the symbol are solid, as shown for the capacitor at pin 5 of the SL6442.

Also, the abbreviations for component values are simplified. Thus, μ (Greek letter mu) after a capacitor value represents μF (microfarad), n is nF (nanofarad), and p is pF (picofarad). (The Japanese, and others, often go one step further and use a lowercase u instead of the micro symbol.)

With resistor values, k is thousands of ohms (Ω—Greek letter omega), M is megohms, and the absence of a unit of measure means that ohm is the unit. If μ, n and p are used with an inductance, they represent μH (microhenry), nH (nanohenry), and pH (picohenry). Examples are the 18-nH and 82-nH coils at pin 6 of the SL6442.

For a decimal value, the letter for the unit of measure is sometimes placed at the location of the decimal point. Thus, 3k3 is 3.3 kilohms (kΩ), or 3300 ohms. 2M2 is 2.2 megohms (MΩ), 7μ7 is 7.7 μF, 0μ1 is 0.1 μF, and 3n7 is 3.7 nF.

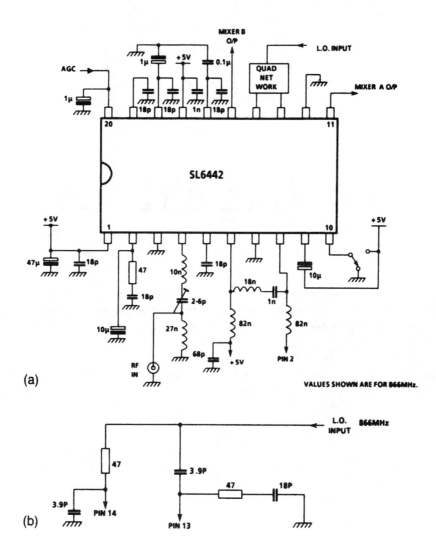

(a)

VALUES SHOWN ARE FOR 866MHz.

(b)

Finding circuits

The circuits are arranged by type or function, with each group assigned to a separate chapter. For example, Chapter 1 contains basic amplifier circuits, Chapter 2 contains switching power supplies, and so on. The Contents lists each chapter in order. The List of figures shows the title of each circuit in order.

To find a particular circuit, begin by noting the chapter in which the circuit is likely to appear. Then look for a title that best describes the circuit you want. For example, if you want amplifiers, look in Chapter 1. If you want power supplies, look in Chapter 2 for switching power supplies or in Chapter 3 for linear power supplies. If you want a power supply that is specifically designed for use in micropower and/or battery-operated equipment, look in Chapter 4.

If you want to test or troubleshoot a circuit, begin with the Contents and find the chapter for the appropriate circuit group. For example, if you want to test power supplies, such tests are described in the first sections of Chapters 2 and 3, including coverage of test equipment and procedures. If the circuit fails to perform properly (fails to meet the tests), the second section of Chapters 2 and 3 describe power-supply circuit troubleshooting (Chapter 2 for switching supplies and Chapter 3 for linear supplies).

Notice that many circuits could appear in more than one chapter. For example, Chapter 1 contains only amplifier circuits. However, Chapter 14 contains some amplifier circuits that are programmable, using a digital-to-analog converter to provide digital control of amplifier gain. Chapter 5 describes only oscillator/generator circuits, but Chapter 14 describes a waveform generator under digital control. So, if you do not locate the circuit you want, even after a careful study of the Contents and List of figures, use the Index at the back of the book. Here, the circuits are indexed under the different name by which they are known, or could possibly be classified. Hundreds of cross references in the Index will aid you in this search.

List of figures

=1=

IC amplifier circuits

This chapter is devoted to IC amplifiers and to those circuits that use IC (integrated circuit) amplifiers as the basic element. Such amplifiers include *op amps* (operational amplifiers), Nortons, *OTAs* (operational transconductance amplifiers), *CFAs* (current-feedback amplifiers), chopper-stabilized amplifiers, *WTAs* (wideband transconductance amplifiers), audio amplifiers, and various special-purpose amplifiers. The discussions assume that you already are familiar with amplifier basics (amplification principles, bias operating points, etc.), practical considerations (heatsinks, power dissipation, component-mounting techniques), and simplified amplifier design (frequency limitations). If you are not familiar with these basics, read *Lenk's Audio Handbook*, McGraw-Hill, 1991, and *Lenk's RF Handbook*, McGraw-Hill, 1991. The following paragraphs summarize both testing and troubleshooting of amplifier circuits. This information is included so readers not familiar with electronic procedures can both test the circuits described here and localize problems if the circuits fail to perform as shown.

Amplifier tests

This section covers the basic tests for IC amplifiers. The section begins with a review of typical amplifier test equipment and then goes on to describe test procedures that can be applied to the circuits of this chapter. Actual circuit test results are used where practical. If the circuits pass these basic tests, use the circuits immediately. If not, use the tests as a starting point for the troubleshooting procedures that are described in this chapter. Keep in mind that all amplifier circuits need not be subjected to all tests described here. However, if an amplifier circuit produces the desired results for all of the tests, you can consider the circuit a successful design.

Amplifier test equipment

The tests described in this chapter can be performed using meters, oscilloscopes, generators, power supplies, assorted clips, patch cords, and so on. So, if you have a good set of test equipment that is suitable for other electronic work, you can probably get by. A possible exception is a distortion meter (especially if you are interested in audio amplifiers). Here are some points that you should consider when selecting and using test equipment.

Matching test equipment to the circuits No matter what test instrument is involved, try to match the capabilities of the test equipment to the circuit. For example, if you are going to measure pulses, square waves, or complex waves (as you might for any IC amplifier test), a peak-to-peak meter can provide meaningful indications, but a scope is the logical instrument.

Voltmeters/multimeters In addition to making routine voltage and resistance checks, the main functions of a meter in amplifier work are to measure frequency response and trace signals from input to output. Many technicians prefer scopes for these procedures. The reason for the preference is that scopes also show distortion of the waveform during measurement or signal tracing. Other technicians prefer the simplicity of a meter, particularly in such procedures as voltage-gain and power-gain measurements.

You can sometimes get by with any ac meter (even a basic multimeter, analog or digital) for all amplifier work. However, for accurate measurements, use a wideband meter, preferably a dual-channel model. (Obviously, the meter must have a bandwidth greater than the amplifier circuit being tested!) The dual-channel feature makes it possible to monitor both channels of a stereo circuit simultaneously. This feature is particularly important for stereo frequency-response and crosstalk measurements but is of no great importance for nonstereo amplifiers.

Scopes If you have a good scope for TV and VCR work, use that scope for all amplifier-circuit measurements. If you are considering a new scope, remember that a *dual-channel* instrument lets you monitor both channels of a stereo circuit (as is the case with a dual-channel voltmeter). Of greater importance, a dual-channel scope lets you monitor the input and output of an amplifier simultaneously. A scope also has the advantage over a meter in that the scope can display such common IC-amplifier conditions as distortion, hum, ripple, overshoot, and oscillation. It is an indispensable tool for measuring such characteristics as settling time, slew rate, and noise. (However, the meter is easier to read when you are measuring only gain.)

Distortion meters If you are already in audio/stereo work, you probably have distortion meters (and know how to use them effectively). There are two types of distortion measurements: *harmonic* and *intermodulation*. No particular meter is described here. Instead, descriptions are included of how harmonic and intermodulation distortion measurements are made.

Decibel measurement basics

The *decibel*, or dB, is widely used in amplifier work to express logarithmically the ratio between two power or voltage levels. For example, a typical IC op-amp data sheet lists voltage gain, power gain, and common-mode rejection ratio in dB. The decibel is one-tenth of a bel. (The bel is too large for most practical applications.)

Although there are many ways to express a ratio, the decibel is used in amplifiers for two reasons: (1) the decibel is a convenient unit to use for all types of amplifiers and (2) the decibel is related to the reaction of the human ear and is thus well suited for use with audio amplifiers.

Humans can listen to ordinary conversation comfortably and can hear thunder (which is 100,000 times louder than conversation) without damage to the ear. This capability exists because the response of the human ear to sound waves is approximately proportional to the logarithm of the sound-wave energy and is not directly proportional to the energy.

The common logarithm ($\log_{10}$) of a number is the number of times 10 must be multiplied by itself to equal that number. For example, the logarithm of 100 (that is 10×10), or (10^2) is 2. Likewise, the logarithm of 100,000 (10^5) is 5. The relationship is written:

$$\log_{10} 100{,}000 = 5$$

In comparing two powers, it is possible to use the bel (which is the logarithm of the ratio of the two powers). For example, in comparing the power of ordinary conversation with that of thunder, the increase in sound is equal to:

$$\log_{10} \frac{power\ of\ thunder}{power\ of\ conversation} \text{ or } \log_{10} \frac{100{,}000}{1}$$

Using the more convenient decibel, the increase in sound from ordinary conversation to thunder is equal to:

$$10 \log_{10} \frac{100{,}000}{1} \text{ or 50 decibels (or 50 dB)}$$

For convenience, the same method is used in measuring the increase in amplifier power, whether the amplifiers are used with audio frequencies or not. The increase in power of any amplifier can be expressed as:

$$gain\ (in\ dB) = 10 \log_{10} \frac{power\ output}{power\ input}$$

This relationship also can be expressed as:

$$gain\ (in\ dB) = 10 \log_{10} \frac{P_2}{P_1}$$

Usually, P_2 represents power output and P_1 represents power input. If P_2 is greater than P_1, there is a power gain, expressed in positive decibels (+dB). With P_1 greater than P_2, there is a power loss, expressed in negative decibels (–dB).

Whichever is the case, the ratio of the two powers (P_1 and P_2) is taken, and the logarithm of this ratio is multiplied by 10. As a result, power ratio of 10 = 10-dB gain, power ratio of 100 = 20-dB gain, power ratio of 1000 = 30-dB gain, and so on.

Doubling power ratios

Doubling the power of an amplifier produces a power gain of +3 dB. For example, if the volume control of an amplifier is turned up so the power rises from 4 to 8 W (watts), the gain is up +3 dB. If the power output is reduced from 4 to 2 W, the gain is down –3 dB.

If the original 4 W is increased to 8 W, the power gain is +3 dB. Increasing the power output further to 16 W produces another gain of +3 dB, with a total power gain of +6 dB. At 40 W, the power is increased 10 times (from the original 4 W), and the total power gain is +10 dB, and so on.

Adding decibels

There is another convenience in using decibels for amplifier work. When several amplifier stages are connected so that one works into another (stages connected in *cascade*, as is the usual case in IC amplifiers), the gains of each stage are multiplied. For example, if three stages each with a gain of 10, are connected in cascade, there is a total power gain of $10 \times 10 \times 10$, or 1000.

In the decibel system, the decibel gains are added. Using the example, the decibel power gain is 10 + 10 + 10, or +30 dB. Similarly, if two amplifier stages are connected, one of which has a gain of +30 dB, and the other a loss of –10 dB, the net result is +30 – 10, or +20 dB.

Using decibels to compare voltages and currents

The decibel system also is used to compare the voltage input and output of an amplifier. (Decibels can be used to express current ratios. However, this is generally not practical in amplifiers.) When voltages (or currents) are involved, the decibel is a function of:

$$20 \log \frac{output\ voltage}{input\ voltage}, \ 20 \log \frac{output\ current}{input\ current}$$

The ratio of the two voltages (or currents) is taken, and the logarithm of this ratio is multiplied by 20.

Note that, although power ratios are independent of source and load impedance values, *voltage* and *current ratios* in those equations hold true only when the source and load impedances are equal. In amplifiers where input and output impedances differ, voltage and current ratios are calculated as:

$$20 \log \frac{E_1 \sqrt{R_2}}{E_2 \sqrt{R_1}}, \ 20 \log \frac{I_1 \sqrt{R_2}}{I_2 \sqrt{R_1}}$$

IC amplifier circuits

where R_1 is the source or input impedance and R_2 the load or output impedance ($E_1 \sqrt{R_2}$ and $I_1 \sqrt{R_2}$ are always higher in value than $E_2 \sqrt{R_1}$ and $I_1 \sqrt{R_1}$).

As is true for the power relationship, if the voltage output is greater than the input, there is a decibel gain (+dB). If the output is less than the input, there is a voltage loss (–dB).

Note that doubling the voltage produces a gain of +6 dB. Conversely, if the voltage is cut in half, there is a loss of –6 dB. To get the net effect of several voltage-amplifier stages working together, add the decibel gains (or losses) of each.

Decibels and reference levels

When an amplifier has a power gain of +20 dB, this has no numerical meaning in actual power output. Instead, it means that the power output is 100 times as great as the power input. For this reason, decibels are often used in specific reference levels.

The most common reference levels for audio amplifiers are the *VU* (volume unit) and the *dBm* (decibel meter).

When VU is used, it is assumed that the zero level is equal to 0.001 W (1 milliwatt or 1 mW) across a 600-Ω (ohm) impedance. Thus:

$$VU = 10 \log \frac{P_2}{0.001} = 10 \log \frac{P_2}{10^{-3}} = 10 \log 10^3\, P_2$$

where P_2 is the output power.

Both the dBm and VU have the same zero level base. A dBm scale is normally found on meters when the signal to be measured is a sine wave (normally 1 kHz—kilohertz), and the VU is used for complex audio waveforms.

Frequency response

You can measure amplifier frequency response with a generator and a meter or scope. Tune the generator to various frequencies and measure the resultant output response at each frequency. Plot the results in the form of a graph or response curve. Figure 1-A shows the test connections to measure *open-loop gain* (A_{OL}) for a typical op amp (a Harris CA3450). *Open-loop gain* applies to the IC-amplifier gain without feedback. Figure 1-B shows the plot (sometimes called a *Bode* plot) or graph for the IC when frequency response is measured in the open-loop condition. (Note that Fig. 1-B also shows the phase shift that occurs at various frequencies.)

Figure 1-C shows the test connections for the same IC when closed-loop gain is being measured (with direct feedback between pins 3 and 6 for unity gain and with resistors at pins 3 and 6 for a voltage gain, or A_V, of 10). Figures 1-D and 1-E show the graphs for closed-loop gain.

The frequency at which the output begins to drop is called the *rolloff point*. The specifications for some IC amplifiers consider the rolloff point to start when the output drops 3 dB below the flat portion of the curve. In the graph of Fig. 1-E, rolloff begins at about 10 MHz and drops 3 dB at about 20 MHz.

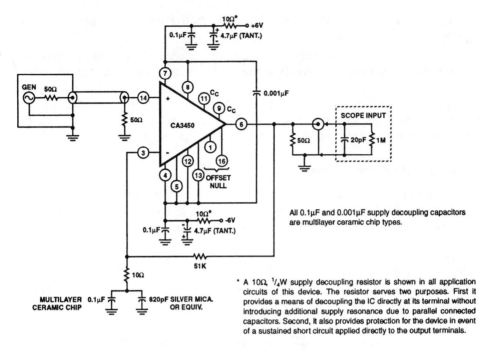

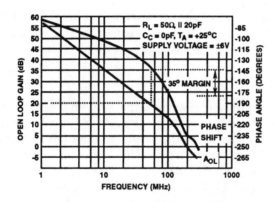

Fig. 1-A Open-loop gain test connections. Harris Semiconductor, Linear & Telecom ICs, 1994, p. 2-217.

Fig. 1-B Bode plot for CA3450. Harris Semiconductor, Linear & Telecom ICs, 1993, p. 2-216.

Some IC amplifiers provide for connection of an external compensation circuit (usually a capacitor but sometimes a capacitor-resistor combination). Such compensation circuits alter both the rolloff point and the gain-frequency relationship. No external compensation is provided in the circuit of Fig. 1-A. The circuit of Fig. 1-C shows an external compensating capacitor connected at pins 9 and 11. Figure 1-D shows the effect of different capacitor (C_c) values on gain and phase shift. Many IC amplifiers have built-in compensation, so open-loop gain-phase characteristics cannot be altered.

IC amplifier circuits

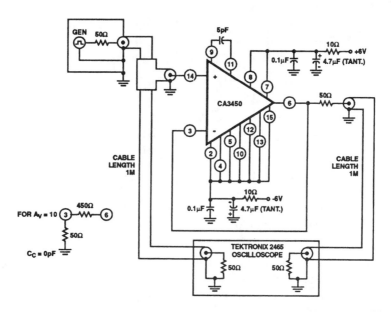

Fig. 1-C   Closed-loop gain test connections. HARRIS SEMICONDUCTOR, LINEAR & TELECOM ICS, 1994, P. 2-218.

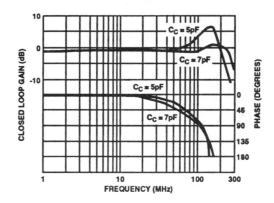

Fig. 1-D   Phase/frequency plot for unity gain.
HARRIS SEMICONDUCTOR, LINEAR & TELECOM ICS, 1994, P. 2-216.

The procedure for measuring frequency response is to apply a constant-amplitude signal while monitoring the output. Vary the input signal frequency (but not amplitude) across the entire operating range of the amplifier. Plot the voltage output at various frequencies across the range on a graph as follows.

1. Connect the equipment as shown in Figs. 1-A or 1-B. Keep in mind that these illustrations show the connections for a specific IC amplifier. However, the circuits have all of the elements of typical gain/frequency test circuit. The test signal is applied to the noninverting input, with the inverting input connected to

Amplifier tests

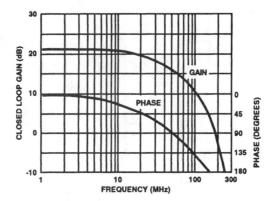

Fig. 1-E   Phase/frequency plot for $A_V = 10$. Harris Semiconductor, Linear & Telecom ICs, 1994, p. 2-216.

the output (or possibly to ground for the open-loop test of some IC amplifiers). The output should be an amplified, noninverted version of the input. Both the input and output are terminated at some specific test value (typically 50 Ω). The input should be terminated in an impedance equal to that of the signal source, with the output terminated at the impedance of the test instrument (scope) input.

Note that the IC amplifier characteristics will change (sometimes drastically) with changes in output load (and external compensation). So, if the data-sheet test values are not close to those of the real load, try testing the IC with real-world values at the output, in addition to the tests with data-sheet values. (You might find that the IC will not meet your particular frequency/gain or rolloff requirements!)

This same precaution applies to tests with and without external compensation. For example, the test circuit of Fig. 1-A shows no external compensation (no connection between pins 9 and 11). The test circuit of Fig. 1-C shows an external capacitor at pins 9/11. By comparing the graphs of Figs. 1-B, 1-D, and 1-E, note that the gain/frequency characteristics are different with and without compensation. For example, the open-loop gain (Fig. 1-B) is about 57 dB at a frequency of 1 MHz but drops to about 10 dB at 100 MHz. When external compensation is used, and the feedback resistors are set for a voltage gain of 10 (20 dB), the voltage gain is flat up to about 10 MHz and then drops off to 10 dB at 100 MHz.

2. Initially, set the generator frequency to the low end of the range—about 1 MHz (megahertz) for the IC amplifier. Then set the generator amplitude to the desired input level. In the absence of a realistic test input voltage, set the generator output to an arbitrary value.

A simple method of finding a satisfactory input level is to monitor the circuit output and increase the generator amplitude at the amplifier center frequency (at 20 or 30 MHz for the IC) until the amplifier is overdriven. The amplifier is overdriven when further increases in generator output do not cause further increases in meter reading (or the output waveform peaks begins to flatten on the scope display). Set the generator output just below this point. Then, return the meter or scope to monitor the generator voltage

IC amplifier circuits

(at the circuit input) and measure the voltage. Keep the generator at this voltage throughout the test.

When making any voltage-output tests, be sure you do not exceed the IC limits. For example, as shown in Fig. 1-F, the IC can deliver an output voltage of about 5 V_{p-p} (volts, peak-to-peak) at a frequency of 20 MHz. If the closed-loop gain is set for 10, and an input signal of 0.6 V (volts) is applied at a frequency of 20 MHz, the output will be 6 V. This condition is beyond the IC capability and will probably result in distortion of the output waveform.

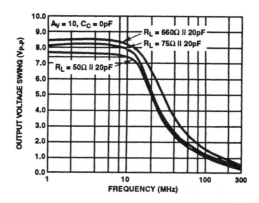

Fig. 1-F   Output voltage versus frequency. Harris Semiconductor, Linear & Telecom ICs, 1994, p. 2-217.

3. If the circuit is provided with any operating or adjustment controls (volume, loudness, gain, treble, bass, balance, and so on), set the controls to some arbitrary point when making the initial frequency-response measurement. The response measurements can then be repeated at different control settings if desired. Although there are no controls as such in this test circuit, the IC is provided with an offset-null feature at pins 1 and 16. If there is an unbalance in the differential input of the IC or if there is a level shift in the stages following the input, the output might not be zero with a zero input. A voltage applied at pins 1 and 16 can be adjusted to correct this condition.

Figure 1-G shows the open-loop frequency-response test circuit for another IC amplifier (the Harris CA3094A) where an offset null is used. In this circuit the input is shorted, and the voltage applied at pin 2 is adjusted for a zero output. Note that the IC of Fig. 1-G is an *OTA* (operational transconductance amplifier). OTAs are discussed further in this chapter.

4. Record the amplifier output voltage on the graph. Without changing the generator amplitude, increase the generator frequency by some fixed amount and record the new amplifier output voltage. The amount of frequency increase is arbitrary and usually depends on the frequency range of the IC. In the IC, an increase of 1 MHz between measurements (up to about 10 MHz) is realistic. At frequencies higher than 10 MHz, an increase of 10 MHz between measurements should be satisfactory. Of course, small increases between measurements will

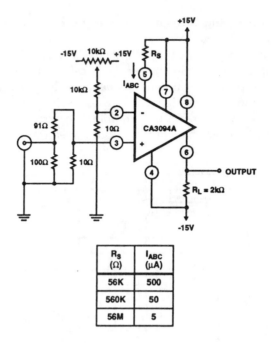

Fig. 1-G Open-loop gain test connections for an OTA.
HARRIS SEMICONDUCTOR, LINEAR & TELECOM ICS, 1994, P. 2-94.

show up any abnormalities (such as *peaking*). Note that the gain peaks at frequencies between 100 and 200 MHz for the IC (Fig. 1-D).

5. After the initial frequency-response measurement, check the effect of operating or adjustment controls (if any). For example, in an audio amplifier, the volume, loudness, and gain controls should have the same effect across the entire frequency range. Treble and bass controls might have some effect on all frequencies. However, a treble control should have the greatest effect at the high end, but bass controls should be most effective at the low end.

6. Remember that generator output can vary with changes in frequency (a fact that is possibly overlooked in making frequency-response tests). Monitor the generator output amplitude after each change in frequency. It is essential that the generator output amplitude remain constant over the entire frequency range of the test.

Voltage gain

Voltage gain for an amplifier is measured in the same way as frequency response. The ratio of output voltage to input voltage (at any given frequency across the entire frequency range) is the voltage gain. Because the input voltage (generator output) is held constant for a frequency-response test, a voltage-gain curve should be identical to a frequency-response curve (such as shown in Figs. 1-B, 1-D, and 1-E). Keep in mind that the voltage gain shown by Fig. 1-B depends primarily on the IC, but the gain shown in Figs. 1-D and 1-E is set by external factors (feedback resistors).

Power output and power gain

Power output of an amplifier is found by noting the output voltage across the load resistance, at any frequency across the entire frequency range. For example, in the circuit of Fig. 1-A, if the output voltage is 7 V, the power is:

$$P = \frac{E^2}{R} = \frac{7^2}{50} = 0.98 \text{ W}$$

As a practical matter, never use a wire-wound component (or any component that has reactance) for the load resistance. Reactance changes with frequency, and that causes the load to change. Use a composition resistor or potentiometer for the load.

To find the *power gain* of an amplifier, begin by finding both the input and output power. You can find input power in the same way as output, but you must know or calculate the input impedance. Calculating input impedance is not always practical in some circuits, especially in designs where input impedance depends on transistor gain. (The procedure for finding dynamic input impedance is described further in this chapter.) With input power known (or estimated), the power gain is the ratio of output power to input power. When the input is terminated in a resistance that is far lower than the amplifier input impedance, you can use the value of the input terminating resistance to find input power.

Input sensitivity

In some IC amplifiers, an input-sensitivity specification is used in place of or in addition to power-output/gain specifications. *Input sensitivity* implies a minimum power output with a given voltage input, such as 3-W output with a 100-mV (milliwatt) input. Input sensitivity usually applies to power IC amplifiers. To find input sensitivity, simply apply the specified input and note the actual power output.

Bandwidth

Some specifications require that the IC amplifier deliver a given voltage or power output across a given frequency range. Usually, the voltage bandwidth is not the same as the power bandwidth. For example, an amplifier might produce full-power output up to 1 MHz, even though the frequency response is flat up to 10 MHz. That is, voltage (without a load) remains constant up to 10 MHz, whereas power output (with a load) remains constant up to 1 MHz. Figure 1-H shows the test connections and procedures to measure bandwidth at –3 dB points for a typical op amp (the Harris CA3020/CA3020A).

Load sensitivity

Most amplifiers, especially power amplifiers, are sensitive to changes in load. An amplifier produces maximum power gain when the output impedance is the same as the load impedance. The test circuit for load-sensitivity measurement is the same as for frequency response (Figs. 1-A and 1-G) except that the load resistance

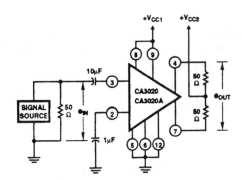

PROCEDURES:

1. Apply desired value of V_{CC1} and V_{CC2}
2. Apply 1kHz input signal and adjust for e_{IN} = 5mV (rms)
3. Record the resulting value of e_{OUT} in dB (reference value)
4. Vary input-signal frequency, keeping e_{IN} constant at 5mV, and record frequencies above and below 1kHz at which e_{OUT} decreases 3dB below reference value
5. Record bandwidth as frequency range between -3dB points

Fig. 1-H Measurement of bandwidth at –3-dB points.

HARRIS SEMICONDUCTOR, LINEAR & TELECOM ICs, 1994, P. 2-50.

is variable. Again, never use a wire-wound load resistance; the reactance can result in considerable error.

To find load sensitivity, measure the power output at various load-impedance and output-impedance ratings. That is, set the load resistance to various values (including a value equal to the supposed amplifier-output impedance). Record the voltage and/or power gain at each setting. Repeat the test at various frequencies. Figure 1-I shows a typical load-sensitivity response curve. Notice that if the load is twice the output impedance (as indicated by a 2:1 ratio, or a normalized load impedance of 2), the output power is reduced to about 50%.

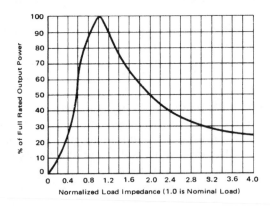

Fig. 1-I Typical load-sensitivity response curve.

Dynamic output impedance or resistance

The load-sensitivity test can be reversed to find the dynamic output impedance or resistance of an amplifier. The connections (Figs. 1-A and 1-G) and the procedures are the same except that the load resistance is varied until *maximum power output* is found. Power is removed, the load resistance is disconnected from the circuit, and the resistance is measured with an ohmmeter. This resistance is equal to the dynamic output impedance of the amplifier (but only at that measurement frequency). The test can be repeated across the entire frequency range as required.

Figure 1-J shows output resistance versus frequency for a typical IC amplifier (a Harris CA3450). Note that the output resistance remains below about 10 Ω at frequencies up to about 80 MHz and then rises rapidly to more than 80 Ω as the frequency increases from 80 MHz to about 120 MHz.

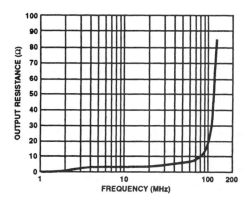

Fig. 1-J   Output resistance versus frequency.
Harris Semiconductors, Linear & Telecom ICs, 1994, p. 2-217.

Dynamic input impedance or resistance

Use the circuit and procedures of Fig. 1-K to find the dynamic input impedance of an amplifier. Note that the IC shown in Fig. 1-K has two inputs to be measured. Also, note that the accuracy of this impedance measurement (and the output impedance measurement) depends on the accuracy with which the resistance R is

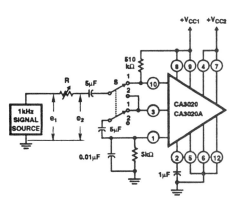

PROCEDURES:

Input Resistance Terminal 10 to Ground (R_{IN10})

1. Apply desired value of V_{CC1} and V_{CC2} and set S in Position 1
2. Adjust 1-kHz input for desired signal level of measurement
3. Adjust R for $e_2 = e_1/2$
4. Record resulting value of R as R_{IN10}

Input Resistance Terminal 3 to Ground (R_{IN3})

1. Apply desired value of V_{CC1} and V_{CC2} set S in Position 2
2. Adjust 1-kHz input for desired signal level of measurement
3. Adjust R for $e_2 = e_1/2$
4. Record resulting value of R as R_{IN3}

Fig. 1-K   Measurement of input resistance. Harris Semiconductors, Linear & Telecom ICs, 1994, p. 2-51.

Amplifier tests

measured. Again, a noninductive (not wire-wound) resistance must be used for R. The impedance found by this method applies only to the frequency used during the test. Current drain, power output, efficiency, and sensitivity. Figure 1-L shows a circuit and the procedures for measuring zero-signal dc current drain, maximum-signal dc current drain, maximum power output, circuit efficiency, sensitivity, and transducer power gain. Again, the circuit of Fig. 1-L applies to a specific IC amplifier (Harris CA3020/3020A), but a similar circuit can be used for most IC amplifiers.

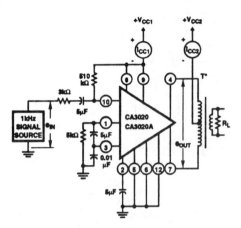

PROCEDURES:
Zero-Signal DC Current Drain

1. Apply desired value of V_{CC1} and V_{CC2} and reduce e_{IN} to 0V
2. Record resulting values of I_{CC1} and I_{CC2} in mA as Zero-Signal DC Current Drain

Maximum-Signal DC Current Drain, Maximum Power Output, Circuit Efficiency, Sensitivity, and Transducer Power Gain

1. Apply desired value of V_{CC1} and V_{CC2} and adjust e_{IN} to the value at which the Total Harmonic Distortion in the output of the amplifier = 10%
2. Record resulting value of I_{CC1} and I_{CC2} in mA as Maximum Signal DC Current Drain
3. Determine resulting amplifier power output in watts and record as Maximum Power Output (P_{OUT})
4. Calculate Circuit Efficiency (η) in % as follows:

$$\eta = 100 \frac{P_{OUT}}{V_{CC1}I_{CC1} + V_{CC2}I_{CC2}}$$

where P_{OUT} is in watts, V_{CC1} and V_{CC2} are in volts, and I_{CC1} and I_{CC2} are in amperes.

5. Record value of e_{IN} in mV (rms) required in Step 1 as Sensitivity (e_{IN})
6. Calculate Transducer Power Gain (G_p) in dB as follows:

$$G_p = 10\log_{10} \frac{P_{OUT}}{P_{IN}}$$

where P_{IN} (in mW) $= \dfrac{e_{IN}^2}{3000 + R_{IN(10)}**}$

*T: PUSH-PULL OUTPUT TRANSFORMER; LOAD RESISTANCE (R_L) SHOULD BE SELECTED TO PROVIDE INDICATED COLLECTOR-TO-COLLECTOR LOAD IMPEDANCE (R_{CC})

Fig. 1-L Measurement of current drain, power output, efficiency, and sensitivity. Harris Semiconductors, Linear & Telecom ICs, 1994, p. 2-50.

Sine-wave analysis

All amplifiers are subject to distortion. That is, the output signal might not be identical to the input signal. Theoretically, the output should be identical to the input except for the amplitude. You can check by applying a sine wave at the amplifier input (using a circuit similar to Fig. 1-A or 1-G) and monitoring both the input and output with a scope. If there is no change in the scope display except for amplitude, there is no distortion.

In practical test or troubleshooting, analyzing sine waves to pinpoint amplifier problems that produce distortion is a difficult job. Unless distortion is severe, it might pass unnoticed. Sine waves are best used where harmonic-distortion or intermodulation-distortion meters are combined with the scope for distortion analysis. (Distortion meters are discussed further in this chapter.) If a scope is used alone for distortion analysis, square waves provide the best results. (The reverse is true for frequency-response or power measurements.)

Square-wave analysis

Distortion analysis is more effective with square waves because of the high odd-harmonic content in square waves (and because it is easier to see a deviation from a straight line with sharp corners than from a curving line). The procedure for checking distortion with square waves is essentially the same as that used with sine waves. Square waves are introduced into the amplifier input, and the output is measured with a scope (Fig. 1-M). The primary concern is deviation of the output waveform from the input waveform (which also is monitored on the scope). If the scope has a dual-trace feature, the input and output can be monitored simulta-

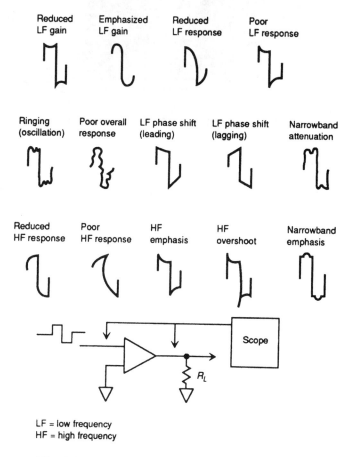

LF = low frequency
HF = high frequency

Fig. 1-M   Basic square-wave distortion analysis.

Amplifier tests

neously. Also, if the scope has an invert function, the output can be inverted from the input for a better comparison of input and output.

If there is a change in the waveform, the nature of the change can sometimes reveal the cause of the distortion. Notice that the drawings of Fig. 1-M are generalized and that the same waveform can be produced by different causes. For example, poor *LF* (low-frequency) response appears to be the same as (high-frequency) emphasis.

Figure 1-N shows the waveforms that are produced by an actual circuit. Notice that the output (trace B) does a good job of following the input (trace A) at a gain of –1. That is, the output is inverted from the input, and there is no gain (unity gain). Also notice that there is some reduced high-frequency response in the output trace B but not the exaggerated response shown in Fig. 1-M.

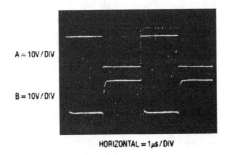

A = 10V / DIV

B = 10V / DIV

HORIZONTAL = 1μs / DIV

Fig. 1-N   Amplifier response to square waves.

The third, fifth, seventh, and ninth harmonics of a clean square wave are emphasized. If an amplifier passes a given frequency and produces a clean square-wave output, it is reasonable to assume that the frequency response is good up to at least nine times the square-wave frequency.

Harmonic distortion

No matter what IC amplifier is used or how well the IC is designed, there is a possibility of odd or even harmonics being present with the fundamental signal. These harmonics combine with the fundamental and produce distortion, as is the case when any two or more signals are combined. The effects of second- and third-harmonic distortion are shown in Fig. 1-O.

Harmonic-distortion meters operate on the *fundamental-suppression principle*. A sine wave is applied to the amplifier input, and the output is measured on a scope or meter. The output is then applied through a filter that suppresses the fundamental frequency. Any output from the filter is then the result of harmonics. Figure 1-P shows typical connections and procedures for measurement of harmonic distortion, where a Hewlett-Packard Type 302A, or equivalent, analyzer is used to measure the *THD* (total harmonic distortion) of a Harris CA3020/CA3020A. This circuit also is used for *S/N* (signal-to-noise) measurements as described further in this chapter.

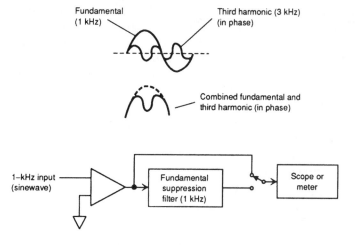

Fig. 1-O Basic harmonic-distortion analysis.

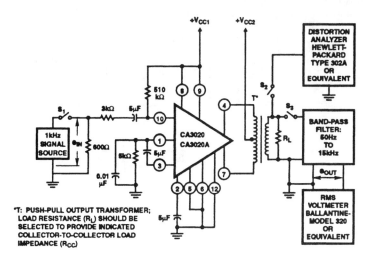

PROCEDURES:

Signal-to-Noise Ratio

1. Close S_1 and S_3; open S_2
2. Apply desired values of V_{CC1} and V_{CC2}
3. Adjust e_{IN} for an amplifier output of 150mW and resulting value of E_{OUT} in dB as e_{OUT1} (reference value)
4. Open S_1 and record resulting value of e_{OUT} in dB as e_{OUT2}
5. Signal-to-Noise Ratio $(S/N) = 20\log_{10}\dfrac{e_{OUT1}}{e_{OUT2}}$

Total Harmonic Distortion

1. Close S1 and S2; open S3
2. Apply desired values of V_{CC1} and V_{CC2}
3. Adjust e_{IN} for desired level amplifier output power
4. Record Total Harmonic Distortion (THD) in %

Fig. 1-P Measurement of THD and S/N ratio.

HARRIS SEMICONDUCTORS, LINEAR & TELECOM ICS, 1994, P. 2-51.

Amplifier tests

In some tests, particularly in audio-amplifier tests, a scope is combined with a harmonic-distortion meter to find the harmonic frequency. For example, if the input is 1 MHz and the output (after filtering) is 3 MHz, third-harmonic distortion is indicated. (Reduce the scope horizontal sweep so you can see one input cycle. If there are three cycles at the output for the same time period as one input cycle, this indicates third-harmonic distortion.)

The percentage of harmonic distortion also is determined by this method. For example, if the output is 100 mV (millivolts) without the filter and 3 mV with the filter, this indicates a 3% harmonic distortion. Notice that total harmonic distortion varies with the power output of the amplifier. For that reason, it is generally necessary to adjust the input voltage for a given power output, as shown in Fig. 1-P. Also note that THD depends on load.

Intermodulation distortion

When two signals of different frequencies are mixed in an amplifier, it is possible that the lower-frequency signal will modulate the amplitude of the higher-frequency signal. This modulation produces a form of distortion that is known as *IMD* (intermodulation distortion). Figure 1-Q shows the basic elements of IMD meters (a signal generator and a highpass filter). The generator portion produces a

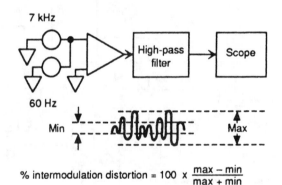

$$\% \text{ intermodulation distortion} = 100 \ \times \ \frac{\text{max} - \text{min}}{\text{max} + \text{min}}$$

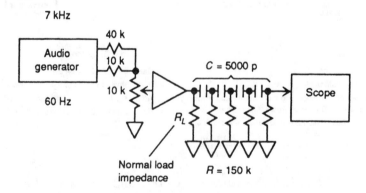

Fig. 1-Q   Basic intermodulation-distortion analysis.

IC amplifier circuits

higher-frequency signal (usually 7 kHz for standard recording-industry testing) that is modulated by a low-frequency signal (usually 60 Hz).

The mixed signals are applied to the amplifier input, with the output connected through a highpass filter to a scope. The highpass filter removes the low-frequency (60 Hz) signal. The only signal that appears on the scope should be the 7 kHz. If any 60-Hz signal is present on the scope, the 60-Hz signal is being passed through as modulation on the 7-kHz signal.

Figure 1-Q also shows a simple IMD test circuit that can be made up in the shop. The highpass filter is designed to pass signals that are about 200 Hz and higher. The purpose of the fixed 40- and 10-kΩ resistors is to set the 60-Hz signal at four times the amplitude of the 7-kHz signal (assuming that both signals leave the generator at the same amplitude). Adjust the 10-kΩ potentiometer that controls the mixed 60-Hz/7-kHz signals to some level that does not overdrive the amplifier being tested.

Calculate the percentage of IMD using the equation shown in Fig. 1-Q. For example, if the maximum output (shown on the scope) is 1 V, and the minimum is 0.99 V, the percentage of IMD is about:

$$\frac{1.0 - 0.99}{1.0 + 0.99} = 0.005 \times 100 = 0.5\%$$

Background noise

If a scope is sufficiently sensitive, it can be used to check and measure the background-noise level of an amplifier, as well as to check for the presence of hum, oscillation, and so on. The scope should be capable of measurable deflection with an input below 1 mV (and considerably less if an IC amplifier is involved).

The basic procedure consists of measuring amplifier output with the volume or gain controls (if any) at maximum but without an input signal. A meter can be used, but the scope is better because the frequency and nature of the noise (or other signals) are displayed visually. Scope gain must be increased until there is a noise or *hash* indication.

A noise indication might be caused by pickup in the leads between the amplifier and scope. If in doubt, disconnect the leads from the amplifier but not the scope. If you suspect that 60-Hz power-line hum is present in the amplifier output (picked up from the power supply or other source), set the scope sync controls to the line position. If a stationary signal pattern appears, the signal is the result of line hum getting into the circuit. If a signal appears that is not at the line frequency, the signal can be the result of oscillation in the amplifier or stray pickup. Short the amplifier input terminals. If the same signal remains, suspect oscillation in the amplifier circuits.

With present-day IC amplifiers, the internal or background noise is considerably less than 1 mV, and it is impossible to measure directly—even with a sensitive scope. You must use a circuit that amplifies the output of the IC under test before the output is applied to the scope. Figure 1-R is such a circuit (and is used for noise tests of an OP-77 IC). The IC under test is connected for high voltage gain, as is the

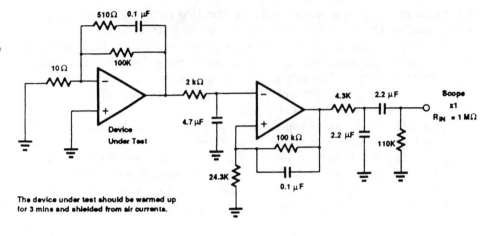

Fig. 1-R   0.1- to 10-Hz noise test circuit.

following amplifier. (The total voltage gain is 50,000.) This gain makes it possible to monitor (and record) noise on a chart recorder (Fig. 1-S). Noise is measured over a 10-s (second) interval, noting the peak-to-peak value, which is about 25 nV (nanovolts) in Fig. 1-S.

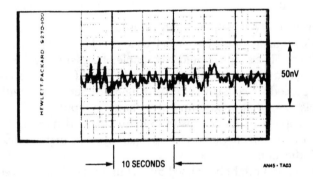

Fig. 1-S   Typical amplifier background noise as measured on a chart recorder over a 10-s interval.

Signal-to-noise ratio

Some IC amplifiers are tested for signal-to-noise (S/N) ratio instead of (or in addition to) background noise. Figure 1-P shows the circuit connections and procedures for S/N measurement of the CA3020/CA3020A. (This circuit is the same as for THD except that the distortion analyzer is not connected when S/N is measured.)

A signal-to-noise test shows the relationship of background noise to signal amplitude, when the amplifier is operated under specific conditions. For example,

IC amplifier circuits

in the circuit of Fig. 1-P, the input signal is increased in amplitude until the output is 150 mW, and the output voltage is recorded in dB. Then the input signal is removed, but the input terminals remain connected together through resistors and capacitors, so the only output is the noise voltage within the IC. This background-noise voltage also is recorded in dB, and the S/N ratio is calculated as shown.

Slew rate (transient response)

Amplifier slew rate is the maximum rate of change in output voltage that the amplifier can produce when maintaining linear characteristics (symmetrical output without clipping). Slew rate is often listed under the heading of *transient response* in data sheets. Other transient response characteristics include *rise time, settling time, overshoot,* and possibly *error band.* All these topics are discussed further in this chapter.

Slew rate is expressed by the difference in output voltage divided by difference in time, d_{Vo}/d_t. Usually, slew rate is listed in volts per microsecond. For example, if the output voltage from an amplifier can change 7 V in 1 μs (microseconds), the slew rate is 7 (which can be listed as 7 V/μs). The major effect of slew rate on circuit performance is that (all other factors being equal) a higher slew rate results in higher power output.

You can estimate the *approximate power bandwidth* of an amplifier if you know the slew rate. The equation is:

$$\text{full power bandwidth in megahertz} = \frac{(\text{slew rate})}{(6.28 \times \text{peak output voltage})}$$

For example, the slew rate for a Harris HA-2529 is listed as 150 (typical) when the peak-to-peak output voltage is ±10 V (a 10-V peak output voltage). Using the equation, the power bandwidth is:

$$\frac{150}{62.8} = 2.39 \text{ MHz}$$

The data sheet for the HA-2529 shows a full-power bandwidth of 2.1 MHz minimum and 2.6 MHz typical.

A simple way to find amplifier slew rate is to measure the slope of the output waveform when a square-wave input is applied, as shown in Fig. 1-T. The input square wave must have a rise time that exceeds the slew-rate capability of the amplifier. As a result, the output does not appear as a square wave but as an integrated wave. In the example shown, the output voltage rises (and falls) about 40 V in 1 μs. Notice that slew rate is usually measured in the closed-loop condition (with negative feedback) and that slew rate increases with higher gain.

Figure 1-U shows the slew-rate and transient-response test circuits for a typical IC op amp (Harris HA-2529). Figure 1-U also includes some definitions for slew rate, settling time, rise time, overshoot, and error band, which are discussed in the following section.

Amplifier tests

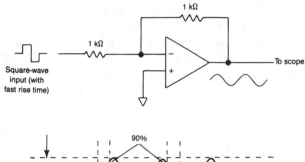

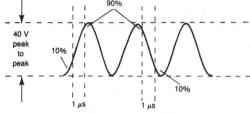

Example shows a slew rate of about 40 (40 V/μs) at unity gain

Fig. 1-T   Slew-rate measurement.

Test Circuits

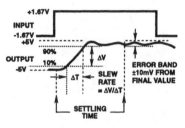

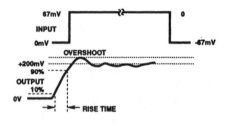

NOTE: Measured on both positive and negative transitions from 0V to +200mV and 0V to -200mV at the output.

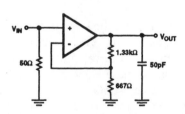

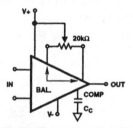

SUGGESTED V$_{OS}$ ADJUSTMENT AND COMPENSATION HOOK UP

Tested offset adjustment range is |V$_{OS}$ + 1mV| minimum referred to output. Typical ranges are +28mV to -18mV with R$_T$ = 20kΩ

Fig. 1-U   Slew-rate and transient-response tests.
HARRIS SEMICONDUCTORS, LINEAR & TELECOM ICs, 1994, P. 2-311.

IC amplifier circuits

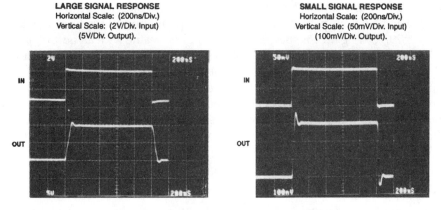

Fig. 1-U   Continued.

Rise time, settling time, and overshoot

There are many ways to measure these transient-response characteristics. Figure 1-V shows the settling time test circuit.

Settling Time Circuit

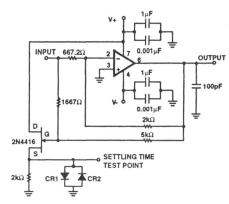

- $A_V = -3$
- Feedback and summing resistor ratios should be 0.1% matched.
- Clipping diodes CR1 and CR2 are optional. HP5082-2810 recommended.

Fig. 1-V   Settling time test circuit. HARRIS SEMICONDUCTORS, LINEAR & TELECOM ICS, 1994, P. 2-310.

For this IC amplifier (HA-2529), rise time is specified with an output of 200 mV and a gain of 3. Thus, the small-signal response scope displays must be used. As shown, the rise time (measured from the 10% point to the 90% point, Fig. 1-U) is about 40 ns (nanoseconds). The data sheet specifies a typical rise time of 20 ns and a maximum of 50 ns.

Settling time is the total length of time from input-step application until the output remains within a specified error band or point around the final value. For the HA-2529, settling time is specified to 0.1% of final value with an output of 10 V and a gain of –3. Thus, the large-signal response must be used. As shown, settling time is somewhat less than 150 ns (from the start of the input, through the overshoot, and back to where the output levels to 10 V). The data sheet specifies a typical settling time of 200 ns. From a circuit-performance standpoint, an increase in rise time, settling time, and overshoot lowers the frequency response and bandwidth. If the IC amplifier is used in pulse applications, rise times and settling times can distort the output pulse.

Phase shift

The phase shift between input and output of some IC amplifiers is not a significant design factor but is a critical factor in others, particularly for op amps (and other ICs operating as op amps). This is because an op amp generally uses the principle of feeding back output signals to the input. Under ideal open-loop conditions, the output should be exactly 180° out of phase with the inverting input and in phase with the noninverting input. Any substantial deviation from this condition can cause op-amp circuit problems.

For example, assume that an op-amp circuit uses the inverting input (with the noninverting input grounded) and the circuit output is fed back to the inverting input. If the output is not shifted the full 180° (for example, if the shift is only a few degrees), the circuit might oscillate (because the output being fed back is almost in phase with the input). Even if there is no oscillation, the amplifier gain will not be stabilized, and the circuit will not operate properly.

A dual-trace scope, connected as shown in Fig. 1-W, is the ideal tool for phase measurement. For the most accurate results, the cables that connect the input and output should be of the same length and characteristics. At higher frequencies, a difference in cable length or characteristics can introduce a phase shift. For simplicity, adjust the scope controls until one cycle of the input signal occupies exactly nine screen divisions (typically 9 centimeters) horizontally. Then, find the phase/division factor of the input signal. For example, if 9 centimeters represents one cycle (360°), 1 centimeter represents 40° (360/9 = 40).

With the phase/division factor established, measure the horizontal distance between corresponding points on the two waveforms (input and output signals). Then multiply the measured distance by the phase/division factor of 40°/cm (degrees per centimeter) to find the phase difference. For example, if the horizontal distance is 0.6 cm with a 40°/centimeter factor, the phase difference is:

$$0.6 \times 40° = 24°$$

If the scope has speed magnification, you can get more accurate results. For example, if the sweep rate is increased 10 times, the magnified phase factor is:

$$40°/\text{cm} \times 0.1 = 4°C$$

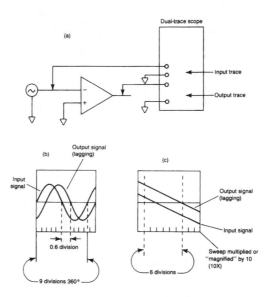

Fig. 1-W Phase-shift measurement.

Figure 1-W shows the same signal with and without sweep magnification. With a 10× magnification, the horizontal distance is 6 centimeters, and the phase difference is:

$$6 \times 4 = 24°$$

Feedback measurement

Because IC-amplifier circuits can include feedback (particularly op-amp circuits), it is sometimes necessary to measure feedback voltage at a given frequency with given operating conditions. The basic feedback-measurement connections are shown in Fig. 1-X. Although it is possible to measure the feedback voltage as shown in Fig. 1-X(A), a more accurate measurement is made when the feedback lead is terminated in the normal operating impedance, as shown in Fig. 1-X(B).

If an input resistance is used in the circuit, and this resistance is considerably lower than the IC input resistance, use the circuit-resistance value. If in doubt, measure the input impedance of the IC and terminate the feedback lead in that value (to measure open-loop feedback voltage). Remember that open-loop voltage gain must be substantially higher than the closed-loop voltage gain for most op-amp circuits to perform properly.

Input-bias current

IC-amplifier input-bias current is the average value of the two input-bias currents of the differential-input stage. In circuit performance, the significance of input-bias current is that the resultant voltage drops across input resistors (such as the resistors at pin 3 of the IC in Fig. 1-G) restrict the input common-mode voltage range at

Amplifier tests

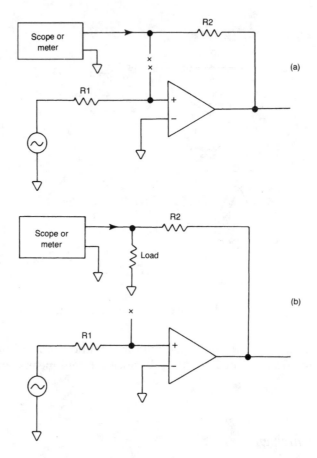

Fig. 1-X Feedback measurement.

higher impedance levels. The input-bias current produces a voltage drop across the input resistors. This voltage drop must be overcome by the input signal (which can be a problem if the input signal is low and the input resistors are large).

Input-bias current can be measured using the circuit of Fig. 1-Y. Any resistance value for R1 and R2 can be used, provided that the value produces a measurable voltage drop and that the resistance values are equal. A value of 1 kΩ (with a tolerance of 1% or better) for both R1 and R2 is realistic for typical op amps.

If it is not practical to connect a meter in series with both inputs (as shown), measure the voltage drop across R1 and R2, and calculate the input-bias current. For example, if the voltage is 3 mV across 1-kΩ resistors, the input-bias current is 3 μA (microamperes). Try interchanging R1 and R2 to see if any difference is the result of a difference in resistor values. In theory, the input-bias current should be the same for both inputs. In the real world, the bias currents should be almost equal. Any great difference in input bias is the result of unbalance in the input differential amplifier of the IC, and it can seriously affect circuit operation (and it usually indicates a defective IC).

IC amplifier circuits

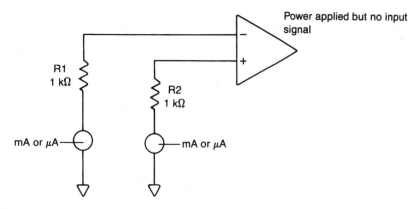

Power applied but no input signal

Fig. 1-Y   Input-bias current measurement.

Input-offset voltage and current

Input-offset voltage is the voltage that must be applied at the input terminals to get zero output voltage, whereas input-offset current is the difference in input-bias current at the amplifier input. Offset voltage and current are usually referred back to the input because the output voltages depend on feedback.

From a circuit performance standpoint, the effect of input-offset is that the input signal must overcome the offset before an output is produced. Also, with no input, there is a constant shift in output level. For example, if an IC amplifier has a 1-mV input-offset voltage, and a 1-mV signal is applied, there is no output. If the signal is increased to 2 mV, the amplifier produces only the peaks in excess of 1 mV.

Input-offset voltage and current can be measured using the circuit of Fig. 1-Z. As shown, the output is alternately measured with R3 shorted and with R3 in the circuit. The two output voltages are recorded as E_1 (R1 closed, R3 shorted) and E_2 (S1 open, R3 in the circuit).

With the two output voltages recorded, the input-offset voltage and current can be calculated using the equations of Fig. 1-Z. For example, assume that R1, R2, and R3 are at the values shown, that E_1 is 83 mV, and that E_2 is 363 mV:

$$\text{Input-offset voltage} = \frac{83 \text{ mV}}{100} = 0.83 \text{ mV}$$

$$\text{Input-offset current} = \frac{280 \text{ mV}}{100 \text{ k}\Omega \, (1 + 100)} = 27.7 \text{ nA}$$

Common-mode rejection

There are many definitions for *CMR* (common-mode rejection), which is sometimes listed as *CMRR* or *CM*$_{rej}$). No matter what definition is used, the first step to measure CMR is to find the open-loop gain of the IC at the desired operating frequency. Then connect the IC in the common-mode test circuit of Fig. 1-AA. In-

Amplifier tests

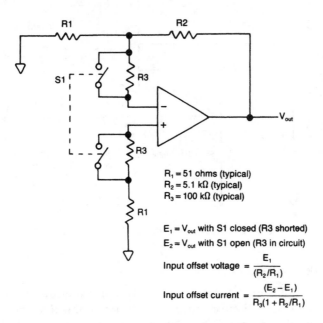

R₁ = 51 ohms (typical)
R₂ = 5.1 kΩ (typical)
R₃ = 100 kΩ (typical)

$E_1 = V_{out}$ with S1 closed (R3 shorted)
$E_2 = V_{out}$ with S1 open (R3 in circuit)

Input offset voltage $= \dfrac{E_1}{(R_2/R_1)}$

Input offset current $= \dfrac{(E_2 - E_1)}{R_3(1 + R_2/R_1)}$

Fig. 1-Z   Input-offset voltage and current measurement.

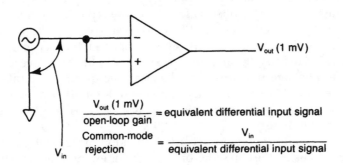

$\dfrac{V_{out}\,(1\ mV)}{\text{open-loop gain}}$ = equivalent differential input signal

$\dfrac{\text{Common-mode}}{\text{rejection}} = \dfrac{V_{in}}{\text{equivalent differential input signal}}$

Fig. 1-AA   Common-mode rejection measurement.

crease the common-mode voltage (at the same frequency used for the open-loop gain test) until you get a measurable output. Do not exceed the maximum input common-mode voltage specified in the data sheet. If no such value is available, do not exceed the normal input voltage of the IC.

To simplify the calculation, increase the input voltage until the output is at some exact value, such as the 1 mV shown. Divide this value by the open-loop gain to find the equivalent differential input signal. For example, with an open-loop gain of 100 and an output of 1 mV, the equivalent differential is:

$$\dfrac{0.001}{100} = 0.00001$$

IC amplifier circuits

Now measure the input voltage that produced the 1-mV output and divide the input by the equivalent differential to find the common-mode rejection ratio. In the example, simply find the input voltage that produces the 1-mV output and move the decimal point over five places. For example, if the output is 1 mV with a 10-V input and a gain of 100, the ratio is 0.0001. This ratio can be converted to decibels (a voltage ratio of 80 dB).

Power-supply sensitivity (or PSS)

PSS (power-supply sensitivity) is the ratio of change in input-offset voltage to the change in power-supply voltage that produces the change. On some data sheets, the term is expressed in millivolts or microvolts per volt (mV/V or μV/V), which represents the change of input-offset voltage (in microvolts or millivolts) to a change (in volts) of the power supply. In other data sheets, the term *PSRR* (power-supply rejection ratio) is used instead, and it is given in decibels.

No matter what it is called, the characteristic can be measured using the circuit of Fig. 1-Z (the same test circuit as for input-offset voltage). The procedure is the same as for measurement of input-offset voltage except that the supply voltage is changed (in 1-V steps). The amount of change in input-offset voltage for a 1-V power-supply change is the PSRR. The ratio of change can be converted to decibels if required. The circuit of Fig. 1-Z also can be used when the amplifier is operated from two power supplies. One supply voltage is changed (in 1-V steps) while the other supply voltage is held constant.

IC amplifier types

This chapter includes circuits for a number of different IC-amplifier types. Following is a summary of these types.

Operational amplifiers (op amps)

The designation *op amp* was originally used for a series of high-performance direct-coupled amplifiers in analog computers. These amplifiers performed mathematical operations (summing, scaling, subtraction, integration, and so on). Today, the availability of inexpensive IC amplifiers has made the packaged op amp useful as a replacement for any amplifier.

Figure 1-BB shows some classic op-amp functions. As shown by the equation for the *inverting amplifier*, the output voltage V_{OUT} is equal to the input voltage V_{IN} multiplied by the ratio R_2/R_1. If R_1 and R_2 are the same value, there is no gain (*unity gain* or a gain of 1). If R_2 (the feedback resistance) is 100 kΩ and R_1 (the input resistance is 10 kΩ), the voltage gain is 10. Most op amps are operated in this closed-loop configuration, with feedback. The main purpose of the feedback is to stabilize gain at some fixed values. With the inverting amplifier, the output voltage is inverted from the input voltage. This inversion happens because the input is applied to the inverting input (–) of the op amp at pin 2.

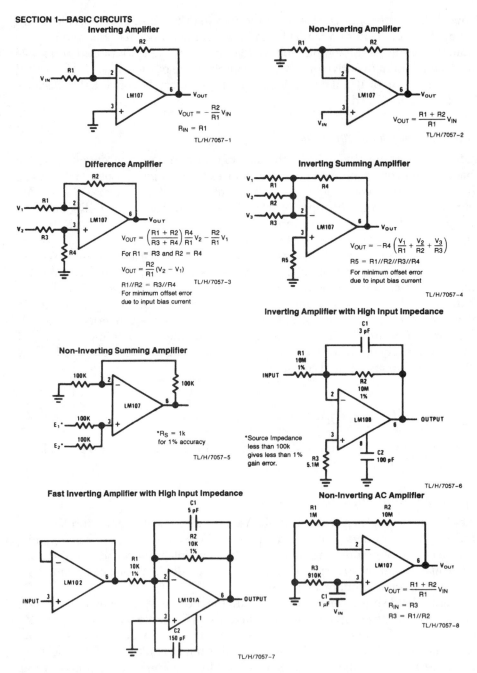

Fig. 1-BB   Classic op-amp functions. NATIONAL SEMICONDUCTOR, LINEAR APPLICATIONS HANDBOOK, 1994, P. 70.

IC amplifier circuits

The *noninverting amplifier* is used when the input and output voltages must be in phase (input applied to the noninverting or + input at pin 3). With noninverting, the V_{OUT} is equal to V_{IN}, multiplied by the ratio of $(R_1 + R_2)/R_1$.

In the *difference amplifier*, V_{OUT} depends on the difference between the two input voltages $(V_2 - V_1)$, multiplied by the ratio of the resistances. For the *inverting summing amplifier*, V_{OUT} is the sum of voltages V_1, V_2, and V_3. For example, if all three input voltages are 5 V, the output is 15 V (provided that R1 through R4 are all the same value).

The circuits of Fig. 1-BB require multiple power supplies (typically ±5 V, ±10 V, and ±15 V). This requirement is typical for most op-amp circuits and is one of the limitations of most op amps. (However, it is possible to operate some op amps with a single supply.)

Notice that in the inverting circuits, the noninverting input of the op amp is returned to ground, possibly through a resistor. This grounded input is typical for op amps operated as single-input circuits (even though op amps have a differential input). As a guideline, when a grounding resistor is used, the resistor value is equal to the parallel resistance of the input and feedback resistors (as shown by the equations for the inverting summing amplifier).

Notice that there is no external compensation for many of the op-amp circuits in this chapter. Early op amps often required external compensation circuits (usually capacitors or resistors, or both) to provide a given frequency-response characteristic. Many present-day op amps have internal compensation and do not require external compensation. There are exceptions. For example, the inverting amplifier with high input impedance circuit of Fig. 1-BB requires an external compensating capacitor (pin 8 to ground). From a troubleshooting standpoint if an op-amp circuit is working, but the frequency response is not as required, look for compensation problems.

OTA (operational transconductance amplifier)

An *OTA* (sometimes called a programmable amplifier) is similar to an op amp. However, OTAs and op amps are not necessarily interchangeable. The OTA not only includes the usual differential inputs of an op amp but also has an additional control input in the form of an amplifier bias current or IABC. (Figure 1-G shows an IC amplifier controlled by an I_{ABC} current at pin 6. The amount of IABC current is set by resistor Rs.)

The control input increases the flexibility of the OTA for use in a wide range of applications. For example, if low power consumption, low input bias, and low offset current, or high input impedance are desired, then select low IABC. On the other hand, if operation into a moderate load impedance is the main consideration, then use higher levels of IABC.

The second major difference between an op amp and an OTA is that the OTA output impedance is extremely high (most op amps have very low output impedance). Because of this difference, the output signal of an OTA is best described in current that is proportional to the difference between the voltages at the two inputs (inverting and noninverting).

IC amplifier types

The OTA transfer characteristics (or input/output relationship) are best defined in *transconductance* rather than voltage gain. Transconductance (usually listed as g_m or g_{21}) is the ratio between the difference in current out (I_{OUT}) for a given difference in voltage input (E_{IN}). Except for the high output impedance and the definition of input/output relationships, OTA characteristics are similar to those of a typical op amp.

Figure 1-CC shows a basic OTA circuit, complete with external circuit components. The following summarizes the procedure for finding the I_{ABC} required to produce a given transconductance. Assume that an open-loop gain of 100 is required. Open-loop gain is related directly to load resistance R_L and transconductance. However, the actual load resistance is the parallel combination of R_L and R_F or about 18 kΩ ($R_L \times R_F)/R_L + R_F$). With an open-loop gain of 100 and an actual load of 18 kΩ, the g_m should be 100/18,000, or about 5.5 millimho (mmho).

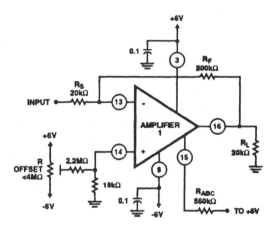

Fig. 1-CC Basic OTA circuit. Harris Semiconductors, Linear & Telecom ICs, 1994, p. 2-58.

The transconductance is set by I_{ABC}. With a data-sheet curve similar to that of Fig. 1-DD, select an I_{ABC} from the minimum-value curve to assure that the OTA provides sufficient gain. As shown in Fig. 1-DD, for a G_m of 5.5 mmho, the required I_{ABC} is approximately 20 μA.

Current-feedback amplifiers (CFAs or Nortons)

CFAs (current-feedback amplifiers) are sometimes called Norton amplifiers. CFAs or CFBs are similar to OTAs in that their characteristics are controlled by an external current or voltage. However, the internal circuits and functions of CFAs are quite different from those of OTAs (and from op amps). With a CFA, there is no separate pin for control current. Any change in amplifier characteristics is set by current that is applied at the inverting and noninverting inputs.

Some manufacturers use a modified amplifier symbol such as shown in Fig. 1-EE (which also shows a simplified version of the classic National Semiconductor

IC amplifier circuits

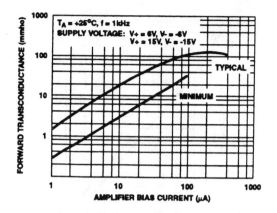

Fig. 1-DD   Forward transconductance curve for OTA.

HARRIS SEMICONDUCTORS, LINEAR & TELECOM ICS, 1994, P. 2-56.

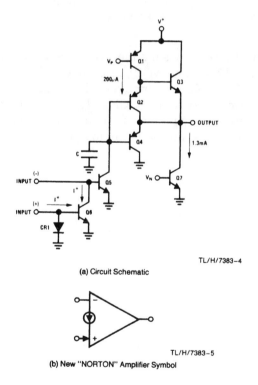

TL/H/7383–4

(a) Circuit Schematic

TL/H/7383–5

(b) New "NORTON" Amplifier Symbol

Fig. 1-EE   Basic current-feedback (Norton) amplifier.

NATIONAL SEMICONDUCTOR, LINEAR APPLICATIONS HANDBOOK, 1994, P. 171.

LM3900). The current arrow between the inputs implies a current-mode of operation. The symbol also signifies that current is removed from the (–) input and that the (+) input is a current input (which can control amplifier gain). The signal can be applied at either the (+) or (–) inputs.

IC amplifier types

Figure 1-FF shows open-loop gain characteristics of the CFA (compared to those of the classic 741 op amp).

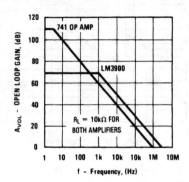

Fig. 1-FF Open-loop gain characteristics of a CFA.
NATIONAL SEMICONDUCTOR, LINEAR APPLICATIONS HANDBOOK, 1994, P. 171.

Figures 1-GG and 1-HH show simplified versions of two classic CFAs (National Semiconductor LM3900 and LM359, respectively). Note that the manufacturer describes these ICs as Norton amplifiers. Two input parameters are very important in circuit performance for these CFAs: I_{BIAS} (the input bias current) and A_I (mirror gain constant). These parameters cannot be measured in the same way as for op amps or OTAs (there is no equivalent to A_I in op amps and OTAs).

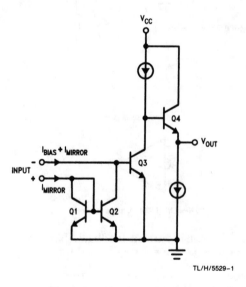

Fig. 1-GG Simplified schematic of the LM3900.
NATIONAL SEMICONDUCTOR, LINEAR APPLICATIONS HANDBOOK, 1994, P. 1138.

IC amplifier circuits

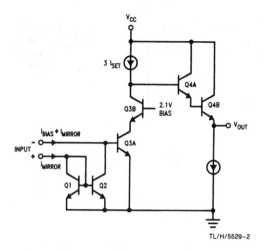

TL/H/5529-2

Fig. 1-HH   Simplified schematic of the LM359.
NATIONAL SEMICONDUCTOR, LINEAR APPLICATIONS HANDBOOK, 1994, P. 1138.

Figures 1-II and 1-JJ show test circuits for measurement of I_{BIAS} in the LM3900 and LM359, respectively. In the circuit of Fig. 1-II, two voltage measurements are made at the output, one with S1 closed and one with S1 open. The output-voltage increase is equal to the voltage appearing across the 1-MΩ resistor, multiplied by the output gain A_V. For the values shown, the output-voltage increase multiplied by 200 gives the bias current in nA, or:

$$I_{\text{BIAS}} \text{ (nA)} = 200 \, \Delta V_{\text{OUT}} = \left(\frac{10^9}{A_V \times 1 \text{ M}\Omega} \right) \Delta V_{\text{OUT}}$$

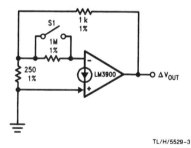

TL/H/5529-3

Fig. 1-II   IBIAS test circuit for the LM3900.
NATIONAL SEMICONDUCTOR, LINEAR APPLICATIONS HANDBOOK, 1994, P. 1138.

The circuit of Fig. 1-JJ is similar except that RSET is added (to provide an I_{SET} of 5 μA), and a 27-pF capacitor is added for circuit stability.

Figures 1-KK and 1-LL show test circuits for measurement of mirror gain AI in the LM3900 and LM359, respectively. In the circuit of Fig. 1-KK, resistors R are se-

IC amplifier types

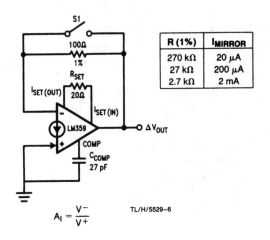

R (1%)	I_{MIRROR}
270 kΩ	20 μA
27 kΩ	200 μA
2.7 kΩ	2 mA

$$A_I = \frac{V^-}{V^+}$$

TL/H/5529-6

Fig. 1-JJ   IBIAS test circuit for the LM359.

NATIONAL SEMICONDUCTOR, LINEAR APPLICATIONS HANDBOOK, 1994, P. 1139.

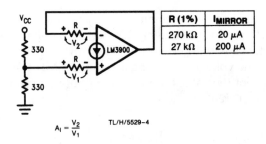

R (1%)	I_{MIRROR}
270 kΩ	20 μA
27 kΩ	200 μA

$$A_I = \frac{V_2}{V_1}$$

TL/H/5529-4

Fig. 1-KK   Mirror-gain test circuit for the LM3900.

NATIONAL SEMICONDUCTOR, LINEAR APPLICATIONS HANDBOOK, 1994, P. 1139.

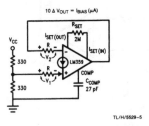

TL/H/5529-5

Fig. 1-LL   Mirror-gain test circuit for the LM359.

NATIONAL SEMICONDUCTOR, LINEAR APPLICATIONS HANDBOOK, 1994, P. 1139.

lected to provide the desired mirror current I_{MIRROR} using the values shown. The voltage across each R is measured, and the ratio of the two voltages is equal to A_I. Mirror gain is affected by I_{BIAS}. Where I_{BIAS} is a significant part of the mirror current, the equation for A_I becomes:

IC amplifier circuits

$$A_I = \frac{(V_2 - R_{IBIAS})}{V_1}$$

Again, the major difference between the circuits of Figs. 1-KK and 1-LL is the addition of R_{SET} and the compensating capacitor.

All the CFA test circuits assume a V_{CC} of 12 V. Test accuracy is only as good as the resistors and meters used (which is generally the case!). Matching is very important for the two resistors R in Figs. 1-KK and 1-LL. A 1% tolerance is recommended. The feedback resistors in Figs. 1-II and 1-JJ also should be 1%. Most 3½ digit DVMs (digital volt meters) have sufficient accuracy for the voltage measurements. The meter input impedance should be at least 10 MΩ (megohms) to prevent circuit loading in the mirror-gain tests.

Chopper-stabilized op amps

Chopper stabilization is often used where stability is essential over time and with variations in temperature and supply voltage. Figures 1-MM and 1-NN show the internal functions and pinouts, respectively, of a chopper-stabilized op amp (the Harris ICL7650S), which is a direct replacement for the industry standard ICL7650 but with improved input-offset voltage, lower input-offset temperature coefficient, reduced input-bias current, and wider common-mode voltage range.

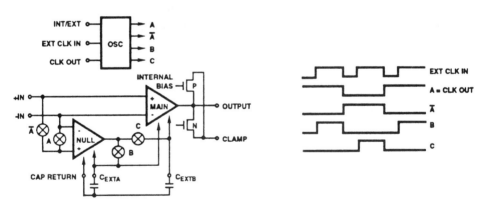

Fig. 1-MM   Internal functions of chopper-stabilized amplifier.
HARRIS SEMICONDUCTOR, LINEAR & TELECOM ICS, 1994, P. 2-695.

As shown in Fig. 1-MM, there are two amplifiers: the main amplifier and the nulling amplifier. Both amplifiers have offset-null capacity. The main amplifier is connected continuously from the input to the output. The nulling amplifier, under control of the chopping oscillator and clock circuit, alternately nulls itself and the main amplifier. The two external capacitors C_{EXTA} and C_{EXTB}, provide the required storage of the nulling potentials. The clock oscillator, and all other control circuits are self-contained. However, the 14-lead version (Fig. 1-NN) includes a provision for an external clock, if required. All of the internal functions are user-transparent,

IC amplifier types

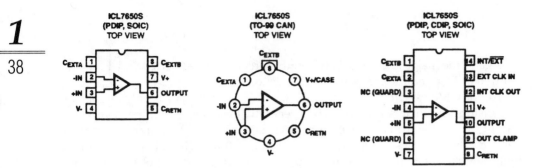

Fig. 1-NN Pinouts of chopper-stabilized amplifier.
HARRIS SEMICONDUCTOR, LINEAR & TELECOM ICs, 1994, P. 2-694.

eliminating common chopper-amplifier problems (intermodulation, spikes, over-range lock-up).

As shown in Fig. 1-OO, the chopper-stabilized IC uses the same basic connections for an inverting amplifier, noninverting amplifier, and voltage follower as a conventional op amp. However, the null/storage capacitors must be connected to the C_{EXTA} and C_{EXTB} pins, with a common connection to the C_{RETN} pin. This connection should be made directly by either a separate wire, or PC trace to avoid injection load-current IR drops into the capacitor circuits. The outside foil by each capacitor (where available) should be connected to the C_{RETN} pin.

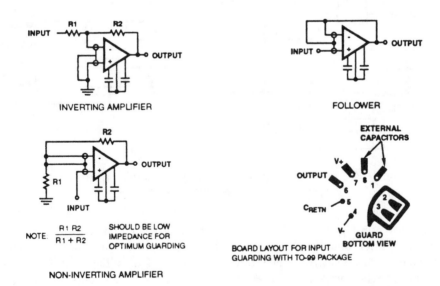

Fig. 1-OO Connections for chopper-stabilized amplifier.
HARRIS SEMICONDUCTOR, LINEAR & TELECOM ICs, 1994, P. 2-703.

IC amplifier circuits

To take full advantage of a chopper-stabilized op amp, the inputs should be guarded. Input guarding of the 8-pin TO-99 package is done with a 10-lead-pin circuit as shown in Fig. 1-OO. With this configuration, the holds adjacent to the inputs are empty when the IC is inserted into the board. The pin configuration of the 14-pin PDIP package (Fig. 1-NN) is designed to facilitate guarding, in that the pins adjacent to the inputs are not used.

Wideband transconductance amplifiers (WTAs)

WTAs (wideband transconductance amplifiers) are similar to OTAs and CFAs (Nortons) in that internal characteristics (such as gain) can be controlled by external components. However, unlike OTAs and CFAs, the WTA does not require any feedback components to produce accurate gain. Without any negative feedback, there is no closed-loop phase shift (the primary cause of oscillation in conventional op amps). Figures 1-PP and 1-QQ show the pinouts and typical operating circuits, respectively, for two WTAs (the Maxim MAX435 and MAX436). As shown, the MAX435 has a differential output, whereas the MAX436 is single ended. However, both ICs have a differential input.

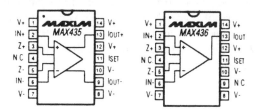

Fig. 1-PP   Pinouts for WTAs. Maxim New Releases DataBook, 1994, p. 8-5.

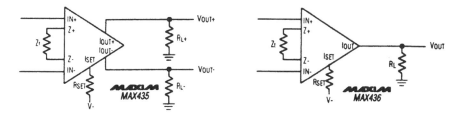

Fig. 1-QQ   Operating circuit for WTAs. Maxim New Releases DataBook, 1994, p. 8-5.

The output of these WTAs is a current that is proportional to the applied differential input. This provides inherent short-circuit protection for the outputs. Circuit gain is set by the ratio of two impedances and an internally set current-gain factor (K).

In effect, WTAs are voltage-controlled current sources, as shown in Fig. 1-RR. Signal gain is set by the ratio of two impedances: the user-selected transconduc-

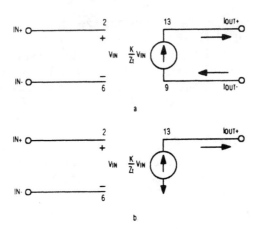

Fig. 1-RR Operation of WTAs from an input/output standpoint.

MAXIM NEW RELEASES DATABOOK, 1994, P. 8-13.

tance element or network (Z_t) and the output-load impedance (Z_L). A differential input voltage V_{IN} applied between the input terminals causes current to flow in output-load Z_L. This current is equal to V_{IN}/Z_t. The current in Z_t is multiplied by a preset current gain K of the WTA and appears at the output terminals as a current equal to $K \times (V_{IN}/Z_t)$. The current flows through the load impedance to produce an output voltage according to:

$$V_{OUT} = K \left(\frac{Z_L}{Z_t} \right) V_{IN}$$

where K = WTA current-gain ratio (factory trimmed)

Z_L = output-load impedance
Z_t = Transconductance element impedance
V_{IN} = differential input voltage

The theoretical voltage gain A_V is set by the impedance of Z_t and Z_L, according to:

$$A_V = K \frac{Z_L}{Z_t}$$

As shown in Fig. 1-SS, the WTA output impedance R_{OUT} (typically 3.5 kΩ) is paralleled with the circuit load-impedance R_L, reducing the equivalent load impedance (the real world load impedance). After accounting for R_{OUT}, the actual circuit gain is calculated as:

$$A_V = \frac{V_{OUT}}{V_{IN}} = \left(\frac{K}{R_t} \right) \left[\frac{(R_{OUT})(R_L)}{R_{OUT} + R_L} \right]$$

IC amplifier circuits

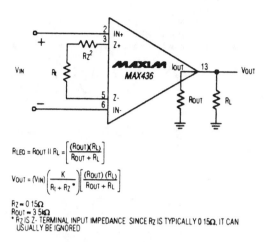

$$R_{LEQ} = R_{OUT} \parallel R_L = \left[\frac{(R_{OUT})(R_L)}{R_{OUT} + R_L}\right]$$

$$V_{OUT} = (V_{IN})\left(\frac{K}{R_t + R_Z}\cdot\right)\left[\frac{(R_{OUT})(R_L)}{R_{OUT} + R_L}\right]$$

$R_Z \approx 0.15\Omega$
$R_{OUT} \approx 3.5k\Omega$
* R_Z IS Z- TERMINAL INPUT IMPEDANCE SINCE R_Z IS TYPICALLY 0 15Ω, IT CAN USUALLY BE IGNORED

Fig. 1-SS   Finite output-impedance of WTAs. MAXIM NEW RELEASES DATABOOK, 1994, P. 8-15.

The voltage-gain error ΔAV, with respect to the theoretical gain, is equal to:

$$\frac{\Delta A_V}{A_V} + \frac{R_L}{R_L + R_{OUT}}$$

The internal current of the WTA is set by an external current source. In turn, this current source is controlled by an external resistor RSET connected from the ISET pin to the V– supply. A typical value for R_{SET} is 5.9 kΩ, which provides an output-drive capability of ±10 mA per output for the MAX435 and ±20 mA for the single-ended MAX436. For a full discussion of WTAs, read the author's *Simplified design of IC amplifier circuits* (Butterworth-Heineman, 1996).

Basic amplifier troubleshooting approaches

The remainder of this chapter is devoted to troubleshooting basic amplifier circuits. Begin with a troubleshooting approach and some practical notes on analyzing amplifier-circuit problems. The introduction concludes with a step-by-step example of amplifier troubleshooting. Much of the information presented here is basic, and the techniques covered are of the most benefit to those readers who are unfamiliar with electronic troubleshooting. The techniques serve as a basis for understanding the step-by-step example of amplifier troubleshooting.

Signal tracing

The basic troubleshooting approach for an amplifier involves signal tracing, such as shown in Fig. 1-TT. The input and output waveforms of each stage are monitored on a scope or meter. Any stage that shows an abnormal waveform (in amplitude, waveshape, and so on) or the absence of an output (with a known-good input sig-

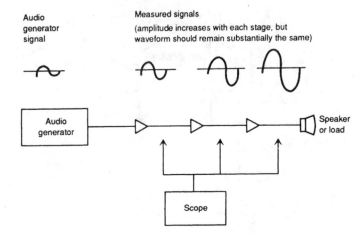

Fig. 1-TT Basic amplifier-circuit signal tracing.

nal) points to a defect in the stage. Voltage and/or resistance measurements on all elements in the transistor (or IC) are then used to pinpoint the problem.

A scope is the most logical instrument to use for checking amplifier circuits (both complete amplifier systems or a single amplifier stage). The scope can duplicate every function of a meter in troubleshooting, and the scope offers the advantage of a visual display. Such a display can reveal common amplifier conditions (hum, distortion, noise, ripple, and oscillation).

When troubleshooting amplifier circuits with signal tracing, use a scope in much the same manner as a meter. Introduce a signal at the input with a generator (Fig. 1-TT). Measure the amplitude and waveform of the input signal on the scope. The input can be sine- or square-wave signals. Move the scope probe to the input and output of each stage, in turn, until the final output is reached. The gain of each stage is measured as voltage on the scope. Also, it is possible to observe any change in waveform from that which is applied to the input. Stage gain and distortion (if any) are established quickly with a scope.

Measuring gain in discrete stages

Take care when measuring the gain of discrete amplifier stages (especially in a circuit where there is feedback). For example, in Fig. 1-UUA if you measure the signal at the base of Q1, the base-to-ground voltage is not the same as the input voltage. To get the correct value of gain, connect the low side of the measuring device (meter or scope) to the emitter and the other lead (high side) to the base. In effect, measure the signal that appears across the base-emitter junctions. This measurement includes the effect of the feedback signal.

As a general safety precaution, never connect the ground lead of a meter or scope to the transistor base unless the lead connects back to an insulated inner chassis or board on the meter or scope. Large ground-loop currents (between the measuring device and the circuit being checked) can flow through the base-emit-

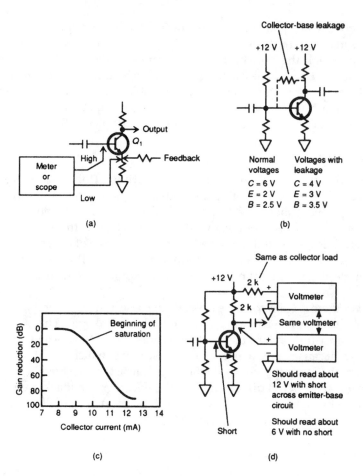

Fig. 1-UU   Basic discrete-circuit amplifier troubleshooting techniques.

ter junction and possibly burn out the transistor. This problem can usually be eliminated by an isolation transformer.

Low-gain problems

Low gain in a feedback amplifier can result in distortion. If gain is normal in a feedback amplifier, some distortion can be overcome. With low gain, the feedback might not be able to bring the distortion within limits. Of course, low gain by itself is sufficient cause to troubleshoot a circuit (with or without feedback).

Most feedback amplifiers have a very high open-loop gain that is set to some specific value by the ratio of resistor values (feedback-resistor value to input load-resistor value, as discussed). If the closed-loop gain is low in an experimental circuit, this usually indicates that the resistance values are incorrect. In an existing amplifier, the problem is usually where the open-loop gain is below the point where the resistors determine gain. When troubleshooting such a situation, if waveforms indicate low gain and transistor (or IC) voltages appear normal, try replacing the transistors (or the IC).

Do not overlook the possibility that the emitter-bypass capacitors (if any) might be open or leaking. If the capacitors are leaking (acting as a resistance in parallel with the emitter resistor), there is considerable negative feedback and little gain. Of course, a completely shorted emitter-bypass capacitor produces an abnormal voltage indication on the transmitter emitter (typically 0 V or ground).

Distortion problems in discrete-amplifier stages

Distortion can be caused by improper bias, overdriving (too much gain), or underdriving (too little gain, preventing the feedback signal from countering the distortion). One problem that is often overlooked in a discrete feedback-amplifier stage with a pattern of distortion is overdriving that results from transistor leakage. Generally, it is assumed that the collector-base leakage of a transistor reduces gain because the leakage is in opposition to signal-current flow. Although this is true for a single stage, it might not be true when more than one feedback stage is involved.

Whenever there is collector-base leakage, the base assumes a voltage nearer to that of the collector (nearer than is the case without leakage). This situation increases both transistor forward bias and transistor current flow. An increase in the transistor current causes a reduction in input resistance (which might or might not cause a gain reduction, depending on where the transistor is located in the circuit). If the feedback amplifier is direct coupled, the effects of feedback are increased. This is because the operating point (set by the base bias) of the following stage is changed, which could possibly result in distortion.

Effects of leakage on discrete circuit performance

When there is considerable leakage in a discrete circuit, the gain is reduced to 0, and/or the signal waveforms are drastically distorted. Such a condition also produces abnormal waveforms and transistor voltage. These indications make troubleshooting relatively easy. The troubleshooting problem becomes very difficult when there is just enough leakage to reduce circuit gain but not enough to distort the waveform seriously (or to produce transistor voltages that are way off).

Collector-base leakage

Collector-base leakage is the most common form of transistor leakage, and it produces a classic condition of low gain (in a single stage). When there is any collector-base leakage, the transistor is forward-biased or the forward bias is increased; see Fig. 1-UU(B). Collector-base leakage has the same effect as a resistance between the collector and base. The base assumes the same polarity as the collector (although at a lower value), and the transistor is forward biased. If leakage is sufficient, the forward bias can be enough to drive the transistor into or near saturation. When a transistor is operated at or near the saturation point, the gain is reduced (for a single stage), as shown in Fig. 1-UU(C).

IC amplifier circuits

Checking transistor leakage in-circuit

If the normal operating voltages are not known, as is the case with all experimental circuits, defective transistors can appear to be good because all of the voltage relationships are normal. The collector-base junction is reverse-biased (collector more positive than base for an npn) and the emitter-base junction is forward-biased (emitter less positive than base for an npn).

A simple way to check transistor leakage is shown in Fig. 1-UU(D). Measure the collector voltage to ground. Then short the base to the emitter and remeasure the collector voltage. If the transistor is not leaking, the base-emitter short turns the transmitter off and the collector voltage rises to the same value as the supply. If there is any leakage, a current path remains (through the emitter resistor, base-emitter short, collector-base leakage path, and collector resistor). Some voltage drop occurs across the collector resistor, and the collector voltage is at some value lower than the supply.

Most meters draw current, and the current passes through the collector resistor when you measure; see Fig. 1-UU(D). The current through the resistor can lead to some confusion, particularly if the meter draws heavy current (has a low ohms-per-volt rating). To eliminate any doubt, connect the meter to the supply through a resistor with the same value as the collector resistor. The voltage drop, if any, should be the same as when the transistor collector is measured to ground. If the drop is much different (lower) when the collector is measured, the transistor is leaking.

For example, assume that the collector measures 4 V with respect to ground; see Fig. 1-UU(D). This voltage means that there is an 5-V drop across the collector resistor and a collector current of 4 mA (8/2000 = 0.004). Normally, the collector is operated at about one-half the supply voltage (at 6 V in this example). Notice that simply because the collector is at 4 V, instead of 6 V, it does not mean that the circuit is faulty. Some circuits are designed that way.

In any event, the transistor should be checked for leakage with the emitter-base short test shown in Fig. 1-UU(D). Now assume that the collector voltage rises to 10 V when the base and emitter are shorted (within 2 V of the 12-V supply). This indicates that the transistor is cutting off, but there is still some current flow through the resistor, about 1 mA (2/2000 = 0.001).

A current flow of 1 mA is high for most present-day meters. To confirm a leaking transistor, connect the same meter through a 2-kΩ resistor (same as the collector-load resistor) to the 12-V supply (preferably at the same point where the collector resistor connects to the power supply). Now, assume that the indication is 11.7 V through the external resistor. This voltage shows that there is some transistor leakage.

The amount of transistor leakage can be estimated as follows: 11.7 − 10 = 1.7-V difference, and 1.7/2000 = 0.00085 = 0.85 mA. However, from a practical troubleshooting standpoint, the presence of any current flow with the transistor supposedly cut off is sufficient cause to replace the transistor.

Example of amplifier-circuit troubleshooting

This step-by-step troubleshooting problem involves locating the defective part (or improperly connected wiring) in a combination discrete-IC audio amplifier. Figure

1-VV shows the schematic diagram. This circuit was chosen as an example because it combines both IC and discrete components. The CA3094B is a programmable amplifier (where gain is set by the resistor at pin 5) that is similar to an OTA.

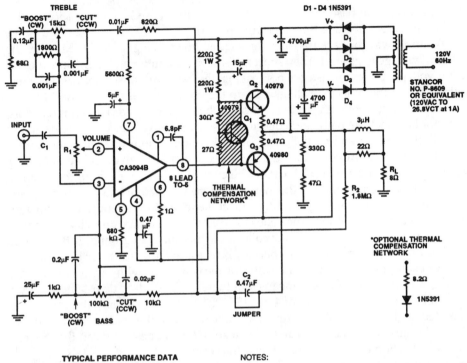

Fig. 1-VV Typical discrete-IC amplifier circuit and performance.

HARRIS SEMICONDUCTOR, LINEAR & TELECOM ICS, 1994, P. 2-100.

TYPICAL PERFORMANCE DATA
For 12W Audio Amplifier Circuit

Power Output (8Ω load, Tone Control set at "Flat")
Music (at 5% THD, regulated supply) 15W
Continuous (at 0.2% IMD, 60Hz and 2kHz
mixed in a 4:1 ratio, unregulated supply)
See Figure 8 in AN6048. . . . . 12W
Total Harmonic Distortion
At 1W, unregulated supply. 0.05%
At 12W, unregulated supply. 0.57%
Voltage Gain 40dB
Hum and Noise (below continuous power output). 83dB
Input Resistance 250kΩ
Tone Control Range. . . . . See Figure 9 in AN6048.

NOTES:
1. For standard input: Short C_2; R_1 = 250kΩ, C_1 = 0.047µF; remove R_2
2. For ceramic cartridge input: C_1 = 0.0047µF, R_1 = 2.5MΩ, remove jumper from C_2; leave R_2

No matter what the trouble symptom, the actual value can eventually be traced to one or more of the circuit parts (transistors, ICs, diodes, capacitors, etc.), unless you have wired the parts incorrectly! Even then, the following waveform, voltage, and resistance checks will indicate which branch within the circuit is at fault.

If you were servicing this circuit in existing equipment, the first step would be to study the literature and test the circuit to confirm the trouble. In this example, the only "literature" is Fig. 1-VV. The circuit description claims an output of 12-W

IC amplifier circuits

into an 8-Ω load. Although the load is shown as R_L, you can assume that the circuit will be used with an 8-Ω speaker. There are no test points or waveforms, the voltage information is incomplete, and there is no resistance-to-ground information. However, with this fragmentary data, you can test the circuit, monitor the signals at various points in the circuit, and localize trouble using the test results.

The first step is to apply a signal at the input and monitor the output. The input can be applied at C1 (as shown). The output is measured at R_L, or at an 8-Ω speaker connected in place of R_L. Use the resistor or the speaker but never operate the circuit without a load. Components Q2 and Q3, and possibly Q1, can be destroyed if they are operated without a load.

The output signal (at the junction of Q2/Q3 emitters and/or the speaker) must be about 10 V to produce 12 W across an 8-Ω load ($9.8^2/8 = 12$ W). If the circuit has a 40-dB voltage gain (as claimed), 0.1 V (100 mV) at the input should be sufficient to fully drive the speaker, depending on the setting of R1.

Connect an audio generator to the input and set the generator to produce 0.1 V at a frequency of 1 kHz. Set both the bass and treble controls to midrange and adjust R1 until you get a good tone on the speaker and/or a readable signal at the Q2/Q3-emitter junction. Adjust R1 for a 10-V signal at the speakers or emitter junctions. The tone will probably burst your eardrums at this point! Adjust R1 until the tone is reasonable and vary both the bass and treble controls. Both tone controls should have some effect on the tone, but the bass control should have the most control. Change the generator frequency to 10 kHz and repeat the tone-control test. Now the treble control should have the most effect.

If the circuit operates as described thus far, it is reasonable to assume that the circuit is good. Quit while you're ahead! If you have access to distortion meters, check distortion against the performance data on Fig. 1-VV, both THD and IMD. Also check the actual voltage input (at pin 2 of the IC) when a 1-V signal appears at the output and determine the true amplifier voltage gain (which should be 40 dB).

If the circuit does not operate as described, set R1 and the bass/treble controls to midrange and monitor the signal voltages at pins 2 and 8 of the IC. You can monitor all test points with an ac voltmeter, dc voltmeter with a rectifier probe, or with a scope. The scope is preferred because any really abnormal distortion at the test points appears on the scope display (as does the voltage).

If there is a signal at pin 2 of the IC but not at pin 8 (or the signal at pin 8 shows little gain over the pin-2 signal), the problem is at the IC portion of the circuit. Check all voltages at the IC. You do not know the exact values, but here are some hints.

The transformer has a 26.8-V center-tapped secondary, so $V+$ and $V-$ should be about 12 to 15 V (and should be substantially the same). In any event, pins 4 and 6 of the IC should be about -12 V, and pin 7 should be about $+12$ V (although pin 7 will probably be lower than pins 4 and 6 because of the 5600-Ω resistor at pin 7).

If the IC voltages appear to be good, but the gain at pin 8 is low, it is possible that the IC is bad, that the resistor at pin 5 is not of the correct value (this resistor determines IC gain, as discussed for OTAs and programmable amplifiers), or that there is too much feedback at pin 3 (from the output through C2 and the tone controls). Remember that most of the voltage gain for this amplifier is in the IC portion of the circuit. Q2 and Q3 provide power output.

Basic amplifier troubleshooting approaches

If there is a good signal at pin 8 of the IC but not at the output (speaker or R_L), the problem is at the discrete portion of the circuit (Q1, Q2, and Q3). Check the collector voltages of Q2 and Q3. These voltages should be about 12 V and should be substantially the same except of opposite polarity. Also, the Q2/Q3 voltages should be substantially the same as at pins 4, 6, and 7 of the IC.

The waveform (signal) and voltage checks that are described here should be sufficient to locate any major defect in the circuit (including an improperly wired experimental circuit!). Of course, if the circuit operates, but performance is not as claimed, it is possible that the problem is one of poor physical layout, wrong component values, and so on. Also, the basic techniques described here can be applied to the other circuits of this chapter.

IC amplifier circuit titles and descriptions

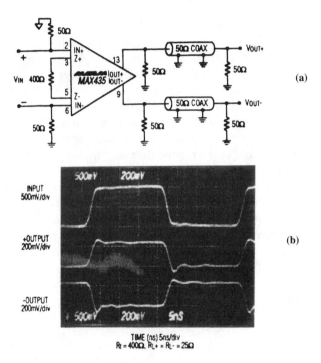

(a)

(b)

WTA coaxial-cable driver

Fig. 1-1 The circuit in the illustration shows a MAX435 WTA connected as a dual-output coaxial-cable driver. Figure 1-1B shows typical waveforms. See Fig. 1-PP for pin connections. Use a 5.9-kΩ resistor for R_{SET} (connected at I_{SET} pin 11). Bypass each power supply to ground with a 0.22-μF ceramic capacitor. Bypass I_{SET} pin 11 to V+ with a 0.22-μF ceramic. MAXIM NEW RELEASES DATA BOOK, 1994, PP. 8-15, 8-16.

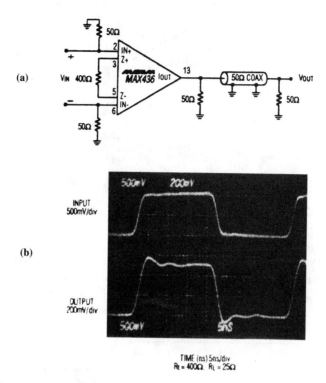

(a)

(b)

WTA coaxial-cable driver (single-ended)

Fig. 1-2   The circuit in the illustration shows a MAX436 WTA connected as a single-output coaxial-cable driver. Figure 1-2B shows typical waveforms. See Fig. 1-PP for pin connections. Use a 5.9-kΩ resistor for R_{SET} and 0.22-μF capacitors as bypass, as described for Fig. 1-1. MAXIM NEW RELEASES DATA BOOK, 1994, P. 8-16.

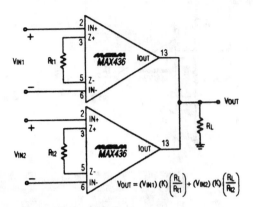

WTA summing amplifier

Fig. 1-3 Two (or more) signals can be summed together by tying the output terminals of the WTAs together as shown. The output voltage V_{OUT} is then the sum of the two (or more) input voltages, as shown by the equation. See Fig. 1-PP for pin connections. Use a 5.9-kΩ resistor for R_{SET} and 0.22-µF capacitors as bypass, as described for Fig. 1-1. MAXIM NEW RELEASES DATA BOOK, 1994, P. 8-16.

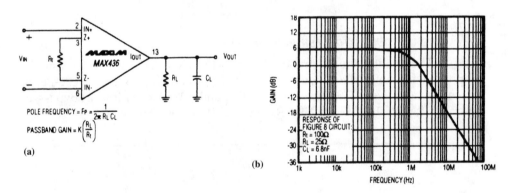

WTA lowpass amplifier

Fig. 1-4 Figure 1-4A shows the single-ended MAX436 connected as a lowpass amplifier. Figure 1-4B shows the gain/frequency response. Note that the reference to Figure 8 in Fig. 1-4B refers to Fig. 1-4A in this book. The pole frequency is determined by R_L and C_L as shown. The passband gain is set by K, R_L, and R_t. MAXIM NEW RELEASES DATA BOOK, 1994, P. 8-17.

IC amplifier circuits

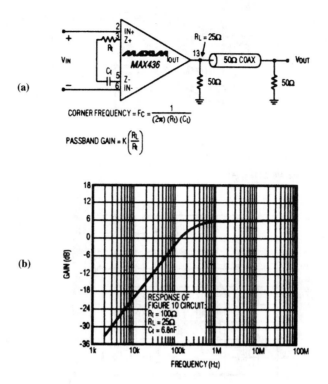

$$\text{CORNER FREQUENCY} = Fc = \frac{1}{(2\pi)(R_t)(C_t)}$$

$$\text{PASSBAND GAIN} = K\left(\frac{R_L}{R_t}\right)$$

WTA *highpass amplifier*

Fig. 1-5 Figure 1-5A shows the single-ended MAX436 connected as a highpass amplifier. Figure 1-5B shows the gain/frequency response. Note that the reference to Figure 10 in Fig. 1-5B refers to Fig. 1-5A in this book. The corner frequency is determined by R_t and C_t as shown. The passband gain is set by K, R_L, and R_t. See Fig. 1-1 for bypass/R_{SET} data. MAXIM NEW RELEASES DATA BOOK, 1994, P. 8-18.

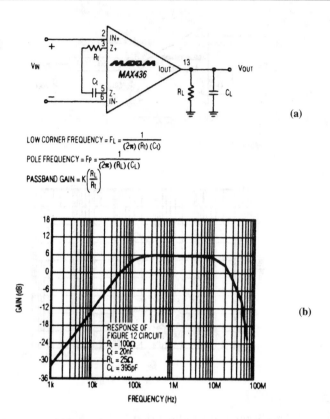

(a)

$$\text{LOW CORNER FREQUENCY} = F_L = \frac{1}{(2\pi)\,(R_t)\,(C_t)}$$

$$\text{POLE FREQUENCY} = F_P = \frac{1}{(2\pi)\,(R_L)\,(C_L)}$$

$$\text{PASSBAND GAIN} = K\left(\frac{R_L}{R_t}\right)$$

(b)

WTA bandpass amplifier

Fig. 1-6 Figure 1-6A shows the single-ended MAX436 connected as a bandpass amplifier. Figure 1-6B shows the gain/frequency response. Note that the reference to Figure 12 in Fig. 1-6B refers to Fig. 1-6A in this book. The low corner frequency is set by R_t/C_t, and the pole frequency is set by R_L/C_L. The passband gain is set by K, R_L, and R_t. Because there is no interaction between the transconductance-network impedance and the output-load impedance, poles, and zeros in the WTA transfer function can be independently set by the two impedance networks. See Fig. 1-1 for bypass/R_{SET} data. MAXIM NEW RELEASES DATA BOOK, 1994, P. 8-18.

IC amplifier circuits

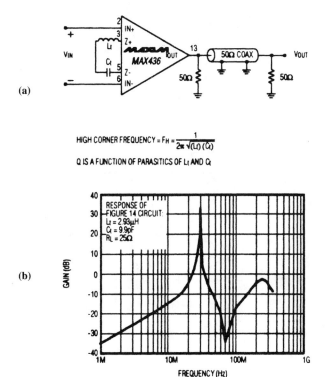

(a)

HIGH CORNER FREQUENCY = $F_H = \dfrac{1}{2\pi \sqrt{(L_t)(C_t)}}$

Q IS A FUNCTION OF PARASITICS OF L_t AND C_t

(b)

WTA tuned amplifier

Fig. 1-7 Figure 1-7A shows the single-ended MAX436 connected as a tuned amplifier. Figure 1-7B shows the gain/frequency response. Note that the Figure 14 referenced in Fig. 1-7B refers to Fig. 1-7A in this book. The high corner frequency is set by the transconductance network of L_t/C_t. The resonant frequency of the transconductance network is determined by the equation shown. The network impedance is minimum at the resonant frequency. This impedance provides maximum amplifier gain at that frequency. The Q (or tuning sharpness) of the amplifier is a function of the parasitic component associated with the L_t/C_t network. If the L_t/C_t network contains much resistance, the tuning will be broad (low Q). A minimum of resistance in the L_t/C_t network produces sharp tuning (high Q). See Fig. 1-1 for bypass/R_{SET} data. MAXIM NEW RELEASES DATA BOOK, 1994, P. 8-18.

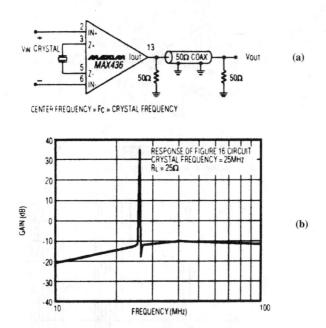

CENTER FREQUENCY = Fc = CRYSTAL FREQUENCY

(a)

(b)

WTA crystal-tuned amplifier

Fig. 1-8 Figure 1-8A shows the single-ended MAX436 connected as a crystal-tuned amplifier. Figure 1-8B shows the gain/frequency response. Note that the Figure 16 referenced in Fig. 1-8B refers to Fig. 1-8A of this book. The circuit of Fig. 1-8A provides a higher Q and more accurate control of the tuned-amplifier frequency. The crystal replaces the LC network shown in Fig. 1-7A. The crystal impedance is minimum at the resonant frequency (25 MHz), resulting in maximum gain for the amplifier at that frequency. See Fig. 1-1 for bypass/R_{SET} data. MAXIM NEW RELEASES DATA BOOK, 1994, P. 8-19.

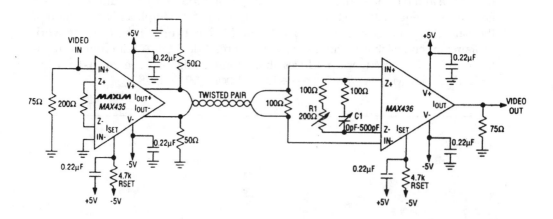

IC amplifier circuits

WTA video twisted-pair driver/receiver

Fig. 1-9 The circuit in the illustration uses the MAX435 and MAX436 connected as a driver/receiver for transmission of video signals over a twisted-pair line (instead of a coax cable). The circuit is good for distances up to 5000 feet (1500 meters), where a single channel of baseband (composite) video is to be transmitted. The twisted-pair must be balanced and properly terminated as shown. R1 is adjusted for proper brightness (to boost overall gain). C1 is adjusted for best color (to extend the bandwidth). Because these equalization adjustments are performed at the receiver end, you simply view the screen (TV, monitor, etc.) and adjust R_1/C_1 for best picture. MAXIM NEW RELEASES DATA BOOK, 1994, P. 8-19.

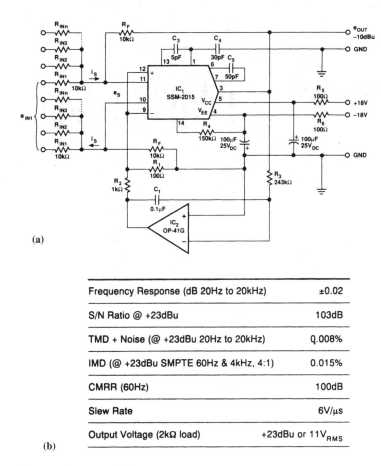

(a)

Frequency Response (dB 20Hz to 20kHz)	±0.02
S/N Ratio @ +23dBu	103dB
TMD + Noise (@ +23dBu 20Hz to 20kHz)	0.008%
IMD (@ +23dBu SMPTE 60Hz & 4kHz, 4:1)	0.015%
CMRR (60Hz)	100dB
Slew Rate	6V/µs
Output Voltage (2kΩ load)	+23dBu or 11V$_{RMS}$

(b)

Balanced summing amplifier

Fig. 1-10 Figure 1-10A shows two IC amplifiers connected as a balanced summing amplifier. Figure 1-10B shows the performance characteristics. IC2 serves as a dc servo amplifier that is referenced to signal ground. The circuit functions as an in-

IC amplifier circuit titles and descriptions

tegrator with a long time constant that retains the integrity of low-frequency audio signals down to 5 Hz and keeps the output voltage at 0 V ±10 mV. Because IC1 is a bipolar device, the noise is low (1.3 nV/$\sqrt{\text{Hz}}$). ANALOG DEVICES, APPLICATIONS REFERENCE MANUAL, 1993, PP. 4-3, 4-4.

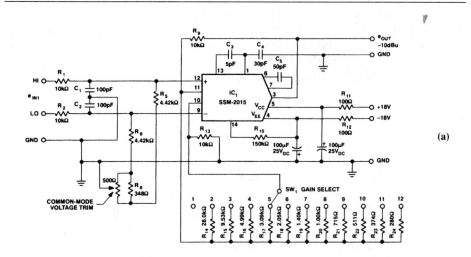

(a)

SW	G_{dB}	e_{IN}(dB)	R_G	VALUE (Ω)
1	10.0	0	R	ω
2	12.5	−2.5	R_{14}	28.0k
3	15.0	−5.0	R_{15}	9.53k
4	17.5	−7.5	R_{16}	4.99k
5	20.0	−10.0	R_{17}	3.09k
6	22.5	−12.5	R_{18}	2.05k
7	25.0	−15.0	R_{19}	1.40k
8	27.5	−17.5	R_{20}	1.00k
9	30.0	−20.0	R_{21}	715
10	32.5	−22.5	R_{22}	511
11	35.0	−25.0	R_{23}	374
12	37.5	−27.5	R_{24}	280

(b)

Specific gain can be calculated from the equation:

$$\text{Gain}_{dB} = 20 \log \left[3.5 + \left(\frac{20 \times 10^3}{R_G} \right) \right] \text{ for } R_8, R_{18} = 10.0\text{k}\Omega$$

IC amplifier circuits

Frequency Response (dB 20Hz to 20kHz)	±0.1
S/N Ratio @ +23dBu	103dB
THD + Noise (20Hz to 20kHz) @ +23dBu	0.008%
IMD (SMPTE 60Hz & 4kHz, 4:1) @ +23dBu	0.015%
CMRR (60Hz)	100dB
Slew Rate	6V/μs
Output Voltage (2kΩ load)	+23dBu or 11V$_{RMS}$

(c)

Balanced-input high-level amplifier

Fig. 1-11 Figure 1-11A shows a single IC connected as a balanced-input high-level amplifier, with selectable gain. Figure 1-11B shows the gain for various positions of the gain-selector switch SW1. Figure 1-11C shows the performance characteristics. The circuit can accept normal audio signals from –27.5 dBu to +0 dBu, with more than 30 dB of headroom. You can select gains other than those given using the equation of Fig. 1-11B. The circuit is well suited for use as the input amplifier in audio-distribution amplifiers, for a balanced-input audio-routing switcher, as the input buffer ahead of an A/D (analog-to-digital) in digital-recording and mixer equipment, or for the low-noise/high-level input of mixing consoles. (The equivalent input noise or E_{IN} is –124 dB.) The common-mode voltage trim potentiometer is included for maximum common-mode noise reduction, and allows the use of low-cost components. Capacitors C1 and C2 should be matched for 1% tolerance. ANALOG DEVICES, APPLICATIONS REFERENCE MANUAL, 1993, PP. 4-5, 4-6.

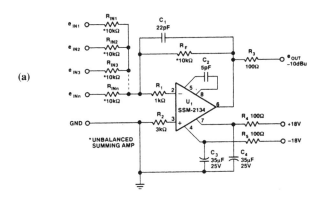

(a)

IC amplifier circuit titles and descriptions

Continued

Frequency Response (20Hz to 20kHz)	±0.02dB
S/N Ratio (@ +23dBu)	104dB
THD + Noise (@ +23dBu, 20Hz to 20kHz)	0.007%
IMD (SMPTE 60Hz and 4kHz, 4:1)	0.015%
Slew Rate	10V/μs
Output Voltage (2kΩ load)	+23.3dBu or 11.3V$_{RMS}$

(b)

Unbalanced, virtual-ground summing amplifier

Fig. 1-12 Figure 1-12A shows a single IC connected as an unbalanced, virtual-ground summing amplifier, with very low noise (2.8 nV/$\sqrt{\text{Hz}}$). Figure 1-12B shows the performance characteristics. The low noise is possible because the IC is a bipolar device, instead of the usual FET. ANALOG DEVICES, APPLICATIONS REFERENCE MANUAL, 1993, P. 4-7.

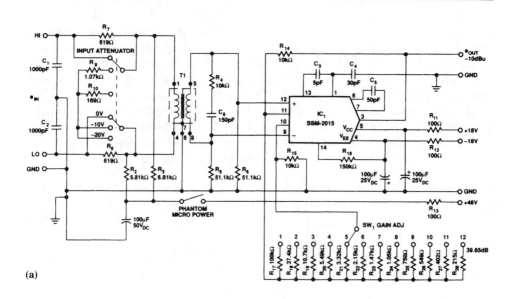

(a)

SW	G_{dB}	*e_{IN}(dB)	R_G	VALUE (Ω)
1	9.6	−37.5	R_{17}	100k
2	12.1	−40.0	R_{18}	37.4k
3	14.6	−42.5	R_{19}	10.7k
4	17.1	−45.0	R_{20}	5.49k
5	19.6	−47.5	R_{21}	3.32k
6	22.1	−50.0	R_{22}	2.15k
7	24.6	−52.5	R_{23}	1.47k
8	27.1	−55.0	R_{24}	1.05k
9	29.6	−57.5	R_{25}	750
10	32.1	−60.0	R_{26}	549
11	34.6	−62.5	R_{27}	402
12	39.6	−65.0	R_{28}	215

*Input attenuator set to the 0dB position.

Unspecified overall circuit gain can be calculated from the equation:

(b)

$$G_{dB} = 20 \log\left[3.5 + \left(\frac{20 \times 10^3}{R_G}\right)\right] + 17.9$$

Frequency Response (20Hz to 20kHz, −60dBu, 50dB gain)	±0.15dB
THD + Noise (20Hz to 20kHz, −60dBu, 50db gain)	0.045%
IMD (+23dBu, SMPTE 60Hz and 4kHz, 4:1)	0.05%
EIN (Equivalent Input Noise, 150Ω source)	−127dB
Input Impedance (20Hz to 5kHz)	1500Ω
Source Impedance	150Ω
CMR at 1kHz (common-mode rejection at 1kHz)	120dB
CMVR (common-mode voltage range)	±150V_{DC}
Slew Rate (overall circuit)	6V/µs
Gain Range (overall circuit)	17.5dB to 36dB
Output Voltage SSM-2015 (±18V_{DC}, 2kΩ load)	+23dBu or 11V_{RMS}
SSM-2016 (±24V_{DC}, 2kΩ load)	+25.7dBu or 15V_{RMS}
Output Headroom (SSM-2015, 2kΩ load, −10dBu nominal)	33dB

(c)

High-performance, transformer-coupled microphone preamplifier

Fig. 1-13 Figure 1-13A shows a single IC connected as a transformer-coupled microphone preamplifier, with selectable gain. Figure 1-13B shows the gain for various positions of the gain-selector switch SW1. Figure 1-13C shows the performance characteristics. You can select gains other than those given using the equation of Fig. 1-13B. Microphone input loading is 1.5 kΩ. The input circuit contains a three-position input attenuator using optimized source levels versus amplifier headroom. A phantom microphone-power circuit is included for condenser microphones that require 24 to 48 V. Where additional headroom is required, use the SSM-2016, which can be powered up to ±36 V. When powered by ±24 V, headroom increases to 35.7 dB with E_{IN} of –127 dB and an IC dissipation of 600 mW (no signal) and 725 mW (worst case). When powered by ±36 V, headroom increases to 39.3 mW with an IC dissipation of 1.2 W (no signal) and 1.5 W (worst case). ANALOG DEVICES, APPLICATIONS REFERENCE MANUAL, 1993, PP. 4-9, 4-10.

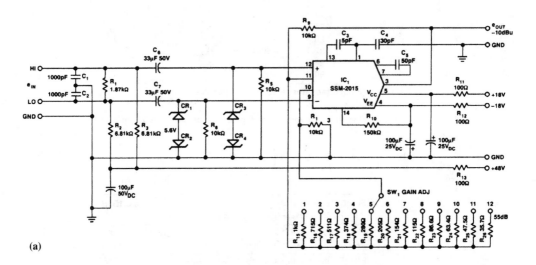

(a)

SW	G_{dB}	$e_{IN}(dB)$	R_G	VALUE (Ω)
1	27.5	–37.5	R_{15}	1.00k
2	30	–40	R_{16}	715
3	32.5	–42.5	R_{17}	511
4	35	–45	R_{18}	374
5	37.5	–47.5	R_{19}	280
6	40	–50	R_{20}	205
7	42.5	–52.5	R_{21}	154
8	45	–55	R_{22}	115
9	47.5	–57.5	R_{23}	86.6
10	50	–60	R_{24}	63.4
11	52.5	–62.5	R_{25}	47.5
12	55	–65	R_{26}	35.7

(b)

IC amplifier circuits

Continued

Frequency Response (20Hz to 20kHz)	±0.1dB
THD + Noise (@ +23dBu, 20Hz to 20kHz)	0.03%
IMD (@ +23dBu, SMPTE 60Hz & 4kHz, 4:1)	0.05%
EIN (Equivalent Input Noise, 150Ω source)	−124dB
CMR (Common-Mode Rejection at 1kHz)	105dB
Slew Rate	6V/μs
Output Voltage (2kΩ load)	+23dBu or 11V$_{RMS}$
Output Headroom (2kΩ load, −10dBu nominal)	33dB

(c)

Balanced low-noise microphone preamplifier

Fig. 1-14 Figure 1-14A shows a single IC connected as a balanced low-noise microphone preamplifier, with selectable gain. Figure 1-14B shows the gain for various positions of the gain-selector switch SW1. Figure 1-14C shows the performance characteristics. You can select gains other than those given with the equation:

$$\left[3.5 + \left(\frac{20 \times 10^3}{RG}\right)\right]$$

where *RG = gain resistance.*

A phantom microphone-power circuit is included for condenser microphones that require 24 to 48 V. Capacitors CR1–CR4 protect the input transistors of the IC when connecting the microphone to the preamplifier circuit. Common-mode voltage range is ±5.5 V. ANALOG DEVICES, APPLICATIONS REFERENCE MANUAL, 1993, PP. 4-11, 4-12.

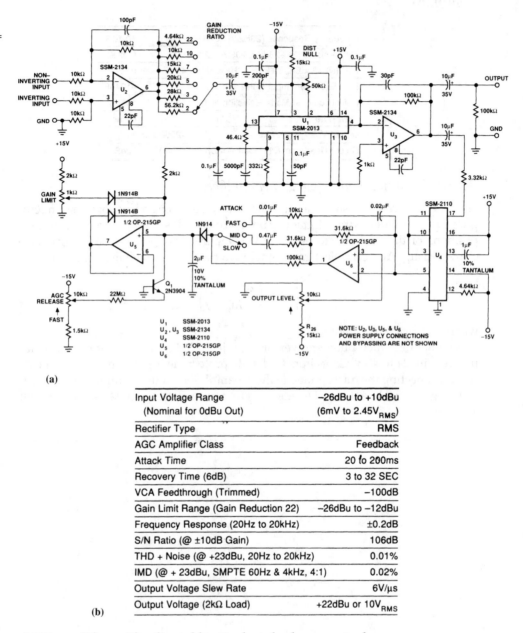

(a)

Input Voltage Range	−26dBu to +10dBu
(Nominal for 0dBu Out)	(6mV to 2.45V_{RMS})
Rectifier Type	RMS
AGC Amplifier Class	Feedback
Attack Time	20 to 200ms
Recovery Time (6dB)	3 to 32 SEC
VCA Feedthrough (Trimmed)	−100dB
Gain Limit Range (Gain Reduction 22)	−26dBu to −12dBu
Frequency Response (20Hz to 20kHz)	±0.2dB
S/N Ratio (@ ±10dB Gain)	106dB
THD + Noise (@ +23dBu, 20Hz to 20kHz)	0.01%
IMD (@ + 23dBu, SMPTE 60Hz & 4kHz, 4:1)	0.02%
Output Voltage Slew Rate	6V/μs
Output Voltage (2kΩ Load)	+22dBu or 10V_{RMS}

(b)

AGC amplifier with adjustable attack and release control

Fig. 1-15 Figure 1-15A shows an *AGC* (automatic gain control) amplifier with selectable gain-reduction compression ratios and time-domain adjustable AGC attack
and release. Figure 1-15B shows the performance characteristics. Six gain-reduction
slope ratios can be selected. The AGC output level is set by the rectified signal voltage compared to the reference voltage from the OUTPUT LEVEL control. The

IC amplifier circuits

three-position ATTACK switch allows selection of fast, medium, and slow compression and AGC response. The gain recovery is linear and time adjustable and has maximum gain-limiting (gating) to preclude input source-noise floor rise. The AGC release rate is controlled by the constant-current discharge of the integrator capacitor. Recovery time is linear and can be adjusted by changing the integrator discharge current supplied by Q1 and regulated by the AGC RELEASE control. ANALOG DEVICES, APPLICATION REFERENCE MANUAL, 1993, PP. 4-13, 4-14.

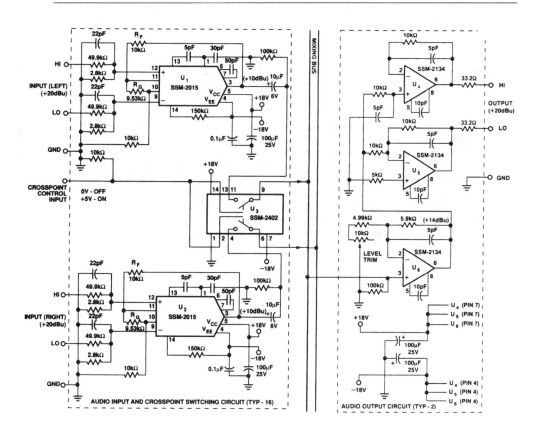

(a)

Continued

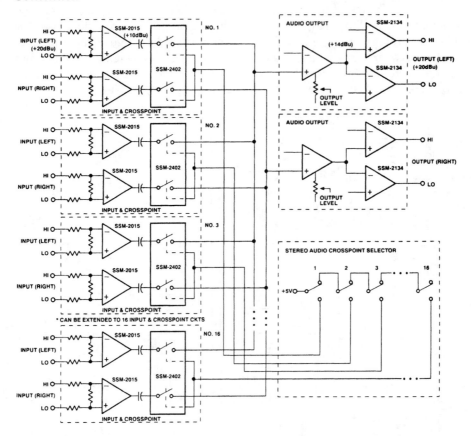

(b)

Max Input Level	+30dBu
Input Impedance, Unbalanced	100kΩ
Input Impedance, Balanced	200kΩ
Common-Mode Rejection (20Hz to 20kHz)	>70dB
Common-Mode Voltage Limit	±98V Peak
Max Output Level	+30dBu/dBm
Output Impedance	67Ω
Gain Control Range	±2dB
Output Voltage Slew Rate	6V/µs
Frequency Response (±0.05dB)	20Hz to 20kHz
Frequency Response (±0.5dB)	10Hz to 50kHz
THD + Noise (20Hz to 20kHz, +8dBu)	0.005%
THD + Noise (20Hz to 20kHz, +24dBu)	0.03%
IMD (SMPTE 60Hz & 4kHz, 4:1, +24dBu)	0.02%
Crosstalk (20Hz to 20kHz)	>80dB
S/N Ratio @ 0dB Gain	135dB

(c)

IC amplifier circuits

High-performance stereo routing switcher

Fig. 1-16 Figure 1-16A shows a 16-to-1 (channels) stereo-audio routing switcher. Figure 1-16B shows a functional diagram of the switcher. Figure 1-16C shows the performance characteristics. Switching control of the SSM-2402 can be activated by mechanical switches, or 5-V *TTL/CMOS* (transistor-transistor logic/complementary metal-oxide semiconductor) logic, using many X/Y control schemes, destination control, or computer control, as needed. The output level of each channel can be set individually. ANALOG DEVICES, APPLICATIONS REFERENCE MANUAL, 1993, PP. 4-15, 4-16, 4-17.

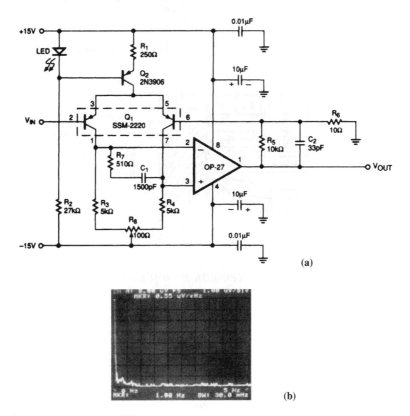

(a)

(b)

Ultra low-noise preamplifier

Fig. 1-17 Figure 1-17A shows a preamplifier suitable for low-output transducers such as audio microphones, magnetic pickups or low-impedance strain gauges. Figure 1-17B shows a spectrum analyzer display of total output noise when driven through a 10-Ω source impedance. The circuit can provide a gain of 1000, over a 200-kHz bandwidth, with a noise of 0.5 nV/$\sqrt{\text{Hz}}$. The input stage current is set by the current source of Q2, R1, and a GaAsP *LED* (light-emitting diode) used as a 1.6-V zener with a temperature coefficient almost identical to that of the Q2 base-emitter junction. Potentiometer R8 provides for input-offset adjustment. THD is less than 0.005% (with a 10-V signal from 20 Hz to 20 kHz). ANALOG DEVICES, APPLICATIONS REFERENCE MANUAL, 1993, PP. 4-55, 4-56.

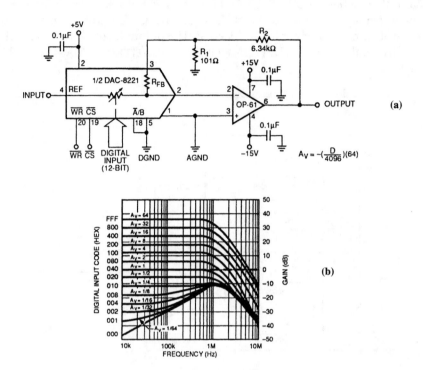

(a)

$$A_V = -\left(\frac{D}{4096}\right)(64)$$

(b)

Digitally programmable gain/attenuation amplifier

Fig. 1-18 Figure 1-18A shows a digitally programmable amplifier that can produce gain as well as attenuation in the range of ⅟₆₄ to 64, using a 12-bit D/A (digital-to-analog) converter (Chapter 9). Figure 1-18 shows the gain versus frequency response for each of the 12 binary gain settings. As shown by the equation, the circuit has a transfer function of:

$$A_V = -\left(\frac{D}{4096}\right)(64)$$

where *D* represents the binary-weighted digital code of the D/A converter. ANALOG DEVICES, APPLICATIONS REFERENCE MANUAL, 1993, P. 8-83.

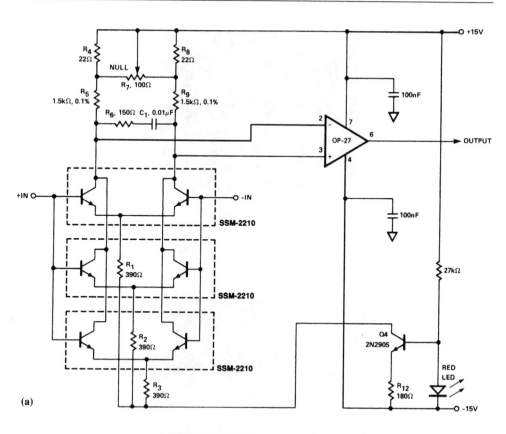

(a)

Input Noise Voltage Density at 1kHz		500pV/√Hz
Input Noise Voltage from 0.1Hz to 10Hz		40nV$_{p-p}$
Input Noise Current at 1kHz		1.5pA/√Hz
Gain-Bandwidth	G = 10	3MHz
	G = 100	600kHz
	G = 1000	150kHz
Slew Rate		2V/μs
Open-Loop Gain		3×10^7
Common-Mode Rejection		130dB
Input Bias Current		3μA
Supply Current		10mA
Nulled TCV$_{OS}$		0.1μV/°C Max
T.H.D. at 1kHz	G = 1000	0.002%

(b)

IC amplifier circuit titles and descriptions

Continued

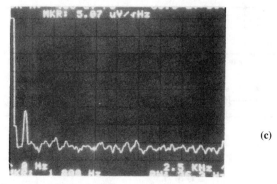

(c)

Spectrum analyzer display of broadband noise with a
gain of 10,000.
Horizontal axis = 0 to 2.5kHz.
Normalized vertical axis = $830pV/\sqrt{Hz}$ R.T.I.
$e_n = 507pV/\sqrt{Hz}$ at 1kHz.

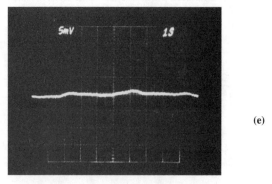

(d)

Low frequency noise spectrum at a gain of 10,000
showing a low 1.5Hz noise corner.

(e)

Peak-to-peak noise from 0.1 to 10Hz.
Overall gain is 100,000.

IC amplifier circuits

Low-noise op amp

Fig. 1-19 Figure 1-19A shows an op-amp circuit suitable for high-precision instrumentation, or for use as a strain-gage amplifier, tape-head preamplifier, microphone preamplifier, and any application where low-output, low-impedance transducers are used. Such transducers require very low-voltage noise to maintain a good signal-to-noise ratio. Figure 1-19B shows the performance characteristics. Figures 1-19C through 1-19E show spectrum analyzer and oscilloscope displays. The circuit achieves the low-noise capability using three SSM-2210 dual transistors at the input. Current for the input stage is set by Q4 and a GaAsP LED. The voltage difference between the forward voltage of the LED and the base-emitter voltage of Q4 (about 1 V) is applied across R12 to produce a temperature-stable emitter current. Potentiometer R7 provides for input-offset adjustment. You can get input-offset drives of less than 0.1 μV/°C when the offset is nulled close to zero. If this null is not required, omit R4, R7, R8, and connect R5/R9 directly to V+. ANALOG DEVICES, APPLICATIONS REFERENCE MANUAL, 1993, PP. 13-36, 13-37.

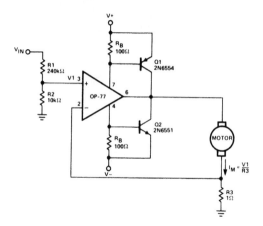

Servomotor amplifier

Fig. 1-20 In this op-amp circuit, R_B sets the bias point for transistors Q1/Q2. Because $V_{BE(ON)}$ varies greatly with temperature, a guard band is required to prevent Q1 and Q2 from conducting simultaneously. Select R_B such that the transistors do not conduct until I_M equals the op-amp quiescent supply current I_{SY}. Q1 and Q2 will begin to conduct at about $V_{BE(ON)} = 0.5$ V. In this circuit:

$$R_B = \frac{\left[V_{BE(ON)}\right]}{\left(I_{SY} + I_M\right)} = \frac{0.5}{(0.0025 + 0.0025)} = 100 \ \Omega$$

IC amplifier circuit titles and descriptions

To maximize voltage swing across the motor, V_1 must be minimized. If at full load $V_1 = 0.2$ V with $V+ = 15$ V and $V_{BE1} = 0.8$ V, the voltage across the motor will be:

$$V_M = (V+ - 2) - V_{BE1} - V_1 = (15 - 2) - 0.8 - 0.2 = 12 \text{ V}$$

V_{IN} can be scaled with a resistive divider as:

$$\frac{V_{IN}}{V_1} = \frac{(R_1 + R_2)}{R_2}$$

with $R_1 = 240$ kΩ and $R_2 = 10$ kΩ, $V_{IN} = 5$ will produce $I_M = 200$ mA. ANALOG DEVICES, APPLICATIONS REFERENCE MANUAL, 1993, P. 13-49.

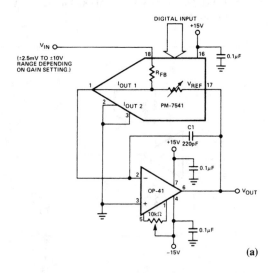

(a)

DIGITAL IN	GAIN (A_V)
4095	−1.00024
2048	−2
1024	−4
512	−8
256	−16
128	−32
64	−64
32	−128
16	−256
8	−512
4	−1024
2	−2048
1	−4096
0	OPEN LOOP

(b)

IC amplifier circuits

Precision programmable-gain amplifier

Fig. 1-21 In this op-amp circuit, the digitally programmable gain has 12-bit accuracy over the range of –1 to –1024 and 10-bit accuracy to –4096. The low bias current of the OP-41 JFET (junction field-effect transistor) input maintains this accuracy. Capacitor C1 limits the noise-voltage bandwidth allowing accurate measurements down to microvolt levels. The gain/digital relationships are shown in Fig. 1-21B. ANALOG DEVICES, APPLICATIONS REFERENCE MANUAL, 1993, P. 13-50.

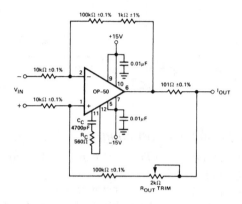

Bilateral current source

Fig. 1-22 In this op-amp circuit, the compliance is better than ±11 V at an output current of 20 mA, and the trimmed output resistance is typically 2 MΩ with R_L ≤500 Ω. For the resistor values shown, the maximum V_{IN} is 200 mV. ANALOG DEVICES, APPLICATIONS REFERENCE MANUAL, 1993, P. 13-50.

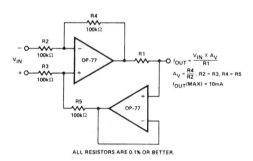

ALL RESISTORS ARE 0.1% OR BETTER.

Precision current pump

Fig. 1-23 In this op-amp circuit, the accuracy of I_{OUT} is improved with a noninverting voltage follower in the feedback loop. To maximize voltage compliance of I_{OUT}, keep R_1 to a minimum. ANALOG DEVICES, APPLICATIONS REFERENCE MANUAL, 1993, P. 13-50.

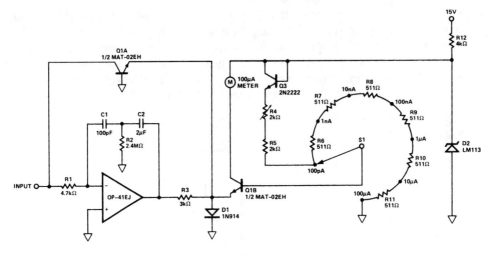

Op-amp, wide-range, low-current ammeter

Fig. 1-24 The ammeter shown in Fig. 1-24 can measure currents from 100 pA to 100 μA without high-value resistors. Accuracy is 1% over most of the range, depending upon accuracy of the divider resistor and the input-bias current of the op-amp. Using the OP-41 as the input amplifier allows low-end measurement down to a few pA (because of the 5-pA input-bias current). Because the voltage across the inputs of the inverting amplifier is forced to virtual zero, the current meter's effective series voltage drop is less than 500 μV at any current level. R4 is used to adjust full-scale deflection with a 1-μA input current, which gives maximum accuracy over the range of currents. The low V_{OS} and exceptionally good log performance of the MAT-02 assure high accuracy over the full six-decade operating range. ANALOG DEVICES, APPLICATIONS REFERENCE MANUAL, 1993, P. 13-53.

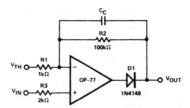

Op-amp precision threshold detector/amplifier

Fig. 1-25 When V_{IN} is less than V_{TH}, the amplifier output swings negative, reverse-biasing diode D1. As a result $V_{OUT} = V_{TH}$. When V_{TH} is equal to or greater than V_{TH}, the loop closes and:

$$V_{OUT} = V_{TH} + (V_{IN} - V_{TH})\left(1 + \frac{R_2}{R_1}\right)$$

Capacitor C_C is selected to smooth the loop response. ANALOG DEVICES, APPLICATIONS REFERENCE MANUAL, 1993, P. 13-53.

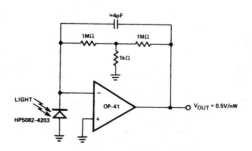

Op-amp wide-dynamic-range light detector

Fig. 1-26 The circuit in the illustration produces an output voltage proportional to light input over a 60-dB range. The 5-pA input-bias current of the OP-41 assures a low-output voltage offset. ANALOG DEVICES, APPLICATIONS REFERENCE MANUAL, 1993, P. 13-53.

IC amplifier circuit titles and descriptions

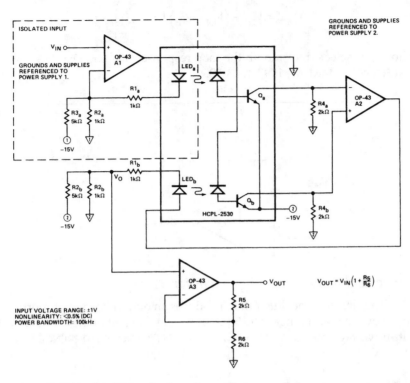

Op-amp isolation amplifier

Fig. 1-27 This isolation amplifier operates on the principle that the nonlinearities of one optocoupler will be tracked by the nonlinearities of another, if they are well matched. By using an optocoupler in the feedback loop of A2, nonlinearities of the isolating optocoupler will be canceled. V_O will equal V_{IN}, plus an offset created by imperfect matching between a and b side resistors and optocouplers. With stable supplies, the circuit displays less than 0.5% dc nonlinearity with a 2-V_{p-p} signal. Power bandwidth is 100 kHz. The dual optocoupler provides isolation against 600-Vdc common-mode voltages. ANALOG DEVICES, APPLICATIONS REFERENCE MANUAL, 1993, P. 13-54.

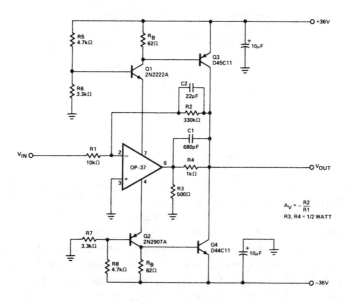

±36-V low-noise op amp

Fig. 1-28 An OP-37 provides a low-noise front end for this amplifier, which can deliver more than ±200 mA to a load with a 70-V peak-to-peak output swing. Transistors Q1 and Q2 act as series regulators, which are used to step down the supply voltage for the OP-37 to ±15 V. Transistors Q3 and Q4 provide the high current output drive. Resistors R3 and R4 provide for an output voltage gain of 3. This gain is reduced to unity at high frequencies by C_1 (to maintain stability). ANALOG DEVICES, APPLICATIONS REFERENCE MANUAL, 1993, P. 13-55.

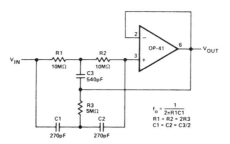

Op-amp high-Q notch filter

Fig. 1-29 The low bias current and high input impedance of the OP-41 permit small-value capacitors and large resistors to be used in this 60-Hz notch filter. The 5-pA bias current develops 100 μV across R1 and R2. Other notch frequencies can be produced by proper selection of R_1 and C_1, as shown by the equation. ANALOG DEVICES, APPLICATIONS REFERENCE MANUAL, 1993, P. 13-56.

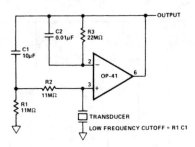

Piezoelectric transducer amplifier

Fig. 1-30 Piezoelectric transducers often require a high-input-resistance amplifier. The OP-41 can provide input resistance in the range of 10^{12} Ω. However, a dc return for the bias current is needed. To maintain a high R_{IN}, large-value resistors greater than 22 MΩ often are required, making the circuit impractical. Using this circuit, input resistances are reduced by bootstrapping the resistors to the output. With this arrangement, the lower cutoff frequency is determined more by the *RC* product of R_1 and C_1 than by the resistor values and the equivalent capacitance of the transducer. ANALOG DEVICES, APPLICATIONS REFERENCE MANUAL, 1993, P. 13-56.

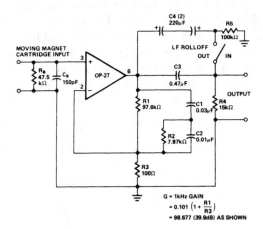

RIAA phono pre-amplifier

Fig. 1-31 With a 1-kΩ source, circuit noise measures 63 dB below a 1-mV reference level, unweighted, in a 20-kHz bandwidth. Precision metal-film resistors and film capacitors of polystyrene or polypropylene are recommended. Avoid high-*K* ceramics. C3 and R4 form a simple –6-dB per octave rumble filter (with a corner at 22 Hz), which can be bypassed by C4 (a nonpolarized electrolytic). Circuit distortion is generally below 0.01% at levels up to 7 V_{rms}. At 3-V output, the circuit produces less than 0.03% THD at frequencies up to 20 kHz. ANALOG DEVICES, APPLICATIONS REFERENCE MANUAL, 1993, P. 13-57.

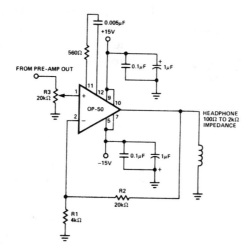

Headphone amplifier

Fig. 1-32 The circuit in the illustration provides amplification for one headphone. Two identical circuits are required for stereo headphones. For low-level pre-amp output signals, the amplifier gain can be increased by reducing R1 according to:

$$R_1 = \frac{20 \text{ k}\Omega}{(AV - 1)}$$

Circuit performance (with $V_{OUT} = 6$ V$_{rms}$ and $R_1 = 4$ kΩ) is as follows: THD @ 100 Hz = 0.0025%, @ 1 kHz = 0.003%, @ 10 kHz = 0.011%; signal-to-noise ratio ≥80 dB; response flatness = ±0.4 dB from 10 Hz to 20 kHz; bandwidth = –3 dB at 56 kHz. ANALOG DEVICES, APPLICATIONS REFERENCE MANUAL, 1993, P. 13-57.

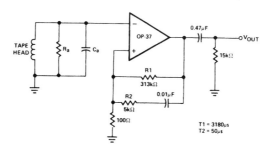

NAB tape-head pre-amplifier

Fig. 1-33 The circuit in the illustration produces 50 dB gain at 1 kHz and a dc gain greater than 70 dB. Worst-case output offset is just higher than 500 mV. To minimize the tape head offset, head dc resistance should be below 1 kΩ. A single 47-μF output capacitor can block the final output offset without affecting the dynamic range. The tape head can be coupled directly to the amplifier input because the worst-case bias current of 80 nA with a 400-mH, 100-μin (micro-inch) head (such as the PRB2H7K) will not be troublesome. Avoid transients at power-up and power-down, which can magnetize the head. ANALOG DEVICES, APPLICATIONS REFERENCE MANUAL, 1993, P. 13-58.

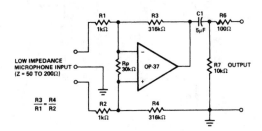

Microphone pre-amplifier

Fig. 1-34 The transformerless circuit in Fig. 1-34 amplifies differential signals from low-impedance of 2 kΩ. Bandwidth is 110 kHz. A dummy resistor might be necessary if the microphone is to be unplugged (to prevent feedback and amplifier oscillation). Either close-tolerance (0.1%) bridge resistors should be used, or R4 should be trimmed for best CMRR. All resistors should be metal-film types. Total circuit noise should be less than 6 nV/$\sqrt{\text{Hz}}$, which is equivalent to 0.9 μV in a 20-kHz noise bandwidth, or nearly 61 dB below a 1-mV input signal. ANALOG DEVICES, APPLICATIONS REFERENCE MANUAL, 1993, P. 13-58.

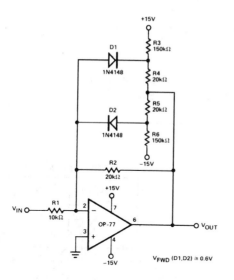

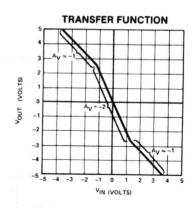

Piecewise-linear amplifier (decreasing gain)

Fig. 1-35 The circuit in the illustration is used to linearize a nonlinear input signal or to create a nonlinear output function from a linear input, as shown by the transfer function. At $V_{OUT} = 0$ V, both D1 and D2 are reverse biased and

$$\frac{V_{OUT}}{V_{IN}} = -\frac{R_2}{R_1}$$

As V_{IN} goes positive, V_{OUT} becomes negative according to that gain until a threshold is reached,

$$\frac{-(V_{OUT} + V_{FWD})}{R_4 4} = \frac{15\ V}{R_3}$$

where D1 becomes forward biased. For more positive values of V_{IN}, the gain is:

$$\frac{V_{OUT}}{V_{IN}} = -\frac{(R_2 \bullet R_4)}{R_2 + R_4)\ R_1} - \frac{R_2 \bullet V_{FWD}}{(R_2 + R_5)\ V_{IN}}$$

A similar action occurs as V_{IN} goes negative. Beyond the point where D2 becomes forward biased,

$$\frac{(V_{OUT} - V_{FWD})}{R_5} = \frac{15\ V}{R_6}$$

IC amplifier circuit titles and descriptions

the gain is:

$$\frac{V_{OUT}}{V_{IN}} = -\frac{(R_2 \bullet R_5)}{(R_2 + R_5)R_1} - \frac{R_2 \bullet V_{FWD}}{(R_2 + R_5)\ V_{IN}}$$

Additional diode/resistor combinations can be added to further contour the gain. ANALOG DEVICES, APPLICATIONS REFERENCE MANUAL, 1993, P. 13-59.

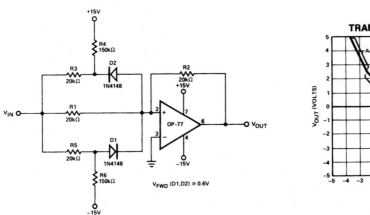

Piecewise-linear amplifier (increasing gain)

Fig. 1-36 The circuit in the illustration performs the linear-to-nonlinear or nonlinear-to-linear transformation by increasing gain beyond fixed thresholds, as shown by the transfer function. At $V_{IN} = 0$ V, both D1 and D2 are reverse biased and

$$\frac{V_{OUT}}{V_{IN}} = \frac{R_2}{R_1}$$

As V_{IN} goes positive beyond the threshold,

$$\frac{(V_{IN} - V_{FWD})}{R_5} = \frac{15\ V}{R_6}$$

D1 conducts, and the gain becomes:

$$\frac{V_{OUT}}{V_{IN}} = \frac{R_2(R_1 + R_5)}{(R_1 \bullet R_5)} + \frac{V_{FWD} \bullet R_2}{V_{IN} \bullet R_5}$$

IC amplifier circuits

As V_{IN} goes negative beyond the threshold,

$$\frac{-(V_{IN} + V_{FWD})}{R_3} = \frac{15\ V}{R_4}$$

D2 conducts, and the gain becomes:

$$\frac{V_{OUT}}{V_{IN}} = -\frac{R_2(R_1 + R_3)}{(R_1 \bullet R_3)} + \frac{V_{FWD} \bullet R_2}{V_{IN} \bullet R_3}$$

Additional diode/resistor combinations can be added to further contour the gain. ANALOG DEVICES, APPLICATIONS REFERENCE MANUAL, 1993, P. 13-60.

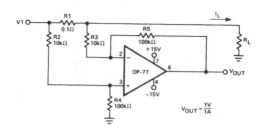

Current monitor

Fig. 1-37 The monitor circuit in the illustration can sense current at any point between the ±15-V supplies ($|V_1|$ less than 14.3 V guaranteed). This capability makes the circuit suitable for sensing current in applications such as full-bridge drivers where bidirectional current is associated with large common-mode voltage changes. The 120-dB CMRR of the OP-77 makes the amplifier's contribution to common-mode error negligible, leaving only the error produced by the resistor-ratio inequality. Ideally, $R_2/R_4 = R_3/R_5$. This ratio is best trimmed by R_4. ANALOG DEVICES, APPLICATIONS REFERENCE MANUAL, 1993, P. 13-60.

IC amplifier circuit titles and descriptions

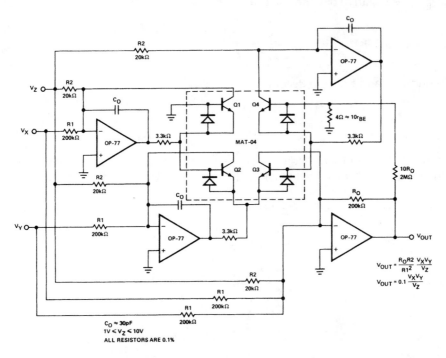

Precision analog multiplier/divider

Fig. 1-38 The multiplier/divider achieves precision because of the low emitter resistance, r_{BE} and superior V_{OS} matching of the MAT-04 quad transistor array. In this circuit, linearity error because of the transistors is less than ±0.1%. The OP-77 helps maintain accuracy with a V_{OS} less than 25 µV. For even higher accuracy, the offset voltages can be nulled. ANALOG DEVICES, APPLICATIONS REFERENCE MANUAL, 1993, P. 13-61.

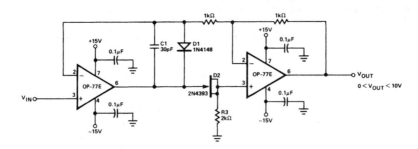

Precision absolute-value amplifier

Fig. 1-39 In the circuit in Fig. 1-39, the signal always appears as a common-mode signal to the op amps. The OP-77E CMRR of 1 µV/V assures errors of less than 2 ppm (pulses per minute). The high gain and low TCV_{OS} assure accurate operation with inputs from microvolts to volts. ANALOG DEVICES, APPLICATIONS REFERENCE MANUAL, 1993, P. 13-61.

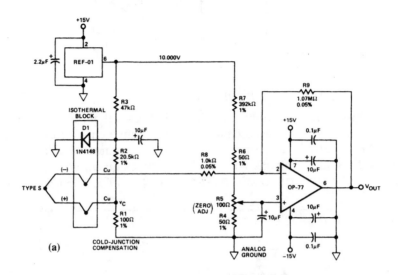

	TYPE	SEEBECK COEFFICIENT, α	R1	R2	R7	R9
(b)	K	39.2μV/°C	110Ω	5.76kΩ	102kΩ	269kΩ
	J	50.2μV/°C	100Ω	4.02kΩ	80.6kΩ	200kΩ
	S	10.3μV/°C	100Ω	20.5kΩ	392kΩ	1.07MΩ

Thermocouple amplifier with cold-junction compensation

Fig. 1-40 The high gain, low noise and low offset drift of the OP-77 are combined to produce a thermocouple amplifier with a total accuracy typically better than ±0.5°C. Cold-junction compensation is performed by R1, R2, and D1, which are mounted isothermally with the thermocouple terminating junctions. Calibration is done using R5, after the circuit has stabilized for about 15 minutes. A copper-wire short is applied across the terminating junctions, simulating a 0°C ice point. Then adjust R5 for 0.000-V output. Remove the short and the amplifier is ready to use. The circuit can be used for type S, J, and K thermocouples, with the appropriate resistor values as shown in Fig. 1-40B. R9 is chosen to give a +10,000-V output for a 1000°C measurement. ANALOG DEVICES, APPLICATIONS REFERENCE MANUAL, 1993, P. 13-62.

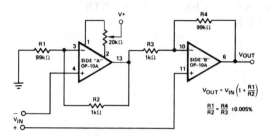

Two op-amp instrumentation amplifier

Fig. 1-41 The differential amplifier in Fig. 1-41 offers high input impedance and a PSRR greater than 100 dB. Because the high circuit gain occurs in side B, side A offset should be adjusted with respect to V_{OUT}. This adjustment simultaneously corrects for side B offset voltage. For the circuit values given, voltage gain is 100. ANALOG DEVICES, APPLICATIONS REFERENCE MANUAL, 1993, P. 13-62.

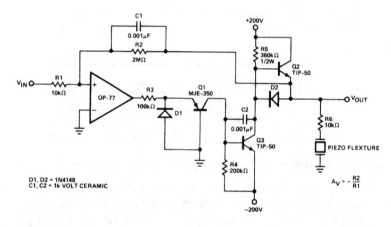

±200-V low-offset op amp

Fig. 1-42 The circuit in the illustration is designed to drive piezo-fixture elements in precise-positioning applications. The high-gain and low-offset voltage of the OP-77 produce an accurate, stable drive voltage to the piezo device, allowing predictable, repeatable submicron movements to be easily controlled. The output of the OP-77 creates a proportional current drive through the common-base-connected Q1 to the base of Q3. Transistor Q3 forms a Class A amplifier with bias resistor R5. Transistor Q2 is an emitter-follower used to boost output current. Resistor R6 prevents oscillation caused by the capacitive loading of the piezo device on the amplifier output. ANALOG DEVICES, APPLICATIONS REFERENCE MANUAL, 1993, P. 13-63.

IC amplifier circuits

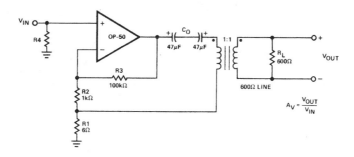

Impedance transforming amplifier

Fig. 1-43 The circuit in the illustration simulates a source impedance R_S equal to the load impedance R_L. If $R_S = R_L$, then the voltage gain A_V = the unloaded voltage gain divided by 2. The unloaded gain is approximately R_3/R_2. When the output is loaded, a second feedback loop is closed. The voltage gain for the second loop is approximately R_L/R_1. This second-loop gain combines with A_V (unloaded) to give:

$$A_{V(\text{loaded})} = \frac{A_{V(\text{unloaded})} \cdot A_V}{A_{V(\text{unloaded})} + A_V}$$

If $R_3/R_2 = R_L/R_1$, then:

$$A_{V(\text{loaded})} = \frac{A_{V(\text{unloaded})}}{2}$$

These approximations assume that $R_1 \ll R_2 \ll R_3$. The OP-50 requires no compensation for the circuit values shown and can easily drive the 600-Ω line. ANALOG DEVICES, APPLICATIONS REFERENCE MANUAL, 1993, P. 13-63.

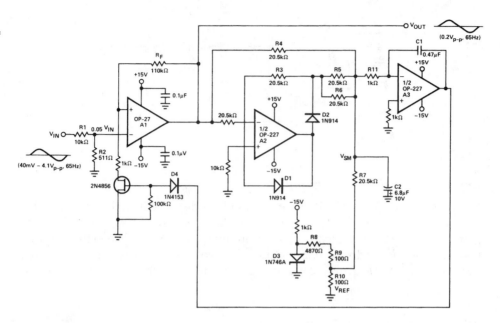

Low-noise AGC amplifier

Fig. 1-44 In this circuit, a JFET is used to control gain of the low-noise OP-27 over a two-decade input-voltage range. For inputs from 40 mV to 4.1 V_{p-p}, the AGC maintains a 0.2 V_{p-p} output. Amplifier A2 performs an absolute-value operation on V_{OUT} and sums the result with a 0.2-V reference on C2. The deviation of this sum, VSM, from zero is amplified by A3 and controls the JFET gate. If the peak-to-peak amplitude of V_{OUT} exceeds 0.2 V, V_{SM} becomes positive and drives the JFET gain negative. This increases the JFET's channel resistance, lowering the gain of A1. The reverse of this occurs if V_{OUT} falls below 0.2 V_{p-p}. The values of C1 and C2 are chosen to optimize the circuit response time for a given input-voltage frequency. This example is designed for 65 Hz. Higher frequencies would justify lower values of C1/C2 to speed the AGC response. ANALOG DEVICES, APPLICATIONS REFERENCE MANUAL, 1993, P. 13-64.

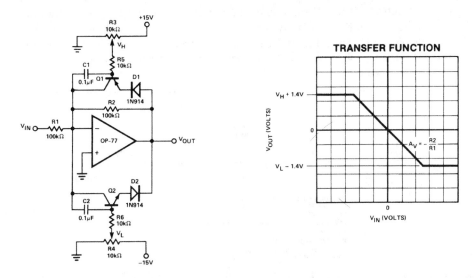

Amplifier with active output clipping

Fig. 1-45 The circuit in the illustration allows adjustment of the op-amp maximum output voltage. Below the clipping levels, circular gain is $A_V = -R_2/R_1$. As V_{OUT} rises above ($V_H + 1.4$ V), the base-emitter of Q1 becomes forward biased, allowing collector current to flow to the summing node, thus clamping V_{OUT}. A similar action occurs as V_{OUT} goes below ($V_L = 1.4$ V) and is clamped by Q2. ANALOG DEVICES, APPLICATIONS REFERENCE MANUAL, 1993, P. 13-65.

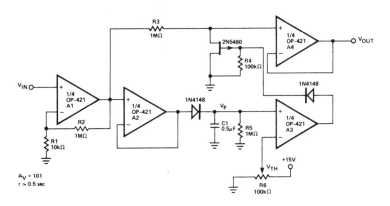

Low-power amplifier with squelch

Fig. 1-46 The OP-421 is the heart of this variable-squelch amplifier, which requires less than 2 mA of supply current (when R_L is infinity). A1 provides a high impedance input amplifier, with $A_V = 1 + R_2/R_1$. The OP-421 output drives a unity-gain output buffer A4 and a peak detector with a time constant set by C_1 and R_5. When the peak-detector output VP exceeds the adjustable threshold V_{TH}, the comparator A3 drives the gate of the FET high, turning the FET off. At lower input-signal levels, V_p falls below V_{TH}, and the FET turns on, clamping the input of A4 to ground. ANALOG DEVICES, APPLICATIONS REFERENCE MANUAL, 1993, P. 13-65.

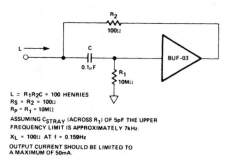

Buffer/amplifier simulated inductor

Fig. 1-47 The circuit in the illustration simulates an inductor of 100 henrys, using the values shown but occupies much less space. A typical inductor of 50 henrys can occupy up to 5 cubic inches. See Fig. 1-49 for pin numbers and power connections. ANALOG DEVICES, APPLICATIONS REFERENCE MANUAL, 1993, P. 13-67.

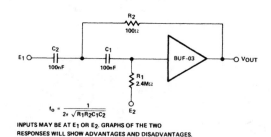

(a)

IC amplifier circuits

Continued

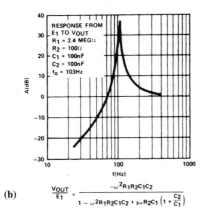

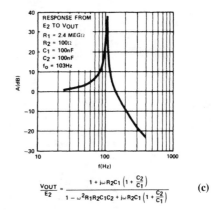

(b) $\dfrac{V_{OUT}}{E_1} = \dfrac{-\omega^2 R_1 R_2 C_1 C_2}{1 - \omega^2 R_1 R_2 C_1 C_2 + j\omega R_2 C_1 \left(1 + \frac{C_2}{C_1}\right)}$

$\dfrac{V_{OUT}}{E_2} = \dfrac{1 + j\omega R_2 C_1 \left(1 + \frac{C_2}{C_1}\right)}{1 - \omega^2 R_1 R_2 C_1 C_2 + j\omega R_2 C_1 \left(1 + \frac{C_2}{C_1}\right)}$ **(c)**

Buffer/amplifier tuned circuits

Fig. 1-48 The tuned circuit shown in Fig. 1-48A uses the simulated inductor of Fig. 1-47 (R1, R2, C1), plus capacitor C2. Depending on whether the circuit is driven at E_1 or E_2, the responses of Figs. 1-48B or 1-48C result. The resonant response in both cases is +38 dB at 103 Hz. The Fig. 1-48B response (E1) is +2.5 dB at 200 Hz and –10 dB at 50 Hz. On the other hand, the Fig. 1-49C response (E2) is –9 dB at 200 Hz and + 2.5 dB at 50 Hz. See Fig. 1-49 for pin numbers and power connections. ANALOG DEVICES, APPLICATIONS REFERENCE MANUAL, 1993, P. 13-67.

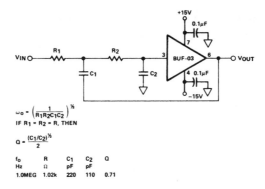

$\omega_0 = \left(\dfrac{1}{R_1 R_2 C_1 C_2}\right)^{\frac{1}{2}}$

IF $R_1 = R_2 = R$, THEN

$Q = \dfrac{(C_1/C_2)^{\frac{1}{2}}}{2}$

f_0 Hz	R Ω	C_1 pF	C_2 pF	Q
1.0MEG	1.02k	220	110	0.71

Buffer/amplifier lowpass filter (high frequency)

Fig. 1-49 Unlike most op amps (which typically cannot handle 1-MHz signals in the 5- to 10-V range with full-power bandwidth), the BUF-03 has a greater than 4-MHz full-power bandwidth for a 20 V_{p-p} sine wave. ANALOG DEVICES, APPLICATIONS REFERENCE MANUAL, 1993, P. 13-67.

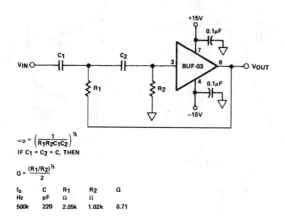

$$\omega_0 = \left(\frac{1}{R_1 R_2 C_1 C_2}\right)^{\frac{1}{2}}$$

IF $C_1 = C_2 = C$, THEN

$$Q = \frac{(R_1/R_2)^{\frac{1}{2}}}{2}$$

f_0 Hz	C pF	R_1 Ω	R_2 Ω	Q
500k	220	2.05k	1.02k	0.71

Buffer/amplifier highpass filter (high frequency)

Fig. 1-50 As in the Fig. 1-49 circuit, the bandwidth of the BUF-03 permits high-frequency operation with full-power bandwidth. ANALOG DEVICES, APPLICATIONS REFERENCE MANUAL, 1993, P. 13-68.

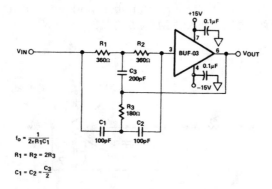

$$f_0 = \frac{1}{2\pi R_1 C_1}$$

$$R_1 = R_2 = 2R_3$$

$$C_1 = C_2 = \frac{C_3}{2}$$

Buffer/amplifier notch filter

Fig. 1-51 The circuit in the illustration is designed as a 4.5-MHz trap (or notch filter) for use in TV receivers. The elements are chosen so that no capacitor is less than 100 pF. Other trap or notch frequencies can be produced by proper selection of components, using the equation shown. Note that larger values of C (in relation to R) produce a sharper notch (or trap frequency). ANALOG DEVICES, APPLICATIONS REFERENCE MANUAL, 1993, P. 13-68.

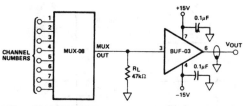

NOTE 1: STRAY CAPACITANCE AT MULTIPLEXER OUTPUT
NODE SHOULD BE MINIMIZED TO REDUCE
CHANNEL-TO-CHANNEL CROSSTALK.
NOTE 2: A BUFFER WHOSE SLEW RATE IS TOO SMALL WILL
INCREASE CHANNEL-TO-CHANNEL CROSSTALK.

Buffer/amplifier high-speed line driver for multiplexers

Fig. 1-52 The circuit in the illustration shows the BUF-03 used as a data line driver, providing full-power bandwidth to up to 4 MHz at 20 V_{p-p}. ANALOG DEVICES, APPLICATIONS REFERENCE MANUAL, 1993, P. 13-68.

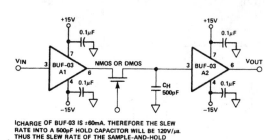

ICHARGE OF BUF-03 IS ±60mA. THEREFORE THE SLEW
RATE INTO A 500pF HOLD CAPACITOR WILL BE 120V/µs.
THUS THE SLEW RATE OF THE SAMPLE-AND-HOLD
CIRCUIT IS LIMITED BY THE CAPACITOR CHARGING TIME.

Buffer/amplifier high-speed sample and hold

Fig. 1-53 The circuit in the illustration provides high speeds because there are no feedback loops to slow down the settling times. Typically, this circuit is followed by a successive-approximation analog-to-digital converter (Chapter 9). ANALOG DEVICES, APPLICATIONS REFERENCE MANUAL, 1993, P. 13-68.

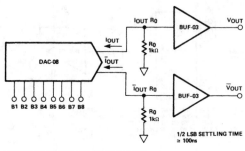

SYSTEM WILL DRIVE CABLES OR TWISTED PAIRS.

Buffer/amplifier high-speed voltage-output DAC

Fig. 1-54 The circuit in the illustration develops both V_{OUT} and $\overline{V_{OUT}}$. Note that the output capacitance of the DAC-08 is approximately 15 pF. As R_O increases in value, so does the settling time for V_{OUT} and $\overline{V_{OUT}}$. See Fig. 1-49 for pin numbers and power connections. ANALOG DEVICES, APPLICATIONS REFERENCE MANUAL, 1993, P. 13-68.

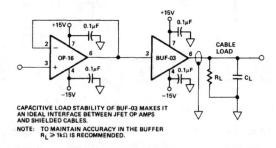

CAPACITIVE LOAD STABILITY OF BUF-03 MAKES IT AN IDEAL INTERFACE BETWEEN JFET OP AMPS AND SHIELDED CABLES.

NOTE: TO MAINTAIN ACCURACY IN THE BUFFER $R_L \geqslant 1k\Omega$ IS RECOMMENDED.

Buffer/amplifier line or cable driver

Fig. 1-55 The circuit in the illustration uses a BUF-03 to convert a FET op amp into a cable or line driver. (FET drivers have speed but often lack stability or current capability.) ANALOG DEVICES, APPLICATIONS REFERENCE MANUAL, 1993, P. 13-69.

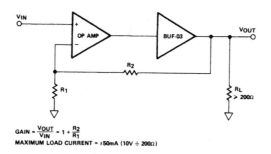

$$\text{GAIN} = \frac{V_{OUT}}{V_{IN}} = 1 + \frac{R_2}{R_1}$$

MAXIMUM LOAD CURRENT = ±50mA (10V ÷ 200Ω)

Buffer/amplifier line driver with current boost

Fig. 1-56 The circuit in the illustration is an alternate to that shown in Fig. 1-55 and is used when greater accuracy is required. Note that the value of R_L should normally be greater than 1 kΩ. However, in the Fig. 1-56 circuit, this limitation does not apply (because the error caused by lower impedances is inside the op-amp feedback loop). ANALOG DEVICES, APPLICATIONS REFERENCE MANUAL, 1993, P. 13-69.

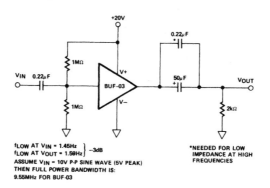

$\left.\begin{array}{l}\text{f\,LOW AT V}_{IN} = 1.45\text{Hz}\\ \text{f\,LOW AT V}_{OUT} = 1.59\text{Hz}\end{array}\right\}$ −3dB

ASSUME V_{IN} = 10V P-P SINE WAVE (5V PEAK)
THEN FULL POWER BANDWIDTH IS:
9.55MHz FOR BUF-03

*NEEDED FOR LOW
IMPEDANCE AT HIGH
FREQUENCIES

Buffer/amplifier with single supply

Fig. 1-57 The circuit in Fig. 1-57 requires only a single ±20-V supply for the BUF-03. See Fig. 1-49 for pin connections. ANALOG DEVICES, APPLICATIONS REFERENCE MANUAL, 1993, P. 13-69.

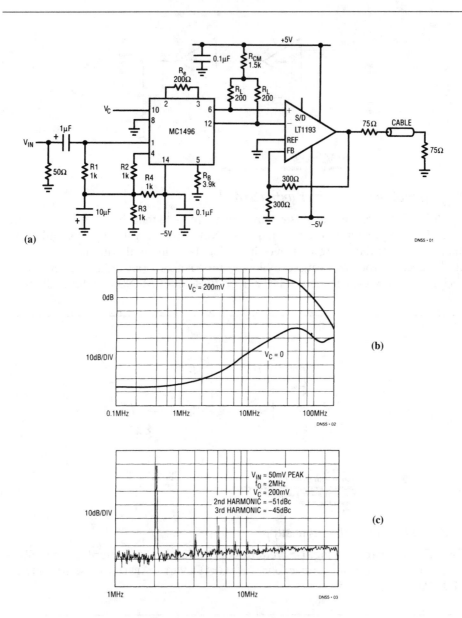

Low-cost 50-MHz voltage-controlled amplifier (VCA)

Fig. 1-58 The circuit in the illustration combines an LT1193 video-difference amplifier with an MC1496 balanced modulator to make a low-cost 50-MHz VCA. The input signal at pin 1 of the MC1496 is multiplied by the control voltage at pin 10 and appears as a differential output current at pins 6/12. The LT1193 acts to level shift the differential signal and convert it to a single-ended output. Positive V_C causes positive gain, negative V_C produces a phase inversion ($-A_V$), and zero V_C

provides maximum attenuation. The voltage gain of the VCA can be increased (at the expense of bandwidth) by changing the value of load resistors R_L. Shorting R_{CM} and increasing R_L to 2 kΩ will increase the maximum gain by 20 dB, and the –3-dB bandwidth will drop to about 10 MHz. Figures 1-58B and 1-58C show the circuit characteristics. LINEAR TECHNOLOGY, LINEAR APPLICATIONS HANDBOOK, 1993, PP. DN55-1, -2.

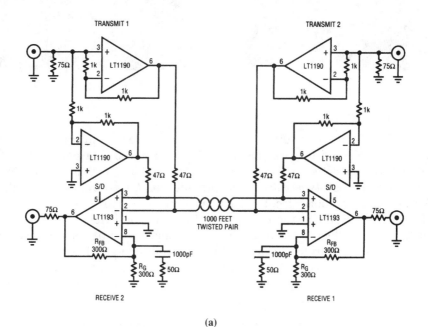

(a)

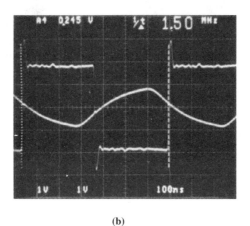

(b)

IC amplifier circuit titles and descriptions

Continued

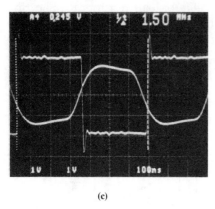

(c)

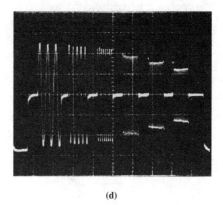

(d)

Bidirectional video bus

Fig. 1-59 The circuit in the illustration makes it possible to send color video 1000 feed over low-cost twisted-pair line instead of coaxial cable (such as RG-59/U). The cost of coax is between 25¢ and 50¢ per foot, but a PVC twisted-pair line is only pennies per foot. Twisted-pair also provides for *drops* or receiver taps along the line. When the amplifier feedback point is compensated (or equalized) with the 50 kΩ and 1000 pF combination as shown, the –3-dB bandwidth is boosted from 750 kHz to 4 MHz. Figure 1-59B shows the uncompensated performance when a 1.5-MHz square wave is fed through 1000 feet of twisted-pair line. Figure 1-59C shows the compensated performance. Figure 1-59D shows a typical TV multiburst pattern when passed through 1000 feet of twisted pair. LINEAR TECHNOLOGY, LINEAR APPLICATIONS HANDBOOK, 1993, PP. DN65-1, -2.

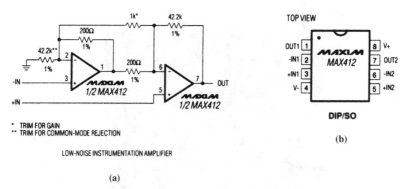

(a)

(b)

Low-noise instrumentation amplifier

Fig. 1-60 The circuit in the illustration uses both sections of a MAX412 to form an instrumentation amplifier with less than 2.4 nV/√Hz of noise. With a ±5-V supply, the output voltage swing is ±3.6 V into 2 kΩ. Figure 1-60B shows the pin configurations. MAXIM NEW RELEASES DATA BOOK, 1992, P. 3-31.

IC amplifier circuits

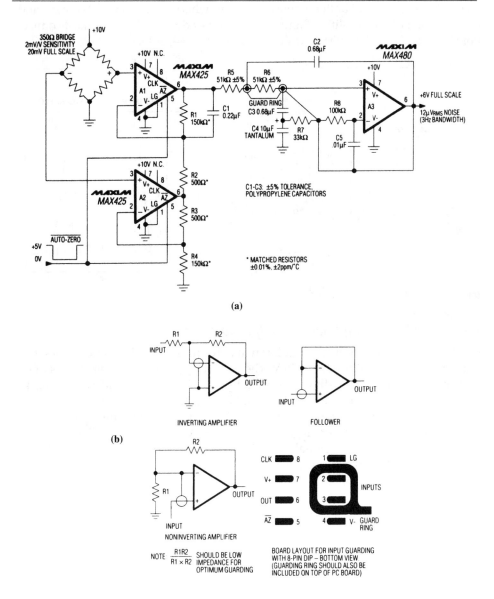

(a)

(b)

Weigh-scale amplifier (single supply)

Fig. 1-61 In this circuit, A1 and A2 (with appropriate gain resistors R1 through R4) amplify the differential 20-mV full-scale signal of the strain-gauge bridge by a factor of 300. The signal is then filtered and buffered by A3. Scale measurements to within a few parts per million are possible. Amplifier A1 and A2 switching clocks run free to minimize noise. The circuit assumes that the strain gauge is loaded with a tare weight, so the output signal is always positive. This permits the circuit to be operated from a single 10-V supply. Guard traces on the top and bottom of the PC

IC amplifier circuit titles and descriptions

board should surround both the sensitive nodes where R5, R6, and C2 connect and the IN+ node of the amplifier. Figure 1-61B shows additional input-guard connections. MAXIM NEW RELEASES DATA BOOK, 1992, PP. 3-44, 3-45.

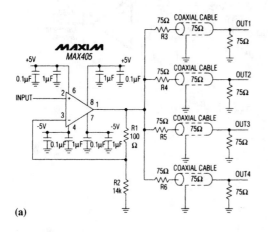

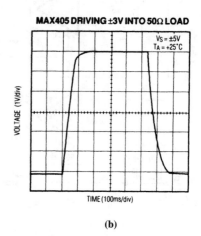

(a)

(b)

Video distribution amplifier with adjustable gain

Fig. 1-62 The circuit in the illustration will drive four 150-Ω loads (four back-terminated 75-Ω coax cables) to ±2.25 V. The receiving end shows a signal loss of –6 dB because of the voltage divider formed by the 75-Ω output load and the back-terminating resistor. The value of R1 should be selected to provide the desired signal amplitude at the output of the MAX405. The range of values for R_1 is from 0 to 1.4 kΩ, which corresponds to signal gains of 0.99 V/V to 1.10 V/V. The MAX405 will drive a 24-MHz 6-V_{p-p} signal into a 50-Ω load as shown in Fig. 1-62B. MAXIM NEW RELEASES DATA BOOK, 1992, P. 8-13.

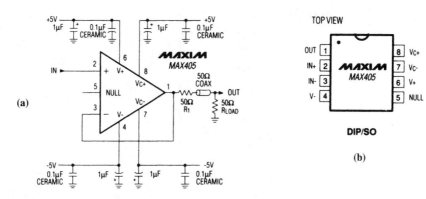

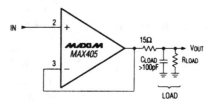

(a)

(b)

Precision video buffer amplifier

Fig. 1-63 The circuit in the illustration shows the MAX405 connected as a basic video buffer/amplifier. The MAX405 has a 180-MHz bandwidth, 650-V/μs slew rate and 35-ns settling time to within 0.1%. Figure 1-63B shows the pin configuration. MAXIM NEW RELEASES DATA BOOK, 1992, P. 8-29.

Driver for large capacitive loads

Fig. 1-64 The circuit in the illustration shows the MAX405 connected to drive large capacitive loads, using a series resistor at the output. See Fig. 1-63 for power connections. MAXIM NEW RELEASES DATA BOOK, 1992, P. 8-34.

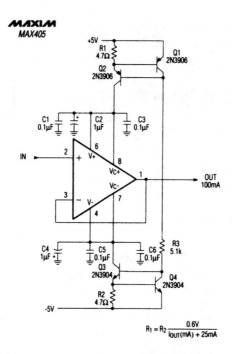

MAXIM
MAX405

$$R_1 = R_2 \frac{0.6V}{I_{OUT}(mA) + 25mA}$$

Buffer/amplifier with continuous-output short-circuit protection

Fig. 1-65 The circuit in the illustration shows the external connections required to provide short-circuit protection when the MAX405 is used at temperatures above 50°C. These external connections are not required at temperatures below 50°C because the MAX405 is provided with internal short-circuit protection. Typical values of output short-circuit current are 180 mA for positive polarity and 150 mA for negative polarity using ±5-V supplies. MAXIM NEW RELEASES DATA BOOK, 1992, P. 8-34.

IC amplifier circuits

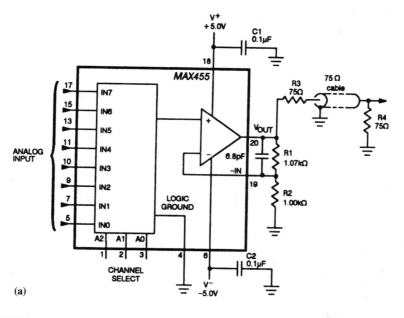

(a)

MAX453		MAX454			MAX455			
A0	Channel	A1	A0	Channel	A2	A1	A0	Channel
L	0*	L	L	0*	L	L	L	0
H	1	L	H	1	L	L	H	1
		H	L	2	L	H	L	2
		H	H	3	L	H	H	3
					H	L	L	4*
					H	L	H	5
					H	H	L	6
					H	H	H	7

*Default channel if selection pins are left floating.

(b)

Video multiplexer/amplifier

Fig. 1-66 The circuit in the illustration shows a MAX455 combination CMOS video multiplexer/amplifier used to drive a back-terminated 75-Ω cable. Resistors R3 and R4 terminate the cable at both ends. R3 also attenuates the signal by a factor of two. To make up for the signal loss, the amplifier is run at a gain of 2 V/V, providing unity gain from signal input to cable output. Amplifier closed-loop gain is set by R_1 and R_2 as follows:

$$\frac{V_{OUT}}{V_{IN}} = \frac{G \times (RL + R_2)}{(G \times R_2) + (R_1 + R_2)}$$

where G is the open-loop gain (about 70 V/V with a 150-Ω load). Multiplexer channels are selected by the A0, A1, and A2 pins, as shown by the logic table of Fig. 1-66B. The GND pin (which is a logic ground, not an analog ground) is connected to digital ground. MAXIM NEW RELEASES DATA BOOK, 1992, P. 8-42.

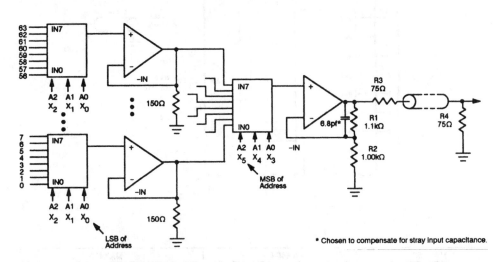

Multiplexer/amplifier for 64 channels

Fig. 1-67 The circuit in the illustration shows how 64 channels can be multiplexed using nine MAX455s (Fig. 1-66). Eight MAX455s select eight out of 64 channels, and a final MAX455 selects one of the eight intermediate channels. The first eight MAX455s are connected as a unity-gain amplifier with 150-Ω load resistors. This arrangement results in a voltage gain of about 0.99 V/V. The 150-Ω loads also cause the unity-gain amplifiers to peak around 40 MHz, which tends to cancel the rolloff of the final amplifier running at a gain of 2 V/V. The overall gain is adjusted by R_1. The –3-dB frequency is about 35 MHz. MAXIM NEW RELEASES DATA BOOK, 1992, P. 8-43.

IC amplifier circuits

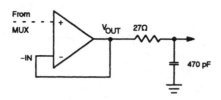

Multiplexer/amplifier used to drive large capacitive loads

Fig. 1-68 The circuit in the illustration shows the MAX455 connected to drive large capacitive loads, using a series resistor at the output. See Fig. 1-66 for power connections. The series resistor minimizes signal peaking at high frequencies. As a guideline, the resistor should be chosen such that the *RC* product is 10 ns or longer. The circuit of Fig. 1-68 should not be used if *R* is greater than 150 Ω (or *C* is less than 100 pF). The MAX455 can drive 100 pF directly without an isolation resistor. MAXIM NEW RELEASES DATA BOOK, 1992, P. 8-44.

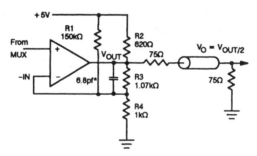

*Chosen to compensate for stray input capacitance.

Multiplexer/amplifier connected for minimum phase distortion

Fig. 1-69 The circuit in the illustration is used when video signals are of one polarity (say from 0 to +1 V), resulting in phase distortion. Such distortion can be minimized by biasing the output stage of the video amplifier. In this circuit, a signal is driven from 0 to +2 V into a 150-Ω load. Resistor R2 provides 6.5 mA of drive to the load at midscale (1 V). The amplifier (instead of supplying 0 to 13 mA) supplies a more symmetric ±8 mA, which reduces phase distortion to about 1 degree at 4 MHz. Because of the amplifier's finite gain of 0.5 mA/mV, the current from R2 introduces an offset voltage. Adding R1 compensates for this offset. Resistors R3 and R4 set the closed-loop gain of the amplifier. See Fig. 1-66 for power connections. MAXIM NEW RELEASES DATA BOOK, 1992, P. 8-44.

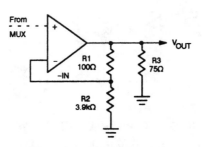

Multiplexer/amplifier connected for unity gain

Fig. 1-70 The circuit in the illustration shows the connections for operating the amplifier portion of the MAX455 at unity gain. Resistors R1 and R2 adjust the gain to be nominally 1.00 V/V. R3 is the load resistor. If precise unity gain is not needed, R1 and R2 can be omitted, and $-V_{IN}$ can be connected directly to V_{OUT}. MAXIM NEW RELEASES DATA BOOK, 1992, P. 8-43.

=2=

Switching
power-supply circuits

The discussions in this chapter assume you are already familiar with power-supply and regulator basics (such as operation of rectifier diodes, switch-mode regulators and supplies, converters, inverters, etc.) and basic power-supply testing/troubleshooting. If not, read the author's *Simplified Design of Switching Power Supplies* and *Simplified Design of Linear Power Supplies*, both published by Butterworth-Heinemann. The following paragraphs summarize both the testing and troubleshooting of power-supply/regulator circuits. This information is included so readers not familiar with power supplies can both test the circuits described here and localize problems if the circuits fail to perform as shown. Notice that power supplies with linear regulators are covered in Chapter 3, and micropower/single-cell battery supplies are covered in Chapter 4.

Power-supply/regulator testing

This section describes the basic tests for all types of power-supply and regulator circuits. Both simple tests and more advanced tests are described. If the circuits pass these tests, the circuits can be used immediately. If not, the tests provide a starting point for the troubleshooting procedures that are described in the power-supply/regulator troubleshooting portion of this chapter.

Basic tests

This section is devoted to simple, practical test procedures that can be applied to a just-completed power supply during design/experimentation (or to a suspected power supply as part of troubleshooting). More advanced tests are covered in this chapter. However, the following procedures are usually sufficient for practical applications.

The basic function of an off-line power supply is to convert alternating current into direct current. In a dc/dc converter, direct current is converted to direct current but at a different voltage (usually higher but lower in many cases). In any event, you can check the power-supply function simply by measuring the output voltage. However, for a more thorough test of a supply, the output voltage should be measured with a load, without a load, and possibly with a partial load.

If the supply delivers the full-rated output voltage into a full-rated load, the basic supply function is met. In addition, it is often helpful to measure the regulating effect of the supply, the supply internal resistance, and the amplitude of any ripple at the supply output. The following paragraphs describe each of the basic tests.

Output tests   Figure 2-A is the basic power-supply test circuit. This arrangement permits the supply to be tested at no load, half load, and full load, depending on the position of S1. With S1 in position 1, no load is on the supply. At positions 2 and 3, there is half load and full load, respectively.

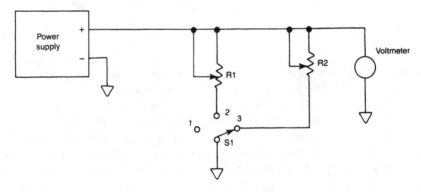

Fig. 2-A   Basic power-supply test circuit.

Using Ohm's law, $R = E \backslash I$, R1 and R2 are chosen on the basis of output voltage and load current (maximum or half load). For example, if the supply is designed for at output of 5 V at 500 mA (full load), the value of R_2 is 5/0.5 = 10 Ω. The value of R_1 is 5/0.25 = 20 Ω.

Where more than one supply is to be tested, R1 and R2 should be variable, and adjusted to the correct value before testing (using an ohmmeter with the power removed). The resistors must be noninductive (not wire wound) and must dissipate the rated power without overheating. For example, using the previous values for R_1 and R_2, the power dissipation of R_1 is 5 × 0.5 = 2.5 W (use at least 5 W), and the dissipation for R_2 is 5 × 0.25 = 1.25 W (use at least 2 W).

1. Connect the equipment (Fig. 2-A).

2. Set R1 and R2 to the correct value.

3. Apply power. Set the input voltage to the correct value. Use the midrange value for the input voltage unless otherwise specified. For example, the input voltage

for a typical switching regulator/supply is between +4 V and +20 V, with certain tests (such as load and line regulation) with the input between +5.8 and +15 V. For dc/dc converters, a separate variable supply is required for the input voltage. (Figure 2-B shows such a supply.) For off-line supplies, the input can be adjusted to the desired voltage with a variable transformer (or variac). If you must make the test with an input, use a 9-V battery for an approximate midrange value.

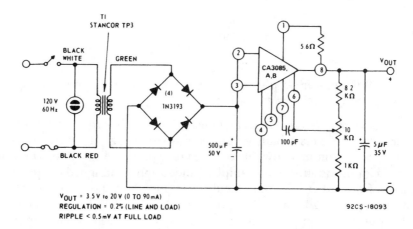

V_{OUT} = 3.5 V to 20 V (0 TO 90 mA)
REGULATION = 0.2% (LINE AND LOAD)
RIPPLE < 0.5 mV AT FULL LOAD

Fig. 2-B Adjustable off-line linear supply.

4. Measure the output voltage at each position of S1.

5. Calculate the current at positions 2 and 3 of S1, using Ohm's law, $I = E/R$. For example, assume that R_1 is 20 Ω and that the output voltage meter indicates 4.8 V at position 2 of S1. The actual load current is 4.8/20, or 240 mA. If the supply output is 5 V at position 1, and it drops to 4.8 V at position 2, the supply is not producing full output with a load. This drop indicates poor regulation, possibly resulting from poor wiring design (in an experimental supply) or from component failure (when discovered as part of troubleshooting).

Load-regulation tests   Load regulation (also known as *load effect* or *output regulation* is usually expressed as a percentage of output voltage and can be determined by:

$$\%regulation = \frac{(no\text{-}load\ voltage) - (full\text{-}load\ voltage)}{full\text{-}load\ voltage} \times 100$$

A low percentage of regulation is desired because it indicates that the output voltage changes very little with load changes. Use the following steps when measuring load regulation.

1. Connect the equipment (Fig. 2-A).

2. Set R2 for the correct value for a full load.

3. Apply power. Measure the output voltage at position 1 (no load) and position 3 (full load).

4. Using the equation, calculate the percentage of regulation. For example, if the no-load voltage is 5 V, and the full-load voltage is 4.999 V, the percentage of regulation is:

$$\left[\frac{(5 - 4.999)}{4.999} \right] \times 100 = 0.2\%$$

5. Notice that the power-supply output regulation is usually poor (high percentage) when the *internal resistance* is high.

Line-regulation tests Line regulation (also known as *line effect, input regulation,* or possibly *source effect* is usually expressed as a percentage of output voltage and represents the maximum allowable output voltage (with a given load) for maximum-rated input variation. For example, if the supply is designed to operate with an ac input from 110 to 120 V, and the dc output is 100 V, the output is measured (1) with an input of 120 V and (2) with an input of 110 V. If there is no change in output, input regulation is perfect (and probably impossible). If the output varies by 1 V, the output variation is 1%. The actual power-supply input regulation can be measured at full load, or half load, as desired, using the test connections shown in Fig. 2-A. However, the input voltage must be varied from maximum to minimum values (with a variac or separate dc supply) and monitored with an accurate voltmeter.

Internal-resistance tests Power-supply internal resistance is determined by the equation:

$$Internal \; resistance = \frac{(no \; load \; voltage) - (full\text{-}load \; voltage)}{current}$$

A low internal resistance is most desirable because this indicates that the output voltage changes very little with load.

1. Connect the equipment (Fig. 2-A).

2. Set R2 to the correct value.

3. Apply power. Measure the actual output voltage at position 1 (no load) and position 3 (full load).

4. Calculate the actual load current at position 3 (full load). For example, if R2 is adjusted to 10 Ω, and the output voltage at position 3 is 4.999 (as in the preceding example), the actual load current is:

$$\frac{4.999}{10} = 499 \; mA$$

Switching power-supply circuits

With no-load voltage, full-load voltage, and actual load current established, find the internal resistance using the equation, for example, with a no-load voltage of 5 V, a full-load voltage of 4.999, and a current of 499 mA (0.499 A), the internal resistance is:

$$\frac{(5 - 4.999)}{0.499} = 0.002 \ \Omega$$

Efficiency tests Power-supply efficiency is usually expressed as a percentage and represents output power divided by input power (times 100 to find percentage). Although the calculation is simple, you must have some means of measuring the input current as well as voltage. If you do not have an ammeter that will measure input current, use a resistor in series with the input supply. Then measure the voltage across the resistor, and calculate current ($I = E/R$). If you use a 1-Ω resistance, the result will be in amperes. A 1000-Ω series resistance will indicate milliamperes (mA). Although the steady-state input current of most switching-regulator ICs is low, the starting-current surge can be high.

Assume that the full-load output voltage is 4.999 V with a load of 75 mA and that the input is 4.5 V with an input current of 20 mA. The input power is 90 mW (4.5×0.02) and the output power is 75 mW, so the efficiency is 83.3% (75/90). This efficiency is typical for most battery-powered switching-regulator circuits.

Ripple tests Any off-line supply, no matter how well regulated or filtered, has some ripple. A battery-operated switching supply also produces an oscillator signal (which has the same effect as ripple). No matter what the source, ripple can be measured with a meter or scope. Usually, the factor of most concern is the ratio between ripple and full-output voltage. For example, if 0.03 V of ripple is measured with a 5-V output, the ratio is 0.03/5, or 0.006 which can be converted to a percentage ($0.006 \times 100 = 0.6\%$).

1. Connect the equipment (Fig. 2-A).

2. Set R2 to the correct value. Ripple is usually measured under full-load power.

3. Apply power. Measure the dc output voltage at position 3 (full load).

4. Set the meter to measure the alternating current. Any voltage measured under these conditions is ac ripple.

5. Find the percentage of ripple, as a ratio between the two voltages (ac/dc).

6. One problem often overlooked in measuring ripple with a meter is that any ripple voltage is not a pure sine wave. Most meters provide accurate ac voltage indications only for pure sine waves. A better way to measure ripple is with a scope (as shown in Fig. 2-C) where peak values can be measured directly.

7. Adjust the scope controls to produce two or three stationary cycles of ripple on the screen. Notice that a full-wave rectifier produces two ripple "humps" per cycle, but a half-wave produces one hump per cycle.

A study of the ripple waveform can sometimes show the source of defects in a power-supply circuit. Here are some examples.

- *If the supply is unbalanced* (one rectifier passing more current than the others), the ripple humps are unequal in amplitude.

- *If there is noise or fluctuations in the supply,* especially where zener diodes are involved, the ripple humps will vary in amplitude and shape.

- *If the ripple varies in frequency,* the ac source is varying. (In switching supplies, the switching frequency is varying.)

- *If a full-wave supply produces a half-wave output,* one rectifier is not passing current.

Advanced tests

The basic tests of the last section are generally sufficient for most practical applications (usually for the experimenters and serious hobbyists). However, you can use many other tests to measure the performance of commercial and lab power supplies. The most common (and most important) of such tests are covered in the following paragraphs.

Figure 2-C shows test connections for measurement of the five most important operating specifications of a power supply: source effect, load effect, *PARD* (periodic and random deviation), drift, and temperature coefficient. There are additional specifications, such as noise-spike measurements and transient-recovery time measurements. However, these measurements require elaborate test setups and are generally applied to lab or commercial supplies. For a thorough discussion

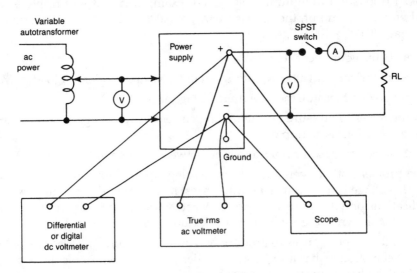

Fig. 2-C Test connections for measuring source effect, load effect, PARD (periodic and random deviation), drift, and temperature coefficient.

of power-supply test and measurements, read the author's *Complete Guide to Electronic Power Supplies*, 1990, Prentice-Hall.

Test equipment All the tests described here can be performed with only four test instruments: a variable autotransformer (variac), a differential or digital ac voltmeter, a true ac voltmeter, and a scope. Of course, on those supplies that are to be battery operated, a separate variable dc supply is required (Fig. 2-B).

Use of a separate supply brings up a problem. If there is any ripple or other output fluctuations in the separate (input) supply, these can pass through the supply under test, making the test supply appear defective. This problem can be checked by substituting a battery (of the same voltage) to power the supply being tested. If the ripple or other fluctuations remain, the problem is in the supply under test.

Make certain that the autotransformer (in off-line supplies) or variable supply (for battery-operated supplies) has an adequate current rating. If not, the input voltage to the supply under test might be severely distorted and the rectifying/regulating circuit might operate improperly.

The dc voltmeter should have a 1 mV (or better) accuracy. The scope should have a sensitivity of 100 µV/cm (microvolts per centimeter) and a bandwidth of at least 10 MHz. Both the scope and meter should have some means of measuring current (some form of current probe, preferably a clip-on type). In switching supplies, a scope display of the indicator current can often pinpoint circuit problems (as covered in this chapter). A nonelectronic *VOM* (volt-ohm-milliammeter), such as the classic Simpson 260 or Tripplet 630, also can be useful in switching supply tests. Some digital and other electronic meters are affected by the switching-oscillator signals.

Proper connections For the most accurate results, the test connections should be permanent (not clip leads) and should be made to the exact point on the supply terminals. Clip-lead connections can produce measurement errors. Instead of measuring pure supply characteristics, you are measuring supply characteristics, plus the resistance between output terminals and point of connection. Even using clip leads to connect the load to the supply terminals can produce a measurement error.

Separate leads All measurement instruments must be connected directly by separate pairs of leads to the monitoring points (Fig. 2-C). This connection avoids the subtle mutual-coupling effects that can occur between measuring instruments (unless all are returned to the low-impedance terminals of the supply). Use twisted pairs or shielded cable to avoid pickup on the measuring loads.

Load resistance Make certain that the load resistance is adequate for the supply and test requirements. Typically, the load resistance and load voltage should permit operation of the supply at maximum-rated output voltage and current.

Pickup and ground-loop effects Always check test connections for possible pickup and/or ground-loop problems. As a simple test, turn off the supply and observe the scope for any unwanted signals (especially at the line frequency, typically 50/60 Hz) with the scope leads connected directly on the supply output terminals. Then connect both scope leads to either terminal (+ or −), whichever is grounded to the board—or to common ground. If there is any noise in either test condition, with the supply off, you have possible pickup and/or ground-loop effects.

Alternating-current (ac) voltmeter connections Connect the ac voltmeter as close as possible to the input ac terminals of an off-line supply. The voltage indication is then a valid measurement of the supply input, without any error introduced by the drop present in the leads that connect the supply input to the ac line. The same is true for the dc input of battery-operated supplies. That is, measure the dc input at the supply terminals, not at the output of the variable supply or battery.

Line regulator Do not use any form of line regulator when testing an off-line supply and when using the supply (unless specifically recommended for that supply). This precaution is especially important for switching supplies and regulators. A line regulator can change the shape of the output waveform in a switching supply and thus offset any improvement produced by a constant line input to the supply.

Source effect or line regulation

No matter what the test is called, the measurement is made by varying the input voltage throughout the specified range from low limit to high limit and noting the change in voltage at the supply output terminals. The test is performed with all other test conditions constant. The supply should stay within specifications for any rated output voltage, combined with any rated output current. The extreme source-effect test is with the maximum output voltage and maximum output current.

Load effect or load regulation

This test is made by closing and opening switch S1 (Fig. 2-C) and noting any change in output voltage. The test is performed with all other test conditions constant. The supply should stay within specifications for any rated output voltage, combined with any rated input voltage. The extreme load-effect test is with maximum output voltage and maximum output current.

Noise and ripple (or PARD)

In many cases, PARD (periodic and random deviation) has replaced the terms *noise* and *ripple* and represents deviation of the dc output voltage from the average value, over a specified bandwidth, with all other test conditions constant.

As an example, in Hewlett-Packard lab supplies, PARD is measured in *rms* (root means square) or peak-to-peak values over a 2-Hz to 2-MHz bandwidth. Fluctuations below 20 Hz are considered to be drift. Peak-to-peak measurements are of particular importance for applications where noise spikes can be detrimental (such as in digital logic circuits, Chapter 6). The rms measurement is not ideal for noise because output noise spikes of short duration can be present in ripple but not appreciably increase the rms value. Always use twisted-pair leads (for single-ended scopes) or shielded two-wire leads (for differential scopes) when making PARD or noise/ripple tests.

Drift (stability)

Drift measurements are made by monitoring the supply output on a differential or digital voltmeter over a stated measurement interval (typically eight hours, after a 30-minute warmup). In some cases, a strip chart is used to provide a permanent record. Place a thermometer near the supply to verify that the ambient temperature remains constant during the period of measurement. The supply should be at a location immune from stray air currents (away from open doors or windows and from air-conditioning vents). If practical, place the supply in an oven and hold the temperature constant. As a guideline, a well-regulated supply will drift less during the eight-hour period than during the 30-minute warmup.

Temperature coefficient (or TC)

TC (also known as *tempco*) measurements are made by placing the supply in an oven and varying the temperature over a given range, following a 30-minute warmup. The supply is allowed to stabilize at each measurement temperature. In the absence of other specifications, the temperature coefficient is the output-voltage changes that result from a 5°C change in temperature. The measuring instrument should be placed outside the oven and must have a long-term stability that is adequate to ensure that any voltmeter drift does not affect measurement accuracy.

Switching power-supply troubleshooting

The remainder of the introduction to this chapter is devoted to troubleshooting switching power-supply circuits. In general, most supply problems are the result of wiring mistakes (which you never make), defective (or inadequate) components, and possibly test errors. All these mistakes can be located by basic voltage checks, resistance checks, and point-to-point wiring checks. However, switch-mode supplies and switching regulators present problems—especially when an experimental circuit is first tested. The following notes describe some of the most common troubleshooting problems for such supplies and regulators.

Ground loops

Figure 2-D shows a typical ground-loop condition. A generator is driving a 5-V signal into 50 Ω on the experimental circuit, which results in a 100-mA current. The return path for this current divides between the ground from the generator (typically, the shield on a BNC cable) and the secondary ground loop. The secondary ground loop is created by the scope-probe ground clip (shield), and the two "third-wire" connections on the generator and scope.

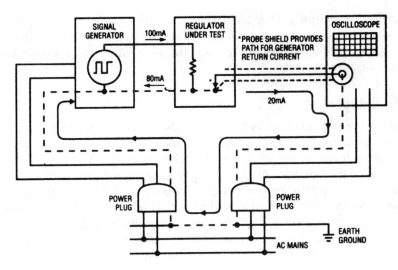

Fig. 2-D Ground-loop errors.

Assume that 20 mA flows in the parasitic ground loop. If the scope ground lead has a resistance of 0.2 Ω, the scope will show a 4-mV bogus signal. The problem gets much worse for higher currents and for fast-signal edges, where the inductance of the scope-probe shield is important. The most practical solution is to use an isolation transformer for the scope. For a quick check, touch the scope-probe tip to the probe ground clip, with the clip connected to the experimental-circuit ground. The scope should show a flat line. Any signal displayed on the scope is a ground-loop problem or pickup problem.

Scope probe compensation

Always check that the scope probe is properly compensated when testing switching supplies. It is especially important for the ac attenuation (on a 10× probe, for example), to match the dc attenuation exactly. If not, low-frequency signals will be distorted and high-frequency signals will have the wrong amplitude. Remember that at typical switching frequencies, the waveshape might appear good because the probe acts purely capacitive, so the wrong amplitude might not be immediately obvious.

Ground-clip pickup

Do not make any test measurements on a switching regulator with a standard (alligator) ground-clip lead. Replace the alligator clip with a special soldered-in probe terminator, which you can get from many probe manufacturers. The standard ground-clip lead can act as an antenna, and it can pick up magnetic and other radiated signals. Make the test described for ground loops if you suspect pickup by the scope probe.

Measuring at the component

Make all measurements (output voltage, ripple, etc.) at the component, not at a wire that is connected to a component. This precaution is necessary because wires are not shorts. For example, switching regulators (such as those of this chapter) produce square waves or pulses. In turn, these pulses are applied to an output capacitor (in most cases). A typical regulator can produce about 2-V per inch spikes in the capacitor leads. The farther you measure from the capacitor, the greater the spike voltage.

EMI (electromagnetic interference) suppression

EMI (electromagnetic interference) is a fact of life with switching regulators. EMI takes two basic forms: conducted (which travels down input and output wiring) and radiated (which produces electric and magnetic fields). Although these fields do not usually cause regulator problems, they can create problems for surrounding circuits. The following guideline is helpful in minimizing EMI problems.

1. Avoid long high-current grounds and feedback nodes. Figure 2-E shows the right and wrong ways to make grounds and feedback connections to switching regulators. Even though low-power switching-regulator ICs are generally easy to use, you must pay some attention to PC (printed circuit) layout and routing—especially at power levels greater than 1 W or when high-speed *PWM* (pulse-width modulation) ICs are used. Trace out the high-current paths and minimize their length—especially the ground trace. Use a *star ground*, in which all grounds are brought to one point. Place any input filter capacitor physically close to the IC. Minimize any stray capacitance at the feedback (FB) pin. Return all compen-

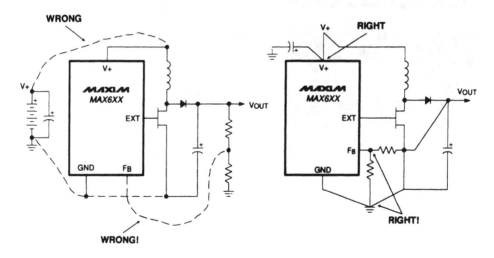

Fig. 2-E   Correct and incorrect ground and feedback connections for switching regulators.

sation capacitors and bypass capacitors to quiet, well-filtered points (such as an analog ground pin).

2. Use inductors or transformers with good EMI characteristics, such as toroids or pot cores. Avoid rod inductors. If you must use rod inductors, keep them in the output filter, where ripple current is low (hopefully). Use the inductor values that are shown in the circuit descriptions. Figure 2-F shows some typical current waveforms that are produced by good and bad inductors. In general, most inductor problems (other than using the incorrect size) can be traced to inadequate saturation (peak current) ratings or excessive dc resistance. If an inductor saturates, the current rises exponentially with time. If there is excessive resistance, a distinct *LR* characteristic is seen. If the waveform takes small (but strange) bends, the inductor might be producing both effects.

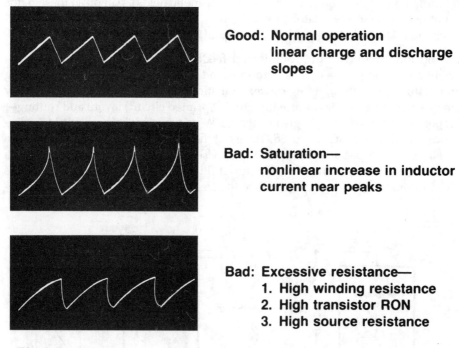

Good: Normal operation linear charge and discharge slopes

Bad: Saturation— nonlinear increase in inductor current near peaks

**Bad: Excessive resistance—
1. High winding resistance
2. High transistor RON
3. High source resistance**

Fig. 2-F Typical current waveforms produced by good and bad inductors.

3. Route all traces carrying high ripple current over a ground plane to minimize radiated fields. This includes the catch-diode leads, input and output capacitor leads, snubber leads, inductor leads, IC input and switch-pin leads, and input power leads. Keep these leads short and keep the components close to the ground plane.

4. Keep sensitive, low-level circuits as far away as possible and use field-canceling tricks, such as twisted-pair differential lines.

5. In very critical applications, add a "spike killer" bead on the catch diode to suppress high harmonics, which can create higher transient switch voltages at switch turn-off. Check each switch waveform carefully.

6. Add an input filter if radiation from input lines could be a problem. Just a few μH (microhenrys) in the output line will allow the regular input capacitor to swallow nearly all the ripple current that is created at the regulator input.

Troubleshooting hints and tips

The following notes apply specifically when troubleshooting an inoperative or poorly performing switch-mode supply or switching regulator, particularly those that are in experimental form.

If the circuit is inoperative, look for such things as transformers wired backwards (always check the polarity dots or color codes on transformers), electrolytic capacitors wired backwards (usually you will find this out shortly after power is applied) and IC pins reversed (check the data sheet and follow the wiring that is given in the circuit schematics).

If the circuit works (no smoke, fire, or explosion), but the performance is not as expected (low efficiency, low output voltage or current, line or load regulation out of tolerance, etc.), one of the first troubleshooting steps is to display the inductor-current waveform on a scope. The most convenient way to monitor the inductor current is with a clip-on type current probe. Figure 2-F shows some typical inductor-current waveforms.

Ideally, both the charge and discharge slopes of the waveforms should be linear, as shown in Fig. 2-F. If so, but the circuit does not perform as desired, look for problems with the inductor, oscillator frequency, output capacitor value, and diode characteristics, in that order. For example, if the inductance has decreased because of saturation, the output will increase, but the inductor-current waveform will be nonlinear near the peaks as shown. On the other hand, if the inductor resistance value is excessive, the waveform might be nonlinear during the entire charge time as shown.

If you suspect that the inductor is faulty, try operating the inductor in a standard test circuit, such as shown in Fig. 2-G. Compare the waveforms obtained with the inductor in the test circuit against the waveforms shown in Figs. 2-H through 2-K. In each case, trace A is the voltage at the VSW pin of the LT1070, and trace B is the current (measured with a clip probe at the VSW pin). When the VSW-pin voltage is low, inductor current flows. With high inductance (Fig. 2-H), current rises slowly with no saturation. As inductance decreases, current rise is steeper (Figs. 2-I and 2-J) but still no saturation. However, the highest inductance (Fig. 2-K) but wound on a low-capacity core, produces extreme saturation and is unsuitable for switching-regulator use. Using this test circuit and procedure, you can narrow the selection of an inductor down to the best unit for your circuit (with regard to cost, size, etc.) Figure 2-L shows an inductor selection kit, which includes 18 inductors of various ratings and size. The kit is available from:

Pulse Engineering, Inc.
PO Box 12235
San Diego, California 92112
(619) 268-2400

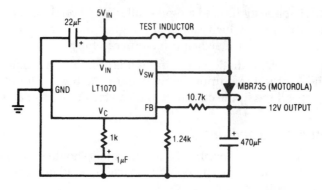

Fig. 2-G   Standard test circuit for inductors used in switching regulators.

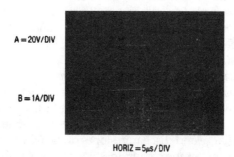

A = 20V/DIV

B = 1A/DIV

HORIZ = 5μs/DIV

Fig. 2-H   Waveforms for 450-nH high-capacity core.

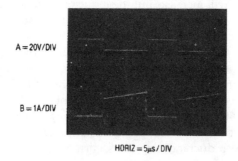

A = 20V/DIV

B = 1A/DIV

HORIZ = 5μs/DIV

Fig. 2-I   Waveforms for 170-nH high-capacity core.

If the input voltage appears to dip, it is possible that the input leads (from the battery to the switching IC for example) are too long when connected in experimental form. (This problem should not occur when the circuit is in PC form). Switching regulators draw current from the input supply in pulses. Long input wires can cause dips in experimental form, try adding input capacitance (perhaps a 1000-μF or higher capacitance close to the IC regulator). If this doesn't cure the problem, it might be necessary to add some input capacitance when the circuit is in final form.

Switching power-supply circuits

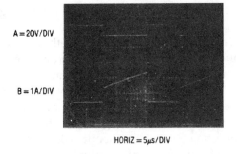

A = 20V/DIV

B = 1A/DIV

HORIZ = 5μs/DIV

Fig. 2-J   Waveforms for 55-μH high-capacity core.

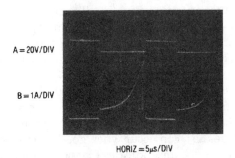

A = 20V/DIV

B = 1A/DIV

HORIZ = 5μS/DIV

Fig. 2-K   Waveforms for 500-μH low-capacity core.

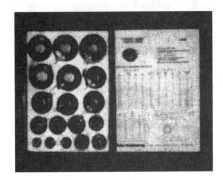

Fig. 2-L   Model 845 inductor-selection kit.

If a nonbattery input supply simply does not come up and switching does not start, with the right components all properly connected, it is possible that the input supply cannot deliver the necessary start-up current. Switching regulators have negative input resistance at start-up and draw high current. The high current can latch some input supplies to a low or off condition. In a battery-input supply, it is possible that the batteries are not "stiff" (not capable of delivering a large momentary current at start-up).

If efficiency is low (much more power going into the supply than power coming out), suspect the inductors (or transformers). Core or copper losses might be the problem. Of course, the problem could be an accumulation of all losses (inductor, capacitor, diode, etc.), which results in an inefficient supply/regulator circuit.

If the switch timing varies, check for excessive ripple at the output and at any feedback or compensation pins of the IC regulator. If the switch timing varies from cycle to cycle, try connecting a capacitor (about 1000 to 3000 pF) across the output capacitor (and/or at the feedback/compensation pin) to ground. If any of these capacitors eliminate the variation in switch timing, you have located the problem area.

If you definitely have high output ripple or noise spikes, suspect the output capacitor. Capacitors have a capacitance value (in μF or pF) and an *ESR* (equivalent series resistance) expressed in ohms. An increase in capacitance value decreases ripple, but an increase in ESR increases ripple. It is possible that your capacitor has high ESR, even though the capacitance value is correct, when there is ripple of unknown origin.

If the IC blows up in an experimental circuit (with the right components, all properly connected, no shorted outputs, and no reversed electrolytics or battery polarities), it is possible that start-up surges are causing momentary large switching voltages. (These voltages also can occur when good experimental circuits are converted to PC form). Generally, this indicates that one or more components were on the borderline (such as excessive leakage in the output capacitor or catch diode).

If the IC runs hot, suspect a problem in the heatsink (not properly connected to the IC regulator). For example, a TO-220 package has a thermal resistance of about 50 to 55°C/W with no heatsink. A 5-V, 3-A output (15-W) regulator with a typical 10% switch loss will dissipate more than 1.5 W in the IC. This energy causes a 75°C temperature rise or 100°C case temperature at 25°C room temperature (which is hot!). In an experimental circuit, simply soldering the TO-220 tab to an enlarged copper pad on the PC board will reduce thermal resistance to about 25°C/W.

If you have poor load or line regulation, check in the following order:

- An output capacitor with high ESR (particularly if the capacitor is outside the feedback loop)
- A ground-loop error in the scope (Fig. 2-D)
- Improper connections of the output-divider resistors to current-carrying lines. (Figure 2-E shows both correct and incorrect connections for switching regulators)
- Excessive output ripple; switching frequency too high. (It is assumed that you have checked the inductor for saturation, as shown in Fig. 2-F.)

If efficiency is poor, do not overlook the diodes. If the diode is not fast enough, or is not rated for the current level, efficiency will drop, even though the circuit might work. If the diode is hopelessly underrated, it will be destroyed. Most switching-regulator diodes are Schottky or ultrafast diodes. The major advantage of Schottky versus ultrafast is efficiency. As a guideline, when the output voltage is below 12 V, a Schottky will provide about 5% improvement in efficiency compared to an ultrafast. If the voltage is greater than 12 V, the advantage is less significant. Of course, the breakdown rating of any diode must exceed the input voltage, with allowances made for input surges.

Switching power-supply circuit titles and descriptions

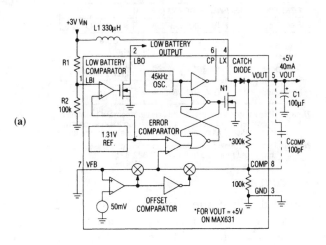

(a)

(b)

V_IN (V)	VOUT (V)	IOUT (mA)	EFF. (%)	INDUCTOR		
				P.N. (Note 2)	μH	Ω
2	5	5	78	CB 6860-21	470	0.4
2	5	10	74	G 1B253	250	0.44
2	5	15	61	G 1B103	100	0.25
3	5	25	82	CB 6860-21	470	0.4
3	5	40	75	CB 7070-29	220	0.55
3	12	5	79	CB 6860-19	330	0.35
3	12	10	79	CB 7070-28	180	0.48
5	12	12	88	CB 6860-21	470	0.4
5	12	25	87	CB 6860-19	330	0.35
3	15	5	73	CB 7070-29	220	0.55
3	15	8	71	CB 7070-27	150	0.43
5	15	10	85	CB 6860-21	470	0.4
5	15	15	85	CB 6860-19	330	0.35
8	15	35	90	G 1B503	500	0.56

Note 2: CB = Cadell-Burns, NY, (516) 746-2310
G = Gowanda Electronics Corp., NY, (716) 532-2234
Other Manufacturers listed in Table 2.

2

122

MANUFACTURER	TYPICAL PART #	DESCRIPTION
BOBBIN INDUCTORS		
Dale	IHA-104	500µH, 0.5Ω
Caddell-Burns	7070-29	220µH, 0.55Ω
Gowanda	1B253	250µH, 0.44Ω
TRW	LL-500	500µH, 0.75Ω
POTTED TOROIDAL INDUCTORS		
Dale	TE-3Q4TA	1mH, 0.82Ω
TRW	MH-1	600µH, 1.9Ω
Gowanda	050AT1003	100µH, 0.05Ω
FERRITE CORES AND TOROIDS (Note 4)		
Siemens	B64290-K38-X38	Tor. Core, $4µH/T^2$
Magnetics	555.130	Tor. Core, $53nH/T^2$
Stackpole	57-3215	Pot Core, 14mm x 8mm
Magnetics	G-41408-25	Pot Core, 14 x 8, $250nH/T^2$

(c)

Note 3: This list does not constitute an endorsement by Maxim Integrated Products and is not intended to be a comprehensive list of all manufacturers of these components.

Note 4: Permag Corp. is a distributor for many of the listed core and toroid manufacturers. (516) 822-3311.

Basic +3 to +5-V converter

Fig. 2-1 The circuit in the illustration uses the MAX631/632/633 to provide basic step-up switching regulation. For operation at one of the preset output voltages (+5 V for the MAX631, +12 V for the MAX632, and +15 V for the MAX633), V_{FB} is connected to ground as shown. For an output voltage other than the preset value, use the circuit of Fig. 2-2. Figure 2-1B shows inductor values for typical input voltages and output current. Figure 2-1C shows coil and core manufacturers for the inductors. The value of low-battery voltage is set by:

$$R_1 = R_2 \left(\frac{V_{LB}}{1.31 \text{ V}} - 1 \right)$$

where V_{LB} is the desired low-battery detection voltage. MAXIM NEW RELEASES DATA BOOK, 1992, PP. 4-59, 60, 61.

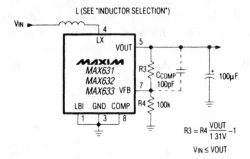

Converter with adjustable output

Fig. 2-2 When the ICs of Fig. 2-1 must provide an adjustable output, an external voltage divider (R3 and R4) is required to set the value of V_{OUT}. With R4 at any value between 10 kΩ and 10 MΩ (100 kΩ typically), the value of R_3 is set by:

$$R_3 = R_4 \left(\frac{V_{OUT}}{1.31 \text{ V}} - 1 \right)$$

The inductors shown in Fig. 2-1B and 2-1C still apply. MAXIM NEW RELEASES DATA BOOK, 1992, P. 4-61.

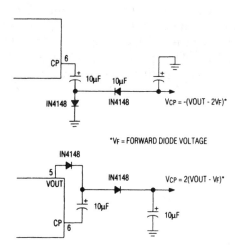

Converter with negative output (charge pump)

Fig. 2-3 The *CP* (charge-pump) output of the IC (Fig. 2-1) is a low-impedance buffer that swings from ground to V_{OUT} at the IC-oscillator frequency. Two external capacitors and diodes can be connected as shown to generate a negative output voltage of $-(V_{OUT} - 1.2\text{ V})$ or a positive output of $2(V_{OUT} - 1.2\text{ V})$. The 1.2 V is the forward drop of two silicon diodes. Both circuits can be used at once if desired. With 10 μF capacitors, the output impedance of V_{CP} is about 30 Ω. If space is critical, the capacitors can be reduced but with a slight increase in output impedance and V_{CP} output ripple. MAXIM NEW RELEASES DATA BOOK, 1992, P. 4-62.

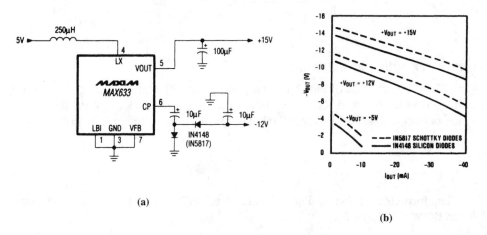

(a)

(b)

Converter with negative output (switching charge pump)

Fig. 2-4   The circuit in the illustration provides about ±10 mA with V_{OUT} at +15 V and ±15 mA if V_{OUT} is –12 V. The magnitude of the negative output is about 3 V less than V_{OUT} because of the forward voltage drop of the 1N4148 diodes and the output impedance of the CP pin. Using Schottky diodes (1N5817) will increase the absolute value of the negative output by about 1 V. The performance of the CP output is shown in Fig. 2-4B. MAXIM NEW RELEASES DATA BOOK, 1992, P. 4-62.

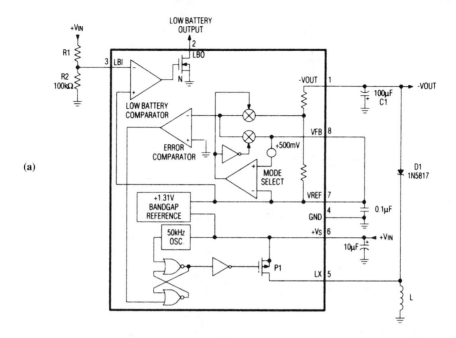

(a)

(b)

V$_{IN}$ (V)	VOUT (V)	IOUT (mA)	Part No.	INDUCTOR	
				μH	Ω
+3	-5	5	7070-27	150μH	0.43
+5	-5	25	7070-27	150μH	0.43
+9	-5	40	7070-31	330μH	0.72
+12	-5	45	7070-33	470μH	0.88
+15	-5	50	7070-35	680μH	1.5
+5	-12	12	7070-26	120μH	0.32
+9	-12	30	7070-31	330μH	0.72
+12	-12	40	7070-33	470μH	0.88
+3	-15	2	7070-27	150μH	0.43
+5	-15	8	7070-27	150μH	0.43
+9	-15	25	7070-31	330μH	0.72

Note 5: Caddell-Burns N.Y. (516) 746-2310.

Switching power-supply circuit titles and descriptions

MANUFACTURER	TYPICAL PART #	DESCRIPTION
ASIA		
TDK Corporation 13-1, Nihonbashl 1-chome Chuo-ku Tokyo 103 Japan		
EUROPE		
Richard Jahre GmbH Luetzowstrasse 90 1000 Berlin 30 Germany		
BOBBIN INDUCTORS		
Dale	IHA-104	500µH, 0.5Ω
Caddell-Burns	7070-29	220µH, 0.55Ω
Gowanda	1B253	250µH, 0.44Ω
TRW	LL-500	500µH, 0.75Ω
POTTED TOROIDAL INDUCTORS		
Dale	TE-3Q4TA	1mH, 0.82Ω
TRW	MH-1	600µH, 1.9Ω
Gowanda	050AT1003	100µH, 0.05Ω
FERRITE CORES AND TOROIDS (Note 4)		
Allen Bradley	T0451S100A	Tor. Core, 500nH/T^2
Siemens	B64290-K38-X38	Tor. Core, 4µH/T^2
Magnetics	555.130	Tor. Core, 53nH/T^2
Stackpole	57-3215	Pot Core, 14mm x 8mm
Magnetics	G-41408-25	Pot Core, 14 x 8, 250nH/T^2

(c)

Note 3: This list does not constitute an endorsement by Maxim Integrated Products and is not intended to be a comprehensive list of all manufacturers of these components.
Note 4: Permag Corp. is a distributor for many of the listed core and toroid manufacturers (516) 822-3311.

Basic inverting switching regulator

Fig. 2-5 The circuit in the illustration uses the MAX635/636/637 to provide basic inverted switching regulation. For operation at one of the preset output voltages (–5 V for the MAX635, –12 V for the MAX636, and –15 V for the MAX637), V_{FB} is connected to V_{REF} as shown. No external resistors are required. For an output voltage other than reset value, use the circuit of Fig. 2-6. Figure 2-5B shows the inductor values for typical input voltages and output currents. Figure 2-5C shows coil and core manufacturers for the inductors. The value of low-battery voltage is set by:

$$R_1 = R_2 \left(\frac{V_{LB}}{1.31 \text{ V}} - 1 \right)$$

where V_{LB} is the desired low-battery detection voltage. MAXIM NEW RELEASES DATA BOOK, 1992, PP. 4-80, 81.

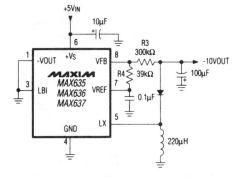

Inverter with adjustable output

Fig. 2-6 When the IC of Fig. 2-5 must provide an adjustable output, an external voltage divider (R3 and R4) is required to set the value of V_{OUT}. With R4 at any value between 10 kΩ and 10 MΩ (100 kΩ typically), the value of V_{OUT} is set by:

$$V_{OUT} = -1.31 \text{ V} \times \left(\frac{R_3}{R_4}\right)$$

MAXIM NEW RELEASES DATA BOOK, 1992, P. 4-81.

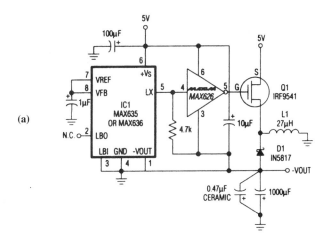

(a)

Continued

V_{IN}	-VOUT	IOUT	EFFICIENCY	IC1	L1
5V	-5V	400mA	70%	MAX635	27µH
5V	-5V	500mA	64%	MAX635	18µH
5V	-12V	150mA	75%	MAX636	27µH
5V	-12V	200mA	70%	MAX636	18µH

(b)

Notes: 18µH Coil = Caddell-Burn's (Mineola, NY) Model 6860-04.
27µH Coil = Caddell-Burn's Model 6860-06.

Medium-power inverter

Fig. 2-7 In the Fig. 2-7 circuit, the MAX635 or MAX 636 is used to control an external P-channel MOSFET (metal-oxide semiconductor field-effect transistor) through a MAX626 MOSFET driver. Figure 2-7B shows the recommended inductor values and part numbers for various output voltages and currents. MAXIM NEW RELEASES DATA BOOK, 1992, P. 4-83.

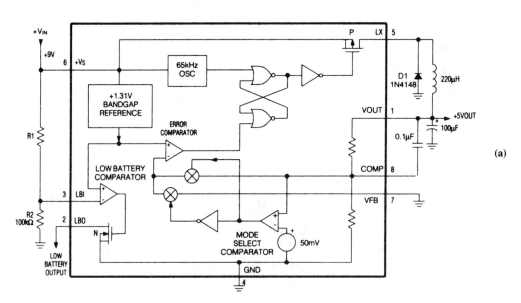

(a)

MAXIM PART NO.	V_{IN} (V)	VOUT (V)	IOUT (mA)	TYP EFF (%)	I_{pk} (mA)	PART NO.*	INDUCTOR (L)	
							µH	Ω
MAX638	7-9.5	5	35	92	200	7070-27	150	0.4
	8-9.5	5	55	89	200	7070-27	150	0.4
	10-14	5	50	92	300	7070-30	270	0.6
	12	5	60	92	250	7070-30	270	0.6
	12	5	75	89	300	7070-28	180	0.5

(b)

* Caddell-Burns, NY, (516) 746-2310

Switching power-supply circuits

Step-down switching regulator

Fig. 2-8 The circuit in Fig. 2-8 shows a MAX638 connected to provide a fixed +5-V output for various input voltages. Figure 2-8B shows the recommended inductor values and part numbers for the input voltages. The value of the low-battery voltage is set by:

$$R_1 = R_2 \left(\frac{V_{LB}}{1.31 \text{ V}} - 1 \right)$$

where V_{LB} is the desired low-battery detection voltage. MAXIM NEW RELEASES DATA BOOK, 1992, P. 4-88.

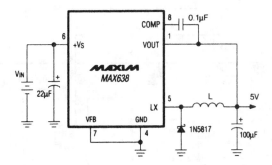

Step-down switching regulator (without low-battery indicator)

Fig. 2-9 The circuit in the illustration is essentially the same as that of Fig. 2-8, but it is used where a low-battery function is not required. The inductors shown in Fig. 2-8B apply to Fig. 2-9. MAXIM NEW RELEASES DATA BOOK, 1992, p. 4-89.

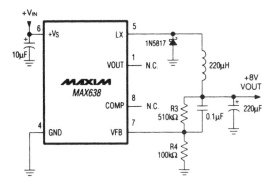

Step-down switching regulator with adjustable output

Fig. 2-10 When the IC of Fig. 2-8 or 2-9 must provide an adjustable output, an external voltage divider (R3 and R4) is required to set the value of V_{OUT}. With R_4 at any value between 10 kΩ and 10 MΩ (100 kΩ typically), the value of R_3 is set by:

$$R_3 = R_4 \left(\frac{V_{OUT}}{1.31 \text{ V}} - 1 \right)$$

The inductors shown in Fig. 2-8B still apply. MAXIM NEW RELEASES DATA BOOK, 1992, P. 4-89.

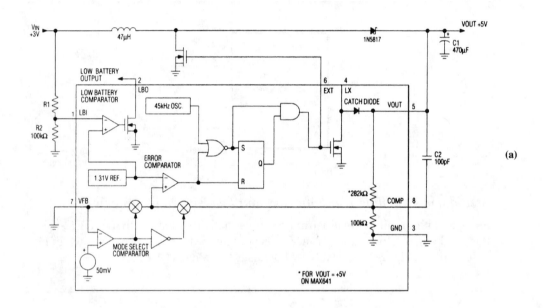

(a)

MAXIM PART NO.	V_{IN} (V)	VOUT (V)	IOUT (mA)	TYP EFF (%)	I_{pk} (A)	PART NO.*	INDUCTOR (L) μH	Ω
MAX641	3	5	200	83	1.3	6860-13	100	.10
	3	5	300	80	2.0	6860-09	47	.05
MAX642	5	12	200	91	1.2	6860-08	39	0.05
	5	12	350	89	2	6860-04	18	0.03
	5	12	550	87	3.5	7200-02	12	0.01
MAX643	5	15	100	92	1.2	6860-08	39	0.05
	5	15	150	89	1.5	6860-06	27	0.04
	5	15	225	89	2	6860-04	18	0.03
	5	15	325	85	3.5	7200-02	12	0.01

(b)

* Ferrite Bobbin Coils from Caddell-Burns, NY (516) 746-2310

Switching power-supply circuits

Continued

PART NUMBER	PKG.	Ron AT (I_{DS}, $V_{GS} = X$)	V(MAX)	MFG.
IRFD121	4p DIP	0.3Ω (1.3A, 10V)	60	H/IR
BUZ71A	TO-220	0.12Ω (6A, 10V)	50	MOT/SI/SM
BUZ21	TO-220	0.1Ω (9A, 10V)	100	MOT/SI/SM
IRF513	TO-220	0.8Ω (2A, 10V)	100	H/IR/MOT/SI
IRF530	TO-220	0.18Ω (8A, 10V)	100	H/IR/MOT/SI
IRF540	TO-220	0.085Ω (8A, 10V)	100	H/IR/MOT/SI
IRF620	TO-220	0.8Ω (2.5A, 10V)	200	H/IR/MOT/SI
IRF640	TO-220	0.18Ω (10A, 10V)	200	H/IR/MOT/SI

Manufacturer Code: H= Harris, IR= International Rectifier, MOT= Motorola, SM= Siemens, SI= Siliconix

(c) **N-Channel Logic-Level Power MOSFETs**

PART NUMBER	PKG.	Ron AT (I_{DS}, $V_{GS} = X$)	V(MAX)	MFG.
RFP25N06L	TO-220	0.85Ω (12.5A, 5V)	50	H
RFP12N10L	TO-220	0.20Ω (6A, 5V)	100	H
PFP15N06L	TO-220	0.14Ω (7.5A, 5V)	50	H
IRL540	TO-220AB	0.11Ω (24A, 4V)	100	IR
IRL734	TO-220AB	0.3Ω (7.8A, 4V)	60	IR
IRZ14	TO-220AB	0.07Ω (23A, 4V)	60	IR
MTM25N05L	TO-220AB	0.1Ω (12.5A, 5V)	50	MOT
MTM15N05L	TO-220AB	0.15Ω (7.5A, 5V)	50	MOT
MTP12N10L	TO-220AB	0.18Ω (6A, 5V)	100	MOT

Manufacturer Code: H= Harris, IR= International Rectifier, MOT= Motorola

Note: This list does not constitute an endorsement by Maxim Integrated Products and is not intended to be a comprehensive list of all manufacturers of these components.

Basic step-up switching regulator (10 W)

Fig. 2-11 The circuit in the illustration uses the MAX641/642/643 to provide basic step-up (+3 V to +5 V) switching operation with 10 W output. For operation at one of the preset output voltages (+5 V for the MAX641, +12 V for the MAX642, and +15 V for the MAX643), the V_{BF} pin is connected to ground. For an output voltage other than the preset value, use the circuit of Fig. 2-12. Figure 2-11B shows inductor values for typical inputs and outputs. Figure 2-11C shows recommended external MOSFET characteristics and part numbers. The value of low-battery voltage is set by:

$$R_1 = R_2 \left(\frac{V_{LB}}{1.31 \text{ V}} - 1 \right)$$

where V_{LB} is the desired low-battery detection voltage. MAXIM NEW RELEASES DATA BOOK, 1992, PP. 4-97, 100, 101.

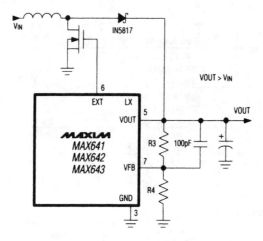

10-W switching regulator with adjustable output

Fig. 2-12 When the ICs of Fig. 2-11 must provide an adjustable output, an external voltage divider (R3 and R4) is required so the the value of V_{OUT}. With R4 at any value between 10 kΩ and 10 MΩ (100 kΩ typically), the value of R5 is set by:

$$R_3 = R_4\left(\frac{V_{OUT}}{1.31\ V} - 1\right)$$

The inductors shown in Fig. 2-11B and MOSFETs shown in Fig. 2-11C still apply. MAXIM NEW RELEASES DATA BOOK, 1992, P. 4-98.

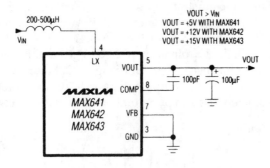

Low-power step-up conversion

Fig. 2-13 In low-power applications, the L_x output and the internal diode of the MAX641/642/643 can be used instead of an external MOSFET. This configuration requires only two capacitors and one inductor. The power-handling capability of this circuit is about 250 mW. Use the inductors shown in Fig. 2-1B and 2-1C. MAXIM NEW RELEASES DATA BOOK, 1992, P. 4-101.

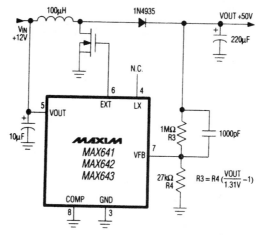

High-voltage step-up regulator

Fig. 2-14 If the external MOSFET or transistor has an adequate voltage rating, the output-voltage range of the MAX641/642/643 can be extended using the circuit of Fig. 2-14. The inductors shown in Fig. 2-11B and the MOSFETs shown in Fig. 2-11C still apply. Maxim New Releases Data Book, 1992, p. 4-102.

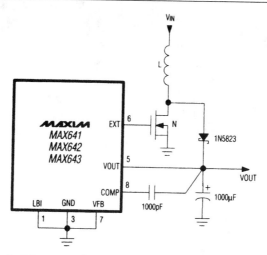

Basic step-up switching regulator (without low-battery indicator)

Fig. 2-15 The circuit in the illustration is essentially the same as that of Fig. 2-11, but it is used where a low-battery function is not required. The inductors and MOSFETs of Figs. 2-11B and 2-11C still apply. The output power is determined by the current ratings of the external MOSFET and inductor, as well as the switching time of the EXT output into the gate capacitance of the MOSFET. Typical switching times are 125 ns when V_{OUT} is +15 V and 160 ns when V_{OUT} = +5 V, with a MOSFET input capacitance of 350 pF. Maxim New Releases Data Book, 1992, p. 4-102.

Switching power-supply circuit titles and descriptions

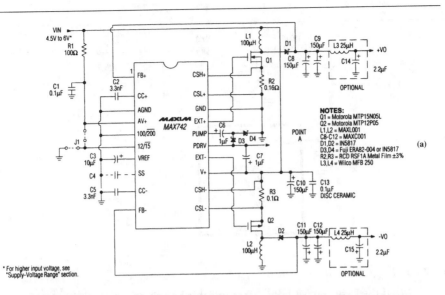

(a)

VIN
4.5V to 6V*

NOTES:
Q1 = Motorola MTP15N05L
Q2 = Motorola MTP12P05
L1,L2 = MAXL001
C8-C12 = MAXC001
D1,D2 = IN5817
D3,D4 = Fuji ERA82-004 or IN5817
R2,R3 = RCD RSF1A Metal Film ±3%
L3,L4 = Wilco MFB 250

POINT
A

* For higher input voltage, see
"Supply-Voltage Range" section.

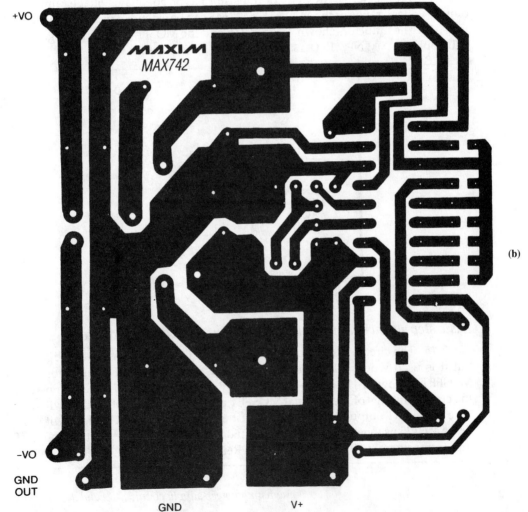

(b)

+VO

-VO

GND
OUT

GND
IN

V+

Continued

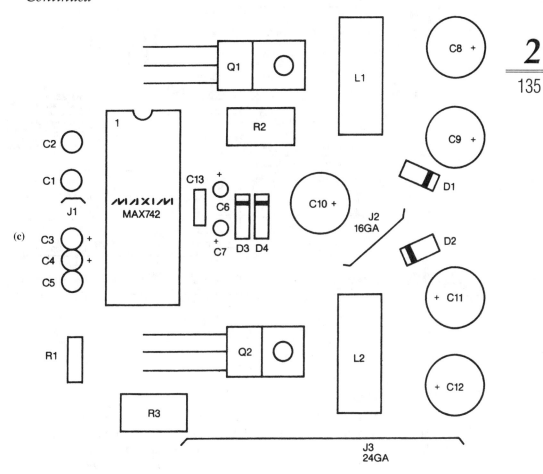

Dual-output, switch-mode regulator (6 W)

Fig. 2-16 The circuit in Fig. 2-16A provides ±200 mA at ±15 V, or ±250 mA at ± 12 V. Although designed for operation from a +3-V logic supply, the MAX742 works well from 4.2 V (the undervoltage lock-out threshold) V to +10 V (absolute maximum plus a safety margin). If the input exceeds 8 V, ground the PDRV pin, and remove C6, D3 and D4. By heatsinking the Q1/Q2, using cores with higher current capability (such as Gowanda #050AT1003), and using higher filter capacitance, the output capability can be increased to 10 W. Figures 2-16B and 2-16C show PC layout and parts placement, respectively. MAXIM NEW RELEASES DATA BOOK, 1992, P. 4-161, 162, 163.

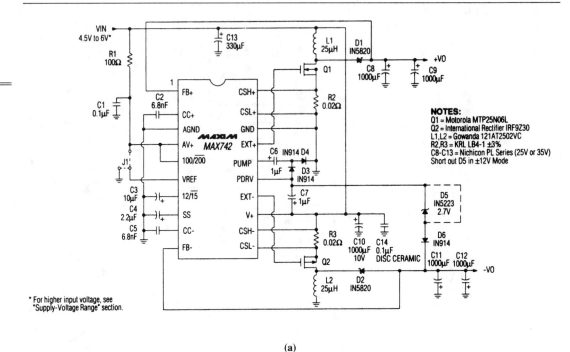

(a)

Dual-output, switch-mode regulator (22 W)

Fig. 2-17 The circuit in Fig. 2-17A provides ±750 mA at ±15 V, or ±950 mA at ±12 V. Both Q1 and Q2 must have heatsinks. D3, D4, and C6 might not be necessary if the circuit load is less than 100 mA on start-up. Although designed for operation from a +5-V input, the input can be from +4.2 to +10 V. Ground the PDRV pin and remove C6, D3, C4 if the input exceeds 8 V. (See Fig. 2-16). Figures 2-17B and 2-17C show PC layout and parts placement, respectively. Maxim New Releases Data Book, 1992, pp. 4-161, 164, 165.

Switching power-supply circuits

Continued

Dual-Output, Switch-Mode Regulator
(+5V to ±12V or ±15V)

MAX742

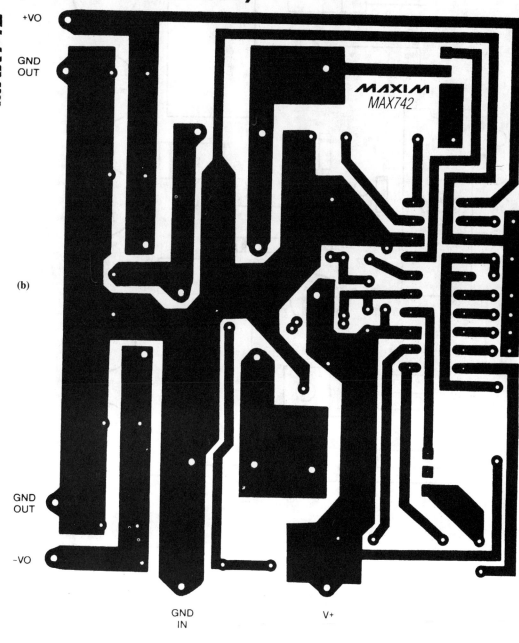

(b)

+VO

GND
OUT

GND
OUT

-VO

GND
IN

V+

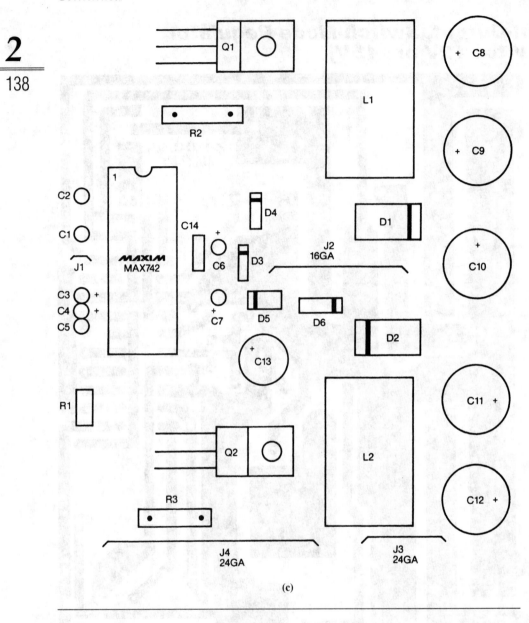

(c)

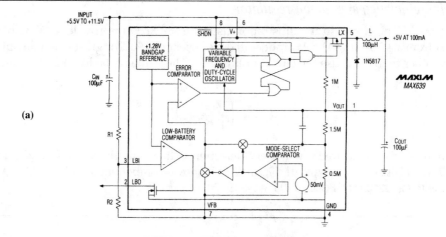

(a)

(b)

Inductors — Through Hole				
P/N	Size	Value (μH)	I$_{MAX}$ (A)	Series-R (Ω)
MAXL001*	0.65 x 0.33" dia.	100	1.75	0.2
7300-13**	0.63 x 0.26" dia.	100	0.89	0.27
7300-15**	"	150	0.72	0.36
7300-17**	"	220	0.58	0.45
7300-19**	"	330	0.47	0.58
7300-21**	"	470	0.39	0.86
7300-25**	"	1000	0.27	2.00

*Maxim Integrated Products

**Caddell-Burns
258 East Second Street
Mineola, NY 11501-3508
(516) 746-2310

Inductors — Surface Mount				
P/N	Size	Value (μH)	I$_{MAX}$ (A)	Series-R (Ω)
CD54	5.2 x 5.8 x 4.5mm	100	0.52	0.63
CD54	"	220	0.35	1.50
CDR74	7.1 x 7.7 x 4.5mm	100	0.52	0.51
CDR74	"	220	0.35	0.98
CDR105	9.2 x 10.0 x 5.0mm	100	0.80	0.35
CDR105	"	220	0.54	0.69

Sumida Electric (USA)
637 East Golf Road
Arlington Heights, IL 60005
(708) 956-0666

Capacitors — Low ESR				
P/N	Size	Value (μF)	ESR (Ω)	V$_{MAX}$ (V)
MAXC001*	0.49 x 0.394" dia.	150	0.2	35
267 Series**	D SM packages	47	0.2	10
267 Series**	E SM packages	100	0.2	6.3

*Maxim Integrated Products

**Matsuo Electronics
2134 Main Street
Huntington Beach, CA 92648
(714) 969-2491

Schottky Diodes — Surface Mount			
P/N	Size	V$_F$(V)	I$_{MAX}$ (A)
SE014	SOT89	0.55	1
SE024	SOT89	0.55	0.95

Collmer Semiconductor
14368 Proton Road
Dallas, TX 75244
(214) 233-1589

NOTE: This list does not constitute an endorsement by Maxim Integrated Products and is not intended to be a comprehensive list of all manufacturers of these components.

Switching power-supply circuit titles and descriptions

High-efficiency step-down regulator

Fig. 2-18 The circuit in the illustration uses the MAX639 to provide +5 V at 100 mA from a +5.5-V to +11.5-V input, with an efficiency of 90% or greater. For an output other than +5 V, use the circuit of Fig. 2-19. The value of the low-battery voltage is set by:

$$R_1 = R_2 \left(\frac{V_{LB}}{1.28} - 1 \right)$$

where V_{LB} is the desired low-battery detection voltage and the value R_2 is the 10 kΩ to 10 MΩ (100 kΩ typical). Figure 2-18B shows component suppliers. MAXIM NEW RELEASES DATA BOOK, 1993, PP. 4-19, 20.

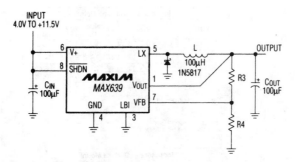

High-efficiency step-down regulator with adjustable output

Fig. 2-19 When the MAX639 of Fig. 2-18 must provide an adjustable output, an external voltage divider (R3 and R4) is required to set the value of V_{OUT}. With R4 at any value between 10 kΩ and 10 MΩ (100 kΩ typical), the value of R3 is set by:

$$R_3 = R_4 \left(\frac{V_{OUT}}{1.28} - 1 \right)$$

MAXIM NEW RELEASES DATA BOOK, 1993, P. 4-20.

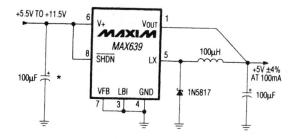

* An input bypass capacitor must be less than 1cm from the V+ pin (pin 6). If the main bypass capacitor is too distant, then a high-frequency 0.1μF ceramic capacitor must be placed within 1cm of the V+ pin.

Basic high-efficiency step-down regulator (without low-battery indicator)

Fig. 2-20 The circuit in the illustration is essentially the same as that of Fig. 2-18, but it is used where a low-battery function is not required. MAXIM NEW RELEASES DATA BOOK, 1993, P. 4-15.

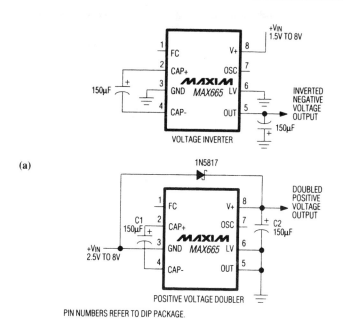

(a)

PIN NUMBERS REFER TO DIP PACKAGE.

Continued

MANUFACTURER	CAPACITOR	CAPACITOR TYPE
Illinois Capacitor	RZS	Aluminum electrolytic
Mallory Capacitor	TDC & TDL	Tantalum
Nichicon	PF & PL	Aluminum electrolytic
Sprague Electric	672D, 673D, 674D, 678D	Aluminum electrolytic
Sprague Electric	135D, 173D, 199D	Tantalum
United Chemi-Con	LXF & SXF	Aluminum electrolytic

(b)

Illinois Capacitor	(708) 675-1760	FAX (708) 673-2850
Mallory Capacitor	(317) 856-3731	FAX (317) 856-2500
Nichicon	(708) 843-7500	FAX (708) 843-2798
Sprague Electric	(508) 339-8900	FAX (508) 339-5063
United Chemi-Con	(708) 696-2000	FAX (708) 640-6341

Simple voltage converter/doubler

Fig. 2-21 The circuits in the illustration use the MAX665 to provide either an inverted negative voltage output or a doubled positive-voltage output from a positive input. The inverter provides a –1.5-V to –8-V output (depending on input). The doubler provides +9.35 V from a +5-V input. Efficiency is generally greater than 90%. An *FC* (frequency-control) pin selects either 10-kHz (FC open), or 45-kHz (FC connected to *V*+) operation. The circuit also can be driven externally, or the oscillator frequency can be lowered by connecting an external capacitor between OSC (pin 13) and ground. Figure 2-21B shows recommended manufacturers for pump capacitors C1 and C2. MAXIM NEW RELEASES DATA BOOK, 1993, PP. 4-23, 31.

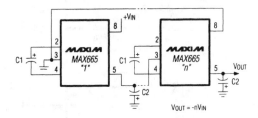

PIN NUMBERS REFER TO DIP PACKAGE.

Cascaded voltage converter/doublers

Fig. 2-22 The circuit in the illustration shows how two or more MAX665s can be cascaded to produce a larger negative multiplication of the initial supply voltage. The resulting output resistance is about equal to the sum of the individual MAX665 R_{OUT} values. The output voltage, where *n* is an integer representing the number of devices cascaded, is defined by $V_{OUT} = -n \, (V_{IN})$. See Fig. 2-21B for capacitor data. MAXIM NEW RELEASES DATA BOOK, 1993, P. 4-31.

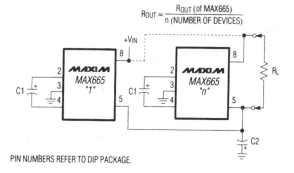

$$R_{OUT} = \frac{R_{OUT} \text{ (of MAX665)}}{n \text{ (NUMBER OF DEVICES)}}$$

PIN NUMBERS REFER TO DIP PACKAGE.

Parallel voltage converter/doublers

Fig. 2-23 The circuit in the illustration shows how two or more MAX665s can be paralleled to reduce output resistance as shown by the equation. Each device requires its own pump capacitor C1, but the reservoir capacitor C2 serves all devices. See Fig. 2-21B for capacitor data. The C2 value must be increased by a factor of n, where n is the number of devices. MAXIM NEW RELEASES DATA BOOK, 1993, P. 4-31.

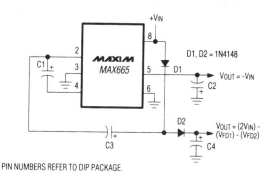

PIN NUMBERS REFER TO DIP PACKAGE.

Combined positive multiplication and negative conversion

Fig. 2-24 The circuit in the illustration shows how a MAX665 can produce both a negative V_{OUT} and a multiplied positive V_{OUT} from a positive V_{IN}. Capacitors C1 and C3 are pump and reservoir, respectively, for the negative output. Capacitors C2 and C4 are pump and reservoir for the multiplied positive output. This circuit produces higher source impedances of the generated supplies because of the common charge-pump driver's finite impedance. See Fig. 2-21B for capacitor data. MAXIM NEW RELEASES DATA BOOK, 1993, P. 4-31.

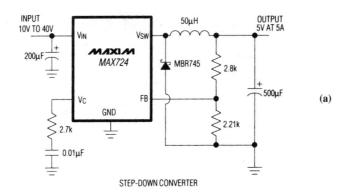

STEP-DOWN CONVERTER

(a)

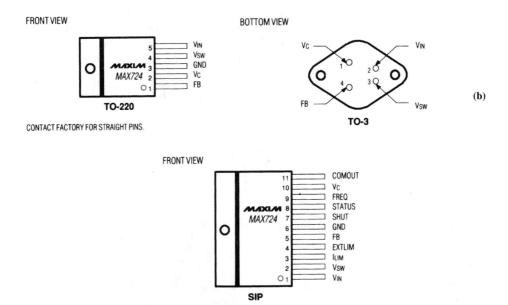

CASE IS CONNECTED TO GROUND. PINS ARE FORMED.

5-A PWM step-down regulator

Fig. 2-25 The circuit in the illustration uses a MAX724 to provide 5 V at 5 A, with a 10-V to 40-V input. The MAX724H (high-voltage version) permits inputs up to 60 V. As shown in Fig. 2-25B, the MAX724 is available in 5-pin TO-220, 4-pin TO-3, and 11-pin *SIP*s (single in-line packages). MAXIM NEW RELEASES DATA BOOK, 1993, P. 4-65.

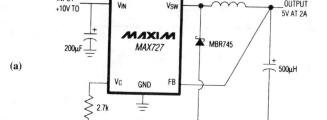

(a)

FRONT VIEW

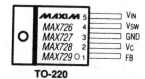

CONTACT FACTORY FOR STRAIGHT PINS.

(b)

2-A PWM step-down regulator

Fig. 2-26 The circuit in the illustration uses a MAX727 to provide 5 V at 2 A, with a 10-V to 40-V input. The H (high voltage) version permits inputs up to 60 V. The MAX726 has a 2.5-V to 40-V adjustable output. The MAX727, 728, and 729 have preset outputs of +5 V, +3.3 V, and +3 V, respectively. As shown in Fig. 2-26B, the devices are available in 5-pin TO-220 and 11-pin SIP packages. MAXIM NEW RELEASES DATA BOOK, 1993, P. 4-67.

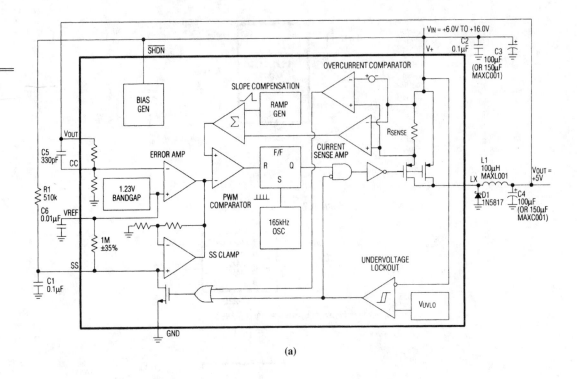

(a)

PART	MIN (V)	MAX (V)
MAX730	5.2	11.0
MAX750	4.0	11.0
MAX738	6.0	16.0
MAX758	4.0	16.0

(b)

MAX730/MAX750 Circuit Conditions				Soft-Start Time (ms) vs. C1 (μF)			
R1 (kΩ)	V+ (V)	I_{OUT} (mA)	C4 (μF)	0.01μF (ms)	0.047μF (ms)	0.1μF (ms)	0.47μF (ms)
510	6	0	100	2	6	11	28
510	9	0	100	1	4	6	15
510	11	0	100	1	2	4	11
510	9	150	100	1	4	8	21
510	9	300	100	1	5	9	27
510	9	150	390	3	6	9	23
510	9	150	680	4	6	9	24
none	6	0	100	16	34	51	125
none	9	0	100	10	22	34	82
none	11	0	100	8	18	28	66
none	9	150	100	34	134	270	1263
none	9	150	390	39	147	280	1275
none	9	150	680	40	152	285	1280

(c)

MAX738/MAX758 Circuit Conditions				Soft-Start Time (ms) vs. C1 (μF)			
R1 (kΩ)	V+ (V)	I_{OUT} (mA)	C4 (μF)	0.01μF (ms)	0.047μF (ms)	0.1μF (ms)	0.47μF (ms)
510	7	0	100	1	4	6	18
510	12	0	100	1	2	3	8
510	16	0	100	1	1	2	6
510	12	300	100	1	3	5	3
510	12	750	100	1	5	8	21
none	7	0	100	12	27	40	100
none	12	0	100	7	16	25	54
none	16	0	100	6	13	20	68
none	12	300	100	27	112	215	1114

(d)

Production Method	Inductors	Capacitors
Surface Mount	Sumida (708) 956-0666 CD54-101KC (MAX730) CD105-101KC (MAX738) Coiltronics (305) 781-8900 CTX100-series	Matsuo (714) 969-2491 267-series
Miniature Through-Hole	Sumida (708) 956-0666 RCH654-101K (MAX730) RCH895-101K (MAX738)	Sanyo (619) 661-6322 OS-CON-series Low ESR Organic Semiconductor
Low-Cost Through-Hole	Maxim MAXL001 100μH Iron-Powder Toroid Renco (516) 586-5566 RL1284-100	Maxim MAXC001 150μF, Low ESR Electrolytic Nichicon (708) 843-7500 PL-series Low ESR Electrolytics United Chemicon (708) 696-2000 LXF-series

2

147

Switching power-supply circuit titles and descriptions

+5-V PWM step-down regulators

Fig. 2-27 The circuit in the illustration uses the MAX730/738 to provide basic PWM step-down regulation. Figure 2-27B shows the input voltage range. Figure 2-27C shows the soft-start times. Figure 2-27D shows recommended component suppliers. For an output other than +5 V, use the circuit of Fig. 2-28. MAXIM NEW RELEASES DATA BOOK, 1993, PP. 4-73, 74, 75, 78.

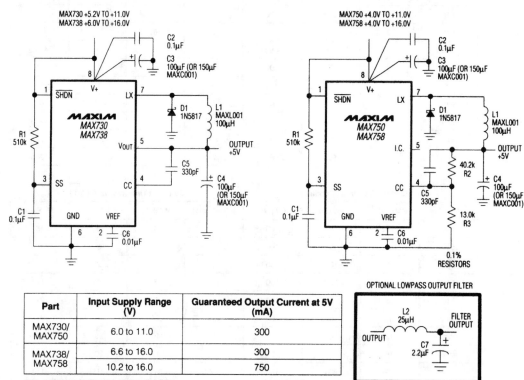

Part	Input Supply Range (V)	Guaranteed Output Current at 5V (mA)
MAX730/ MAX750	6.0 to 11.0	300
MAX738/ MAX758	6.6 to 16.0	300
	10.2 to 16.0	750

NOTE: PIN NUMBERS REFER TO 8-PIN PACKAGES

+5-V and adjustable PWM step-down regulators

Fig. 2-28 These circuits use both the fixed and MAX730/738 and adjustable MAX750/758 to provide a +5-V output. To provide outputs other than +5 V, use the MAX750/758 and alter the values of R_2/R_3. With R_3 at any value in the 10 kΩ to 10-MΩ range, the value of R_2 is set by:

$$R_2 = R_3 \left(\frac{V_{OUT}}{1.23\ V} - 1 \right)$$

See Fig. 2-27 for input voltage range, typical soft-start times, and component suppliers. MAXIM NEW RELEASES DATA BOOK, 1993, P. 4-77.

Switching power-supply circuits

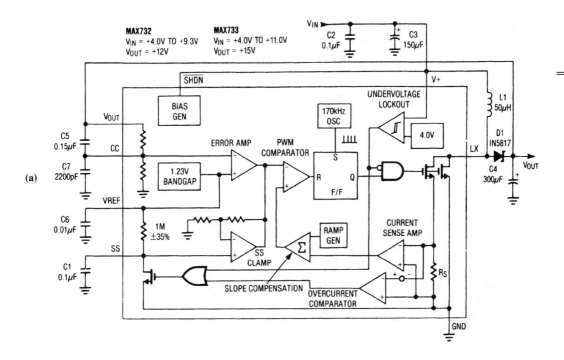

(a)

(b)

MAX732 CIRCUIT CONDITIONS			SOFT-START TIME (ms) vs. C1 (µF)		
V+ (V)	IOUT (mA)	C4 (µF)	0.1µF	0.47µF	1.0µF
4.5	0	300	57 ms	115 ms	123 ms
6.0	0	300	40	80	70
9.0	0	300	29	57	44
4.5	100	300	92	348	780
6.0	100	300	59	209	444
9.0	100	300	29	57	60
4.5	200	300	175	713	1690
6.0	200	300	84	340	756
9.0	200	300	28	76	123

MAX733 CIRCUIT CONDITIONS			SOFT-START TIME (ms) vs. C1 (µF)		
V+ (V)	IOUT (mA)	C4 (µF)	0.1µF	0.47µF	1.0µF
4.5	0	300	90 ms	208 ms	251 ms
6.0	0	300	64	135	148
9.0	0	300	36	67	53
12.0	0	300	28	49	33
4.5	75	300	157	680	1380
6.0	75	300	103	426	882
9.0	75	300	46	162	305
12.0	75	300	28	49	33
4.5	125	300	235	1124	2260
6.0	125	300	133	596	1255
9.0	125	300	54	231	476
12.0	125	300	30	49	41

Note: Soft-start times are ± 35% accurate, C1 is the soft-start capacitor, C4 is the output capacitor.

Switching power-supply circuit titles and descriptions

PRODUCTION METHOD	INDUCTORS	CAPACITORS
Surface Mount	Sumida (708) 956-0666 CD54-470 (47μH) CD54-180 (18μH) for discontinuous mode Coiltronics (305) 781-8900 CTX 100-series	Matsuo (714) 969-2491 267-series
Miniature Through-Hole	Sumida (708) 956-0666 RCH654-470	Sanyo (619) 661-6322 OS-CON-series Low ESR Organic Semiconductor
Low-Cost Through-Hole	Renco (516) 586-5566 RL 1284-47	Maxim MAXC001 150μF, Low ESR Electroyltic Nichicon (708) 843-7500 PL-series Low ESR Electrolytics United Chemicon (708) 696-2000 LXF-series

(c)

+12-V/+15-V PWM step-down regulators

Fig. 2-29 The circuit in the illustration uses the MAX732 and MAX733 to provide fixed +12-V and +15-V outputs, respectively. Figure 2-29B shows typical soft-start times, and Fig. 2-29C shows component suppliers. MAXIM NEW RELEASES DATA BOOK, 1993, PP. 4-97, 98, 100.

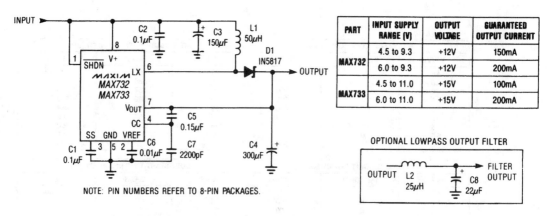

PART	INPUT SUPPLY RANGE (V)	OUTPUT VOLTAGE	GUARANTEED OUTPUT CURRENT
MAX732	4.5 to 9.3	+12V	150mA
	6.0 to 9.3	+12V	200mA
MAX733	4.5 to 11.0	+15V	100mA
	6.0 to 11.0	+15V	200mA

NOTE: PIN NUMBERS REFER TO 8-PIN PACKAGES.

+12V/+15V PWM step-down regulator with output filter

Fig. 2-30 The circuit in the illustration is similar to that of Fig. 2-29, but with an optional lowpass output filter. MAXIM NEW RELEASES DATA BOOK, 1993, P. 4-99.

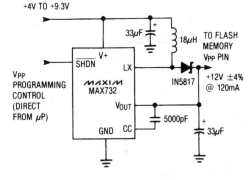

+12-V flash-memory programming power supply

Fig. 2-31 The circuit in the illustration provides +12 V at 120 mA for flash-memory programming. For an input, efficiency is 88%, quiescent current is 1.7 mA, and shutdown current is 70 µA, with 6 µA into the V+ pin. See Fig. 2-29 for soft-start times and component suppliers. See Fig. 2-32 for surface-mount configuration. MAXIM NEW RELEASES DATA BOOK, 1993, P. 4-100.

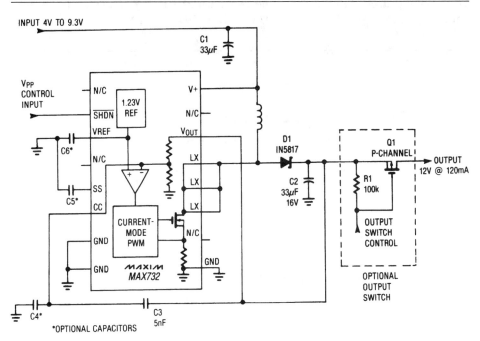

+12-V flash-memory programming evaluation board

Fig. 2-32 The circuit in the illustration is available as a surface-mount evaluation board and is similar to the circuit of Fig. 2-31. Use 18 µH for the inductor across the V+ and L_x pins. MAXIM NEW RELEASES DATA BOOK, 1993, P. 4-103.

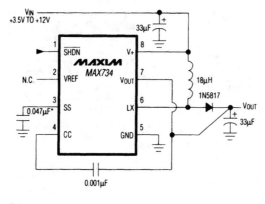

* OPTIONAL

Simple +12-V flash-memory programming supply

Fig. 2-33 The simple circuit in the illustration provides the full +12-V output at 120 mA but requires only 0.3 square inch of space and is readily surface mounted. With a 5-V input, efficiency is 88%, quiescent current is 1.1 mA, and shutdown current is 70 μA. MAXIM NEW RELEASES DATA BOOK, 1993, P. 4-107.

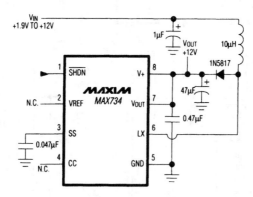

+12-V flash-memory programming supply with 1.9-V input

Fig. 2-34 The circuit in the illustration is similar to that of Fig. 2-33 but is bootstrapped to permit operation with 1.9-V input. MAXIM NEW RELEASES DATA BOOK, 1993, P. 4-107.

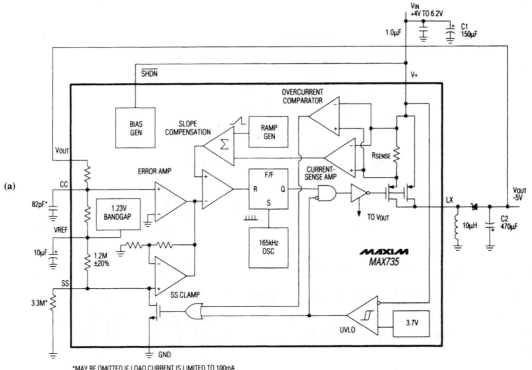

(a)

*MAY BE OMITTED IF LOAD CURRENT IS LIMITED TO 100mA.

(b)

PRODUCTION METHOD	INDUCTORS	CAPACITORS
Surface Mount	Sumida CD54-100 (10µH)	Matsuo 267 series
Miniature Through Hole	Sumida RCH855-100M (10µH)	Sanyo Os-Con series low-ESR organic semiconductor
Low-Cost Through Hole	Renco RL 1284 (10µH)	Nichicon PL series low-ESR electrolytics United Chemicon LXF series

Matsuo USA (714) 969-2491 FAX (714) 960-6492
Matsuo Japan (06) 332-0871
Nichicon (708) 843-7500 FAX (708) 843-2798
Renco (516) 586-5566 FAX (516) 586-5562
Sanyo Os-Con USA (619) 661-6322
Sanyo Os-Con Japan (0720) 70-1005 FAX (0720) 70-1174
Sumida USA (708) 956-0666
Sumida Japan (03) 3607-5111 FAX (03) 3607-5428
United Chemi-Con (708) 696-2000 FAX (708) 640-6311

–5-V inverting PWM regulator (basic)

Fig. 2-35 The circuit in the illustration uses a MAX 735 to provide a –5-V output from a +4-V to +6.2-V input. Output current is 275 mA typical. Figure 2-35B shows component suppliers. MAXIM NEW RELEASES DATA BOOK, 1993, P. 4-114.

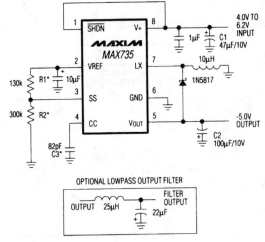

–5-V inverting PWM regulator (surface-mount)

Fig. 2-36 This is the surface-mount version of the circuit in Fig. 2-35, for commercial and extended industrial temperature ranges. MAXIM NEW RELEASES DATA BOOK, 1993, P. 4-113.

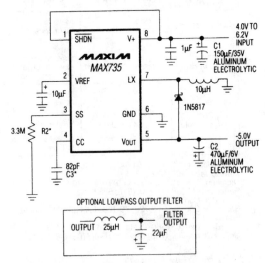

−5-V inverting PWM regulator
(through-hole, commercial temperature ranges)

Fig. 2-37 The circuit in the illustration is the through-hole version of the circuit in Fig. 2-35 for commercial temperature ranges. MAXIM NEW RELEASES DATA BOOK, 1993, P. 4-113.

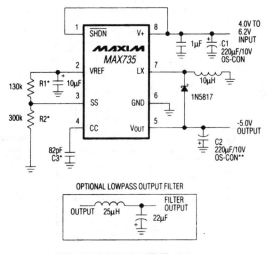

−5-V inverting PWM regulator (through-hole, all temperature ranges)

Fig. 2-38 The circuit in the illustration is the through-hole version of the circuit in Fig. 2-35 for all temperature ranges. MAXIM NEW RELEASES DATA BOOK, 1993, P. 4-113.

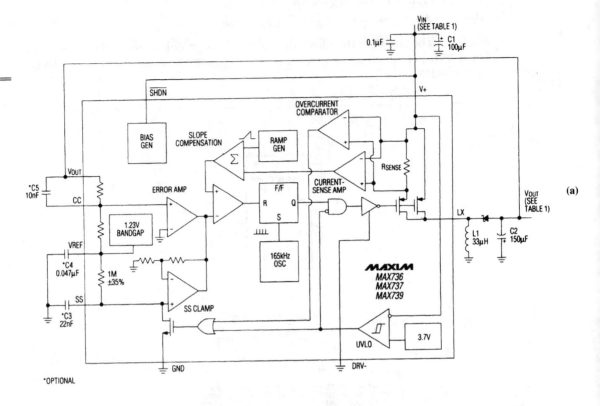

(a)

PRODUCTION METHOD	INDUCTORS	CAPACITORS
Surface Mount	Sumida USA: Phone (708) 956-0666 Japan: Phone (03) 3607-5111 FAX (03) 3607-5428 CD54-330 (33µH) CD54-100 (10µH) Coiltronics Phone (305) 781-8900 FAX (305) 782-4163 CTX 100 series	Matsuo USA: Phone (714) 969-2491 FAX (714) 960-6492 Japan: Phone (06) 332-0871 267 series
Miniature Through-Hole	Sumida USA: Phone (708) 956-0666 Japan: Phone (03) 3607-5111 FAX (03) 3607-5428 RCH654-330 (33µH) RCH108-330 (33µH)	Sanyo Os-Con USA: Phone (619) 661-6322 Japan: Phone (0720) 70-1005 FAX (0720) 70-1174 OS-CON series Low ESR Organic Semiconductor
Low-Cost Through-Hole	Renco Phone (516) 586-5566 FAX (516) 586-5562 RL 1284-33	Nichicon Phone (708) 843-7500 FAX (708) 843-2798 PL series Low ESR Electrolytics United Chemi-Con Phone (708) 696-2000 FAX (708) 640-6311 LXF series

(b)

For wide temperature applications using through-hole components, organic semiconductor capacitors are recommended (C1 and C2 in Figure 1). These capacitors maintain low ESR across their operating temperature range.

Switching power-supply circuits

Device	V+ Range (V)		Output Voltage (V)	Diode D1
	Bootstrapped	Non-Bootstrapped		
MAX736	4 to 8.6	4 to 8.6	-12	1N5818
MAX737	4 to 5.5	4 to 5.5	-15	1N5818
MAX739/759	4 to 11	4 to 15	-5	1N5817/1N5818

(c)

(d)

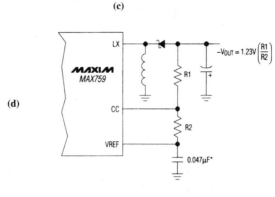

* OPTIONAL

(e)

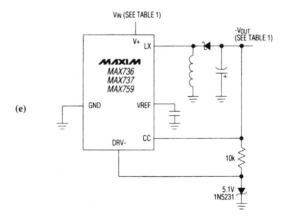

Adjustable inverting PWM regulators (basic)

Fig. 2-39 The circuit in the illustration uses the MAX736/737/739/759 to provide –5 V, –12 V, –15 V and adjustable inverted outputs. See Fig. 2-39B for component suppliers and Fig. 2-39C for the three fixed output voltages. Figure 2-39D shows the circuit for producing an adjustable inverted output. Figure 2-39E shows the circuit connections required for bootstrap operation. MAXIM NEW RELEASES DATA BOOK, 1993, PP. 4-123, 124, 125, 127.

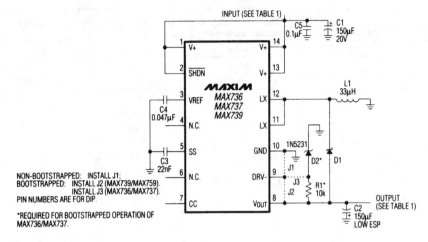

Fixed inverting PWM regulators

Fig. 2-40 The circuit in Fig. 2-40 is the standard application circuit for the MAX736/737/739 of Fig. 2-39. See Fig. 2-39B for components and Fig. 2-39C for fixed input/output voltages. Maxim New Releases Data Book, 1993, p. 4-123.

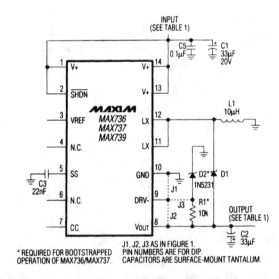

Fixed inverting PWM regulators (surface-mount)

Fig. 2-41 This is the surface-mount discontinuous-conduction version of the Fig. 2-40 circuit. See Fig. 2-39B for components and Fig. 2-39C for fixed input/output voltages. Maxim New Releases Data Book, 1993, p. 4-123.

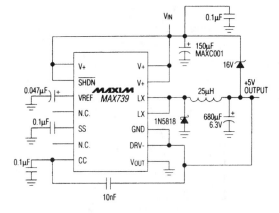

+5-V step-down with a PWM inverter

Fig. 2-42 The circuit in the illustration shows how an inverter can be used to provide a +5-V output, with an input from 9 V to 21 V. The circuit requires a minimum load of 3 mA and can provide up to 1-A full load output, with 60% to 85% efficiency. If the input does not exceed 15 V, ground the DRV pin and remove the zener (for high efficiency). See Fig. 2-40 for pin connections and Fig. 2-39B for components. MAXIM NEW RELEASES DATA BOOK, 1993, PP. 4-126, 127.

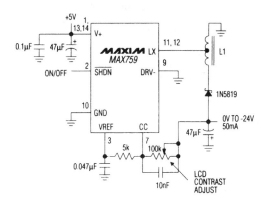

L1 = COILTRONICS CTX100-1 OR MAGNETICS, INC "KOOL-MU" 77030-A7
30 TURNS AND 30 TURNS 26 AWG
PIN NUMBERS ARE FOR DIP

–24-V LCD (liquid-crystal diode) power supply

Fig. 2-43 The circuit in the illustration uses a MAX759 to provide an adjustable negative voltage for power small LCD (liquid-crystal diode) displays and will deliver 30 mA at –24 V with 80% efficiency. See Fig. 2-39B for components. MAXIM NEW RELEASES DATA BOOK, 1993, P. 4-126.

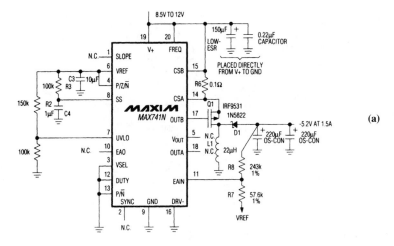

(a)

V$_{IN}$ (V)	I$_{OUT}$ (A)	EFFICIENCY (%)
12	0.1	73
8.5	0.5	81.5
10	0.5	83
12	0.5	85
8.5	1	78
10	1	81
12	1	82

VSEL	P/Z/N̄	OUTPUT	EAIN IMPEDANCE (Ω)
V+	V+	5V	>1k
V+	VREF	Adj. Positive	>50M HiZ
V+	GND	-5V	>13k
VREF	V+	12V	>6.5k
VREF	VREF	Prohibited	NA
VREF	GND	-12V	>13k
GND	V+	15V	>6.5k
GND	VREF	Adj. Negative	>50M HiZ
GND	GND	-15V	>11k

(b)

DUTY	P/N̄	FREQ	OUTA	OUTB	MODE
V+	V+	V+, VREF	N	P	Push-pull
V+	GND	V+, VREF	N	P	Push-pull
V+	V+	GND	GND	GND	Shutdown
V+	GND	GND	V+	V+	Shutdown
VREF	V+	V+, VREF	N	P	Complementary output 50%
VREF	GND	V+, VREF	N	P	Complementary output 50%
VREF	X	GND	GND	V+	Shutdown
GND	V+	V+, VREF	N	P	Complementary output 85% or 95%
GND	GND	V+, VREF	N	P	Complementary output 85% or 95%
GND	X	GND	GND	V+	Shutdown

N = Drives N-Channels (On = V+)
P = Drives P-Channels (On = GND)
X = Don't Care

Switching power-supply circuits

12 V to –5.2 V at 1.5-A ECL power supply

Fig. 2-44 The circuit in the illustration uses a MAX741N controller to provide a fixed –5.2 V at 1.5 A for *ECL* (emitter-coupled logic) applications. The input supply range is from 7.6 V to 15 V and should be bypassed with a low-ESR aluminum electrolytic, such as Nichicon PL series or Sanyo Os-Con series. Bypass capacitors must be placed directly at the *V+* and GND pins as shown. See Fig. 2-44B for efficiency. MAXIM NEW RELEASES DATA BOOK, 1993, PP. 4-133, 134.

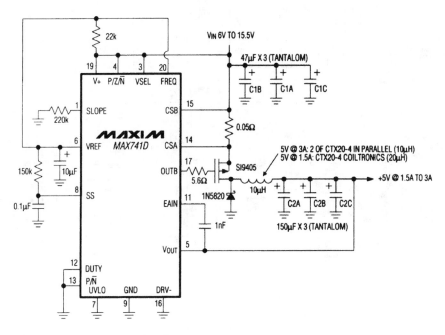

5 V from 6 V to 15.5-V buck converter

Fig. 2-45 The circuit in the illustration uses a MAX741D controller to provide a fixed +5 V at 1.5 A to 3 A. Nichicon PL series capacitors are recommended for commercial temperatures. Use Sanyo OsCon for temperatures below 0°C. GND and *V+* connections should be of the star type. MAXIM NEW RELEASES DATA BOOK, 1993, P. 4-135.

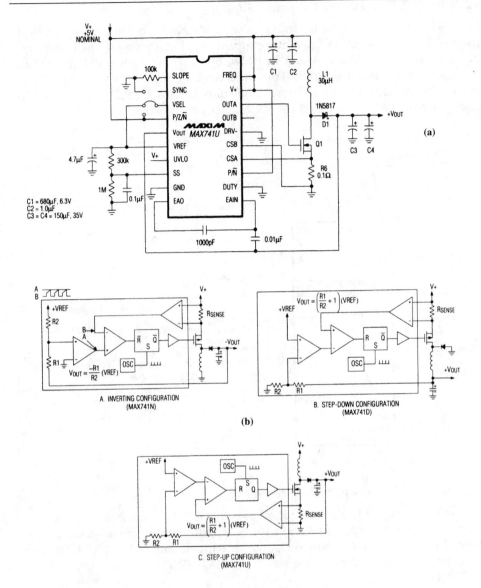

(a)

(b)

A. INVERTING CONFIGURATION
(MAX741N)

B. STEP-DOWN CONFIGURATION
(MAX741D)

C. STEP-UP CONFIGURATION
(MAX741U)

+12 V or +15 V from 5-V step-up converter

Fig. 2-46 The circuit in the illustration uses a MAX741U controller to provide a fixed +12 V or +15 V at 400 mA from a 5-V input. The selection of 12-V or 15-V output depends on pin programming as shown in Fig. 2-44B. For example, for a +12-V output, V_{SEL} (pin 3) is connected to V_{REF} (through jumpers), and P/ZN (pin 4) is connected to V+. For a +15-V output, V_{SEL} is grounded and P/ZN is connected to V+. See Fig. 2-46B for the three basic MAX741 configurations and Fig. 2-45 for pin numbers. MAXIM NEW RELEASES DATA BOOK, 1993, PP. 4-134, 136.

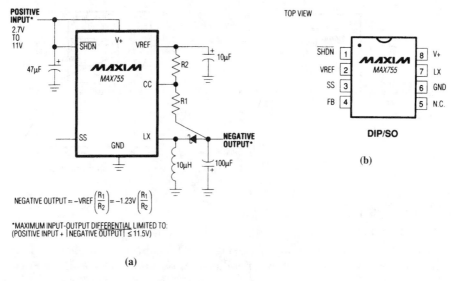

$$\text{NEGATIVE OUTPUT} = -\text{VREF}\left(\frac{R_1}{R_2}\right) = -1.23V\left(\frac{R_1}{R_2}\right)$$

*MAXIMUM INPUT-OUTPUT DIFFERENTIAL LIMITED TO:
(POSITIVE INPUT + | NEGATIVE OUTPUT| ≤ 11.5V)

(a)

Adjustable inverting PWM regulator

Fig. 2-47 The simple circuit in the illustration uses a MAX755 to produce an adjustable negative output from a positive input. The output is set by the values of R_1 and R_2 as shown by the equations. The input can be from 2.7 V to 11 V, but the input/output differential must not exceed 11.5 V. MAXIM NEW RELEASES DATA BOOK, 1993, P. 4-139.

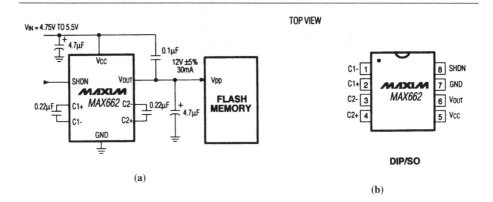

(a)

(b)

+12-V 30-mA flash-memory programming supply

Fig. 2-48 The circuit in the illustration uses a MAX662 to provide +12-V at 30-mA for flash-memory programming. The entire circuit fits into less than 0.2 square inch of PC board space. Figure 2-48B shows capacitor suppliers. MAXIM NEW RELEASES DATA BOOK, 1994, P. 4-47.

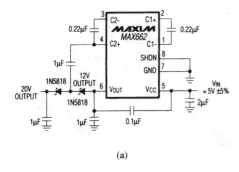

(a)

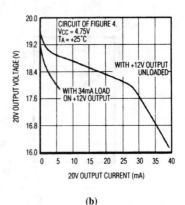

(b)

+12 V and +20 V from a +5-V input

Fig. 2-49 The circuit in the illustration uses a MAX662 to provide both a +12-V and +20-V outputs from a 5-V input. See Fig. 2-48 for capacitor suppliers and Fig. 2-49B for the +20-V output-current capability. MAXIM NEW RELEASES DATA BOOK, 1994, P. 4-47.

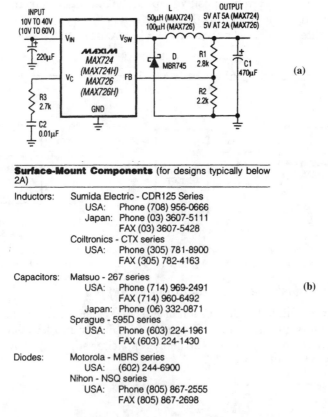

(a)

(b)

Surface-Mount Components (for designs typically below 2A)

Inductors:	Sumida Electric - CDR125 Series
	USA: Phone (708) 956-0666
	Japan: Phone (03) 3607-5111
	FAX (03) 3607-5428
	Coiltronics - CTX series
	USA: Phone (305) 781-8900
	FAX (305) 782-4163
Capacitors:	Matsuo - 267 series
	USA: Phone (714) 969-2491
	FAX (714) 960-6492
	Japan: Phone (06) 332-0871
	Sprague - 595D series
	USA: Phone (603) 224-1961
	FAX (603) 224-1430
Diodes:	Motorola - MBRS series
	USA: (602) 244-6900
	Nihon - NSQ series
	USA: Phone (805) 867-2555
	FAX (805) 867-2698

Switching power-supply circuits

Through-Hole Components	

Inductors:	Sumida - RCH-110 series
	(see above for phone number)
	Cadell-Burns - 7070, 7300, 6860, and 7200 series
	USA: Phone (516) 746-2310
	        FAX (516) 742-2416
	Renco - various series
	USA: Phone (516) 586-5566
	        FAX (516) 586-5562
	Coiltronics - various series
	(see above for phone number)

Capacitors:	Nichicon - PL series low-ESR electrolytics
	USA: Phone (708) 843-7500
	        FAX (708) 843-2798
	United Chemi-Con - LXF series
	USA: Phone (708) 696-2000
	        FAX (708) 640-6311
	Sanyo - OS-CON low-ESR organic semiconductor
	USA: Phone (619) 661-6322
	Japan: Phone (0720) 70-1005
	        FAX (0720) 70-1174

Diodes:	General Purpose - 1N5820-1N5825 Motorola - MBR and MBRD series (see above for phone number)

5-A/2-A step-down PWM regulator

Fig. 2-50 The circuit in the illustration shows the MAX724/726 connected in the basic step-down configuration. Figure 2-50B shows suppliers for both surface-mount and through-hole components. Resistors R1 and R2 set the output voltage. By letting R_2 equal the nearest standard resistor value (so that 1 mA flows through R_1/R_2), the value of R_1 can be found by:

$$R_1 = \left(\frac{V_{OUT} \times R_2}{2.21\ V} \right) - R_2$$

You can use other values of R_2, but they should not exceed 4 kΩ. MAXIM NEW RELEASES DATA BOOK, 1994, P. 4-85.

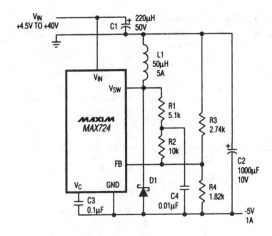

D1 - MOTOROLA MBR745
C1 - NICHICON UPL1C221MRH6
C2 - NICHICON UPL1A102MRH6
L1 - COILTRONICS CTX25-5-52
ALL RESISTORS HAVE 1% TOLERANCE

1-A positive-to-negative inverter

Fig. 2-51 The circuit in the illustration shows the MAX724 connected to provide a −5 V from a positive input. Figure 2-50B shows suppliers for both surface-mount and through-hole components. Both the MAX724 and 726 can convert positive input voltages to negative outputs if sum of input and output voltage is greater than 8 V, and the minimum positive supply is 4.75 V. If V_{IN} does not fall below $2V_{OUT}$, then R1, R2, and C4 can be omitted, and only R3 and R4 set the output voltage. MAXIM NEW RELEASES DATA BOOK, 1994, P. 4-92.

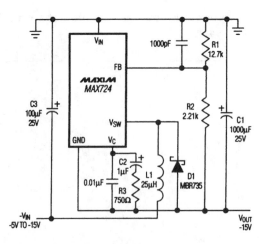

Negative step-up converter

Fig. 2-52 The circuit in the illustration shows the MAX724 connected to provide –15 V from a negative input. Figure 2-50B shows suppliers for both surface-mount and through-hole components. Both the MAX724 and 726 can work as a negative boost converter by tying the GND pin to the negative output. This arrangement allows the IC to operate from input voltages as low as –4.75 V if the regulated output is at least –8 V. Resistors R1 and R2 set the output voltage as discussed for the circuit of Fig. 2-50. MAXIM NEW RELEASES DATA BOOK, 1994, P. 4-92.

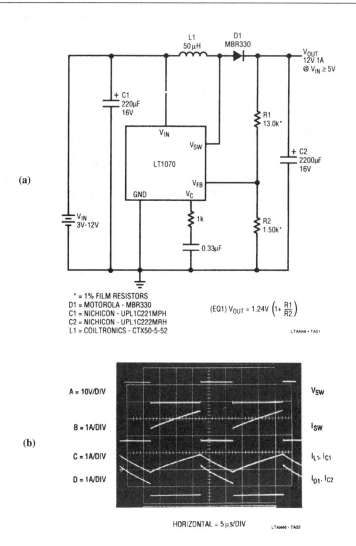

(a)

$$(EQ1) \; V_{OUT} = 1.24V \left(1 + \frac{R1}{R2}\right)$$

* = 1% FILM RESISTORS
D1 = MOTOROLA - MBR330
C1 = NICHICON - UPL1C221MPH
C2 = NICHICON - UPL1C222MRH
L1 = COILTRONICS - CTX50-5-52

LTAN46 · TA01

(b)

HORIZONTAL = 5 μs/DIV   LTAN46 · TA02

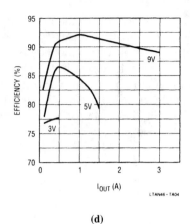

(c)

(d)

Capacitor Manufacturers

1) Nichicion (America) Corporation
   927 East State Parkway
   Schaumburg, IL 60195
   (708)843-7500

2) Sanyo Video Components (USA) Corporation
   1201 Sanyo Avenue
   San Diego, CA 92073
   (619)661-6322

3) United Chemi-Con, Inc.
   9801 West Higgins Road
   Rosemount, IL 60018
   (708)696-2000

4) Sprague Electric Company
   Aluminum Electrolytic Div.
   9800 Kincey Ave. Suite 100
   Huntersville, NC 28078
   (704)875-8070

5) Kemet Electronics
   P.O. Box 5928
   Greenville, SC 29606
   (803)675-1760

6) Marcon
   998Forest Edge Drive
   Vernon Hills, IL 60061
   (708)913-9980

7) Wima
   2269 Saw Mill River Rd
   Bldg. 4C
   P.O. Box 217
   Elmsford, NY 10523
   (914)347-2474

(e)

12 V at 1 A from 3 V to 12 V

Fig. 2-53 The circuit in the illustration uses the LT1070 as a basic step-up regulator. The waveforms, V_{IN}/V_{OUT}, and efficiency are shown in Figs. 2-53B, C, and D, respectively. Fig. 2-53E shows capacitor manufacturers. The Coiltronics address is:

984 S.W. 13th Court
Pompano Beach, FL 33069
Phone: (305) 781-8900
Fax: (305) 782-4163

LINEAR TECHNOLOGY, LINEAR APPLICATIONS HANDBOOK, 1993, PP. AN46-1, 2, 19.

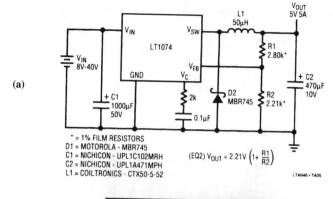

(a)

* = 1% FILM RESISTORS
D1 = MOTOROLA - MBR745
C1 = NICHICON - UPL1C102MRH
C2 = NICHICON - UPL1A471MPH
L1 = COILTRONICS - CTX50-5-52

(EQ2) $V_{OUT} = 2.21V \left(1+ \frac{R1}{R2}\right)$

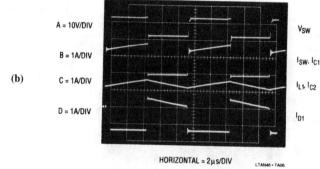

(b)

HORIZONTAL = 2µs/DIV

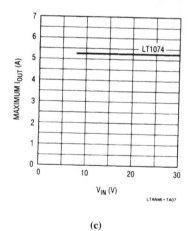

(c)

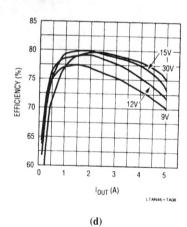

(d)

5 V at 5 A from 8 V to 40 V

Fig. 2-54 The circuit in the illustration uses the LT1074 as a basic step-down regulator. The waveforms, V_{IN}/V_{OUT}, and efficiency are shown in Figs. 2-54B, C and D, respectively. Figure 2-53E shows capacitors manufacturers. See Fig. 2-53 for Coiltronics. LINEAR TECHNOLOGY, LINEAR APPLICATIONS HANDBOOK, 1993, PP. AN46-3, 4.

Switching power-supply circuit titles and descriptions

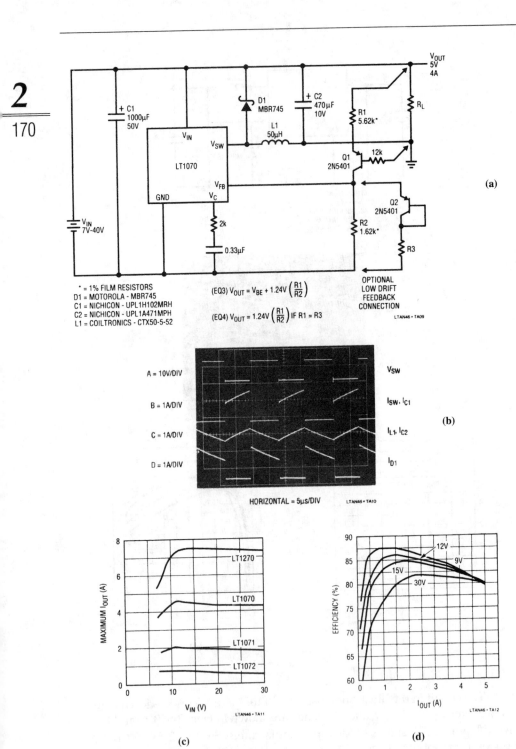

(a)

* = 1% FILM RESISTORS
D1 = MOTOROLA - MBR745
C1 = NICHICON - UPL1H102MRH
C2 = NICHICON - UPL1A471MPH
L1 = COILTRONICS - CTX50-5-52

(EQ3) $V_{OUT} = V_{BE} + 1.24V \left(\dfrac{R1}{R2} \right)$

(EQ4) $V_{OUT} = 1.24V \left(\dfrac{R1}{R2} \right)$ IF R1 = R3

OPTIONAL
LOW DRIFT
FEEDBACK
CONNECTION

LTAN46 • TA09

A = 10V/DIV V_{SW}

B = 1A/DIV I_{SW}, I_{C1}

C = 1A/DIV I_{L1}, I_{C2}

D = 1A/DIV I_{D1}

(b)

HORIZONTAL = 5µs/DIV   LTAN46 • TA10

(c)

LTAN46 • TA11

(d)

LTAN46 • TA12

Switching power-supply circuits

5 V at 4 A from 7 V to 40 V

Fig. 2-55 The circuit in the illustration uses the LT1070 as a step-down regulator with floating input. The waveforms, V_{IN}/V_{OUT}, and efficiency are shown in Figs. 2-55B, C and D, respectively. Figure 2-53E shows capacitor manufacturers. See Fig. 2-53 for Coiltronics. LINEAR TECHNOLOGY, LINEAR APPLICATIONS HANDBOOK, 1993, PP. AN46-5, 6.

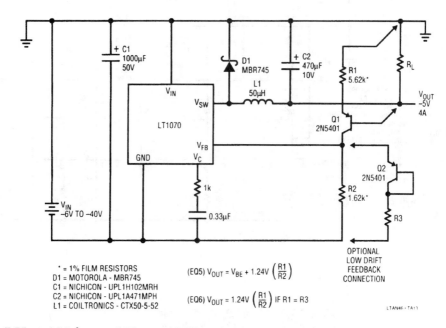

* = 1% FILM RESISTORS
D1 = MOTOROLA - MBR745
C1 = NICHICON - UPL1H102MRH
C2 = NICHICON - UPL1A471MPH
L1 = COILTRONICS - CTX50-5-52

(EQ5) $V_{OUT} = V_{BE} + 1.24V \left(\dfrac{R1}{R2} \right)$

(EQ6) $V_{OUT} = 1.24V \left(\dfrac{R1}{R2} \right)$ IF R1 = R3

LTAN46 - TA13

–5 V at 4 V from –6 V to –40 V

Fig. 2-56 The circuit in the illustration uses the LT1070 as a negative step-down regulator. Figure 2-53E shows capacitor manufacturers. See Fig. 2-53 for Coiltronics. LINEAR TECHNOLOGY, LINEAR APPLICATIONS HANDBOOK, 1993, P. AN46-6.

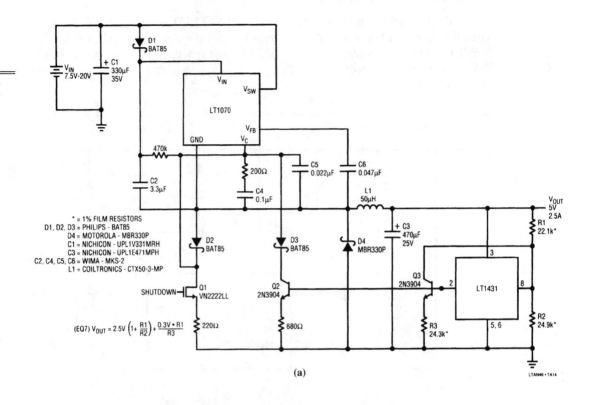

$(EQ7) \ V_{OUT} = 2.5V \left(1 + \dfrac{R1}{R2}\right) + \dfrac{0.3V \cdot R1}{R3}$

(a)

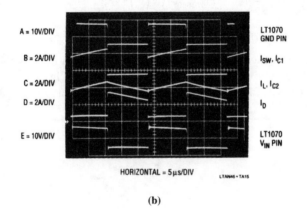

A = 10V/DIV
B = 2A/DIV
C = 2A/DIV
D = 2A/DIV
E = 10V/DIV

LT1070 GND PIN
I_{SW}, I_{C1}
I_L, I_{C2}
I_D
LT1070 V_{IN} PIN

HORIZONTAL = 5μs/DIV

(b)

Switching power-supply circuits

Continued

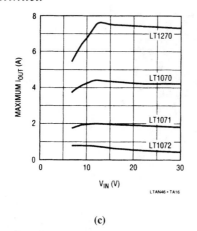

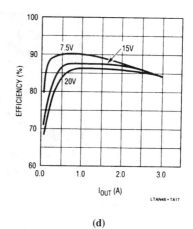

(c)

(d)

5 V at 2.5 A from 7.5 V to 20 V

Fig. 2-57 The circuit in the illustration uses the LT1070 as a high-efficiency buck converter. The LT1431 is a shunt linear regulator (See Chapter 3.) The waveforms, V_{IN}/V_{OUT}, and efficiency are shown in Figs. 2-57B, C, and D, respectively. Figure 2-53E shows capacitor manufacturers. See Fig. 2-53 for Coiltronics. LINEAR TECHNOLOGY, LINEAR APPLICATIONS HANDBOOK, 1993, PP. AN46-7, 8.

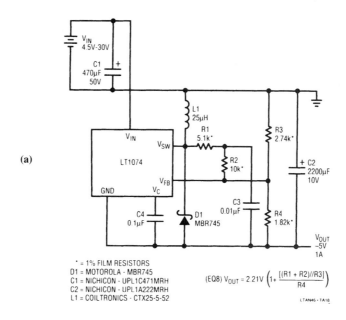

(a)

* = 1% FILM RESISTORS
D1 = MOTOROLA - MBR745
C1 = NICHICON - UPL1C471MRH
C2 = NICHICON - UPL1A222MRH
L1 = COILTRONICS - CTX25-5-52

$$(EQ8) \ V_{OUT} = 2.21V \left(1 + \frac{[(R1 + R2)//R3]}{R4}\right)$$

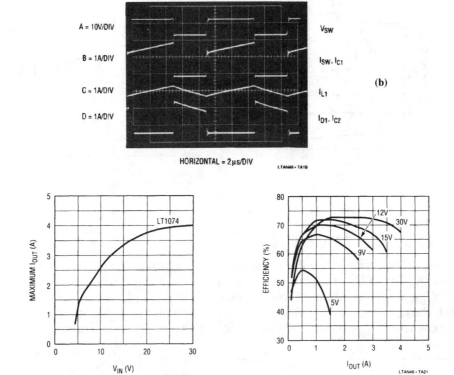

(b)

(c)

(d)

−5 V at 1 A from 4.5 V to 30 V

Fig. 2-58 The circuit in the illustration uses the LT1074 as a positive to negative switch buck-boost. The waveforms, V_{IN}/V_{OUT}, and efficiency are shown in Figs. 2-58B, C and D, respectively. Figure 2-53D shows capacitor manufacturers. See Fig. 2-53 for Coiltronics. LINEAR TECHNOLOGY, LINEAR APPLICATIONS HANDBOOK, 1993, P. AN46-9.

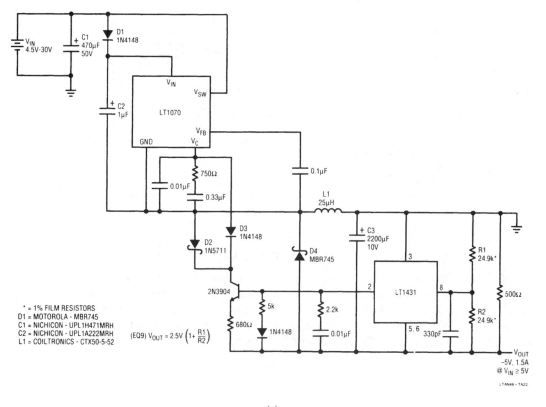

* = 1% FILM RESISTORS
D1 = MOTOROLA - MBR745
C1 = NICHICON - UPL1H471MRH
C2 = NICHICON - UPL1A222MRH
L1 = COILTRONICS - CTX50-5-52

$$(EQ9) \ V_{OUT} = 2.5V \left(1 + \frac{R1}{R2}\right)$$

(a)

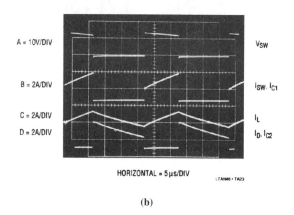

HORIZONTAL = 5μs/DIV

(b)

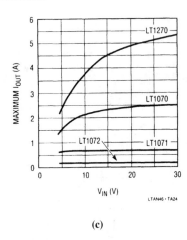

(c)

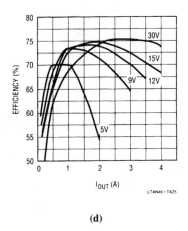

(d)

–5 V at 1.5 A from 4.5 V to 30 V (high efficiency)

Fig. 2-59 The circuit in the illustration uses the LT1070 as a high-efficiency positive to negative regulator. The LT1431 is a shunt linear regulator. (See Chapter 3.) The waveforms, V_{IN}/V_{OUT}, and efficiency are shown in Figs. 2-59B, C, and D, respectively. Figure 2-53E shows capacitor manufacturers. See Fig. 2-53 for Coiltronics. LINEAR TECHNOLOGY, LINEAR APPLICATIONS HANDBOOK, 1993, PP. AN46-10, 11.

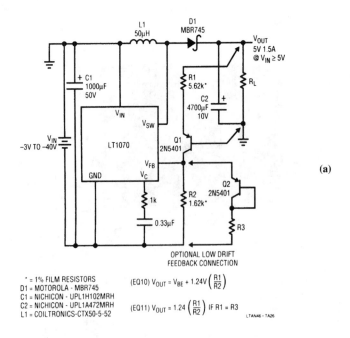

(a)

* = 1% FILM RESISTORS
D1 = MOTOROLA - MBR745
C1 = NICHICON - UPL1H102MRH
C2 = NICHICON - UPL1A472MRH
L1 = COILTRONICS-CTX50-5-52

(EQ10) $V_{OUT} = V_{BE} + 1.24V \left(\frac{R1}{R2}\right)$

(EQ11) $V_{OUT} = 1.24 \left(\frac{R1}{R2}\right)$ IF R1 = R3

Continued

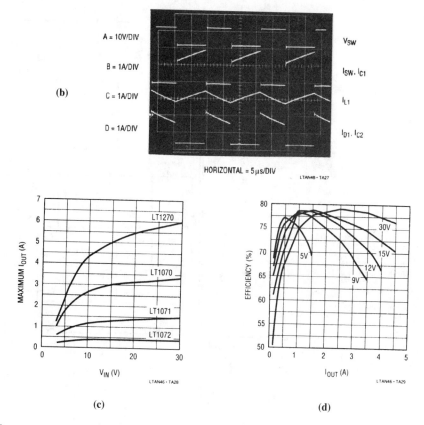

(b)

A = 10V/DIV V_{SW}

B = 1A/DIV I_{SW}, I_{C1}

C = 1A/DIV I_{L1}

D = 1A/DIV I_{D1}, I_{C2}

HORIZONTAL = 5 µs/DIV   LTAN46 · TA27

(c)   **(d)**

5 V at 1.5 A from –3 to –40 V

Fig. 2-60 The circuit in the illustration uses the LT1070 as a negative to positive regulator. The waveforms, V_{IN}/V_{OUT}, and efficiency are shown in Figs. 2-60B, C, and D, respectively. Figure 2-53E shows capacitor manufacturers. See Fig. 2-53 for Coiltronics. LINEAR TECHNOLOGY, LINEAR APPLICATIONS HANDBOOK, 1993, P. AN46-12.

Switching power-supply circuit titles and descriptions

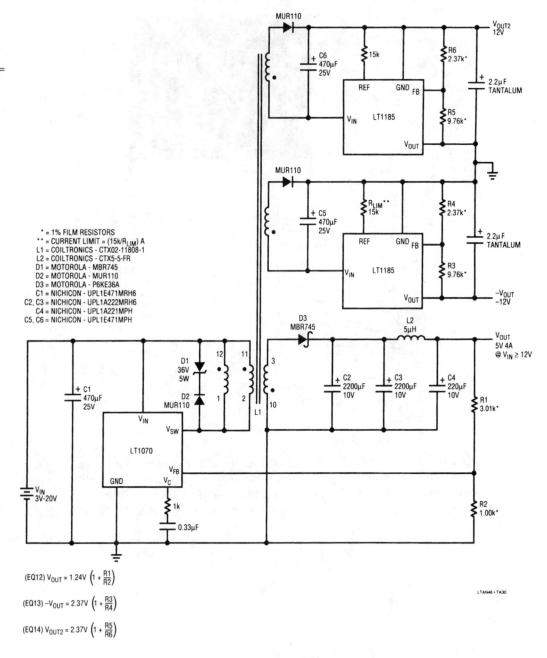

* = 1% FILM RESISTORS
** = CURRENT LIMIT = $(15k/R_{LIM})$ A
L1 = COILTRONICS - CTX02-11808-1
L2 = COILTRONICS - CTX5-5-FR
D1 = MOTOROLA - MBR745
D2 = MOTOROLA - MUR110
D3 = MOTOROLA - P6KE36A
C1 = NICHICON - UPL1E471MRH6
C2, C3 = NICHICON - UPL1A222MRH6
C4 = NICHICON - UPL1A221MPH
C5, C6 = NICHICON - UPL1E471MPH

(EQ12) $V_{OUT} = 1.24V \left(1 + \dfrac{R1}{R2}\right)$

(EQ13) $-V_{OUT} = 2.37V \left(1 + \dfrac{R3}{R4}\right)$

(EQ14) $V_{OUT2} = 2.37V \left(1 + \dfrac{R5}{R6}\right)$

LTAN46 • TA30

(a)

Switching power-supply circuits

Continued

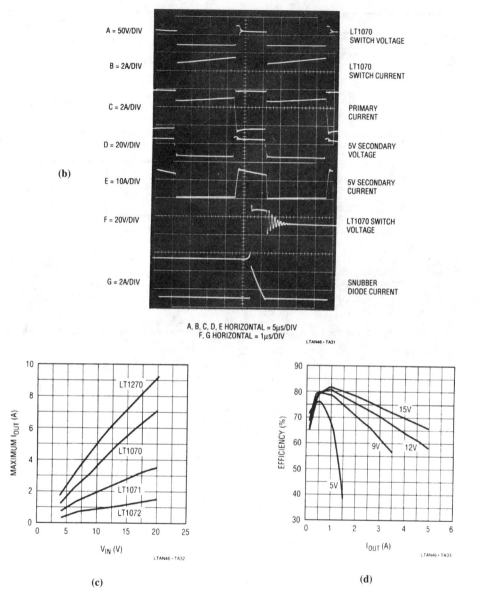

(b)

A = 50V/DIV — LT1070 SWITCH VOLTAGE

B = 2A/DIV — LT1070 SWITCH CURRENT

C = 2A/DIV — PRIMARY CURRENT

D = 20V/DIV — 5V SECONDARY VOLTAGE

E = 10A/DIV — 5V SECONDARY CURRENT

F = 20V/DIV — LT1070 SWITCH VOLTAGE

G = 2A/DIV — SNUBBER DIODE CURRENT

A, B, C, D, E HORIZONTAL = 5μs/DIV
F, G HORIZONTAL = 1μs/DIV

LTAN46 · TA31

(c)

LTAN46 · TA32

(d)

LTAN46 · TA33

5 V and ±12 V from 3 V to 20 V

Fig. 2-61 The circuit in the illustration uses the LT1070 as a flyback regulator. The LT1185s are linear regulators. (See Chapter 3.) The waveforms, V_{IN}/V_{OUT}, and efficiency are shown in Figs. 2-61B, C and D, respectively. Figure 2-53E shows capacitor manufacturers. See Fig. 2-53 for Coiltronics. LINEAR TECHNOLOGY, LINEAR APPLICATIONS HANDBOOK, 1993, PP. AN46-13, 14, 15.

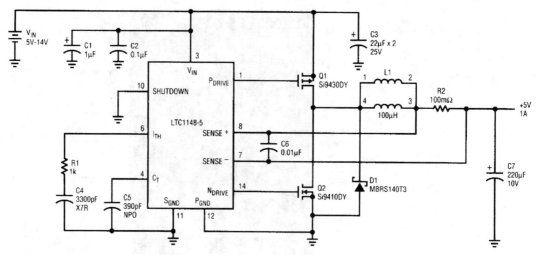

C1 (TA)
C3 AVX (TA) TPSD226K025R0200 ESR = 0.200Ω I_{RMS} = 0.775A
C7 AVX (TA) TPSE227K010R0080 ESR = 0.080Ω I_{RMS} = 1.285A
Q1 SILICONIX PMOS BV_{DSS} = 20V RDS_{ON} = 0.100Ω C_{RSS} = 400pF Q_g = 50nC
Q2 SILICONIX NMOS BV_{DSS} = 30V RDS_{ON} = 0.050Ω C_{RSS} = 160pF Q_g = 30nC
D1 MOTOROLA SCHOTTKY VBR = 40V
R2 KRL SP-1/2-A1-0R100J Pd = 0.75W
L1 COILTRONICS CTX100-4 DCR = 0.175Ω KOOL Mµ CORE

ALL OTHER CAPACITORS ARE CERAMIC

QUIESCENT CURRENT = 180µA
TRANSITION CURRENT (BURST MODE™ OPERATION/CONTINUOUS OPERATION) ≈ 200mA

AN54 • TA01

(a)

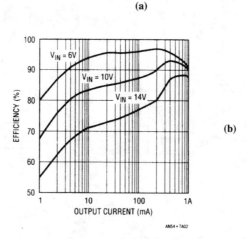

(b)

AN54 • TA02

5 V at 1 A from 5 V to 14 V (surface mount)

Fig. 2-62 The circuit in the illustration uses the LTC1148-5 as a buck converter
with surface-mount components. Figure 2-62B shows the efficiency. LINEAR TECH-
NOLOGY, LINEAR APPLICATIONS HANDBOOK, 1993, P. AN54-3.

Switching power-supply circuits

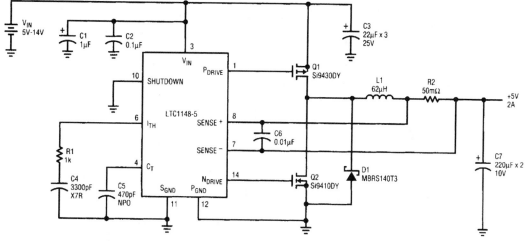

C1 (TA)
C3 AVX (TA) TPSD226K025R0200 ESR = 0.200Ω I_{RMS} = 0.775A
C7 AVX (TA) TPSE227K010R0080 ESR = 0.080Ω I_{RMS} = 1.285A
Q1 SILICONIX PMOS BV_{DSS} = 20V RDS_{ON} = 0.100Ω C_{RSS} = 400pF Q_g = 50nC
Q2 SILICONIX NMOS BV_{DSS} = 30V RDS_{ON} = 0.050Ω C_{RSS} = 160pF Q_g = 30nC
D1 MOTOROLA SCHOTTKY VBR = 40V
R2 KRL SL- 1-C1-0R050J Pd = 1W
L1 COILTRONICS CTX62-2-MP DCR = 0.040Ω MMP CORE (THROUGH HOLE)

ALL OTHER CAPACITORS ARE CERAMIC

QUIESCENT CURRENT = 180μA
TRANSITION CURRENT (BURST MODE™ OPERATION/CONTINUOUS OPERATION) = 400mA

AN54 • TA03

(a)

(b)

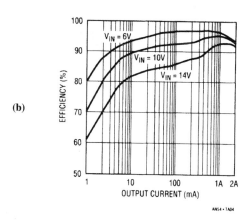

AN54 • TA04

5 V at 2 A from 5 V to 14 V (surface mount)

Fig. 2-63 The circuit in the illustration uses the LTC1148-5 as a buck converter with surface-mount components. Figure 2-63B shows the efficiency. LINEAR TECHNOLOGY, LINEAR APPLICATIONS HANDBOOK, 1993, P. AN54-4.

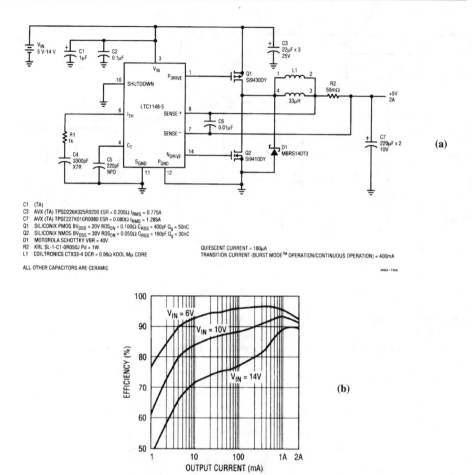

C1 (TA)
C3 AVX (TA) TPSD226K025R0200 ESR = 0.200Ω I$_{RMS}$ = 0.775A
C7 AVX (TA) TPSE227K010R0080 ESR = 0.080Ω I$_{RMS}$ = 1.285A
Q1 SILICONIX PMOS BV$_{DSS}$ = 20V RDS$_{ON}$ = 0.100Ω C$_{RSS}$ = 400pF Q$_g$ = 50nC
Q2 SILICONIX NMOS BV$_{DSS}$ = 30V RDS$_{ON}$ = 0.050Ω C$_{RSS}$ = 160pF Q$_g$ = 30nC
D1 MOTOROLA SCHOTTKY VBR = 40V
R2 KRL SL-1-C1-0R050J Pd = 1W
L1 COILTRONICS CTX33-4 DCR = 0.06Ω KOOL Mμ CORE

ALL OTHER CAPACITORS ARE CERAMIC

QUIESCENT CURRENT = 180μA
TRANSITION CURRENT (BURST MODE™ OPERATION/CONTINUOUS OPERATION) = 400mA

5 V at 2 A from 5 V to 14 V at high frequency (surface mount)

Fig. 2-64 The circuit in the illustration uses the LTC1148-5 as a high-frequency buck converter with surface-mount components. Figure 2-64B shows the efficiency. LINEAR TECHNOLOGY, LINEAR APPLICATIONS HANDBOOK, 1993, P. AN54-5.

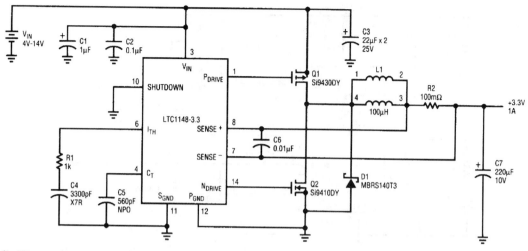

C1 (TA)
C3 AVX (TA) TPSD226K025R0200 ESR = 0.200Ω I_{RMS} = 0.775A
C7 AVX (TA) TPSE227K010R0080 ESR = 0.080Ω I_{RMS} = 1.285A
Q1 SILICONIX PMOS BV_{DSS} = 20V RDS_{ON} = 0.100Ω C_{RSS} = 400pF Q_g = 50nC
Q2 SILICONIX NMOS BV_{DSS} = 30V RDS_{ON} = 0.050Ω C_{RSS} = 160pF Q_g = 30nC
D1 MOTOROLA SCHOTTKY VBR = 40V
R2 KRL SP-1/2-A1-0R100J Pd = 0.75W
L1 COILTRONICS CTX100-4 DCR = 0.175Ω KOOL Mμ CORE

ALL OTHER CAPACITORS ARE CERAMIC

QUIESCENT CURRENT = 180μA
TRANSITION CURRENT (BURST MODE™ OPERATION/CONTINUOUS OPERATION) = 250mA

AN54 • TA07

(a)

(b)

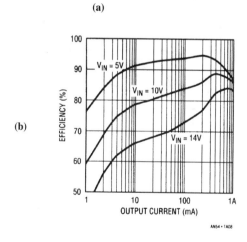

AN54 • TA08

3.3 V at 1 A from 4 V to 14 V (surface mount)

Fig. 2-65 The circuit in the illustration uses the LTC1148-3.3 as a buck converter with surface-mount components. Figure 2-65B shows the efficiency. LINEAR TECHNOLOGY, LINEAR APPLICATIONS HANDBOOK, 1993, P. AN54-6.

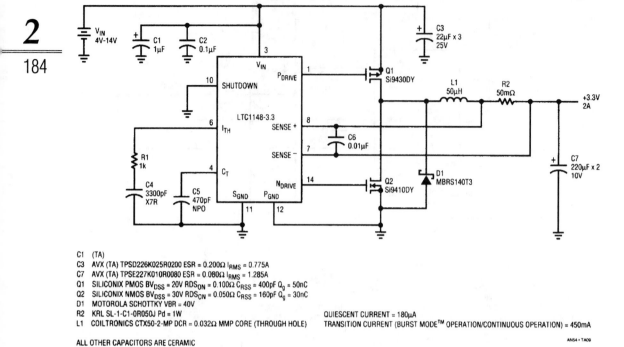

C1 (TA)
C3 AVX (TA) TPSD226K025R0200 ESR = 0.200Ω I$_{RMS}$ = 0.775A
C7 AVX (TA) TPSE227K010R0080 ESR = 0.080Ω I$_{RMS}$ = 1.285A
Q1 SILICONIX PMOS BV$_{DSS}$ = 20V RDS$_{ON}$ = 0.100Ω C$_{RSS}$ = 400pF Q$_g$ = 50nC
Q2 SILICONIX NMOS BV$_{DSS}$ = 30V RDS$_{ON}$ = 0.050Ω C$_{RSS}$ = 160pF Q$_g$ = 30nC
D1 MOTOROLA SCHOTTKY VBR = 40V
R2 KRL SL-1-C1-0R050J Pd = 1W
L1 COILTRONICS CTX50-2-MP DCR = 0.032Ω MMP CORE (THROUGH HOLE)

QUIESCENT CURRENT = 180μA
TRANSITION CURRENT (BURST MODE™ OPERATION/CONTINUOUS OPERATION) = 450mA

ALL OTHER CAPACITORS ARE CERAMIC

AN54 • TA09

(a)

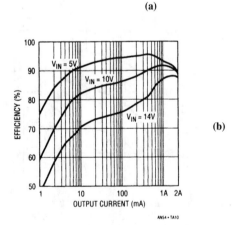

(b)

3.3 V at 2 A from 4 V to 14 V (surface mount)

Fig. 2-66 The circuit in the illustration uses the LTC1148-3.3 as a buck converter with surface-mount components. Figure 2-66B shows the efficiency. LINEAR TECHNOLOGY, LINEAR APPLICATIONS HANDBOOK, 1993, P. AN54-7.

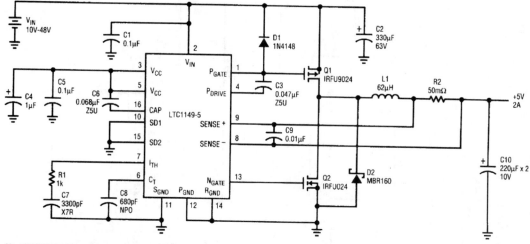

C2 UNITED CHEMI-CON (AL) LXF63VB331M12.5 x 30 ESR = 0.170Ω I_{RMS} = 1.280A
C4 (TA)
C10 SANYO (OS-CON) 10SA220M ESR = 0.035Ω I_{RMS} = 2.360A
Q1 IR PMOS BV_{DSS} = 60V RDS_{ON} = 0.280Ω C_{RSS} = 65pF Q_g = 19nC
Q2 IR NMOS BV_{DSS} = 60V RDS_{ON} = 0.100Ω C_{RSS} = 79pF Q_g = 28nC
D1 SILICON VBR = 75V
D2 MOTOROLA SCHOTTKY VBR = 60V
R2 KRL NP-1A-C1-0R050J Pd = 1W
L1 COILTRONICS CTX62-2-MP DCR = 0.040Ω MMP CORE

QUIESCENT CURRENT = 1.5mA
TRANSITION CURRENT (BURST MODE™ OPERATION/CONTINUOUS OPERATION) = 570mA

ALL OTHER CAPACITORS ARE CERAMIC

AN54 • TA11

(a)

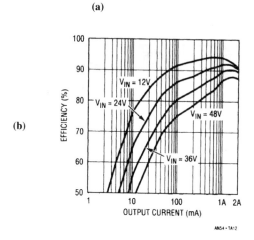

(b)

AN54 • TA12

5 V at 2 A from 10 V to 48 V (high-voltage input)

Fig. 2-67 The circuit in the illustration uses the LTC1149-5 as a buck converter with high-voltage input. Figure 2-67B shows the efficiency. LINEAR TECHNOLOGY, LINEAR APPLICATIONS HANDBOOK, 1993, P. AN54-8.

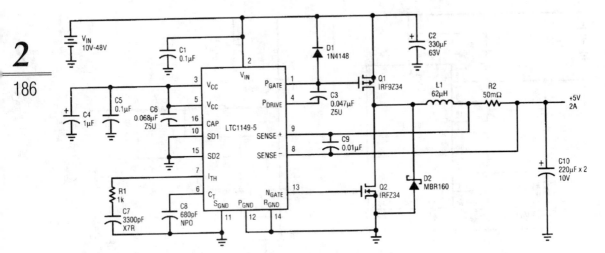

C2   UNITED CHEMI-CON (AL) LXF63VB331M12.5 x 30 ESR = 0.170Ω I_{RMS} = 1.280A
C4   (TA)
C10   SANYO (OS-CON) 10SA220M ESR = 0.035Ω I_{RMS} = 2.360A
Q1   IR PMOS BV_{DSS} = 60V RDS_{ON} = 0.140Ω C_{RSS} = 100pF Q_g = 34nC
Q2   IR NMOS BV_{DSS} = 60V RDS_{ON} = 0.050Ω C_{RSS} = 100pF Q_g = 32nC
D1   SILICON VBR = 75V
D2   MOTOROLA SCHOTTKY VBR = 60V
R2   KRL NP-1A-C1-0R050J Pd = 1W
L1   COILTRONICS CTX62-2-MP DCR = 0.040Ω MMP CORE

QUIESCENT CURRENT = 1.5mA
TRANSITION CURRENT (BURST MODE™ OPERATION/CONTINUOUS OPERATION) = 560mA

AN54 • TA13

ALL OTHER CAPACITORS ARE CERAMIC

(a)

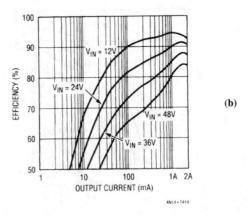

(b)

AN54 • TA14

5 V at 2 A from 10 V to 48 V (with large MOSFETs)

Fig. 2-68 The circuit in the illustration uses the LTC1149-5 as a buck converter
with large MOSFETs. Figure 2-68B shows the efficiency. LINEAR TECHNOLOGY, LIN-
EAR APPLICATIONS HANDBOOK, 1993, P. AN54-9.

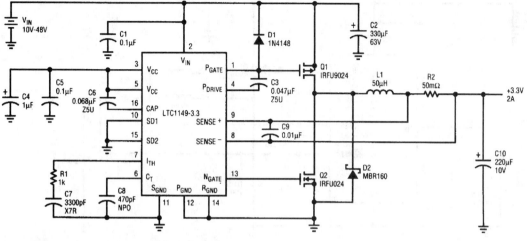

C2 UNITED CHEMI-CON (AL) LXF63VB331M12.5 x 30 ESR = 0.170Ω I_{RMS} = 1.280A
C4 (TA)
C10 SANYO (OS-CON) 10SA220M ESR = 0.035Ω I_{RMS} = 2.360A
Q1 IR PMOS BV_{DSS} = 60V RDS_{ON} = 0.280Ω C_{RSS} = 65pF Q_g = 19nC
Q2 IR NMOS BV_{DSS} = 60V RDS_{ON} = 0.100Ω C_{RSS} = 79pF Q_g = 28nC
D1 SILICON VBR = 75V
D2 MOTOROLA SCHOTTKY VBR = 60V
R2 KRL NP-1A-C1-0R050J Pd = 1W
L1 COILTRONICS CTX50-2-MP DCR = 0.032Ω MMP CORE

ALL OTHER CAPACITORS ARE CERAMIC

QUIESCENT CURRENT = 1.5mA
TRANSITION CURRENT (BURST MODE™ OPERATION/CONTINUOUS OPERATION) = 570mA

AN54 • TA15

(a)

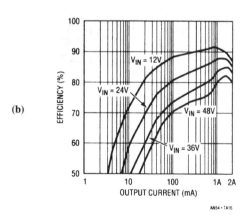

(b)

AN54 • TA16

3.3 V at 2A from 10 V to 48 V (high-voltage input)

Fig. 2-69 The circuit in the illustration uses the LTC1149-3.3 as a buck converter
with high-voltage input. Figure 2-69B shows the efficiency. LINEAR TECHNOLOGY,
LINEAR APPLICATIONS HANDBOOK, 1993, P. AN54-10.

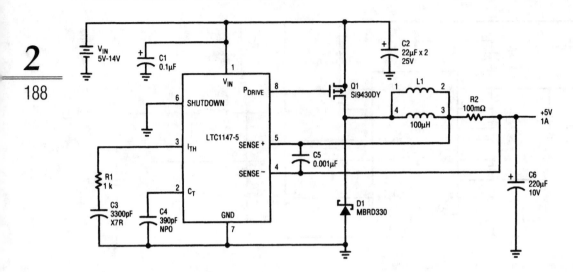

C2 AVX (TA) TPSD226K025R0200 ESR = 0.200Ω I_{RMS} = 0.775A
C5 AVX (TA) TPSE227K010R0080 ESR = 0.080Ω I_{RMS} = 1.285A
Q1 SILICONIX PMOS BV_{DSS} = 20V RDS_{ON} = 0.100Ω C_{RSS} = 400pF Q_g = 50nC
D1 MOTOROLA SCHOTTKY VBR = 30V
R2 KRL SP-1/2-A1-0R100J Pd = 0.75W
L1 COILTRONICS CTX100-4 DCR = 0.175Ω KOOL Mμ CORE

QUIESCENT CURRENT = 190μA
TRANSITION CURRENT (BURST MODE™ OPERATION/CONTINUOUS OPERATION) = 170mA

ALL OTHER CAPACITORS ARE CERAMIC

AN54 • TA17

(a)

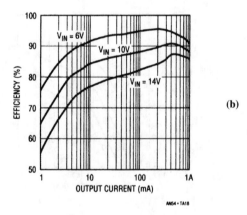

AN54 • TA18

5 V at 1 A from 5 V to 14 V (surface mount)

Fig. 2-70 The circuit in the illustration uses the LTC1147-5 as a buck converter with surface-mount components. Figure 2-70B shows the efficiency. LINEAR TECHNOLOGY, LINEAR APPLICATIONS HANDBOOK, 1993, P. AN54-11.

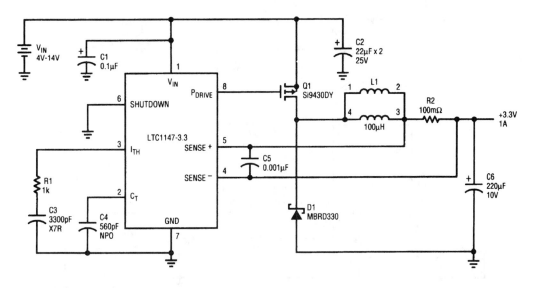

C2 AVX (TA) TPSD226K025R0200 ESR = 0.200Ω I_{RMS} = 0.775A
C6 AVX (TA) TPSE227K010R0080 ESR = 0.080Ω I_{RMS} = 1.285A
Q1 SILICONIX BV_{DSS} = 20V DCR_{ON} = 0.100Ω C_{RSS} = 400pF Q_g = 50nC
D1 MOTOROLA
R2 KRL SP-1/2-A1-0R100 Pd = 0.75W
L1 COILTRONICS CTX100-4 DCR = 0.175Ω KOOL Mµ CORE

QUIESCENT CURRENT = 170µA
TRANSITION CURRENT (BURST MODE™ OPERATION/CONTINUOUS OPERATION) = 170mA

AN54 • TA19

(a)

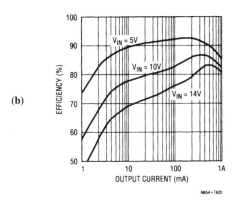

(b)

3.3 V at 1 A from 4 V to 14 V (surface mount)

Fig. 2-71 The circuit in the illustration uses the LTC1147-3.3 as a buck converter with surface-mount components. Figure 2-71B shows the efficiency. LINEAR TECHNOLOGY, LINEAR APPLICATIONS HANDBOOK, 1993, P. AN54-12.

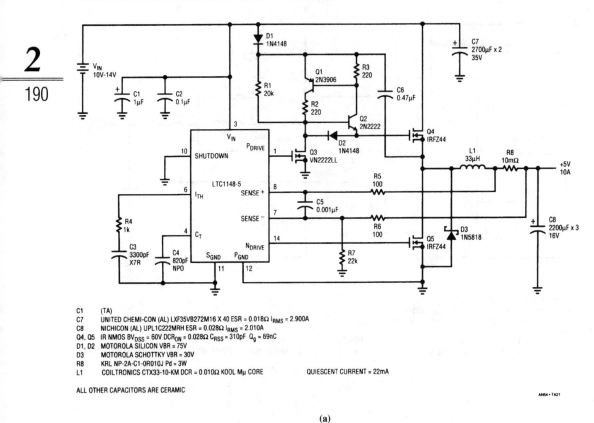

C1 (TA)
C7 UNITED CHEMI-CON (AL) LXF35VB272M16 X 40 ESR = 0.018Ω I_{RMS} = 2.900A
C8 NICHICON (AL) UPL1C222MRH ESR = 0.028Ω I_{RMS} = 2.010A
Q4, Q5 IR NMOS BV_{DSS} = 60V DCR_{ON} = 0.028Ω C_{RSS} = 310pF Q_g = 69nC
D1, D2 MOTOROLA SILICON VBR = 75V
D3 MOTOROLA SCHOTTKY VBR = 30V
R8 KRL NP-2A-C1-0R010J Pd = 3W
L1 COILTRONICS CTX33-10-KM DCR = 0.010Ω KOOL Mμ CORE   QUIESCENT CURRENT = 22mA

ALL OTHER CAPACITORS ARE CERAMIC

AN54 • TA21

(a)

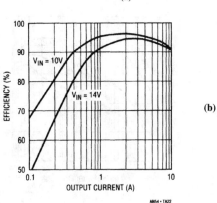

AN54 • TA22

5 V at 10 A from 10 V to 14 V (high current)

Fig. 2-72 The circuit in the illustration uses the LTC1148-5 as a buck converter with high-current output. Figure 2-72B shows the efficiency. LINEAR TECHNOLOGY, LINEAR APPLICATIONS HANDBOOK, 1993, P. AN54-13.

Switching power-supply circuits

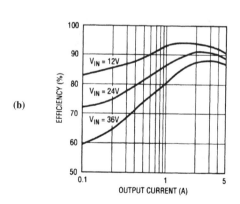

C2   (TA)
C8   NICHICON (AL) UPL1J102MRH ESR = 0.027Ω I_{RMS} = 2.370A
C9   SANYO (OS-CON) 10SA220M ESR = 0.035Ω I_{RMS} = 2.360A
Q1   PNP BV_{CEO} = 30V
Q2   NPN BV_{CEO} = 40V
Q3   SILICONIX NMOS BV_{DSS} = 60V RDS_{ON} = 5.000Ω
Q4   MOTOROLA NMOS BV_{DSS} = 60V RDS_{ON} = 0.050Ω C_{RSS} = 100pF Q_g = 40nC
Q5   IR NMOS BV_{DSS} = 60V RDS_{ON} = 0.050Ω C_{RSS} = 100pF Q_g = 32nC
D1, D2 SILICON VBR = 75V
D3   MOTOROLA SCHOTTKY VBR = 60V
R7   KRL NP-2A-C1-0R020J Pd = 3W
L1   COILTRONICS CTX50-5-52 DCR = 0.021Ω #52 IRON POWDER CORE

ALL OTHER CAPACITORS ARE CERAMIC

(a)

(b)

5 V at 5 A from 12 V to 36 V (high current, high voltage)

Fig. 2-73 The circuit in the illustration uses the LTC1149-5 as a buck converter with high-voltage input and high-current output. Figure 2-73B shows the efficiency. LINEAR TECHNOLOGY, LINEAR APPLICATIONS HANDBOOK, 1993, P. AN54-14.

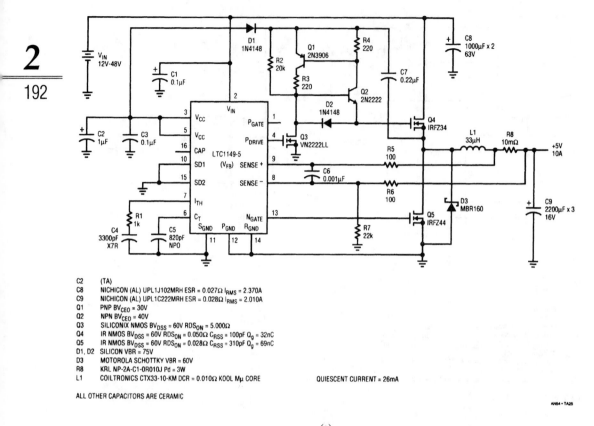

C2	(TA)
C8	NICHICON (AL) UPL1J102MRH ESR = 0.027Ω I_{RMS} = 2.370A
C9	NICHICON (AL) UPL1C222MRH ESR = 0.028Ω I_{RMS} = 2.010A
Q1	PNP BV_{CEO} = 30V
Q2	NPN BV_{CEO} = 40V
Q3	SILICONIX NMOS BV_{DSS} = 60V RDS_{ON} = 5.000Ω
Q4	IR NMOS BV_{DSS} = 60V RDS_{ON} = 0.050Ω C_{RSS} = 100pF Q_g = 32nC
Q5	IR NMOS BV_{DSS} = 60V RDS_{ON} = 0.028Ω C_{RSS} = 310pF Q_g = 69nC
D1, D2	SILICON VBR = 75V
D3	MOTOROLA SCHOTTKY VBR = 60V
R8	KRL NP-2A-C1-0R010J Pd = 3W
L1	COILTRONICS CTX33-10-KM DCR = 0.010Ω KOOL Mμ CORE

QUIESCENT CURRENT = 26mA

ALL OTHER CAPACITORS ARE CERAMIC

AN54 · TA25

(a)

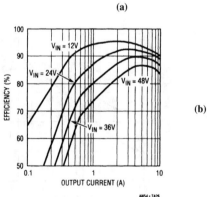

(b)

AN54 · TA26

5 V at 10 A from 12 V to 48 V (high current, high voltage)

Fig. 2-74 The circuit in the illustration uses the LTC1149-5 as a buck converter with high-voltage input and high-current output. Figure 2-74B shows the efficiency. LINEAR TECHNOLOGY, LINEAR APPLICATIONS HANDBOOK, 1993, P. AN54-15.

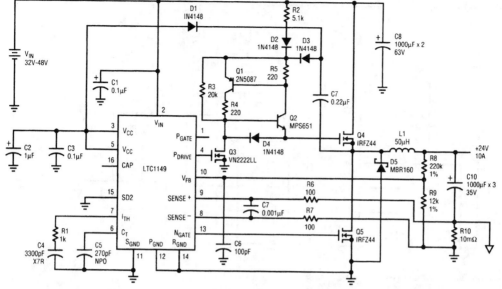

C2	(TA)
C9	NICHICON (AL) UPL1J102MRH ESR = 0.027Ω I$_{RMS}$ = 2.370A
C10	NICHICON (AL) UPL1V102MRH ESR = 0.029Ω I$_{RMS}$ = 1.980A
Q4, Q5	IR NMOS BV$_{DSS}$ = 60V RDS$_{ON}$ = 0.028Ω C$_{RSS}$ = 310pF Q$_g$ = 69nC
Q1	PNP BV$_{CEO}$ = 50V
Q2	NPN BV$_{CEO}$ =
D1, D2, D3, D4	SILICON VBR = 75V
D5	MOTOROLA SCHOTTKY VBR = 60V
R10	KRL NP-2A-C1-0R010J Pd = 3W
L1	COILTRONICS CTX50-10-KM DCR = 0.010Ω KOOL Mµ CORE

ALL OTHER CAPACITORS ARE CERAMIC

V_{OUT} = 1.25V (1 + R8/R9)
QUIESCENT CURRENT = 2mA
TRANSITION CURRENT (BURST MODE™ OPERATION/CONTINUOUS OPERATION) = 1.5A

AN54 • TA27

(a)

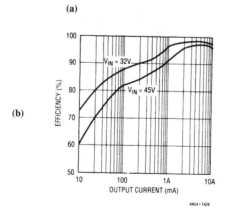

(b)

AN54 • TA28

25 V at 10 A from 32 V to 48 V (high current, high voltage)

Fig. 2-75 The circuit in the illustration uses the LTC1149 as a buck converter with high-voltage input and high-current output. Figure 2-75B shows the efficiency. LINEAR TECHNOLOGY, LINEAR APPLICATIONS HANDBOOK, 1993, P. AN54-16.

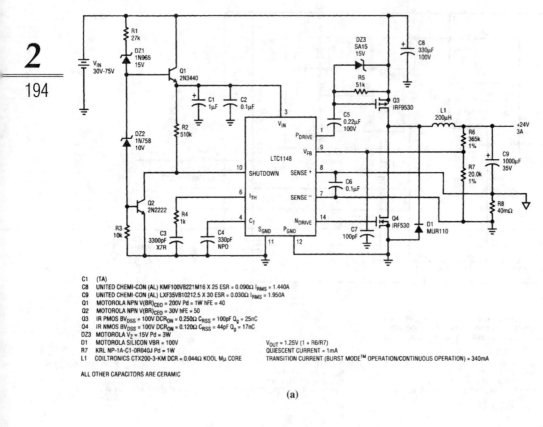

(a)

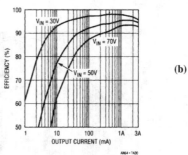

(b)

24 V at 3 A from 30 V to 75 V (high voltage)

Fig. 2-76 The circuit in the illustration uses the LTC1148 as a buck converter with high-voltage input. Figure 2-76B shows the efficiency. LINEAR TECHNOLOGY, LINEAR APPLICATIONS HANDBOOK, 1993, P. AN54-17.

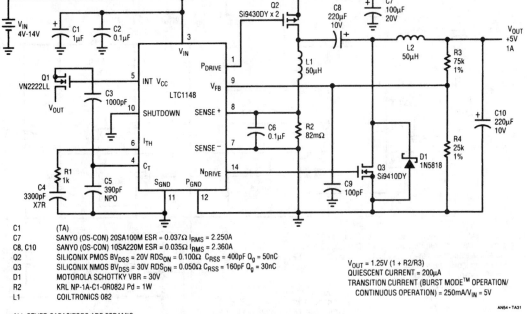

C1   (TA)
C7   SANYO (OS-CON) 20SA100M ESR = 0.037Ω I_{RMS} = 2.250A
C8, C10 SANYO (OS-CON) 10SA220M ESR = 0.035Ω I_{RMS} = 2.360A
Q2   SILICONIX PMOS BV_{DSS} = 20V RDS_{ON} = 0.100Ω C_{RSS} = 400pF Q_g = 50nC
Q3   SILICONIX NMOS BV_{DSS} = 30V RDS_{ON} = 0.050Ω C_{RSS} = 160pF Q_g = 30nC
D1   MOTOROLA SCHOTTKY VBR = 30V
R2   KRL NP-1A-C1-0R082J Pd = 1W
L1   COILTRONICS 082

V_{OUT} = 1.25V (1 + R2/R3)
QUIESCENT CURRENT = 200μA
TRANSITION CURRENT (BURST MODE™ OPERATION/
CONTINUOUS OPERATION) = 250mA/V_{IN} = 5V

AN54 • TA31

ALL OTHER CAPACITORS ARE CERAMIC

(a)

(b)

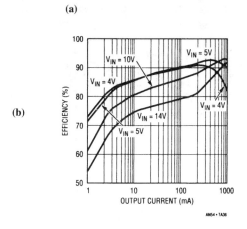

AN54 • TA36

5 V at 1 A from 4 V to 14 V (buck-boost)

Fig. 2-77 The circuit in the illustration uses the LTC1148 as a buck-boost converter. Figure 2-77B shows the efficiency. LINEAR TECHNOLOGY, LINEAR APPLICATIONS HANDBOOK, 1993, P. AN54-18.

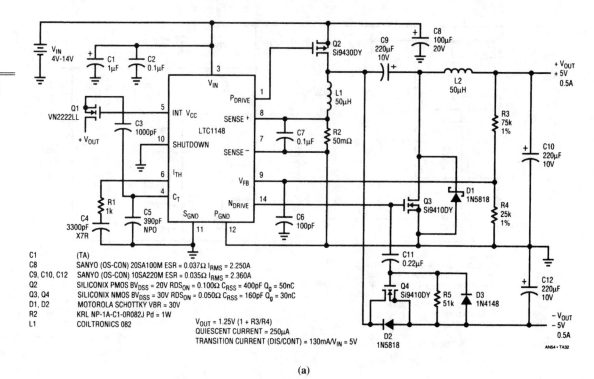

(a)

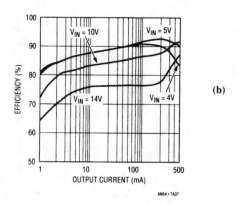

(b)

+5 V at −5 V at 0.5 A from 4 V to 14 V (split supply)

Fig. 2-78 The circuit in the illustration uses the LTC1148 as a split-supply converter. Figure 2-78B shows the efficiency. LINEAR TECHNOLOGY, LINEAR APPLICATIONS HANDBOOK, 1993, P. AN54-19.

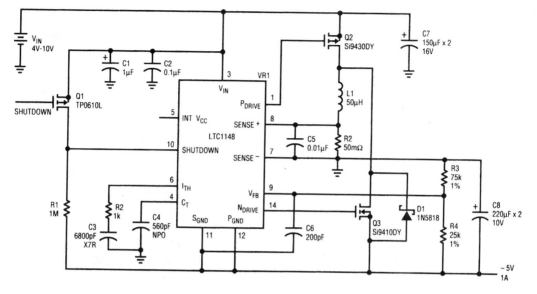

C1   (TA)
C7   SANYO (OS-CON) 16SA150M ESR = 0.035Ω I_{RMS} = 2.280A
C8   SANYO (OS-CON) 10SA220M ESR = 0.035Ω I_{RMS} = 2.360A
Q2   SILICONIX PMOS BV_{DSS} = 20V RDS_{ON} = 0.100Ω C_{RSS} = 400pF Q_g = 50nC
Q3   SILICONIX NMOS BV_{DSS} = 30V RDS_{ON} = 0.050Ω C_{RSS} = 160pF Q_g = 30nC
D1   MOTOROLA SCHOTTKY VBR = 30V
R2   KRL NP-1A-C1-0R050J
L1   COILTRONICS CTX50-2-MP DCR = 0.032Ω MMP CORE

ALL OTHER CAPACITORS ARE CERAMIC

V_{OUT} = 1.25V (1 + R3/R4)

AN54 • TA33

–5 V at 1 A from 4 V to 10 V (inverter)

Fig. 2-79   The circuit in the illustration uses the LTC1148 as a positive-to-negative converter (inverter). LINEAR TECHNOLOGY, LINEAR APPLICATIONS HANDBOOK, 1993, P. AN54-20.

=3=

Linear-supply and voltage-reference circuits

All the testing and troubleshooting procedures described for switching power-supply circuits (in Chapter 2) apply to the circuits in this chapter and are not repeated here. In general, troubleshooting for linear power supplies is simpler than for switching supplies (because you do not have the waveform/inductor problems). However, before you get to the actual circuits, begin with troubleshooting for some typical linear supplies (linear voltage regulators, voltage references, etc.). Although the following examples apply to troubleshooting for an adjustable off-line supply and a dual nonadjustable, pre-regulated off-line supply, the techniques and sequence can generally be applied to any linear supply.

Linear supply troubleshooting

Adjustable off-line supply troubleshooting

The first step in troubleshooting the supply of Fig. 2-B (Chapter 2) is to test the circuit as described in Chapter 2, using both the basic and advanced tests. With a 90-mA load, you should be able to set the output voltage between 3.5 and 20 V, using the 10-kΩ pot (potentiometer) at pin 6 of the CA3085 positive voltage regulator.

If you find no voltage, check the neon lamp when the power switch is closed (right after you make sure the power cord is plugged in!). If the lamp is off, suspect the fuse.

If the lamp is on, check for ac at the transformer secondary (about 24 V across the diodes). If ac is absent, suspect the transformer. If it is present, check for dc between pin 3 of the CA3085 and ground. If it is absent, suspect problems with the diodes. It also is possible that the 500-μF capacitor is shorted or leaking badly.

If there is dc at pins 2/3 (about 25 V), but there is no output (pin 8) or you cannot adjust the output across the range, suspect the 5-μF output capacitor, the 100-pF compensation capacitor, and the CA3085 (in that order).

If you get the correct voltage output (across the range) but there is excessive ripple, or if line/load regulation is out of tolerance (alleged to be 0.2%), suspect the CA3085 (although leaking filter capacitors are always a possibility).

Dual pre-regulated supply troubleshooting

Again, the first step in troubleshooting the supply of Fig. 3-A is to test the circuit as described in Chapter 2. Note that you should get both a +12-V and a –12-V output, neither of which is adjustable. (The adjustment pin on the three-terminal regulators is connected to a fixed voltage divider network in both cases.) If you get an output voltage, but not at the correct level, check the values of the 124-Ω and 1.07-kΩ resistors.

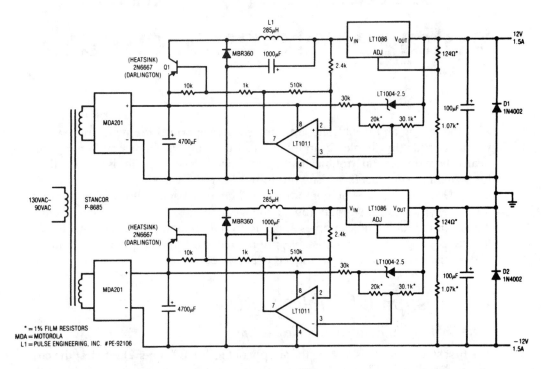

Fig. 3-A Dual pre-regulated off-line linear supply.

If you get +12 V, but not –12 V (or vice versa), you have eliminated half of the circuits as suspects. The same is true if you get excessive ripple or if the two supplies do not show substantially the same line/load regulation. (If both supplies are equally bad, suspect the transformer.)

Assume that the +12-V supply is bad, but the –12-V supply is good (–12-V output with a 1.5-A load, when the line is varied between 90 and 130 Vac). If there is no +12 V, check for ac at the input of the MDA201 and dc at the output. If the ac is absent or is abnormal, suspect the transformer wiring. If there is ac, but not dc, suspect the MDA201 (or the 4700-μF capacitor).

Linear-supply and voltage-reference circuits

If there is dc from the MDA201, compare this to the dc from the MDA201 in the –12-V supply. While you are at it, compare the voltage at V_{IN} of both LT1086 regulators. If V_{IN} for the bad +12-V LT1086 is not substantially the same as for the –12-V regulator, suspect Q1 and the associated parts (such as L1, the LT1011, the MRB360, and the 1000-μF capacitor).

If the dc is the same for both V_{IN} terminals, but the +12-V output is absent or abnormal, suspect the 100-μF output capacitor, diode D1, the LT1086, and the LT1004 zener (in that order). If the zener is leaking badly (or is completely dead), you will get a different voltage at the V_{IN} terminals of the LT1086.

Linear-supply and voltage-reference circuit titles and descriptions

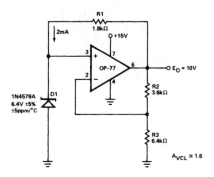

High-stability voltage reference

Fig. 3-1 The simple bootstrapped voltage reference provides a precise 10 V, virtually independent of changes in power-supply voltage, ambient temperature, and output loading. Correct zener operating current of exactly 2 mA is maintained by R1, a selected 5-ppm/°C resistor, connected to the regulated output. Accuracy is primarily determined by three factors: the 5-ppm/°C TC (temperature coefficient) of D1, the 1-ppm/°C ratio-tracking of R2 and R3, and the op-amp V_{OS} errors. The OP-77, with a V_{OS} TC of 0.3 μV/°C contributes only 0.05 ppm/°C of output error, thus effectively eliminating the V_{OS} TC as an error consideration. ANALOG DEVICES, APPLICATIONS REFERENCE MANUAL, 1993, P. 13-56.

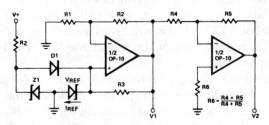

Precision dual-tracking voltage references

Fig. 3-2 In this circuit, R3 is selected to set I_{REF} to operate V_{REF} at its minimum temperature-coefficient current. Proper circuit start up is assured by RZ, Z1, and D1.

$$V_{Z1} \le V_{REF}$$

$$V_1 = V_{REF}\left(1 + \frac{R_2}{R_1}\right)$$

$$R_3 = \frac{(V_1 - V_{REF})}{I_{REF}}$$

$$V_2 = V_1\left(\frac{-R_5}{R_4}\right)$$

ANALOG DEVICES, APPLICATIONS REFERENCE MANUAL, 1993, P. 13-56.

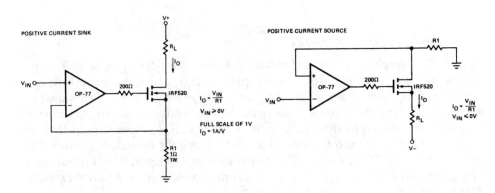

Precision current sinks

Fig. 3-3 These simple, high-current sinks require that the load float between the power supply and the sink. In these circuits, the OP-77's high gain, high CMRR, and low TCV$_{OS}$ assure high accuracy. ANALOG DEVICES, APPLICATIONS REFERENCE MANUAL, 1993, P. 13-64.

Linear-supply and voltage-reference circuits

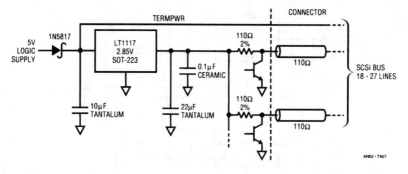

SCSI active termination using a linear regulator

Fig. 3-4 The circuit in the illustration uses an LT1117 low-dropout three-terminal regulator to control the local supply for the active termination of an *SCSI* (small-computer system interface). The LT1117 line regulation makes the circuit immune to variations in *TERMPWR* (termination power signal). The absolute variation in the 2.85-V output is 4% over temperature. In contrast to a passive terminator, two LT1117s require half as many termination resistors and operate at ⅕ the quiescent current, or 20 mA. At these power levels, PC traces provide adequate heatsinking for the LT1117 SOT-223 package. The LT1117 circuit handles fault conditions with short-current limiting, thermal shutdown, and on-chip ESD protection. LINEAR TECHNOLOGY, LINEAR APPLICATIONS HANDBOOK, 1993, P. AN52-6.

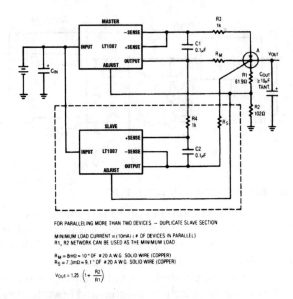

FOR PARALLELING MORE THAN TWO DEVICES — DUPLICATE SLAVE SECTION

MINIMUM LOAD CURRENT = (10mA) (# OF DEVICES IN PARALLEL)
R1, R2 NETWORK CAN BE USED AS THE MINIMUM LOAD

R_M = 8mΩ ~ 10" OF #20 A.W.G. SOLID WIRE (COPPER)
R_S = 7.3mΩ ~ 9.1" OF #20 A.W.G. SOLID WIRE (COPPER)

$V_{OUT} = 1.25 \left(1 + \frac{R2}{R1}\right)$

Linear supply for 3.3-V digital system

Fig. 3-5 This circuit converts 5 V to 3.3 V for use in a digital system, without the use of a switching regulator. Nominal tolerance on the 5-V rail in most systems is ±5% (4.75 to 5.25 V). If the regulator dropout voltage is at the upper extremes of its specification (1.5 V at maximum current and temperature for the LT1083 family), it is still possible to supply 3.25 V to the memory devices when the 5-V rail is at the low end of the specification (4.75 V). This supply is well within the allowable digital-supply voltage range of 3.3 V ±10% (3.0 V to 3.6 V). LINEAR TECHNOLOGY, LINEAR APPLICATIONS HANDBOOK, 1993, P. DN33-2.

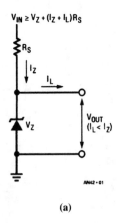

(a)

Continued

CHARACTERISTIC		CARBON COMPOSITION	THIN-FILM CARBON FILM	THICK-FILM CERMET	THIN-FILM NiCr FILM	WIREWOUND	METAL FOIL (MOLDED)	METAL FOIL (HERMETIC)
Ohmic Range		2.7M-100M	1M-4.7M	1M-3M	10M-3M	20k-468k	1k-250k	1k-250k
Absolute Accuracy*	Standard	5%	5.0%	1.0%	0.1%	0.01%	0.01%	0.01%
	Available	20%-5%		5.0%-0.1%	1.0%-0.01%	1.0%-0.05%	1.0%-0.005%	1.0%-0.001%
Temperature Coefficient*	Standard	-5000ppm/°C	-200ppm/°C	100ppm/°C	10ppm/°C	5ppm/°C	2.5ppm/°C-8ppm/°C	0.6ppm/°C
	Available		-100ppm/°C--1500ppm/°C	25ppm/°C-200ppm/°C	5ppm/°C-25ppm/°C	1.0ppm/°C-20ppm/°C		
TCR Tracking*					1: (1-9), 1.0ppm/°C 1: (10-100), 2.0ppm/°C 1: (100-1000), 4.0ppm/°C	1: (1-4), 0.5ppm/°C 1: (5-10), 2.0ppm/°C	1: (1-4), 0.5ppm/°C 1: (5-10), 1.0ppm/°C	1: (1-4), 0.5ppm/°C 1: (5-10), 1.0ppm/°C
Ratio Matching*					1: (1-9), 0.005% 1: (10-100), 0.01% 1: (100-1000), 0.02%	1: (1-4), 0.005% 1: (5-10), 0.1%	1:(1-4), 0.005% 1: (5-10), 0.01%	1: (1-4), 0.005% 1: (5-10), 0.01%
Load-Life Stability*		1KHRS, 6%-4%	3.0%	1.0%	1kHRS, 0.02%	10kHRS, 0.2%	2kHRS, 0.015% 10kHRS, 0.05%	2kHRS, 0.015%
Shelf-Life*		2.0%	0.1%	30ppm/YR	100ppm/YR	25ppm/YR	5ppm/YR	
Voltage Coefficient Of Resistance		-0.02%/V		0.05ppm/V	0.1ppm/V	0.1ppm/V	0.1ppm/V	
Resistor Classification		General Purpose	General Purpose	Semi-Precision	Precision	Precision	Ultra-Precision	Ultra-Precision
Manufacturer's Part Number		Allen-Bradley** CB Series	International** Resistive Co.	International** Resistive Co.	International** Resistive Co. MAR5	Vishay/Ultronix** 105A	Vishay** S102 Series	Vishay** VHP1000

* ± Unless otherwise stated
  % = ppm × 0.0001
  0.0001% = 1ppm
  0.001% = 10ppm
  0.01% = 100ppm
  0.1% = 1000ppm
  1% = 10000ppm
** Parameters may vary between manufacturers

(b)

Linear-supply and voltage-reference circuit titles and descriptions

Continued

Thick-Film

Thick-film resistors are made from a paste mixture of Metal-Oxide (cermet) and binder particles, screen printed onto a ceramic substrate and fired at high temperatures. They are semi-precision components, with standard 1% tolerance and typical TC's of 100ppm/°C to 200ppm/°C.

Carbon Composition/Carbon Film

Carbon composition resistors are made from a large chunk of resistive material. They can handle large overloads for a short period of time. This is their main advantage over the other resistor technologies. They are general purpose components, not precision. Carbon composition resistors do not have constant TC's. TC's can vary anywhere between −2000ppm/°C to −8,000ppm/°C and have shelf-life stabilities of 2% to 5% of resistance value (20,000ppm/Yr to 50,000ppm/Yr).

Carbon film resistors are manufactured using a thin-film process. Initial tolerance and TC are similar to carbon composition. However, they do not have the high overload capability. The sole advantage is their low cost.

Resistor Manufacturers

1. Vishay/Ultronix
   461 North 22nd Street
   P.O. Box 1090
   Grand Junction, CO 81502
   (303) 242-0810

2. Vishay Resistive Systems Group
   63 Lincoln Highway
   Malvern, PA 19355
   (215) 644-1300

3. International Resistive Company
   P.O. Box 1860
   Boone, NC 28607
   (704) 264-8861

4. Julie Research Laboratories
   508 West 26th Street
   New York, NY 10001
   (212) 633-6625

5. Allen-Bradley Company, Inc.
   Electronic Components Division
   1414 Allen-Bradley Drive
   El Paso, TX 79936-4888
   (800) 592-4888

(c)

Basic shunt voltage reference

Fig. 3-6 The circuit in the illustration shows the basic shunt reference using only a zener diode and resistor. The equation shows the relationship of resistance, zener voltage, load current, and zener current. Figures 3-6B and 3-6C show characteristics and manufacturers for typical resistors. LINEAR TECHNOLOGY, LINEAR APPLICATIONS HANDBOOK, 1993, PP. AN42-5, 25, 26.

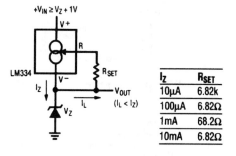

I_Z	R_{SET}
10μA	6.82k
100μA	6.82Ω
1mA	68.2Ω
10mA	6.82Ω

Basic current-stabilized reference

Fig. 3-7 The circuit in the illustration shows the basic current-stabilized reference using an LM334, zener, and resistor. The table shows the relationship of zener current to resistance value. See Fig. 3-6B for resistor characteristics. LINEAR TECHNOLOGY, LINEAR APPLICATIONS HANDBOOK, 1993, P. AN42-5.

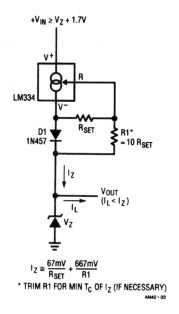

$$I_Z \cong \frac{67mV}{R_{SET}} + \frac{667mV}{R1}$$
* TRIM R1 FOR MIN T_C OF I_Z (IF NECESSARY)
AN42 - 03

Low TC current-stabilized reference

Fig. 3-8 The circuit in the illustration shows the basic current-stabilized reference (with low temperature coefficient, or TC) using an LM334, zener, resistors, and diode. Resistor R1 can be trimmed to provide minimum TC if necessary. See Fig. 3-6B for resistor characteristics. LINEAR TECHNOLOGY, LINEAR APPLICATIONS HANDBOOK, 1993, P. AN42-5.

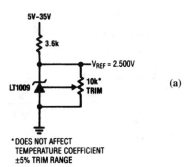

5V–35V

3.6k

V_{REF} = 2.500V

10k*
TRIM

LT1009

(a)

*DOES NOT AFFECT
TEMPERATURE COEFFICIENT
±5% TRIM RANGE

VOLTAGE REFERENCE SELECTION GUIDE*

* COMMERCIAL 0°C to + 70°C
** LTZ1000 requires external control and biasing circuits.

VOLTAGE V_Z (VOLTS)	VOLTAGE TOLERANCE MAXIMUM T_A = 25°C	DEVICE	TEMPERATURE DRIFT, ppm/°C OR mV CHANGE	OPERATING CURRENT RANGE (OR SUPPLY CURRENT)	MAXIMUM DYNAMIC IMPEDANCE (Ω)	MAJOR FEATURE
1.235	± 0.32%	LT1004C-1.2	20ppm (typ)	10μA to 20mA	1.5	Micropower
	± 0.32%	LT1004CS8-1.2	20ppm (typ)	10μA to 20mA	1.5	Micropower
	± 1%	LT1034BC-1.2	20ppm (max)	20μA to 20mA	1.5	Low TC Micropower with 7V Aux. Reference
	± 1%	LT1034C-1.2	40ppm (max)	20μA to 20mA	1.5	Low TC Micropower with 7V Aux. Reference
	± 2%	LM385-1.2	20ppm (typ)	15μA to 20mA	1.5	Micropower
	± 1%	LM385B-1.2	20ppm (typ)	15μA to 20mA	1.5	Micropower
2.5	± 0.5%	LT1004C-2.5	20ppm (typ)	20μA to 20mA	1.5	Micropower
	± 0.8%	LT1004CS8-2.5	20ppm (typ)	20μA to 30mA	1.5	Micropower
	± 0.2%	LT1009C	6mV (max)	400μA to 10mA	1.4	Precision
	± 2.5%	LT1009S8	25ppm (max)	400μA to 20mA	0.6	Precision
	± 0.2%	LT1019C-2.5	20ppm (max)	1.2mA	N/A	Precision Bandgap
	± 4%	LM336-2.5	6mV (max)	400μA to 10mA	1.4	General Purpose
	± 2%	LM336B-2.5	6mV (max)	400μA to 10mA	1.4	General Purpose
	± 3%	LM385-2.5	20ppm (typ)	20μA to 20mA	1.5	Micropower
	± 1.5%	LM385B-2.5	20ppm (typ)	20μA to 20mA	1.5	Micropower
	± 3%	LT580J	85 (max)	1.5mA	N/A	3 Terminal Low Drift
	± 1%	LT580K	40 (max)	1.5mA	N/A	3 Terminal Low Drift
	± 0.4%	LT580L	25 (max)	1.5mA	N/A	3 Terminal Low Drift
	± 0.4%	LT580M	10 (max)	1.5mA	N/A	3 Terminal Low Drift
4.5	± 0.2%	LT1019C-4.5	20ppm (max)	1.2mA	N/A	Precision Bandgap
5.0	± 0.2%	LT1019C-5	20ppm (max)	1.2mA	N/A	Precision Bandgap
	± 1%	LT1021BC-5	5ppm (max)	1.2mA	0.1	Very Low Drift
	± 0.05%	LT1021CC-5	20ppm (max)	1.2mA	0.1	Very Tight Initial Tolerance
	± 1%	LT1021DC-5	20ppm (max)	1.2mA	0.1	Low Cost, High Performance
	± 1%	LT1021CS8	20ppm (max)	1.2mA	0.1	Low Cost, High Performance
	± 0.02%	LT1027A	2ppm (max)	2.0mA	N/A	Low Drift, Tight Tolerance
	± 0.05%	LT1027B	2ppm (max)	2.0mA	N/A	Low Drift, Tight Tolerance
	± 0.05%	LT1027C	3ppm (max)	2.0mA	N/A	Low Drift, Tight Tolerance
	± 0.05%	LT1027D	5ppm (max)	2.0mA	N/A	Low Drift, Tight Tolerance
	± 0.1%	LT1027E	7.5ppm (max)	2.0mA	N/A	Low Drift, Tight Tolerance
	± 0.2%	LT1029AC	20ppm (max)	700μA to 10mA	0.6	Precision Bandgap
	± 1%	LT1029C	34ppm (max)	700μA to 10mA	0.6	Precision Bandgap
	± 0.3%	REF02E	8.5ppm (max)	1.4mA	N/A	Precision Bandgap
	± 0.5%	REF02H	25ppm (max)	1.4mA	N/A	Precision Bandgap
	± 1%	REF02C	6.5ppm (max)	1.6mA	N/A	Precision Bandgap
	± 2%	REF02E	250ppm (max)	2.0mA	N/A	Bandgap
6.9	± 3%	LM329A	10ppm (max)	600μA to 15mA	1.0 (typ)	Low Drift
	± 5%	LM329B	20ppm (max)	600μA to 15mA	1.0 (typ)	Low Drift
	± 5%	LM329C	50ppm (max)	600μA to 15mA	1.0 (typ)	General Purpose
	± 5%	LM329D	100ppm (max)	600μA to 15mA	1.0 (typ)	General Purpose
	± 4%	LTZ1000	0.1ppm/°C	4mA	20.0	Ultra Low Drift, 2ppm Long Term Stability**

(b)

Linear-supply and voltage-reference circuits

Continued

VOLTAGE V_Z (VOLTS)	VOLTAGE TOLERANCE MAXIMUM T_A = 25°C	DEVICE	TEMPERATURE DRIFT, ppm/°C OR mV CHANGE	OPERATING CURRENT RANGE (OR SUPPLY CURRENT)	MAXIMUM DYNAMIC IMPEDANCE (Ω)	MAJOR FEATURE
6.95	±5%	LM399	2ppm (max)	500μA to 10mA	1.5	Ultra Low Drift
	±5%	LM399A	1ppm (max)	500μA to 10mA	1.5	Ultra Low Drift
7.0	±0.7%	LT1021BC-7	5ppm (max)	1.0mA	0.2	Low Drift/Noise, Exc. Stability
	±0.7%	LT1021DC-7	20ppm (max)	1.0mA	0.2	Low Cost, High Performance
10.0	±0.2%	LT1019C-10	20ppm (max)	1.2mA	N/A	Precision Bandgap
	±0.5%	LT1021BC-10	5ppm (max)	1.7mA	0.25	Very Low Drift
	±0.05%	LT1021CC-10	20ppm (max)	1.7mA	0.25	Very Tight Initial Tolerance
	±0.5%	LT1021DC-10	20ppm (max)	1.7mA	0.25	Low Cost, High Performance
	±0.5%	LT1031BC	5ppm (max)	1.7mA	0.25	Very Low Drift
	±0.1%	LT1031CC	15ppm (max)	1.7mA	0.25	Very Tight Initial Tolerance
	±0.2%	LT1031DC	25ppm (max)	1.7mA	0.25	Low Cost, High Performance
	±0.3%	LT581J	30ppm (max)	1.0mA	N/A	3 Terminal Low Drift
	±0.1%	LT581K	15ppm (max)	1.0mA	N/A	3 Terminal Low Drift
	±0.3%	REF01E	8.5ppm (max)	1.4mA	N/A	Precision Bandgap
	±0.5%	REF01H	25ppm (max)	1.4mA	N/A	Precision Bandgap
	±1%	REF01C	65ppm (max)	1.6mA	N/A	Precision Bandgap

(c)

2.5-V reference

Fig. 3-9 The circuit in the illustration shows a 2.5-V reference using the LT1009. See Fig. 3-9B for LT1009 characteristics and Fig. 3-6B for resistor characteristics. The 10-kΩ pot provides for a ±5% trim range. LINEAR TECHNOLOGY, LINEAR APPLICATIONS HANDBOOK, 1993, P. AN42-5.

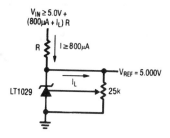

5-V reference

Fig. 3-10 The circuit in the illustration shows a 5-V reference using the LT1009. See Fig. 3-9B for LT1009 characteristics and Fig. 3-6B for resistor characteristics. The 25-kΩ pot provides for a ±5% and –13% trim range. LINEAR TECHNOLOGY, LINEAR APPLICATIONS HANDBOOK, 1993, P. AN42-5.

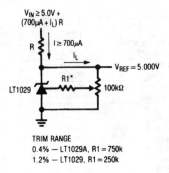

TRIM RANGE
0.4% — LT1029A, R1 = 750k
1.2% — LT1029, R1 = 250k

5-V reference (narrow trim range)

Fig. 3-11 The circuit in the illustration shows a 5-V reference using the LT1029. See Fig. 3-9B for LT1029 characteristics and Fig. 3-6B for resistor characteristics. The 100-kΩ pot provides a narrow trim range (depending on the value of R_1) as shown. LINEAR TECHNOLOGY, LINEAR APPLICATIONS HANDBOOK, 1993, P. AN42-5.

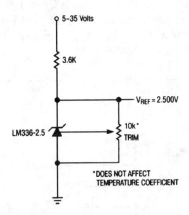

2.5-V reference (temperature-independent trim)

Fig. 3-12 The circuit in the illustration shows a 2.5-V reference using the LM336-2.5. See Fig. 3-9B for LM336 characteristics and Fig. 3-6B for resistor characteristics. The 10-kΩ trimmer potentiometer does not affect the temperature coefficient. LINEAR TECHNOLOGY, LINEAR APPLICATIONS HANDBOOK, 1993, P. AN42-5.

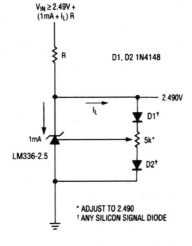

$V_{IN} \geq 2.49V +$
$(1mA + I_L) R$

R   D1, D2 1N4148

2.490V

I_L

D1†

1mA   5k*

LM336-2.5   D2†

* ADJUST TO 2.490
† ANY SILICON SIGNAL DIODE

2.490-V reference (minimum temperature coefficient)

Fig. 3-13 The circuit in the illustration shows a 2.490-V reference using the LM336-2.5. See Fig. 3-9B for LM336 characteristics and Fig. 3-6B for resistor characteristics. LINEAR TECHNOLOGY, LINEAR APPLICATIONS HANDBOOK, 1993, P. AN42-5.

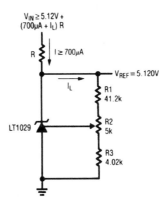

$V_{IN} \geq 5.12V +$
$(700\mu A + I_L) R$

R $I \geq 700\mu A$

$V_{REF} = 5.120V$

I_L

R1
41.2k

LT1029   R2
5k

R3
4.02k

5.120-V reference

Fig. 3-14 The circuit in the illustration shows a 5.120-V reference using the LM1029. See Fig. 3-9B for LM1029 characteristics and Fig. 3-6B for resistor characteristics. Use R2 to trim for desired output. LINEAR TECHNOLOGY, LINEAR APPLICATIONS HAND-BOOK, 1993, P. AN42-6.

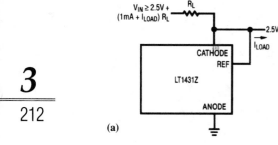
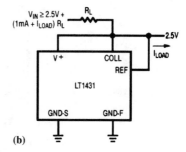

(a)

(b)

2.5-V reference (fixed)

Fig. 3-15 These circuits show alternate connections for a fixed 2.5-V reference using the LT1431 and LT1431Z. See Fig. 3-6B for resistor characteristics. LINEAR TECHNOLOGY, LINEAR APPLICATIONS HANDBOOK, 1993, P. AN42-6.

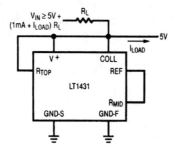

5-V reference (fixed)

Fig. 3-16 The circuit in the illustration shows a 5-V reference using the LT1431. See Fig. 3-6B for resistor characteristics. LINEAR TECHNOLOGY, LINEAR APPLICATIONS HANDBOOK, 1993, P. AN42-6.

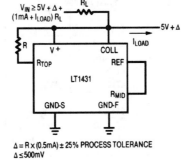

$\Delta = R \times (0.5\text{mA}) \pm 25\%$ PROCESS TOLERANCE
$\Delta \leq 500\text{mV}$

5-V-plus reference

Fig. 3-17 The circuit in the illustration is similar to that of Fig. 3-16 except that the 5-V output can be increased up to 500 mV by connecting resistor R between $V+$ and R_{TOP}. See Fig. 3-6B for resistor characteristics. LINEAR TECHNOLOGY, LINEAR APPLICATIONS HANDBOOK, 1993, P. AN42-6.

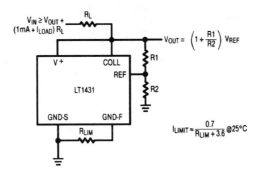

Programmable reference with adjustable current limit

Fig. 3-18 The circuit in the illustration uses the LT1431 to provide a programmable reference output. As shown by the equations, V_{OUT} is set by R_1/R_2 ($V_{\text{REF}} = 5$ V), and the current limit is set by R_{LM}. See Fig. 3-6B for resistor characteristics. LINEAR TECHNOLOGY, LINEAR APPLICATIONS HANDBOOK, 1993, P. AN42-6.

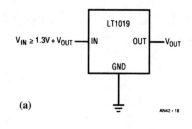

(a)

AN42 · 18

LT1031 PERFORMANCE		
DEVICE	V_{OUT}	TC IN ppm/°C (TYP/MAX)
LT1031B	10V ±5mV	3/5
LT1031C	10V ±10mV	6/15
LT1031D	10V ±20mV	10/25
LT1021 PERFORMANCE		
DEVICE	V_{OUT}	TC IN ppm/°C (TYP/MAX)
LT1021C-5	5V ±2.5mV	3/20
LT1021B-5	5V ±50mV	2/5
LT1021B-7	7V ±50mV	2/5
LT1021D-7	7V ±50mV	3/20
LT1021C-10	10V ±5mV	5/20
LT1021B-10	10V ±50mV	2/5
LT1019 PERFORMANCE		
DEVICE	V_{OUT}	TC IN ppm/°C (TYP/MAX); C = COM, M = MIL
LT1019A-2.5	2.5V ±1.25mV	3/5 (C), 5/10 (M)
LT1019-2.5	2.5V ±5mV	5/20 (C), 8/25 (M)
LT1019A-5	5V ±2.5mV	3/5 (C), 5/10 (M)
LT1019-5	5V ±10mV	5/20 (C), 8/25 (M)
LT1019A-10	10V ±5mV	3/5 (C), 5/10 (M)
LT1019-10	10V ±20mV	5/20 (C), 8/25 (M)
LT1027 PERFORMANCE		
DEVICE	V_{OUT}	TC IN ppm/°C (TYP/MAX)
LT1027A	5V ±1mV	1/2
LT1027B	5V ±2.5mV	1/2
LT1027C	5V ±2.5mV	2/3
LT1027D	5V ±2.5mV	3/5
LT1027E	5V ±5mV	5/7.5

(b)

Basic series reference (LT1019)

Fig. 3-19 The circuit in the illustration shows the basic connections for a series voltage reference. See Fig. 3-19B for LT1019 characteristics. LINEAR TECHNOLOGY, LINEAR APPLICATIONS HANDBOOK, 1993, P. AN42-7.

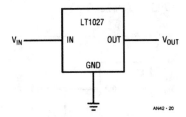

Basic series reference (LT1021)

Fig. 3-20 The circuit in the illustration shows the basic connections for a series voltage reference. See Fig. 3-19B for LT1021 characteristics. LINEAR TECHNOLOGY, LINEAR APPLICATIONS HANDBOOK, 1993, P. AN42-7.

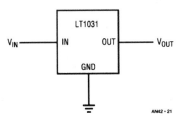

Basic series reference (LT1027)

Fig. 3-21 The circuit in the illustration shows the basic connections for a series voltage reference. See Fig. 3-19B for LT1027 characteristics. LINEAR TECHNOLOGY, LINEAR APPLICATIONS HANDBOOK, 1993, P. AN42-7.

Basic series reference (LT1031)

Fig. 3-22 The circuit in the illustration shows the basic connections for a series voltage reference. See Fig. 3-19B for LT1031 characteristics. LINEAR TECHNOLOGY, LINEAR APPLICATIONS HANDBOOK, 1993, P. AN42-7.

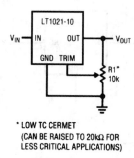

* LOW TC CERMET
(CAN BE RAISED TO 20kΩ FOR
LESS CRITICAL APPLICATIONS)

10-V reference with full trim range

Fig. 3-23 The circuit in the illustration shows the LT1021-10 connected to provide a 10-V output with a ±0.7% trim range. See Figs. 3-9B and 3-19B for LT1021 characteristics and Fig. 3-6B for resistor characteristics. LINEAR TECHNOLOGY, LINEAR APPLICATIONS HANDBOOK, 1993, P. AN42-7.

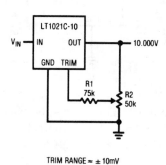

TRIM RANGE ≈ ± 10mV

10-V reference with restricted trim range

Fig. 3-24 The circuit in the illustration shows the LT1021C-10 connected to provide a 10-V output with a ±10-mV trim range. See Figs. 3-9B and 3-19B for LT1021 characteristics and Fig. 3-6B for resistor characteristics. LINEAR TECHNOLOGY, LINEAR APPLICATIONS HANDBOOK, 1993, P. AN42-7.

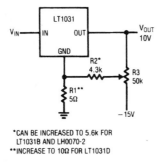

*CAN BE INCREASED TO 5.6k FOR
LT1031B AND LH0070-2
**INCREASE TO 10Ω FOR LT1031D

10-V output with trimmed reference

Fig. 3-25 The circuit in the illustration shows the LT1031 connected to provide a 10-V output with a trimmed reference. See Figs. 3-9B and 3-19B for LT1031 characteristics and Fig. 3-6B for resistor characteristics. LINEAR TECHNOLOGY, LINEAR APPLICATIONS HANDBOOK, 1993, P. AN42-8.

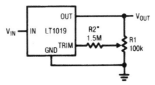

Reference with wide trim range

Fig. 3-26 The circuit in the illustration shows the LT1031 connected to provide a wide trim range (±5%). As shown in Fig. 3-19B, the LT1019 can provide 2.5-V, 5-V, and 10-V reference voltages. See Fig. 3-6B for resistor characteristics. LINEAR TECHNOLOGY, LINEAR APPLICATIONS HANDBOOK, 1993, P. AN42-8.

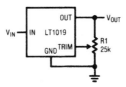

*INCREASE TO 4.7MΩ FOR LT1019A (±0.05%)

Reference with narrow trim range

Fig. 3-27 The circuit in the illustration shows the LT1031 connected to provide a narrow trim range (±0.2%). As shown in Fig. 3-18B, the LT1019 can provide 2.5-V, 5-V, and 10-V reference voltages. See Fig. 3-6B for resistor characteristics. LINEAR TECHNOLOGY, LINEAR APPLICATIONS HANDBOOK, 1993, P. AN42-8.

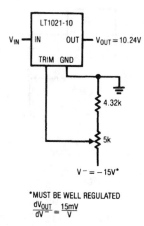

10-V reference trimmed to 10.24 V

Fig. 3-28 The circuit in the illustration shows the LT1021-10 trimmed to provide a 10.24-V output. See Figs. 3-9B and 3-19B for LT1021 characteristics and Fig. 3-6B for resistor characteristics. LINEAR TECHNOLOGY, LINEAR APPLICATIONS HANDBOOK, 1993, P. AN42-8.

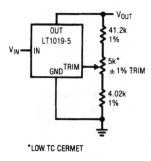

5-V reference trimmed to 5.120 V

Fig. 3-29 The circuit in the illustration shows the LT1019-5 trimmed to provide a 5.120-V output. See Figs. 3-9B and 3-19B for LT1019 characteristics and Fig. 3-6B for resistor characteristics. LINEAR TECHNOLOGY, LINEAR APPLICATIONS HANDBOOK, 1993, P. AN42-8.

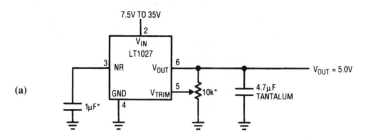

*LOW TC CERMET

10-V reference trimmed to 10.240 V (alternate)

Fig. 3-30 The circuit in the illustration is an alternate to the Fig. 3-28 circuit and is trimmed to provide a 10.240-V output. See Figs. 3-9B and 3-19B for LT1019 characteristics and Fig. 3-6B for resistor characteristics. LINEAR TECHNOLOGY, LINEAR APPLICATIONS HANDBOOK, 1993, P. AN42-8.

(a)

* NOISE REDUCTION CAP AND TRIM POTENTIOMETER OPTIONAL.

AN42 - 30

CHARACTERISTIC	ALUMINUM SOLID ELECTROLYTIC	POLYESTER FILM	SOLID TANTALUM ELECTROLYTIC	MULTILAYER CERAMIC	ALUMINUM ELECTROLYTIC	UNIT
Capacitance	0.47	0.47	0.47	0.47	0.47	µF
ESR* 100kHz	0.198	0.456	4.5	0.062	5.4	Ω
Leakage Current* @ 5V	20	0.03	30	0.16	175	nA
Manufacturer's	SANYO	SANYO	KEMET	KEMET	SANYO	

*Typical

(b)

Linear-supply and voltage-reference circuit titles and descriptions

Capacitor Manufacturers

1. Nichicon (America) Corporation
   927 East State Parkway
   Schaumburg, IL 60195
   (708) 843-7500

2. Sanyo Video Components (USA) Corporation
   1201 Sanyo Avenue
   San Diego, CA 92073
   (619) 661-6322

3. United Chemi-Con, Inc.
   9801 West Higgins Road
   Rosemount, IL 60018
   (312) 696-2000

4. Illinois Capacitor, Inc.
   3757 West Touhy Avenue
   Lincolnwood, IL 60645
   (312) 675-1760

5. Kemet Electronics
   P. O. Box 5928
   Greenville, SC 29606
   (803) 963-6300   (c)

5-V fast-settling trimmed reference

Fig. 3-31 The circuit in the illustration provides a 5-V trimmed output with fast settling time. See Figs. 3-31B and 3-31C for capacitor characteristics and manufacturers, respectively. See Figs. 3-9B and 3-19B for LT1027 characteristics and Fig. 3-6B for resistor characteristics. LINEAR TECHNOLOGY, LINEAR APPLICATIONS HANDBOOK, 1993, P. AN42-8.

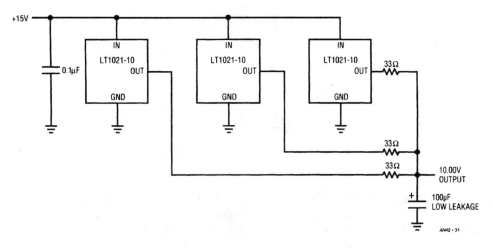

Low-noise statistical-voltage standard

Fig. 3-32   The circuit in the illustration uses three LT1021-10s to provide a 10-V statistical-voltage standard. See Figs. 3-6, 3-9, 3-19, and 3-31 for LT1021, resistor, and capacitor characteristics. LINEAR TECHNOLOGY, LINEAR APPLICATIONS HANDBOOK, 1993, P. AN42-9.

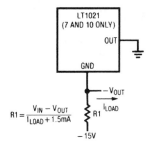

Shunt-mode operation of series reference

Fig. 3-33   The circuit in the illustration shows the LT1021-7 or –10 connected in the shunt mode to provide a –7-V or –10-V reference. See Figs. 3-6 and 3-19 for resistor and LT1021 characteristics. LINEAR TECHNOLOGY, LINEAR APPLICATIONS HANDBOOK, 1993, P. AN42-9.

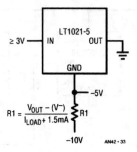

−5-V reference

Fig. 3-34   The circuit in the illustration shows the LT1021-5 connected to provide a −5-V reference. See Figs. 3-6 and 3-19 for resistor and LT1021 characteristics. LINEAR TECHNOLOGY, LINEAR APPLICATIONS HANDBOOK, 1993, P. AN42-9.

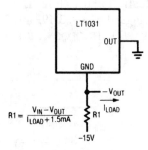

−10-V reference

Fig. 3-35   The circuit in the illustration shows the LT1031 connected to provide a −10-V reference. See Figs. 3-6 and 3-19 for resistor and LT1031 characteristics. LINEAR TECHNOLOGY, LINEAR APPLICATIONS HANDBOOK, 1993, P. AN42-9.

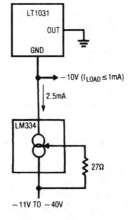

–10-V reference with wide input range

Fig. 3-36 The circuit in the illustration uses an LT1031 and LM334 to provide a –10-V output with an input from –11-V to –40 V. See Figs. 3-6 and 3-19 for resistor and LT1031 characteristics. LINEAR TECHNOLOGY, LINEAR APPLICATIONS HANDBOOK, 1993, P. AN42-9.

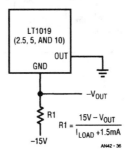

–2.5, –5-V, and –10-V reference

Fig. 3-37 The circuit in the illustration uses an LT1019 series reference connected in the shunt mode to provide negative reference voltages. See Figs. 3-6 and 3-19 for resistor and LT1019 characteristics. LINEAR TECHNOLOGY, LINEAR APPLICATIONS HANDBOOK, 1993, P. AN42-9.

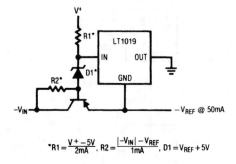

$$^*R1 = \frac{V + -5V}{2mA}, \quad R2 = \frac{|-V_{IN}| - V_{REF}}{1mA}, \quad D1 = V_{REF} + 5V$$

Negative reference with current boost

Fig. 3-38 The circuit in the illustration uses an LT1019 to provide negative reference voltages (–2.5, –5, or –10 V) with a 50-mA current capability. See Figs. 3-6 and 3-19 for resistor and LT1019 characteristics. LINEAR TECHNOLOGY, LINEAR APPLICATIONS HANDBOOK, 1993, P. AN42-9.

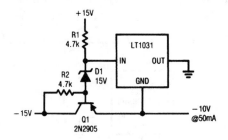

–10-V reference with current boost

Fig. 3-39 The circuit in the illustration uses an LT1031 to provide –10-V reference with a 50-mA current capability. See Figs. 3-6 and 3-19 for resistor and LT1031 characteristics. LINEAR TECHNOLOGY, LINEAR APPLICATIONS HANDBOOK, 1993, P. AN42-10.

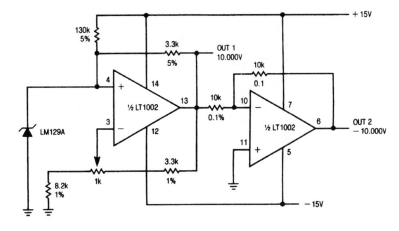

±10-V precision reference

Fig. 3-40 The circuit in the illustration provides both +10 V and –10 V, which can be trimmed to 1 mV. See Fig. 3-6 for resistor characteristics. LINEAR TECHNOLOGY, LINEAR APPLICATIONS HANDBOOK, 1993, P. AN42-10.

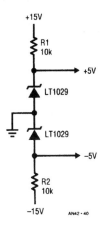

±5-V reference

Fig. 3-41 The circuit in the illustration provides both +5 V and –5 V. See Fig. 3-6 for resistor characteristics. LINEAR TECHNOLOGY, LINEAR APPLICATIONS HANDBOOK, 1993, P. AN42-10.

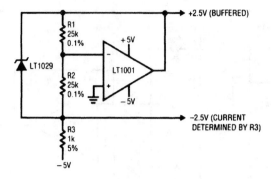

±2.5-V reference

Fig. 3-42 The circuit in the illustration provides both +2.5 V and −2.5 V. See Fig. 3-6 for resistor characteristics. LINEAR TECHNOLOGY, LINEAR APPLICATIONS HANDBOOK, 1993, P. AN42-10.

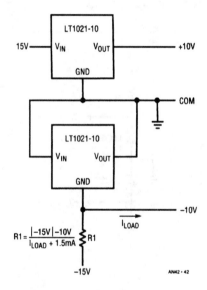

±10-V reference

Fig. 3-43 The circuit in the illustration provides both +10 V and −10 V. See Figs. 3-6 and 3-19 for resistor and LT1021 characteristics. LINEAR TECHNOLOGY, LINEAR APPLICATIONS HANDBOOK, 1993, P. AN42-10.

Linear-supply and voltage-reference circuits

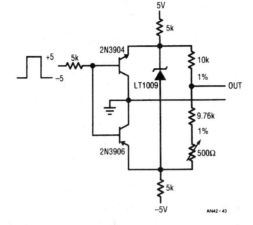

±1.25-V logic-programmable reference

Fig. 3-44 The circuit in the illustration can be adjusted for reference voltages between −1.25 V and +1.25 V and can be programmed on or off by a 5-V logic pulse. See Fig. 3-6 for resistor characteristics. LINEAR TECHNOLOGY, LINEAR APPLICATIONS HANDBOOK, 1993, P. AN42-10.

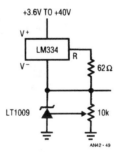

2.5-V reference with wide input range

Fig. 3-45 The circuit in the illustration uses an LT1009 and LM334 to provide a 2.5-V output with an input from 3.6 V to 40 V. See Fig. 3-6 for resistor characteristics. LINEAR TECHNOLOGY, LINEAR APPLICATIONS HANDBOOK, 1993, P. AN42-11.

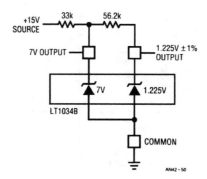

7-V and 1.225-V pre-regulated reference

Fig. 3-46 The circuit in the illustration provides both 7 V and 1.225 V. See Fig. 3-6 for resistor characteristics. LINEAR TECHNOLOGY, LINEAR APPLICATIONS HANDBOOK, 1993, P. AN42-11.

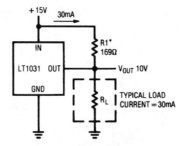

*SELECT R1 TO DELIVER TYPICAL LOAD CURRENT.
LT1031 WILL THEN SOURCE OR SINK AS NECESSARY
TO MAINTAIN PROPER OUTPUT. DO NOT REMOVE LOAD
AS OUTPUT WILL BE DRIVEN UNREGULATED HIGH. LINE
REGULATION IS DEGRADED IN THIS APPLICATION.

10-V reference with shunt resistor for greater current

Fig. 3-47 The circuit in the illustration provides 10-V output with a typical load of 30 mA. See Figs. 3-6 and 3-19 for resistor and LT1031 characteristics. LINEAR TECHNOLOGY, LINEAR APPLICATIONS HANDBOOK, 1993, P. AN42-12.

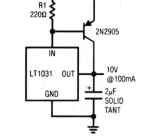

10-V reference with boosted output

Fig. 3-48 The circuit in the illustration provides 10-V output with a typical load of 100 mA. See Figs. 3-6, 3-19, and 3-31 for resistor, capacitor, and LT1031 characteristics. LINEAR TECHNOLOGY, LINEAR APPLICATIONS HANDBOOK, 1993, P. AN42-12.

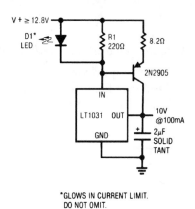

*GLOWS IN CURRENT LIMIT.
DO NOT OMIT.

10-V reference with boosted output and current limit

Fig. 3-49 The circuit in the illustration provides a 10-V output with an LED to limit current at 100 mA. See Figs. 3-6, 3-19, and 3-31 for resistor, capacitor, and LT1031 characteristics. LINEAR TECHNOLOGY, LINEAR APPLICATIONS HANDBOOK, 1993, P. AN42-12.

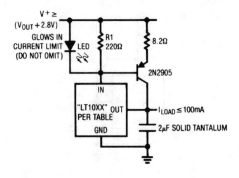

DEVICE	V$_{OUT}$
LT1019-2.5	2.5V
LT1019-5 LT1021-5 LT1027	5V
LT1021-7	7V
LT1019-10 LT1021-10	10V

2.5-V, 5-V, 7-V and 10-V reference with boosted output

Fig. 3-50 The circuit in the illustration provides outputs shown in the table with an LED to limit current at 100 mA. See Figs. 3-6, 3-19, and 3-31 for resistor, capacitor, and IC characteristics. LINEAR TECHNOLOGY, LINEAR APPLICATIONS HANDBOOK, 1993, P. AN42-12.

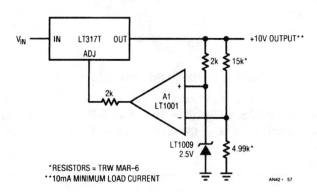

10-V high-current reference

Fig. 3-51 The circuit in the illustration provides a 10-V output with a current of 1.5 A. See Fig. 3-6 for resistor characteristics. LINEAR TECHNOLOGY, LINEAR APPLICATIONS HANDBOOK, 1993, P. AN42-12.

Linear-supply and voltage-reference circuits

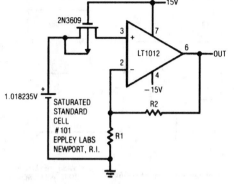

THE TYPICAL 30pA BIAS CURRENT OF THE LT1012 WILL DEGRADE THE
STANDARD CELL BY ONLY 1ppm/YEAR. NOISE IS A FRACTION OF A
ppm. UNPROTECTED GATE MOSFET ISOLATES STANDARD CELL ON
POWER DOWN.

Buffered standard cell

Fig. 3-52 The circuit in the illustration provides a buffer for a standard reference
cell. LINEAR TECHNOLOGY, LINEAR APPLICATIONS HANDBOOK, 1993, P. AN42-13.

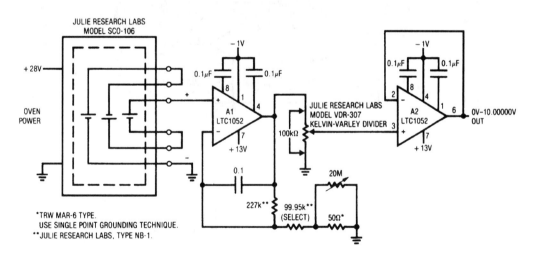

Standard-grade variable voltage reference

Fig. 3-53 The circuit in the illustration can provide an output of 0 V to 10.00000 V
controlled by a Kelvin-Varley divider. The circuit can be trimmed with the 20-MΩ
pot. LINEAR TECHNOLOGY, LINEAR APPLICATIONS HANDBOOK, 1993, P. AN42-13.

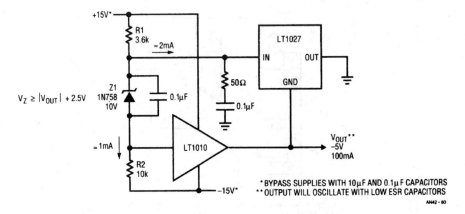

–5-V current-boosted negative reference with overload protection

Fig. 3-54 The circuit in the illustration provides –5 V at 100 mA with overload protection. See Figs. 3-6, 3-19, and 3-31 for resistor, capacitor, and LT1027 characteristics. LINEAR TECHNOLOGY, LINEAR APPLICATIONS HANDBOOK, 1993, P. AN42-13.

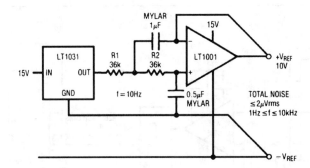

10-V low-noise reference

Fig. 3-55 The circuit in the illustration provides a 10-V output with a minimum of noise. See Figs. 3-6, 3-19, and 3-31 for resistor, capacitor, and LT1031 characteristics. LINEAR TECHNOLOGY, LINEAR APPLICATIONS HANDBOOK, 1993, P. AN42-14.

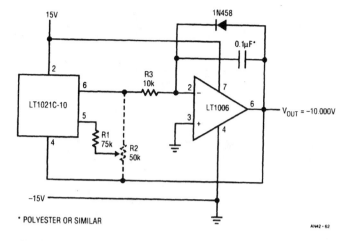

10-V low-noise reference

Fig. 3-56 The circuit in the illustration provides a −10-V output with a minimum of noise. See Figs. 3-6, 3-19, and 3-31 for resistor, capacitor, and LT1031 characteristics. LINEAR TECHNOLOGY, LINEAR APPLICATIONS HANDBOOK, 1993, P. AN42-14.

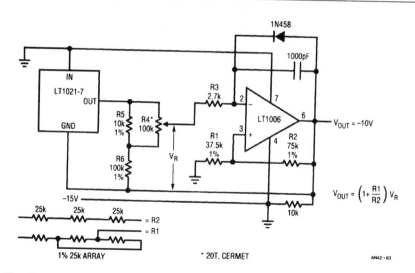

−10-V trimmed, low-noise, low-TC reference

Fig. 3-57 The circuit in the illustration provides a −10-V output with a minimum of noise and low temperature-coefficient, from a single supply. See Figs. 3-6, 3-19, and 3-31 for resistor, capacitor, and LT1021 characteristics. LINEAR TECHNOLOGY, LINEAR APPLICATIONS HANDBOOK, 1993, P. AN42-14.

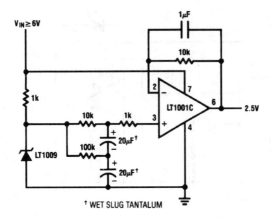

2.5-V low-noise reference

Fig. 3-58 The circuit in the illustration provides a 2.5-V output with a minimum of noise. See Figs. 3-6 and 3-31 for resistor and capacitor characteristics. LINEAR TECHNOLOGY, LINEAR APPLICATIONS HANDBOOK, 1993, P. AN42-15.

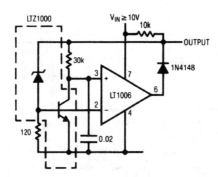

7-V low-noise, low-drift reference

Fig. 3-59 The circuit in the illustration provides a 7-V output with a minimum of noise and very low drift. See Figs. 3-6 and 3-31 for resistor and capacitor characteristics. LINEAR TECHNOLOGY, LINEAR APPLICATIONS HANDBOOK, 1993, P. AN42-15.

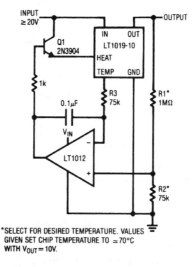

*SELECT FOR DESIRED TEMPERATURE. VALUES
GIVEN SET CHIP TEMPERATURE TO $\approx 70°C$
WITH V_{OUT} = 10V.

10-V temperature-stabilized reference

Fig. 3-60 The circuit in the illustration provides a 10-V output with temperature stabilization. See Figs. 3-6 and 3-31 for resistor and capacitor characteristics. LINEAR TECHNOLOGY, LINEAR APPLICATIONS HANDBOOK, 1993, P. AN42-15.

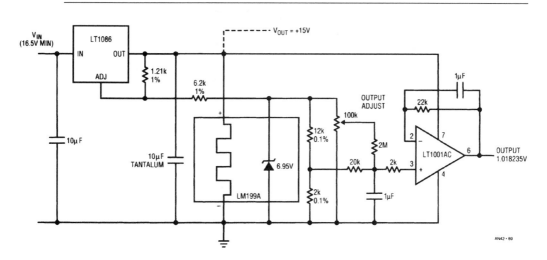

Buffered standard-cell replacement

Fig. 3-61 The circuit in the illustration provides a substitute for a standard cell (see Fig. 3-52) and can be trimmed to 1.018235 V with the 100-kΩ pot. See Figs. 3-6 and 3-31 for resistor and capacitor characteristics. LINEAR TECHNOLOGY, LINEAR APPLICATIONS HANDBOOK, 1993, P. AN42-17.

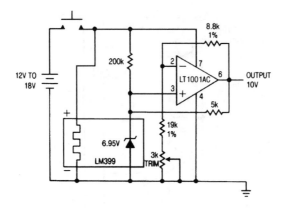

10-V self-biased temperature-stabilized reference

Fig. 3-62 The circuit in the illustration provides a temperature-stabilized reference that can be trimmed to 10-V with the 3-kΩ pot. See Fig. 3-6 for resistor characteristics. LINEAR TECHNOLOGY, LINEAR APPLICATIONS HANDBOOK, 1993, P. AN42-17.

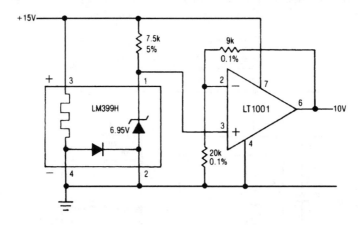

10-V temperature-stabilized reference

Fig. 3-63 The circuit in the illustration provides a fixed 10-V temperature-stabilized reference. See Fig. 3-6 for resistor characteristics. LINEAR TECHNOLOGY, LINEAR APPLICATIONS HANDBOOK, 1993, P. AN42-18.

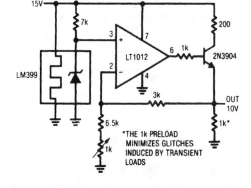

10-V buffered temperature-stabilized reference

Fig. 3-64 The circuit in the illustration provides a buffered, temperature-stabilized reference that can be trimmed to 10-V with the 1-kΩ pot. See Fig. 3-6 for resistor characteristics. LINEAR TECHNOLOGY, LINEAR APPLICATIONS HANDBOOK, 1993, P. AN42-18.

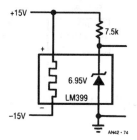

6.95-V temperature-stabilized reference

Fig. 3-65 The simple circuit in the illustration provides a 6.95-V temperature-stabilized output with a minimum of components. See Fig. 3-6 for resistor characteristics. LINEAR TECHNOLOGY, LINEAR APPLICATIONS HANDBOOK, 1993, P. AN42-19.

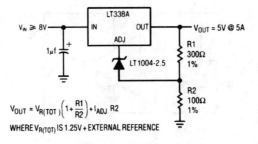

5-V at 5-A reference

Fig. 3-66 The simple circuit in the illustration provides 5 V at 5 A with high stability. See Figs. 3-6 and 3-31 for resistor and capacitor characteristics. LINEAR TECHNOLOGY, LINEAR APPLICATIONS HANDBOOK, 1993, P. AN42-19.

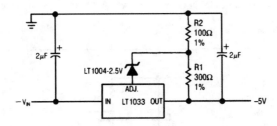

–5-V reference

Fig. 3-67 The simple circuit in the illustration provides –5 V with high stability. See Figs. 3-6 and 3-31 for resistor and capacitor characteristics. LINEAR TECHNOLOGY, LINEAR APPLICATIONS HANDBOOK, 1993, P. AN42-19.

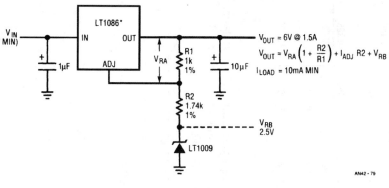

DEVICE	FEATURES
LT1086	1.5A, Low Dropout
LT1085	3A, Low Dropout
LT1084	5A, Low Dropout
LT1083	7.5A, Low Dropout
LT317A	1.5A
LT350	3A
LT338A	5A
LT1038	10A

Regulator with a reference output

Fig. 3-68 The circuit in the illustration provides a regulated output (set by the value of R_1/R_2) in addition to a fixed 2.5-V reference. The current capability is set by device selection (see table). See Figs. 3-6 and 3-31 for resistor and capacitor characteristics. LINEAR TECHNOLOGY, LINEAR APPLICATIONS HANDBOOK, 1993, P. AN42-20.

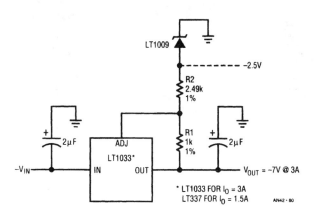

Negative-output regulator with reference

Fig. 3-69 The circuit in the illustration provides a regulated output of –7 V at 3 A (set by the value of R_1/R_2) in addition to a fixed –2.5-V reference. The current capability is set by device selection (see table). See Figs. 3-6 and 3-31 for resistor and capacitor characteristics. LINEAR TECHNOLOGY, LINEAR APPLICATIONS HANDBOOK, 1993, P. AN42-20.

Z1/Z2	Vz	VR	VOUT
LT1034	1.225	1.225	2.475
LT1004	1.235	1.235	2.485
LT1009	2.500	2.500	3.750
LT1034 + LT1009	1.225 + 2.5	3.725	4.975
LT1004 + LT1009	1.235 + 2.5	3.735	4.985
LT1029	5	5	6.250
LT1034 + LT1029	6.225	6.225	7.475
LT1004 + LT1029	6.235	6.235	7.485
LM329	6.9	6.9	8.150
LT1009 + LT1029	2.5 + 5.0	7.500	8.750
LT1034 + LM329	1.225 + 6.9	8.125	9.375
LT1004 + LM329	1.235 + 6.9	8.135	9.385
LT1009 + LM329	2.5 + 6.9	9.400	10.650
2 × LT1029	5 + 5	10.000	11.250
LT1029 + LM329	5 + 6.9	11.900	13.250
2 × LM329	6.9 + 6.9	13.800	15.050

$V_{IN} \geq V_{OUT} + 1.5V$

C_{IN} 10 µF

LT1086 — IN — ADJ — OUT

R1 1.2k

C_{OUT} 150 µF

V_{OUT}

V_R

Z1, Z2 = LM329

$V_R = 13.8V$

$(I_L = 10mA \, MIN)$
$V_{OUT} = 1.25V + V_R$

WHERE V_R = TOTAL EFFECTIVE REFERENCE VOLTAGE
$= V_{Z1} + V_{Z2}$, ETC.

AN42 · 81

Simple stacked reference/regulator

Fig. 3-70 The circuit in the illustration provides both a reference and a regulated output. The voltages are set by the values of Z_1/Z_2 as shown in the table. See Figs. 3-6 and 3-31 for resistor and capacitor characteristics. LINEAR TECHNOLOGY, LINEAR APPLICATIONS HANDBOOK, 1993, P. AN42-21.

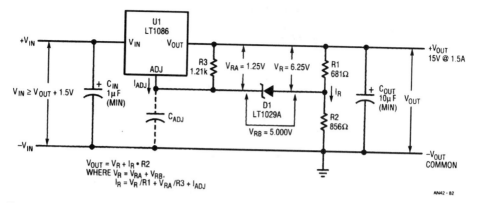

$$V_{OUT} = V_R + I_R \cdot R2$$
WHERE $V_R = V_{RA} + V_{RB}$,
$$I_R = V_R/R1 + V_{RA}/R3 + I_{ADJ}$$

AN42 - 82

Programmable high-stability regulator

Fig. 3-71 The circuit in the illustration provides a regulated output of 15 V at 1.5 A and can be programmed to provide other output voltages as shown by the equations. See Figs. 3-6 and 3-31 for resistor and capacitor characteristics. LINEAR TECHNOLOGY, LINEAR APPLICATIONS HANDBOOK, 1993, P. AN42-21.

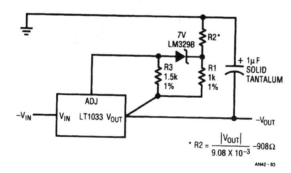

$$* R2 = \frac{|V_{OUT}|}{9.08 \times 10^{-3}} - 908\Omega$$

AN42 - 83

Programmable negative-output high-stability regulator

Fig. 3-72 The circuit in the illustration provides a regulated negative output and can be programmed to provide other output voltages as shown by the equations. See Figs. 3-6 and 3-31 for resistor and capacitor characteristics. LINEAR TECHNOLOGY, LINEAR APPLICATIONS HANDBOOK, 1993, P. AN42-22.

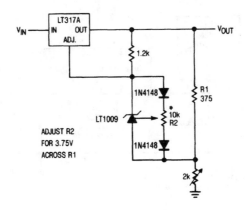

Low-TC regulator

Fig. 3-73 The circuit in the illustration provides an adjustable output with low temperature coefficient. See Fig. 3-6 for resistor characteristics. LINEAR TECHNOLOGY, LINEAR APPLICATIONS HANDBOOK, 1993, P. AN42-22.

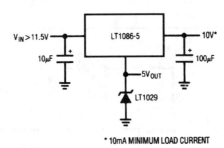

Basic reference/regulator

Fig. 3-74 The circuit in the illustration is similar to that of Fig. 3-70 except that only one reference zener is used. The reference voltage is set by the zener. The output voltage is a combination of the zener voltage and the LT1086 voltage. See Figs. 3-6 and 3-31 for resistor and capacitor characteristics. LINEAR TECHNOLOGY, LINEAR APPLICATIONS HANDBOOK, 1993, P. AN42-23.

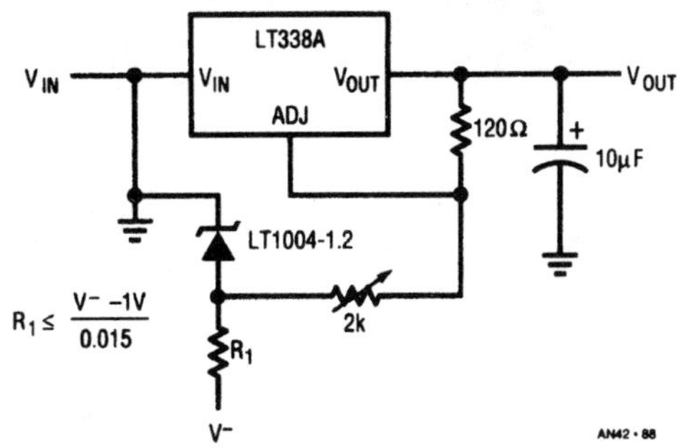

Variable-output regulator (negative supply)

Fig. 3-75 The circuit in the illustration is operated from a negative supply and provides a variable output. The output range is set by the regulator characteristics (see LT1004 data sheet). See Figs. 3-6 and 3-31 for resistor and capacitor characteristics. LINEAR TECHNOLOGY, LINEAR APPLICATIONS HANDBOOK, 1993, P. AN42-23.

=4=

Battery-power and micropower circuits

This chapter is devoted to circuits that can be operated from a battery (often a single 1.5-V cell) and draw a minimum of current (*micropower*). Test and troubleshooting for these circuits is the same as for corresponding circuits in other chapters and are not duplicated here. For example, the circuits of Figs. 4-1 and 4-2 are regulators and voltage references, such as described in Chapter 3. Where it might not be obvious, reference is made (in the circuit description) to the appropriate chapter for testing and troubleshooting. For additional information on micropower and battery circuits, read the author's *Simplified Design of Micropower and Battery Circuits*, 1996, published by Butterworth-Heinemann.

Battery-power/micropower circuit titles and description

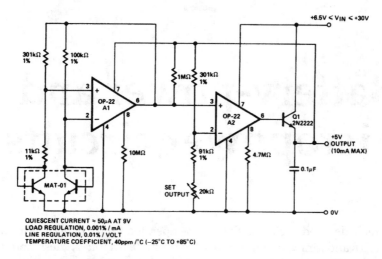

QUIESCENT CURRENT ≈ 50μA AT 9V
LOAD REGULATION, 0.001% / mA
LINE REGULATION, 0.01% / VOLT
TEMPERATURE COEFFICIENT, 40ppm /°C (−25°C TO +85°C)

Micropower 5-V regulator

Fig. 4-1 The 5-V op-amp linear regulator shown is useful for instrumentation requiring good power efficiency. Maximum load current is 10 mA as shown, and it can be increased by changing Q1 to a power transistor and increasing the set current of A2 in proportion. ANALOG DEVICES, APPLICATIONS REFERENCE MANUAL, 1993, P. 13-51.

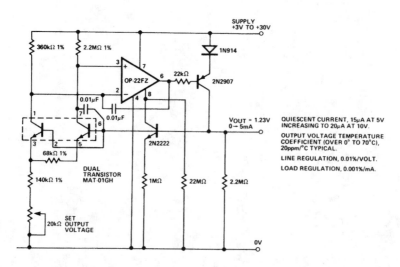

QUIESCENT CURRENT, 15μA AT 5V
INCREASING TO 20μA AT 10V.

OUTPUT VOLTAGE TEMPERATURE
COEFFICIENT (OVER 0° TO 70°C),
20ppm/°C TYPICAL.

LINE REGULATION, 0.01%/VOLT.

LOAD REGULATION, 0.001%/mA.

Micropower 1.23-V band-gap reference

Fig. 4-2 The circuit in the illustration provides a 1.23-V reference with better performance than micropower IC shunt regulators and has the advantage of being a series regulator. ANALOG DEVICES, APPLICATIONS REFERENCE MANUAL, 1993, P. 13-52.

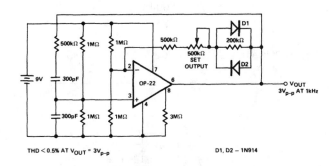

Micropower Wien-bridge oscillator

Fig. 4-3 The circuit in the illustration requires less than 60 μA of current and dissipates less than 500 μA of power, making the circuit ideal for battery-powered applications. Output level is controlled by nonlinear elements D1 and D2. When adjusted for 3-V$_{p-p}$ output, the distortion is lower than 0.5% at 1 kHz. ANALOG DEVICES, APPLICATIONS REFERENCE MANUAL, 1993, P. 13-58.

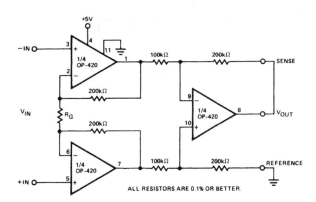

Micropower instrumentation amplifier

Fig. 4-4 The instrumentation amplifier in the illustration requires only 200-μA total quiescent current (with a common-mode voltage of zero) and operates on sin-

Continued

gle-voltage supplies from +1.6 V to +36 V. The output-voltage range is 0 V to (V+ −1.5) volts with a CMRR higher than 100 dB. Differential gain, V, is adjusted with a single resistor, R_G, as given by

$$R_G = \frac{800 \text{ k}\Omega}{(V - 2)}$$

ANALOG DEVICES, APPLICATIONS REFERENCE MANUAL, 1993, P. 13-59.

4

248

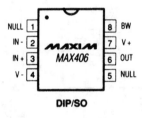

REMOTE-SENSOR BUFFER OPERATES FROM 3V BATTERY

TOP VIEW

NULL	1		8	BW
IN -	2	*MAXIM*	7	V +
IN +	3	MAX406	6	OUT
V -	4		5	NULL

DIP/SO

(b)

Micropower remote-sensor buffer

Fig. 4-5 The circuit in the illustration operates from a single 3-V battery and draws only 1 µA of quiescent current. When pin 8 is connected to V− (ground) as shown, the MAX406 is in the unity-gain mode with a 5 V/ms (volts per millisecond) slew rate and a gain-bandwidth of 8 kHz. When pin 8 is connected to the positive rail, the MAX406 is in the high-speed (uncompensated) mode with a 20 V/ms slew rate and a 40-kHz gain bandwidth. Figure 4-5B shows the pin configurations. MAXIM NEW RELEASES DATA BOOK, 1991, P. 3-23.

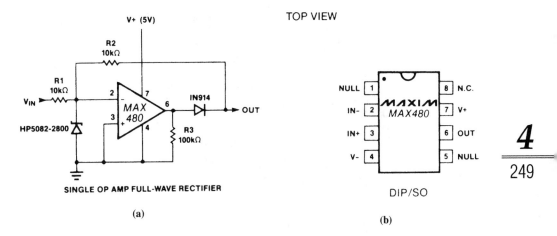

SINGLE OP AMP FULL-WAVE RECTIFIER

(a)

TOP VIEW

DIP/SO

(b)

Micropower full-wave rectifier (single supply)

Fig. 4-6 The MAX480 in this circuit consumes less than 20 μA, allowing operation in excess of 10,000 hours from a 250-mA/hr (milliamps per hour) lithium coin cell. Even with minimum quiescent current, the MAX480 sinks or sources 5 mA from the output. Figure 4-6B shows the pin configuration. MAXIM NEW RELEASES DATA BOOK, 1992, P. 3-47.

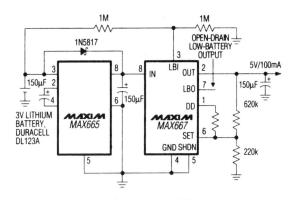

ALL 150μF CAPACITORS ARE MAXC001, AVAILABLE FROM MAXIM.
PIN NUMBERS REFER TO DIP PACKAGE.

+5V at 100 mA from a 3-V lithium battery

Fig. 4-7 The circuit in the illustration generates a +5-V regulated output from a 3-V lithium battery. The circuit can operate continuously for 16 hours with a 40-mA load. MAXIM NEW RELEASES DATA BOOK, 1995, P. 4-32.

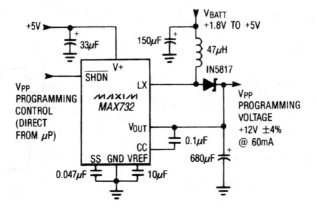

Two-cell to +12-V flash-memory programmer

Fig. 4-8 The circuit in the illustration provides for operation with battery supplies as low as 1.8 V. It is not possible to get output currents greater than 80 mA with battery voltages lower than 2 V, because of the high peak currents required. MAXIM NEW RELEASES DATA BOOK, 1993, P. 4-100.

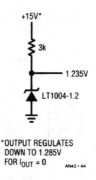

Micropower 1.2-V reference (1.5-V battery)

Fig. 4-9 The circuit in the illustration can provide a 1.2-V output from a 1.5-V battery. LINEAR TECHNOLOGY, LINEAR APPLICATIONS HANDBOOK, 1993, P. AN42-10.

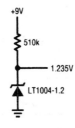

Micropower 1.2-V reference (9-V battery)

Fig. 4-10 The circuit in the illustration can provide a 1.2-V output from a 9-V battery. LINEAR TECHNOLOGY, LINEAR APPLICATIONS HANDBOOK, 1993, P. AN42-10.

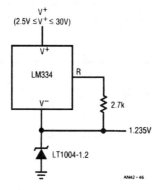

Micropower 1.2-V reference with wide input-voltage range

Fig. 4-11 The circuit in the illustration provides a 1.2-V output with an input from 2.5 V to 30 V. LINEAR TECHNOLOGY, LINEAR APPLICATIONS HANDBOOK, 1993, P. AN42-11.

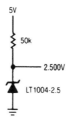

Micropower 2.5-V reference

Fig. 4-12 The circuit in the illustration provides a 2.5-V output with a 5-V input. LINEAR TECHNOLOGY, LINEAR APPLICATIONS HANDBOOK, 1993, P. AN42-11.

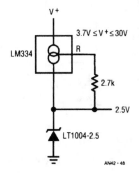

Micropower 2.5-V reference with wide input-voltage range

Fig. 4-13 The circuit in the illustration provides a 2.5-V output with an input from 3.7 V to 30 V. LINEAR TECHNOLOGY, LINEAR APPLICATIONS HANDBOOK, 1993, P. AN42-11.

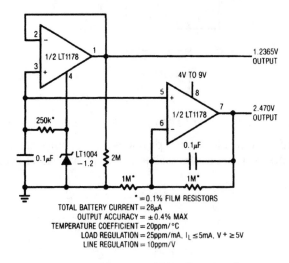

Micropower 1.2264-V and 2.470-V self-buffered reference

Fig. 4-14 The circuit in the illustration provides a 1.2265-V output and a 2.470-V output, both buffered by one-half of an LT1178, with an input from 4 V to 9 V. LINEAR TECHNOLOGY, LINEAR APPLICATIONS HANDBOOK, 1993, P. AN42-11.

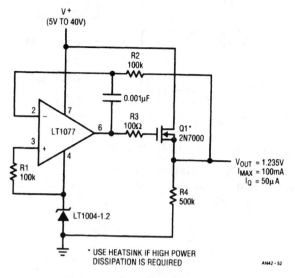

Micropower 1.235-V current-boosted reference

Fig. 4-15 The circuit in the illustration provides a 1.235-V output at 100 mA but has a quiescent current of only 50 μA. LINEAR TECHNOLOGY, LINEAR APPLICATIONS HANDBOOK, 1993, P. AN42-11.

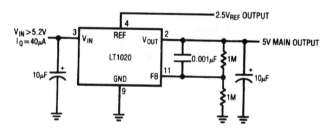

Micropower regulator/reference

Fig. 4-16 The circuit in the illustration provides a 5-V regulated output and a 2.5-V reference with a quiescent current of 40 μA. LINEAR TECHNOLOGY, LINEAR APPLICATIONS HANDBOOK, 1993, P. AN42-22.

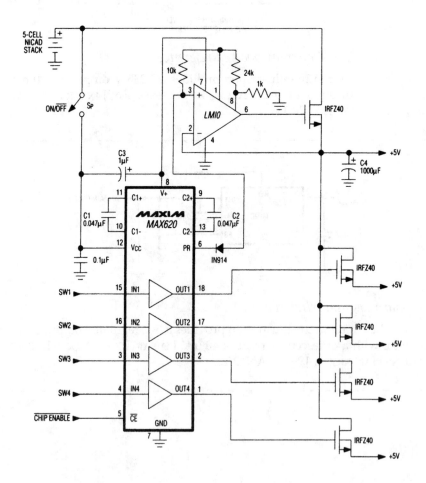

Micropower regulator/reference with shutdown

Fig. 4-17 The circuit in the illustration is identical to that of Fig. 4-16, but it includes a shutdown feature (at pin 3). LINEAR TECHNOLOGY, LINEAR APPLICATIONS HANDBOOK, 1993, P. AN42-23.

Battery-power and micropower circuits

Logic-controlled +5-V regulated power-distribution system

Fig. 4-18 The circuit in the illustration provides a single continuous +5-V output and four switched +5-V supply lines for battery-power equipment. The regulator can supply several amps with a typical dropout voltage of 28 mV at 1 A. MAXIM NEW RELEASES DATA BOOK, 1992, P. 4-29.

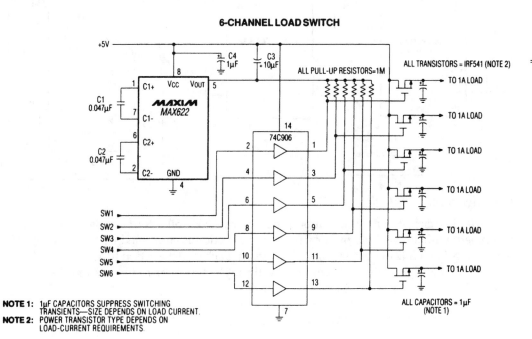

6-CHANNEL LOAD SWITCH

NOTE 1: 1µF CAPACITORS SUPPRESS SWITCHING TRANSIENTS—SIZE DEPENDS ON LOAD CURRENT.
NOTE 2: POWER TRANSISTOR TYPE DEPENDS ON LOAD-CURRENT REQUIREMENTS.

Driver for six high-side switches

Fig. 4-19 The circuit in the illustration is a six-channel load switch driven by a MAX622 to provide high-side (between battery and load) control of battery-operated equipment. The minimum value of the pull-up resistors is determined by:

$$R_{MIN} = \frac{V_{OUT} \times (number\ of\ channels)}{I_{OUT}}$$

where V_{OUT} is the high-side output and I_{OUT} is the MAX622 output current (use 1 mA for most applications). MAXIM NEW RELEASES DATA BOOK, 1992, P. 4-39.

Battery-power/micropower circuit titles and description

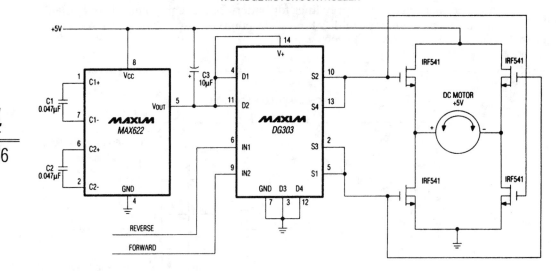

H-BRIDGE MOTOR CONTROLLER

H-bridge motor controller

Fig. 4-20 The circuit in the illustration functions as a controller for battery-power equipment with a reversible motor. The motor direction is controlled by toggling between IN1 and IN2 of the DG303 analog switch. Each switch section turns on the appropriate FET pair, which passes current through the motor in the desired direction. MAXIM NEW RELEASES DATA BOOK, 1992, P. 4-39.

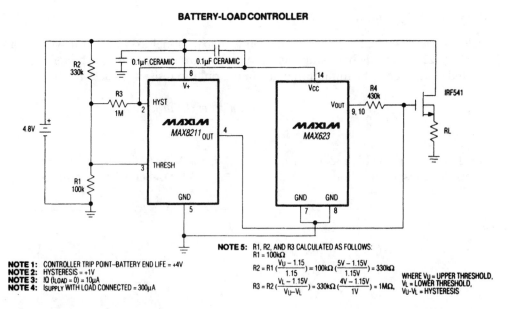

BATTERY-LOAD CONTROLLER

NOTE 1: CONTROLLER TRIP POINT–BATTERY END LIFE = +4V
NOTE 2: HYSTERESIS = +1V
NOTE 3: IQ (ILOAD = 0) = 10µA
NOTE 4: ISUPPLY WITH LOAD CONNECTED = 300µA

NOTE 5: R1, R2, AND R3 CALCULATED AS FOLLOWS:
R1 = 100kΩ

$$R2 = R1 \left(\frac{V_U - 1.15}{1.15} \right) = 100k\Omega \left(\frac{5V - 1.15V}{1.15V} \right) = 330k\Omega$$

$$R3 = R2 \left(\frac{V_L - 1.15V}{V_U - V_L} \right) = 330k\Omega \left(\frac{4V - 1.15V}{1V} \right) = 1M\Omega$$

WHERE V_U = UPPER THRESHOLD,
V_L = LOWER THRESHOLD,
V_U–V_L = HYSTERESIS

Battery-power and micropower circuits

Battery load controller

Fig. 4-21 In this circuit, the MAX8211 under voltage detector detects the battery's end-of-life, and the MAX622 high-side power supply turns the power FET switch on. During normal operation, the MAX8211 HYST pin powers the MAX622, providing gate drive to keep the FET off. When the battery reaches the discharge threshold (end-of-life), the MAX8211 pulls the FET gate low, cutting off current to the load. At the same time, the HYST pin goes low, turning off the MAX622. As a result, supply current is about 10 µA in the load-disconnected condition. MAXIM NEW RELEASES DATA BOOK, 1992, P. 4-40.

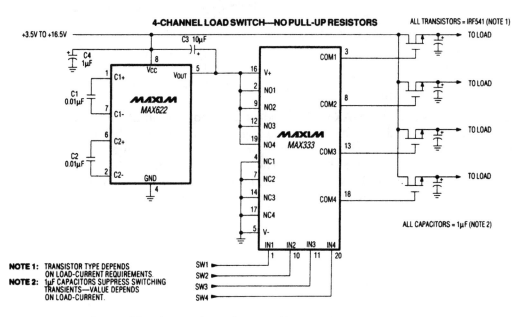

4-CHANNEL LOAD SWITCH—NO PULL-UP RESISTORS

ALL TRANSISTORS = IRF541 (NOTE 1)

ALL CAPACITORS = 1µF (NOTE 2)

NOTE 1: TRANSISTOR TYPE DEPENDS ON LOAD-CURRENT REQUIREMENTS.
NOTE 2: 1µF CAPACITORS SUPPRESS SWITCHING TRANSIENTS—VALUE DEPENDS ON LOAD-CURRENT.

Four-channel load switch with no pull-up resistors

Fig. 4-22 In this circuit, a MAX622 supplies high-side voltage to a MAX333 quad analog switch to control any one of four high-side switches. The FET gates are normally connected to ground when the MAX333 logic inputs are low. MAXIM NEW RELEASES DATA BOOK, 1992, P. 4-41.

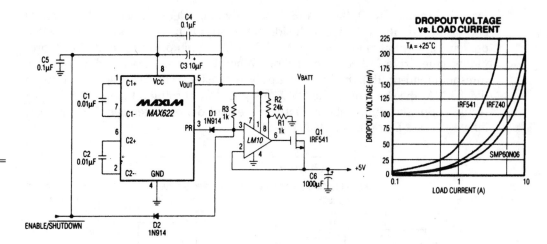

Low-dropout regulator

Fig. 4-23 In this circuit, a MAX622 high-side supply powers, an LM10 reference and op-amp combination, providing sufficient gate drive to turn on the FET. This allows the regulator to have less than 70-mV dropout at 1-A load using an IRF541 and just under 20 mV for an SMP60N06. The regulator is turned on by applying V_{BATT} to the ENABLE/SHUTDOWN connection and turned off by pulling this input to ground. V_{OUT} is set by the ratio of R_1/R_2:

$$R_2 = R_1 \frac{(V_{OUT} - 1)}{0.2}$$

If the application does not require logic shutdown, connect the MAX622 V_{CC} pin directly to the battery and eliminate D2. MAXIM NEW RELEASES DATA BOOK, 1992, P. 4-41.

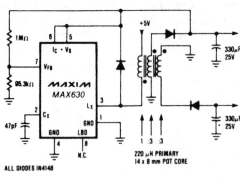

(a)

Continued

MANUFACTURER	TYPICAL PART #	DESCRIPTION
MOLDED INDUCTORS		
Dale	IHA-104	500µH, 0.5 ohms
Nytronics	WEE-470	470µH, 10 ohms
TRW	LL-500	500µH, 0.75 ohms
POTTED TOROIDAL INDUCTORS		
Dale	TE-3Q4TA	1mH, 0.82 ohms
TRW	MH-1	600µH, 1.9 ohms
Torotel Prod.	PT 53-18	500µH, 5 ohms,
FERRITE CORES AND TOROIDS		
Allen Bradley	T0451S100A	Tor. Core, 500nH/T^2
Siemens	B64290-K38-X38	Tor. Core, 4µH/T^2
Magnetics	555130	Tor. Core, 53nH/T^2
Stackpole	57-3215	Pot Core, 14mm x 8mm
Magnetics	G-41408-25	Pot Core, 14 x 8, 250nH/T^2

(b)

Note: This list does not constitute an endorsement by Maxim Integrated Products and is not intended to be a comprehensive list of all manufacturers of these components.

Basic battery-power +5-V to +15-V converter

Fig. 4-24 The circuit in the illustration uses the MAX630 as a basic dc/dc converter but with two additional windings on the inductor. The 1408 (15 mm × 8 mm) pot core specified is an *IEC* (Integrated Electronics Component) standard size available from many manufacturers (see Fig. 4-24B). The –15-V output is semiregulated, typically varying from –13.6 V to 14.4 V as the +15-V load current changes from no load to 20 mA. MAXIM NEW RELEASES DATA BOOK, 1992, PP. 4-51, 53.

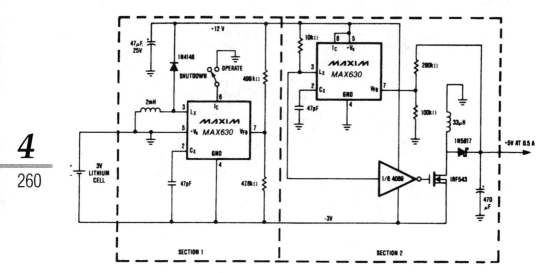

2.5-W 3-V to 5-V converter with low-voltage start-up

Fig. 4-25 The circuit in the illustration provides power for systems (although battery-powered) that need high currents for short periods and then shut down to a low-power state. When shut down, this circuit uses less than 10 μA, and the output falls to 0 V, unlike a standard boost circuit where the output voltage is V_{BATT} −0.6 V when the converter is shut down. Circuit efficiency is 85% with the values shown. Because section 1 powers only section 2, the circuit will start up with battery voltages as low as 1.5 V, independent of the loading on the +5-V output. MAXIM NEW RELEASES DATA BOOK, 1992, P. 4-53.

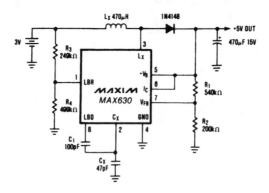

3-V to 5-V converter with low-battery frequency shift

Fig. 4-26 The circuit in the illustration converts 3 V to 5 V at 40 mA, with 85% efficiency, for use by 5-V logic circuits. When pin 6 is driven low, the output voltage will be the battery voltage minus the drop across D1. The optional C1, R3, and R4 circuit lowers the oscillator frequency when the battery voltage falls to 2.0 V. This action maintains the output-power capability by increasing the peak inductor current, compensating for the reduced battery voltage. MAXIM NEW RELEASES DATA BOOK, 1992, P. 4-54.

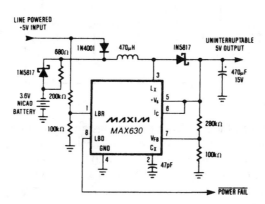

Uninterruptible +5-V supply

Fig. 4-27 The circuit in the illustration provides a continuous supply of regulated +5-V power, with automatic switch-over between line power and battery backup. When the line-powered input voltage is at +5 V, the input provides 4.4 V to the MAX630 and (simultaneously) trickle charges the battery. If the line input falls below battery voltage, the 3.6-V battery supplies power to the MAX630, which boosts the battery voltage up to +5 V, thus maintaining a continuous supply to the uninterruptible +5-V bus. Because the +5-V output is always supplied through the MAX630, there are no power spikes or glitches during power transfer. The MAX630 low-battery detector monitors the line-powered +5 V, and the LBD output can be used to shut down unnecessary sections of the system during power failures. As an alternate, the low-battery detector could monitor the NiCad (nickel-cadmium) battery voltage and provide warning of power loss when the battery is nearly discharged. Unlike battery-backup systems that use 9-V batteries, this circuit does not need +12-V or +15-V to recharge the battery, so it can be used on modules or circuit cards that only have 5 V available. MAXIM NEW RELEASES DATA BOOK, 1992, P. 4-54.

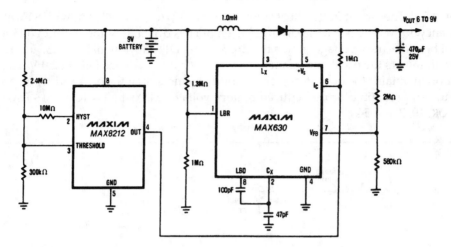

9-V battery life extender

Fig. 4-28 The circuit in the illustration provides a minimum of 7 V until the 9-V battery voltage falls to less than 2 V. When the battery voltage is higher than 7 V, pin 6 of the MAX630 is low, placing the MAX630 in shut down (drawing about 10 mA). When the battery voltage falls to 7 V, pin 4 of the MAX8212 goes high, enabling the MAX630, which then maintains the output at 7 V even as the battery voltage falls below 7 V. The low-battery detector (LBD) is used to decrease the oscillator frequency when the battery voltage falls to 3 V, thereby increasing the output-current capability of the circuit. The circuit in the illustration (with or without the MAX8212) can be used to provide 5 V from four alkaline cells. The initial voltage is about 6 V, and the output is maintained at 5 V even when the battery voltage falls to less than 2 V. MAXIM NEW RELEASES DATA BOOK, 1992, P. 4-55.

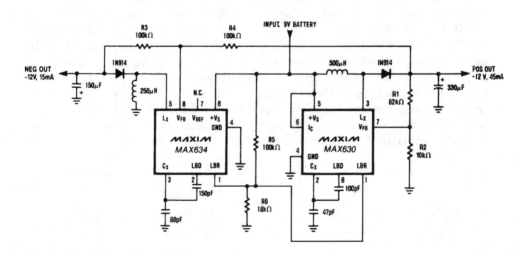

Battery-power and micropower circuits

Dual-tracking regulator for a 9-V battery

Fig. 4-29 The circuit in the illustration provides a dual-tracking ±12-V output from a 9-V battery. The reference for the –12-V output is taken from the positive output through R_3/R_4. Both regulators are set to maximize output power at low-battery voltage by reducing the oscillator frequency (through the LBR input, pin 1) when V_{BATT} falls to 7.2 V. MAXIM NEW RELEASES DATA BOOK, 1992, P. 4-55.

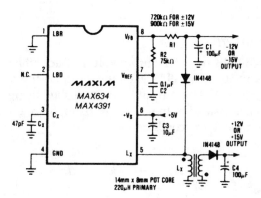

Dual-output inverter

Fig. 4-30 The circuit in the illustration provides dual outputs (±12 V or ±15 V) from a +5-V input. Note that L_X is *bifilar wound* (primary and secondary are wound simultaneously using two wires in parallel). The L_X core is usually a toroid or pot core (see Fig. 4-24B). The negative output is fully regulated, with the positive output semiregulated. MAXIM NEW RELEASES DATA BOOK, 1992, P. 4-72.

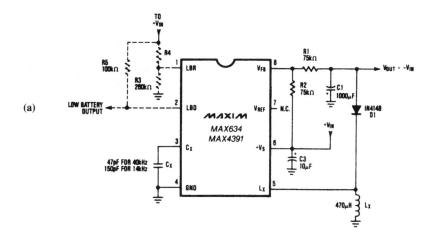

(a)

Battery-power/micropower circuit titles and description

Continued

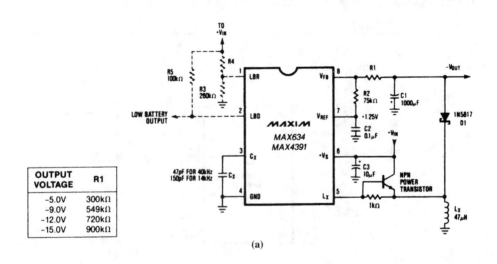

LOW = SHUTDOWN   OR
HIGH = OPERATE

N CHANNEL FET
SUCH AS 2N7000
OR IRFD120

(b)

Regulated voltage inverter

Fig. 4-31 In this circuit, the negative output voltage tracks the positive input voltage and delivers about 50 mA at −9 V when the input is +9 V and about 30 mA at −5 V when the input is +5 V. This output is accomplished using the positive input voltage as the reference instead of the on-board bandgap reference. V_{OUT} is set by the input voltage and the R_1/R_2 combination:

$$V_{OUT} = -\frac{R_2}{R_1} \times +V_S$$

Fig. 4-31B shows a means of selecting low-power or shutdown mode (about 250-µA operating current). The ground pin must be well bypassed, and any voltage drop across the CMOS gate adds to the reference voltage, slightly increasing the regulated output voltage. MAXIM NEW RELEASES DATA BOOK, 1992, P. 4-73.

OUTPUT VOLTAGE	R1
−5.0V	300kΩ
−9.0V	549kΩ
−12.0V	720kΩ
−15.0V	900kΩ

(a)

Continued

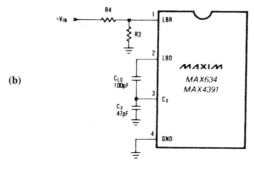

(b)

Voltage inverter with increased output power

Fig. 4-32 The circuit in the illustration provides voltage inversion similar to that of Fig. 4-31 but with an output-current capability greater than 525 mA. All of the npn transistor base current is used to drive the inductor L_X, but the voltage drop across the transistor is about 0.7 V. Figure 4-32B shows a means of compensating for a reduction in input voltage, thus permitting operation over a wider input-voltage range. With the values shown, the oscillator frequency is 40 kHz when the input voltage is higher than 6 V. When the input falls below 6 V, the low-battery detector (pin 2) goes low, placing the 100-pF capacitor in parallel with C_X, reducing the oscillator frequency to 14. This action increases the available output power by a factor of about 3. MAXIM NEW RELEASE DATA BOOK, 1992, PP. 4-74, 75.

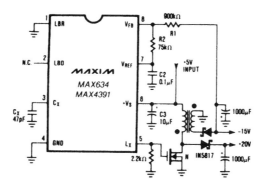

High-power inverter/converter

Fig. 4-33 The circuit in the illustration provides both a –15 V and a +20 V output from a +5-V input. The –15 V is fully regulated for both line and load variations. The +20-V output will vary normally less than 10% with changes in load on either the +20-V or +15-V output, as well as changes in the +5-V input. MAXIM NEW RELEASES DATA BOOK, 1992, P. 4-74.

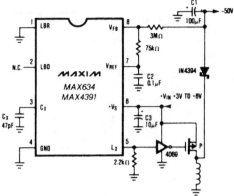

Inverter with increased output voltage

Fig. 4-34   The circuit in the illustration converts any positive voltage input from +3 V to +15 V to any desired output voltage, as long as the voltage breakdown of the external P-channel MOSFET is not exceeded. The circuit also is useful for generating a high-power, high-efficiency, −12-V to −15-V output using a simple one-winding coil. MAXIM NEW RELEASES DATA BOOK, 1992, P. 4-74.

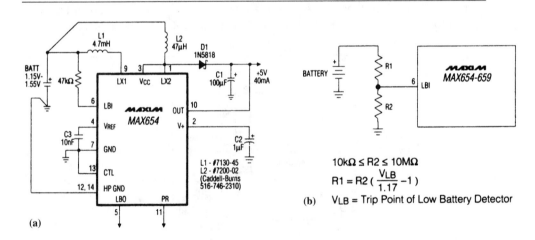

Single-cell step-up converter (5 V at 40 mA)

Fig. 4-35   The circuit in the illustration converts single-cell battery voltage to +5 V at 40 mA. When the battery voltage drops below 1.7 V, the low-battery monitor output (pin 5) goes low and sinks 1.6 mA. With the battery voltage at or greater than 1.7, pin 5 sources 1 µA from *V*+. If a different low-voltage trip point is desired, use the circuit of Fig. 4-35B. Operation of the circuit with some common batteries is shown in Fig. 4-35C. MAXIM NEW RELEASES DATA BOOK, 1992, P. 4-109, 110, 112.

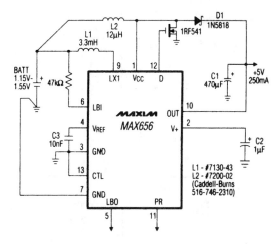

Single-cell step-up converter (5 V at 250 mA)

Fig. 4-36 The circuit in the illustration is similar to that of Fig. 4-35 but with an external switch to carry the additional current. The low-battery circuit of Fig. 4-35C applies. MAXIM NEW RELEASES DATA BOOK, 1992, P. 4-109.

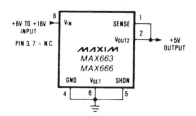

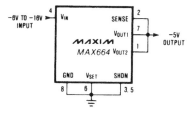

Micropower fixed voltage regulator

Fig. 4-37 The simple circuits in the illustration show the connections for a fixed 5-V output at 40 mA, with a quiescent current of 12 μA maximum. MAXIM NEW RELEASES DATA BOOK, 1992, P. 4-130.

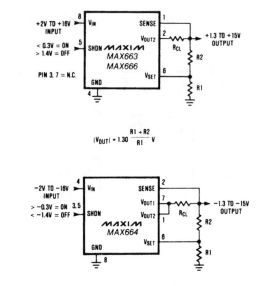

$$|V_{OUT}| = 1.30 \frac{R1 + R2}{R1} \text{ V}$$

Micropower programmable voltage regulator

Fig. 4-38 The circuits in the illustration show the connections for a programmable voltage regulator with a 40-mA output capability. The output voltage is set by R_1/R_2 as shown. Resistor R_{CL} sets the current limit (50 mA maximum) and is found by 0.5 V/current-limit for the MAX663/666, or 0.6/current-limit for the MAX664. External shutdown (if used) is applied to pin 5 for the positive regulator, or pins 3/5 for the negative regulator. MAXIM NEW RELEASES DATA BOOK, P. 4-130.

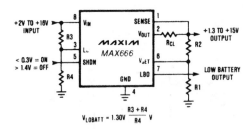

$$V_{LOBATT} = 1.30V \frac{R3 + R4}{R4} \text{ V}$$

Micropower programmable regulator with low-battery detector

Fig. 4-39 The circuit in the illustration shows the connections for a programmable regulator with a 40-mA output and low-battery detection. The equations for current-limit and voltage-output are the same as for Fig. 4-38. MAXIM NEW RELEASES DATA BOOK, 1992, P. 4-131.

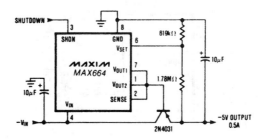

Micropower negative regulator with boosted output

Fig. 4-40 The circuit in the illustration shows the connections for a fixed –5-V output at 0.5 A, using an external transistor. MAXIM NEW RELEASES DATA BOOK, 1992, P. 4-132.

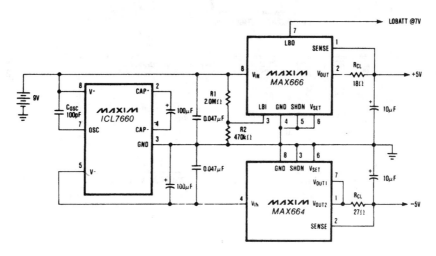

±5-V output from a 9-V battery

Fig. 4-41 The circuit in the illustration uses an ICL7660 charge pump and two micropower regulators to provide ±5-V output at 40 mA from a single 9-V battery. MAXIM NEW RELEASES DATA BOOK, P. 4-132.

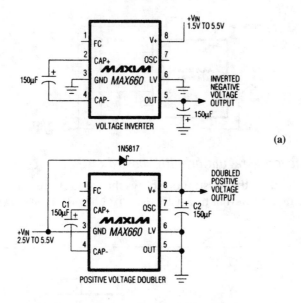

(a)

(b)

(c)

FC	OSC	Oscillator Frequency
Open	Open	10kHz
FC = V+	Open	45kHz
Open or FC = V+	External Capacitor	See Typical Operating Characteristics
Open	External Clock	External Clock Frequency

MANUFACTURER	CAPACITOR	CAPACITOR TYPE
Illinois Capacitor	RZS	Aluminum Electrolytic
Mallory	TDC & TDL	Tantalum
Nichicon	PF & PL	Aluminum Electrolytic
Sprague	672D, 673D, 674D, 678D	Aluminum Electrolytic
Sprague	135D, 173D, 199D	Tantalum
United Chemi-con	LXF & SXF	Aluminum Electrolytic

CMOS voltage converter

Fig. 4-42 The circuit in the illustration shows how a MAX660 can be connected to invert a 1.5-V to 5.5-V battery input, or to double a 2.5-V to 5.5-V battery input. Figure 4-42B shows how to change the oscillator frequency from the normal 10 kHz (with FC and OSC pins open). Figure 4-42C lists some electrolytic capacitors with low ESR (for maximum efficiency, minimum ripple, etc.). MAXIM NEW RELEASES DATA BOOK, 1992, PP. 4-117, 122, 123.

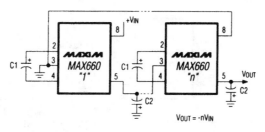

$V_{OUT} = -nV_{IN}$

Cascading for increased output voltage

Fig. 4-43 The circuit in the illustration shows how the ICs of Fig. 4-42 can be cascaded to increase the output voltage. The resulting output resistance is about equal to the sum of the individual MAX660 R_{OUT} values. The output voltage, where *n* represents the number of devices, is equal to $-nV_{IN}$. MAXIM NEW RELEASES DATA BOOK, 1992, P. 4-123.

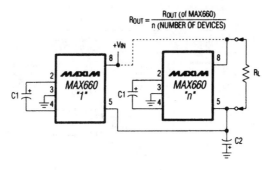

$$R_{OUT} = \frac{R_{OUT} \text{ (of MAX660)}}{n \text{ (NUMBER OF DEVICES)}}$$

Paralleling to reduce output resistance

Fig. 4-44 The circuit in the illustration shows how the ICs of Fig. 4-42 can be connected in parallel to reduce the output resistance. Each device requires a separate pump capacitor C1, but the reservoir capacitor C2 serves all devices. The value of C2 should be increased by a factor of *n* (the number of devices). MAXIM NEW RELEASES DATA BOOK, 1992, P. 4-123.

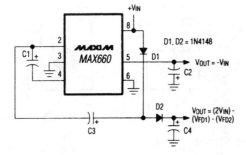

Combined positive multiplier and negative converter

Fig. 4-45 The circuit in the illustration inverts the input to a negative voltage, as well as multiplies the input but produces a higher source impedance. LINEAR TECHNOLOGY, LINEAR APPLICATIONS HANDBOOK, 1992, P. 4-123.

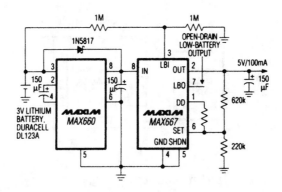

NOTE: ALL 150μF CAPACITORS ARE MAXC001, AVAILABLE FROM MAXIM.

+5-V output from a 3-V lithium battery

Fig. 4-46 The circuit in the illustration can provide 100 mA and will provide continuous operation for 16 hours when the load is 40 mA. MAXIM NEW RELEASES DATA BOOK, 1992, P. 4-124.

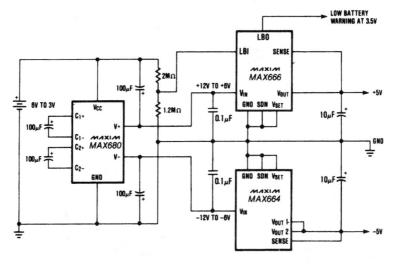

±5-V output from a 3-V or 9-V battery

Fig. 4-47 The circuit in the illustration is an alternate for the circuit of Fig. 4-41 and can be operated with a 3-V battery in place of the 9-V battery. MAXIM NEW RELEASES DATA BOOK, 1992, P. 4-146.

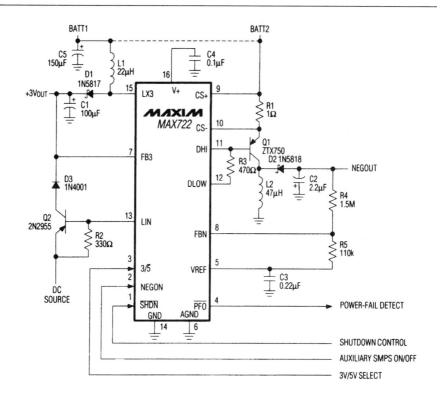

Battery-power/micropower circuit titles and description

Palmtop computer/LCD supply regulator

Fig. 4-48 The circuit in the illustration provides dual, regulated outputs for small, battery-operated microprocessor systems. The MAX722 generates a main output (3 V or 5 V, selectable) and a negative auxiliary output suitable for LCDs. Power can come from two sources (for example from a main battery of two or three alkaline or NiCad cells with a range from 0.9 V to 5.5 V, or an unregulated dc source, such as an ac/dc wall adapter with a range of 7 V to 20 V). MAXIM NEW RELEASES DATA BOOK, 1993, P. 4-51.

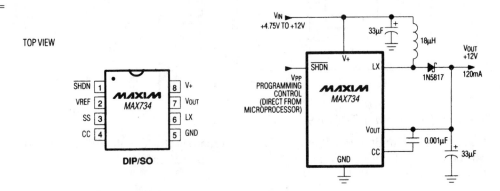

+12-V flash-memory programming supply

Fig. 4-49 The simple circuit in the illustration operates from a battery supply with an input of 1.9-V minimum, has 88% efficiency, an oscillator frequency of 170 kHz, and requires only 0.3 square inches of PC-board space. The operating quiescent current is 1.1 mA, with 70-μA shutdown current. MAXIM NEW RELEASES DATA BOOK, 1993, P. 4-105.

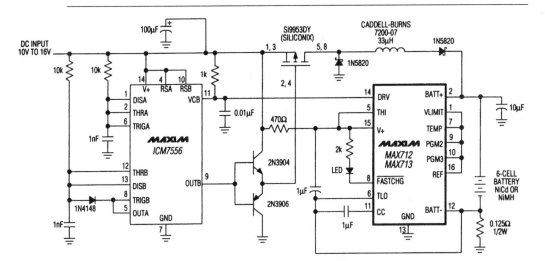

Switch-mode battery charger

Fig. 4-50 The circuit in the illustration is suitable for either NiCad or NiMH (nickel metal hydride) batteries and provides fast charge for up to six cells. MAXIM NEW RELEASES DATA BOOK, 1993, P. 4-65.

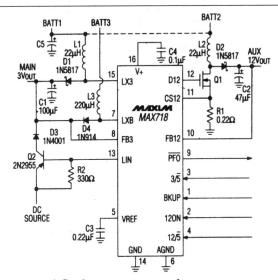

Palmtop-computer and flash-memory regulator

Fig. 4-51 The circuit in the illustration provides regulated dc outputs for small, battery-operated microprocessor systems. The IC generates a main output (3 V or 5 V, selectable) and an auxiliary output for flash memory or PCMCIA (5 V or 12 V, selectable). The circuit accepts up to three input voltages. Power can come from a main battery, or an unregulated dc source such as an ac/dc wall adapter. MAXIM NEW RELEASES DATA BOOK, 1994, P. 4-73.

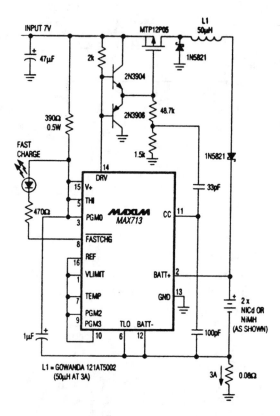

NiCad/NiMH battery charger with step-down regulator

Fig. 4-52   The circuit in the illustration shows a MAX713 connected as a switch-mode regulator to provide a fast charge for both NiCad and NiMH batteries. The input voltage range is battery voltage plus 1.5 V, up to 20 V, with a 7-V minimum. Efficiency with a V_{IN} of 12 V and two cells being charged at 1 A is 80%. The circuit in the illustration is programmed for two cells (PGMO strapped to $V+$; PGM1 open) and for a timeout of 90 minutes (PGM2 and PGM3 strapped to REF). See the MAX713 data sheet for other programming. MAXIM BATTERY MANAGEMENT CIRCUIT COLLECTION, 1994, P. 3.

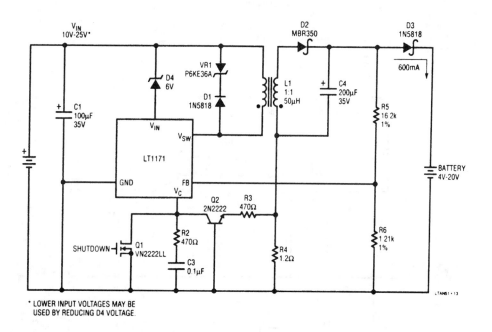

* LOWER INPUT VOLTAGES MAY BE
USED BY REDUCING D4 VOLTAGE.

Constant-current battery charger

Fig. 4-53   The circuit in the illustration uses an LT1171 connected for flyback operation to provide a constant-current battery charger. The circuit can be used for either NiCad or NiMH, but it does not detect when full charge is reached, nor does the circuit indicate a full charge. However, the battery voltage can be higher or lower than the input voltage. For example, a 16-V battery stack could be charged from a 12-V automobile battery. The charge current (sensed by R4) is set at about 600 mA. LINEAR TECHNOLOGY, LINEAR APPLICATIONS HANDBOOK, 1993, P. AN51-11.

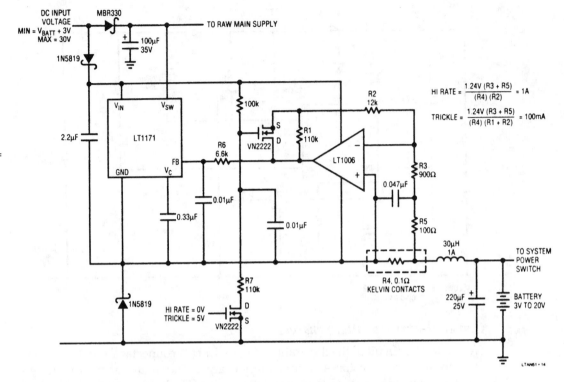

Dual-rate battery charger

Fig. 4-54 The circuit in the illustration shows an LT1171 connected to provide a dual-rate battery charger. Again, the circuit can be used for either NiCad or NiMH but does not detect when full battery charge is reached. However, efficiency is high (90%) so no heatsinks are needed for the LT1171 or diodes. The circuit charges at the high rate until a drop in charging current occurs and then goes to the trickle-charge mode automatically. LINEAR TECHNOLOGY, LINEAR APPLICATIONS HANDBOOK, 1993, P. AN51-12.

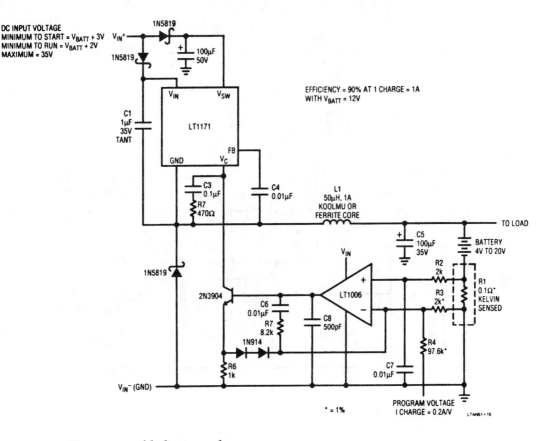

Programmable battery charger

Fig. 4-55 The circuit in the illustration shows an LT1171 connected to provide a programmable battery charger for use with NiCad or NiMH batteries. The circuit does not detect full charge, and input voltage must be higher than battery voltage. Charging current is proportional to the program voltage. (This program voltage can be digitally controlled through D/A converters if desired; see Chapter 9.) A sense resistor in the bottom side (between battery and ground) of the battery senses charging current. This current is compared with the program voltage and a feedback signal is developed to drive the LT1171 VC pin, thus controlling the charge current. Typical efficiency is 90%, eliminating the need for heatsinks (in most cases). LINEAR TECHNOLOGY, LINEAR APPLICATIONS HANDBOOK, 1993, P. AN51-13.

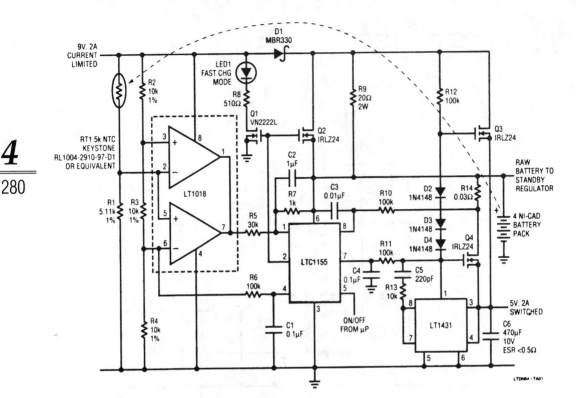

Four-cell NiCad regulator/charger

Fig. 4-56 The circuit in the illustration shows an LTC1155 connected to provide control of a four-cell charger/regulator. The circuit is well suited for a notebook-computer supply, (A four-cell NiCad battery pack can be used to power a 5-V note-book computer.) Little board space is consumed when the LTC1153 and three MOSFETs are housed in SO packages. However, Q3 and Q4 must have heatsinks. Component Q3 can dissipate 7 W if the full 2-A output is required while powered from a wall unit. LINEAR TECHNOLOGY, LINEAR APPLICATIONS HANDBOOK, 1993, P. DN54-1.

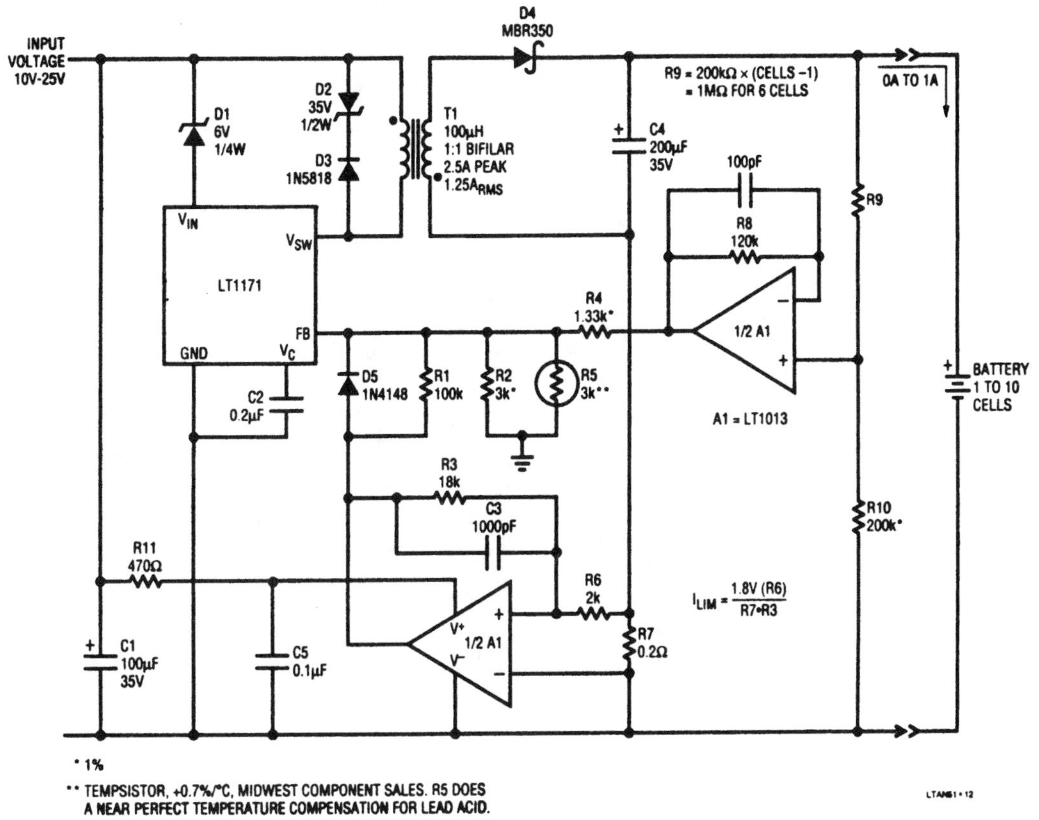

INPUT VOLTAGE 10V-25V

D1 6V 1/4W

D2 35V 1/2W

D3 1N5818

D4 MBR350

T1 100µH 1:1 BIFILAR 2.5A PEAK 1.25A$_{RMS}$

V_{IN}

LT1171

V_{SW}

FB

GND

V_C

C2 0.2µF

D5 1N4148

R1 100k

R2 3k*

R5 3k**

R4 1.33k*

C4 200µF 35V

R8 120k

100pF

1/2 A1

A1 = LT1013

R9 = 200kΩ × (CELLS −1) = 1MΩ FOR 6 CELLS

0A TO 1A

R9

BATTERY 1 TO 10 CELLS

R10 200k*

R3 18k

C3 1000pF

R11 470Ω

C1 100µF 35V

C5 0.1µF

V+

1/2 A1

V−

R6 2k

R7 0.2Ω

$$I_{LIM} = \frac{1.8V \ (R6)}{R7 \cdot R3}$$

* 1%

** TEMPSISTOR, +0.7%/°C, MIDWEST COMPONENT SALES. R5 DOES A NEAR PERFECT TEMPERATURE COMPENSATION FOR LEAD ACID.

LTAN51-12

Lead-acid battery charger

Fig. 4-57 The circuit in the illustration shows a charger suitable for lead-acid rechargeable cells (which require precise nonlinear temperature compensation during charge). The circuit allows operation with input voltages above or below battery voltage. LINEAR TECHNOLOGY, LINEAR APPLICATIONS HANDBOOK, 1993, P. AN51-10.

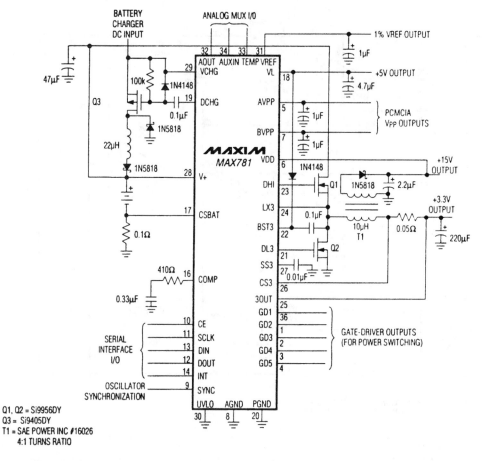

Sub-notebook-computer power controller

Fig. 4-58 The circuit in the illustration is a system-engineered solution for small portable equipment that has relative light (or nonexistent) 5-V requirements. The single MAX781 reduces size, but it produces all outputs normally required by a sub-notebook computer in addition to charging the battery. The circuit is recommended for five-cell to eight-cell operation, with an input voltage range of 5 to 18 V. The quiescent current (with 5-V input) is 1 mA, and the maximum load current is 1.5 A. MAXIM NEW RELEASES BATTERY MANAGEMENT CIRCUIT APPLICATIONS, 1994, P. 35.

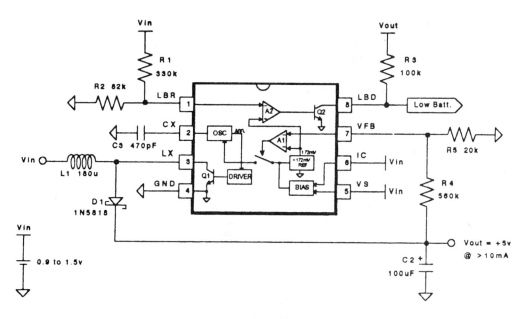

Single-cell step-up switching regulator

Fig. 4-59 The circuit in the illustration shows an EXAR XR-8073 connected for single-cell step-up operation. Using the values shown, the low-battery function trips at 0.86 V. EXAR DATABOOK, 1992, P. 5-147.

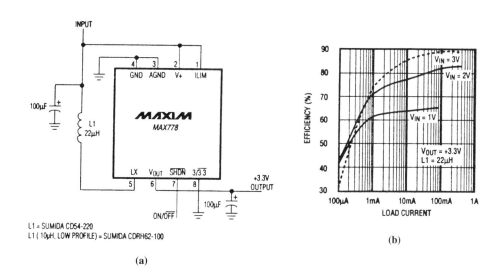

(a)

(b)

Low-voltage-input step-up converter (single cell)

Fig. 4-60 The circuit in the illustration uses a MAX778 to step-up a low voltage (1 V guaranteed) to either 3 V (SEL pin 8 open) or 3.3 V (SEL pin 8 connected to AGND pin 3). Supply voltages can range from 1 V to 6.2 V (1 to 4 cells). With a 2-V input, the circuit delivers a typical 300 mA at 3 V. Figure 4-60B shows the efficiency. MAXIM BATTERY MANAGEMENT CIRCUIT COLLECTION, 1994, P. 8.

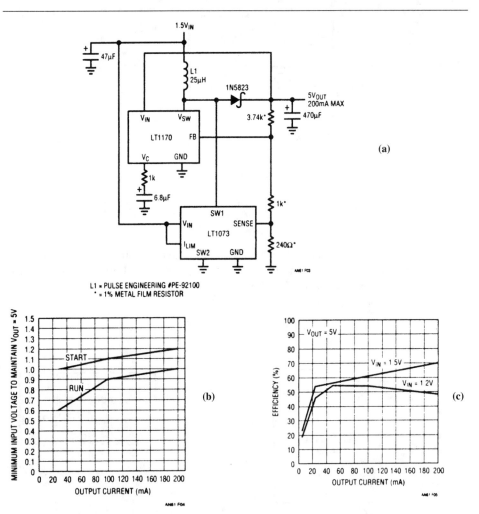

Single-cell 1.5-V to 5-V converter (200 mA)

Fig. 4-61 The circuit shows a single-cell converter with a 200-mA output. Figures 4-61B and 4-61C show the input/output and efficiency, respectively. Typical applications include survival two-way radios, remote transducer-fed date-acquisition systems, etc. LINEAR TECHNOLOGY, 1994, PP. AN61-2, 3.

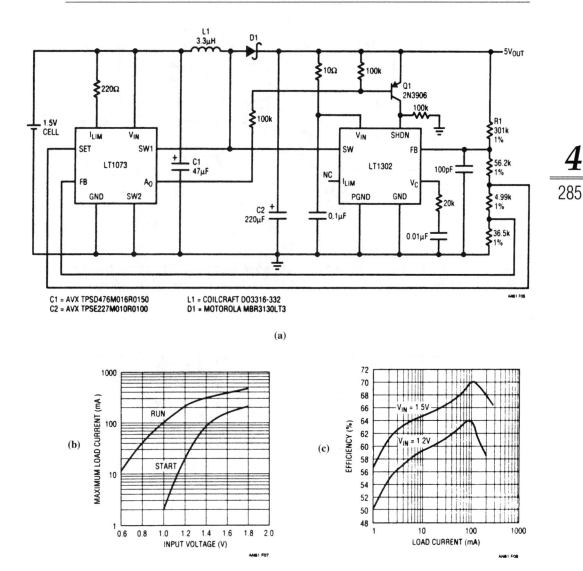

C1 = AVX TPSD476M016R0150   L1 = COILCRAFT DO3316-332
C2 = AVX TPSE227M010R0100   D1 = MOTOROLA MBR3130LT3

AN61-F06

(a)

(b)

(c)

AN61 F07   AN61 F06

Single-cell 1.5-V to 5-V converter (150 mA)

Fig. 4-62   The circuit in the illustration is similar to that of Fig. 4-61, but it provides greater efficiency at lower loads. Figures 4-62A and 4-62B show the maximum permissible loads and efficiency, respectively. LINEAR TECHNOLOGY, 1994, P. AN61-4.

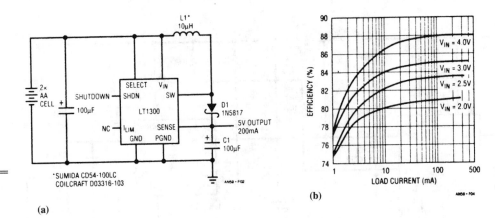

(a)

(b)

Two-AA-cell to 5-V converter (200 mA)

Fig. 4-63   The circuit in the illustration provides a 5-V output at 200 mA from two AA cells. Figure 4-63B shows the efficiency. LINEAR TECHNOLOGY, 1994, PP. AN59-3, 4.

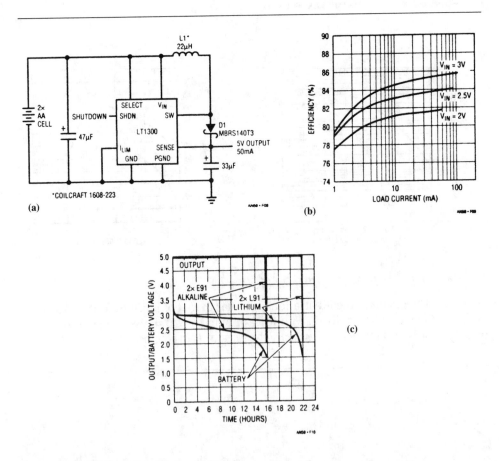

(a)

(b)

(c)

Battery-power and micropower circuits

Two-AA-cell to 5-V converter (50 mA)

Fig. 4-64 The circuit in the illustration provides a 5-V output at 50 mA from two AA cells. Figures 4-64B and 4-64C show the efficiency and battery life, respectively. Linear Technology, 1994, p. AN59-5.

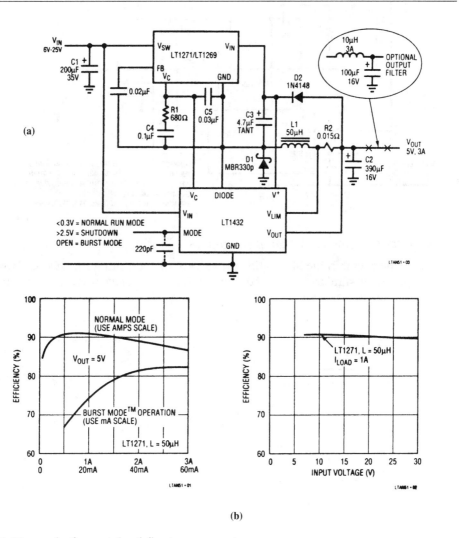

(a)

(b)

5-V supply for notebook/laptop computers

Fig. 4-65 The circuit in the illustration provides a 5-V output for notebook/laptop computers or similar portable equipment. Figure 4-65B shows efficiency. Linear Technology, Linear Applications Handbook, 1993, pp. AN51-1, 3.

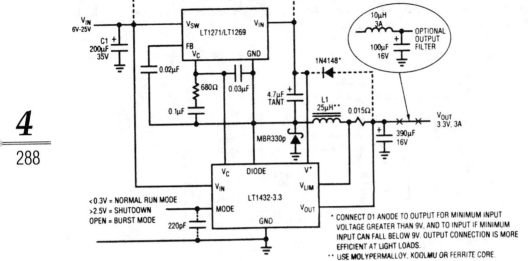

3.3-V supply for notebook/laptop computers

Fig. 4-66 The circuit in the illustration provides a 3.3-V output for notebook/laptop computers or similar portable equipment. LINEAR TECHNOLOGY, LINEAR APPLICATIONS HANDBOOK, 1993, P. AN51-3.

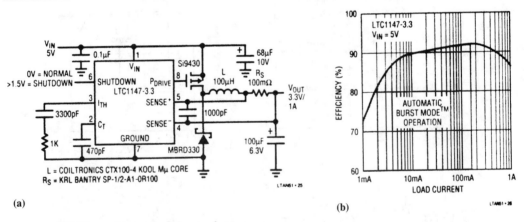

(a)

(b)

5-V to 3.3-V surface-mount regulator

Fig. 4-67 The circuit in the illustration converts 5 V to 3.3 V with high efficiency and is suitable for notebook/laptop-computer or other portable applications. Figure 4-67B shows the efficiency. LINEAR TECHNOLOGY, LINEAR APPLICATIONS HANDBOOK, 1993, PP. AN51-4, 5.

Battery-power and micropower circuits

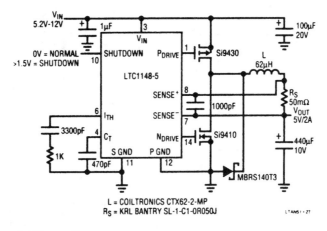

Low-dropout 5-V regulator

Fig. 4-68 The circuit in the illustration can be used with as few as five NiCad or NiMH cells and yet provide a 10-V output because of the low dropout and high efficiency. LINEAR TECHNOLOGY, LINEAR APPLICATIONS HANDBOOK, 1993, P. AN51-5.

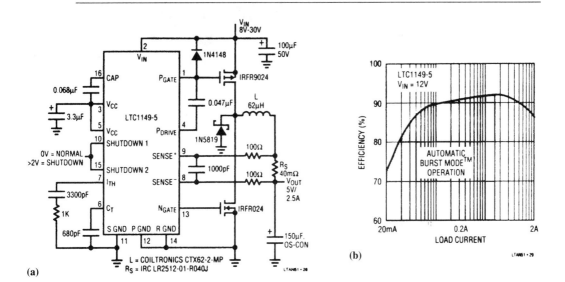

(a) **(b)**

5-V regulator for ac wall adapter

Fig. 4-69 The circuit in the illustration will accept input voltages from 8 V to 30 V to accommodate both notebook/laptop battery packs and ac wall adapters. Figure 4-69B shows the efficiency. LINEAR TECHNOLOGY, LINEAR APPLICATION HANDBOOK, 1995, P. AN51-5.

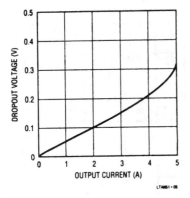

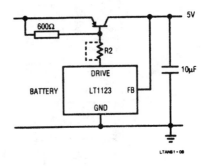

Low-dropout linear regulator for battery supplies

Fig. 4-70 The circuit in the illustration is used with four NiCad batteries to provide a 5-V output. Linear regulators can be used for notebook/laptop computers provided that the input supply has a very narrow voltage range, and the regulator has very low dropout. At full charge, the battery output can be as high as 6 V and is allowed to discharge down to 4.5 V while directly powering the computer. Use an MJE1123 for the transistor, which will produce a dropout of 0.25 V for a 3-V output current. Resistor R2 is used to reduce drive current to the transistor and is optional. LINEAR TECHNOLOGY, LINEAR APPLICATIONS HANDBOOK, 1993, P. AN51-6.

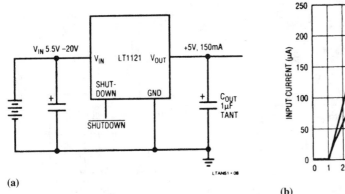

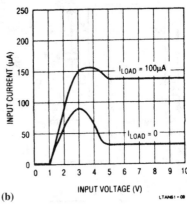

(a)

(b)

Micropower regulator

Fig. 4-71 The circuit in the illustration accepts a wide range of input voltages but draws only 30 μA input current at zero load current. Figure 4-71B shows the input voltage/current relationship. Note that a small (1-μF) output capacitor is used in place of the typical 10-μF capacitor, saving space and money. However, a larger output capacitor can be used without fear of instability. LINEAR TECHNOLOGY, LINEAR APPLICATIONS HANDBOOK, 1993, P. AN51-7.

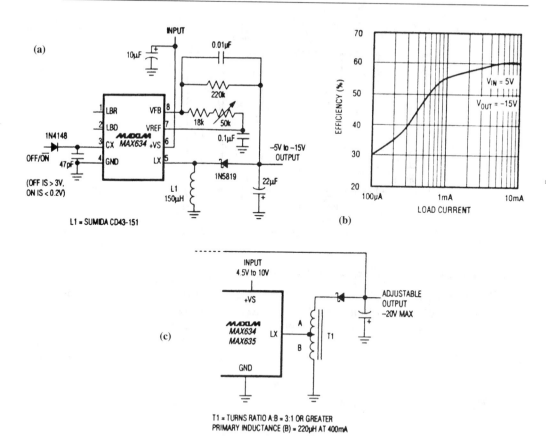

(a)

(b)

(c)

L1 = SUMIDA CD43-151

T1 = TURNS RATIO A:B = 3:1 OR GREATER
PRIMARY INDUCTANCE (B) = 220µH AT 400mA

Micropower LCD-contrast supply

Fig. 4-72 The circuit in the illustration provides an adjustable –5-V to –15-V output for LCD-contrast. Figure 4-72B shows efficiency. This "flea-power" circuit is suitable for small multiplexed LCD displays such as those on cellular telephones. The circuit operates over a 4-V to 6-V range (with a maximum input/output differential of 24 V), draws 500-µA quiescent current, and is capable of a 10 mA output. The maximum input/output of 24 V can be extended by substituting an auto transformer for the inductor, as shown in Fig. 4-72C. MAXIM BATTERY MANAGEMENT CIRCUIT COLLECTION, 1994, P. 48.

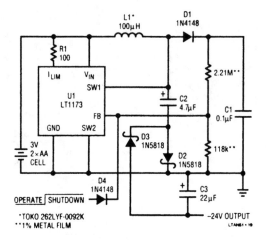

LCD-bias/contrast supply from two AA cells

Fig. 4-73   The circuit in the illustration provides a –24-V output for LCD-bias/contrast using two AA cells. The circuit delivers up to 7 mA from a 2-V input at 73% efficiency. LINEAR TECHNOLOGY, LINEAR APPLICATIONS HANDBOOK, 1993, P. AN51-16.

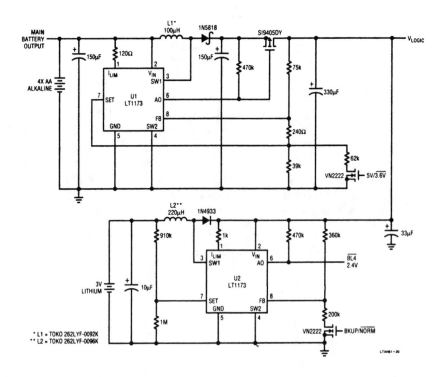

Battery-power and micropower circuits

Four-cell palmtop-computer supply

Fig. 4-74 The circuit in the illustration provides:

- A switchable 3.6 V/5 V output for main logic
- A –24-V output (Fig. 4-75) for LCD display bias/contrast
- A +12-V output (Fig. 4-76) for flash-memory $V_{\text{p-p}}$ generation
- An automatic backup supply

All these are from a four-cell AA input, with a 3-V lithium backup battery. Under no-load conditions, the quiescent current is 380 µA total. The circuit delivers 200 mA at 3.6 V from as little as 2.5-V input, and the four-cell battery can last more than 9.3 hours. The total load for the lithium battery is 5 µA. LINEAR TECHNOLOGY, LINEAR APPLICATIONS HANDBOOK, 1993, P. AN51-17.

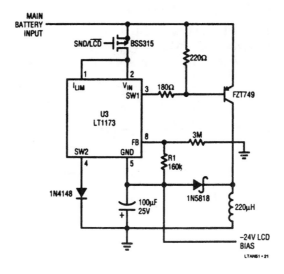

Four-cell LCD bias/contrast

Fig. 4-75 The circuit in the illustration can be added to the circuit of Fig. 4-74 to provide –24 V for LCD bias/contrast. LINEAR TECHNOLOGY, LINEAR APPLICATIONS HANDBOOK, 1993, P. AN51-18.

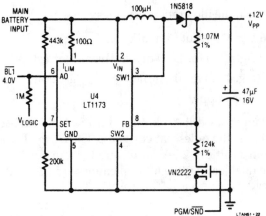

Four-cell flash-memory V$_{p\text{-}p}$ generator

Fig. 4-76 The circuit in the illustration can be added to the circuit of Fig. 4-74 to provide +12 V at 40 mA for flash-memory V$_{p\text{-}p}$ requirements. LINEAR TECHNOLOGY, LINEAR APPLICATIONS HANDBOOK, 1993, P. AN51-19.

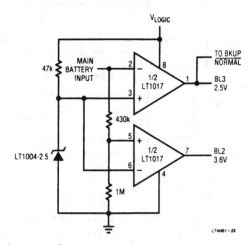

Micropower low-battery detector

Fig. 4-77 The circuit in the illustration can be added to the circuit of Fig. 4-74 to provide for low-battery detection. When the main battery voltage falls below 2.5, or the main battery is removed, BL3 goes high. If the BL3 output is connected to the BKUP/NORM input of U2 (Fig. 4-74), the lithium backup circuit takes over. Comparator BL2 goes low when the main battery voltage falls below 3.6 V and can be used to drive a low-battery indicator (if desired). LINEAR TECHNOLOGY, LINEAR APPLICATIONS HANDBOOK, 1993, P. AN51-19.

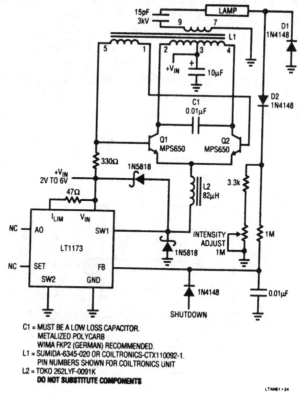

Micropower backlight supply for palmtop computers

Fig. 4-78 The circuit in the illustration provides power to operate a cold-cathode fluorescent tube (CCFL or CCFT, whichever you prefer). Such tubes have generally replaced all other forms of backlight for LCDs (at least for micropower). The LT1173 varies the lamp current as necessary to maintain 1.25 V at the FB pin. When lamp current is low, the LT1173 idles most of the time, drawing about 110-μA of quiescent current. At a 1-mA CCFL/CCFT current, the circuit draws less than 100 mA. LINEAR TECHNOLOGY, LINEAR APPLICATIONS HANDBOOK, 1993, P. AN51-20.

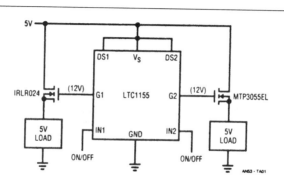

Battery-power/micropower circuit titles and description

Micropower dual high-side switch driver

Fig. 4-79 The circuit in the illustration generates 12 V from a 5-V supply to fully enhance (switch) logic-level N-channel MOSFET switches, with no external components required. Such switches are used in micropower equipment to control power (connect and disconnect loads, batteries, etc.). To be efficient, the switches should be controlled by drivers operated from logic signals and/or sensing elements. The supply current is typically 85 µA with the switch fully enhanced and 8 µA with the LTC1155 in standby. LINEAR TECHNOLOGY, LINEAR APPLICATIONS HANDBOOK, 1993, P. AN53-1.

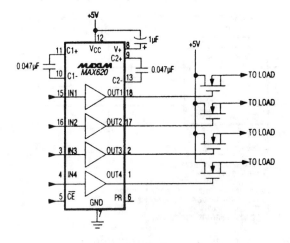

Micropower quad high-side switch driver

Fig. 4-80 To turn on each switch, the MOSFET is forced 11 V above the battery voltage by charge-pump action. (This family of charge-pump, high-side power supplies includes ICs with internal pump capacitors and internal N-channel MOS-FETs.) The input voltage range is 3.5 V to 16.5 V, with a quiescent current of 70 µA and a maximum load current (for each switch) of 5 A. MAXIM BATTERY MANAGEMENT CIRCUIT COLLECTION, 1993, P. 76.

=5=

Oscillator/generator circuits

The discussions in this chapter assume that you are familiar with oscillator basics (such as how discrete-component and IC oscillators operate) and basic oscillator/generator testing and troubleshooting. The following paragraphs summarize both testing and troubleshooting for oscillator/generator circuits. This information is included so readers not familiar with electronic procedures can both test the circuits described here and localize problems if the circuits fail to perform as shown.

For discussion purposes, a circuit is considered to be an oscillator if the primary function is to produce sine waves (such as Fig. 5-A). If the circuit output is a pulse, square wave, triangular wave, ramp, etc. (such as Fig. 5-B), the circuit is considered a generator.

A scope is the logical instrument for both testing and troubleshooting generators because the shape of the circuit output is often critical. However, it is convenient to monitor output amplitude with a meter, and a frequency counter is much easier to use when measuring output frequency (although you can measure both amplitude and frequency with a scope).

Oscillator/generator testing and troubleshooting

No matter how complex or simple the circuits of this chapter appear, they are essentially oscillators (or generators) and can be treated as such for practical testing and troubleshooting. For example, each circuit produces output signals, possibly crystal-controlled but often where frequency depends on RC (resistance-capacitance) LC inductance-capacitance. The output signals must have a given amplitude and must be at a given frequency (or must be capable of tuning across a given frequency range). For generators, the output also must be of a given shape (square, pulse, triangular, etc.). If you measure the signals and find them to be of the correct

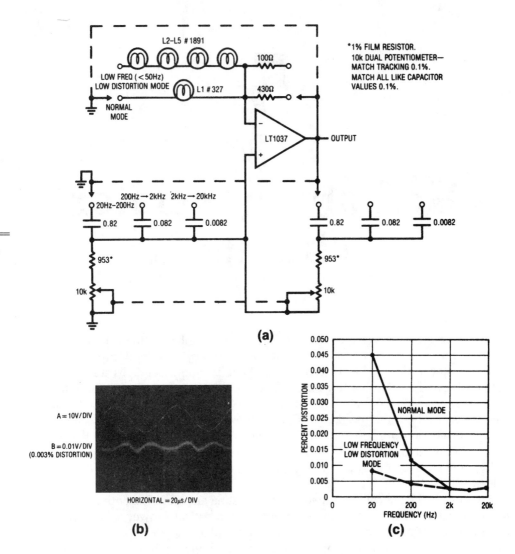

Fig. 5-A   Low-distortion sine-wave oscillator.

frequency, amplitude, and shape, the oscillators or generators are good from a troubleshooting standpoint. If not, the test results provide a good starting point for troubleshooting.

Basic oscillator/generator tests

The first step in troubleshooting oscillator/generator circuits is to measure the amplitude, frequency, and shape of the output signals. Most oscillator/generator circuits have a built-in test point. For example, the sine and cosine outputs of the Fig. 5-C circuits are available at the op-amp outputs. Likewise, triangular and square-wave outputs are available at V4 and V1, respectively, in Fig. 5-D.

Oscillator/generator circuits

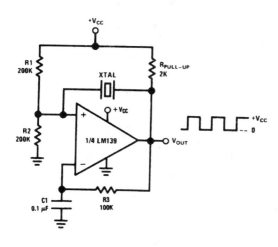

Fig. 5-B Crystal-controlled square-wave oscillator.

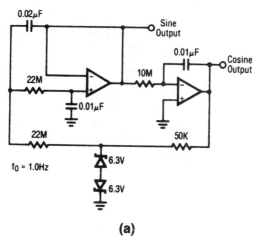

(a)

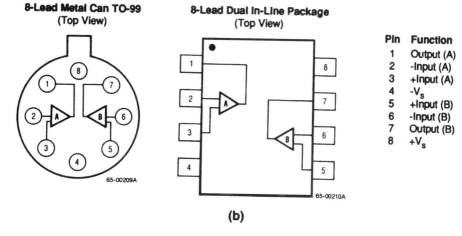

(b)

Fig. 5-C   Low-frequency sine-wave generator.

Oscillator/generator testing and troubleshooting

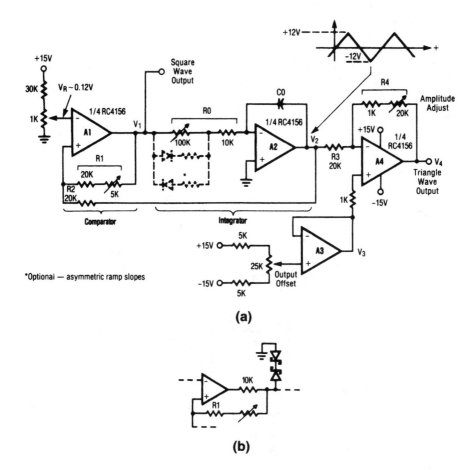

(a)

Optional — asymmetric ramp slopes

(b)

Fig. 5-D Triangle- and square-wave generator.

For an *RF* (radio-frequency) oscillator, the signal can be monitored at the collector or emitter (see Fig. 5-E(A). In this illustration, signal amplitude is monitored with a meter or scope using an RF probe. If you are interested only in the frequency, use a frequency counter.

Oscillator frequency problems

When you measure the oscillator signal, the frequency is (1) right on, (2) slightly off, or (3) way off. If the frequency is slightly off, you can sometimes possibly correct the problem with adjustment. Some oscillators are adjustable (even those with crystal control). The most precise adjustment is made by monitoring the oscillator signal with a frequency counter and adjusting the circuit for exact frequency. However, it also is possible to adjust a crystal oscillator with a meter or scope.

Generally, when a crystal oscillator is adjusted for maximum signal amplitude, the oscillator is at the crystal frequency. However, it is possible (but not likely) that the oscillator is being tuned for a harmonic (multiple or submulti-

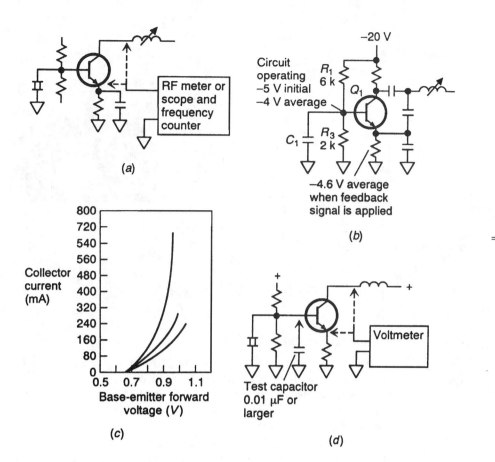

Fig. 5-E   Oscillator testing and troubleshooting.

ple) of the crystal frequency. A frequency counter shows this, but a meter or scope does not.

   If the oscillator frequency is way off, look for a defect rather than improper adjustment. For example, a coil or transformer might have shorted turns, a transistor or capacitor might be leaking badly, an IC or crystal might be defective, or you might have wired the circuit incorrectly (impossible?).

Measuring the frequency with a scope

If you do not have a frequency counter and must measure frequency with a scope, use the following procedure. Adjust the scope controls so you can measure the duration of one complete cycle (along the horizontal trace). For example, if one complete cycle occupies two horizontal divisions, and each division is 10 ns (as shown in Fig. 5-F), each cycle is 20 ns (or 20^{-9} s) in duration. Find the reciprocal of 20 ns, or

$$\frac{1}{20} = 0.05 \times 10^9 = 50^6 = 50 \text{ MHz}$$

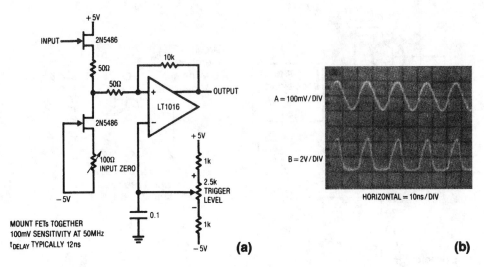

Fig. 5-F   50-MHz trigger.

Oscillator signal-amplitude problems

When you measure the oscillator signal, the amplitude is (1) right on, (2) slightly low, or (3) very low. If the amplitude is right on, leave the oscillator alone (unless the frequency is off). If the amplitude is slightly low, it is sometimes possible to correct the problem with adjustment. Monitor the signal with a meter or scope and adjust the oscillator for maximum signal (or for the desired signal level). For example, in Fig. 5-D, adjust R1 and the V_R voltage (at A1) for a triangle at V_2 with ±12-V amplitude. Then adjust R4 for the desired triangular-output level. With an adjustable crystal oscillator, adjusting for maximum amplitude also locks the oscillator at the correct frequency. In any case, if the adjustment does not correct the problem, look for a leakage in transistors and capacitors, or look for a possible defective IC.

If the amplitude is very low, look for defects such as low power-supply voltages, badly leaking transistors and/or capacitors, defective ICs, and a shorted coil or transformer turns (for RF oscillators). Usually, if the signal output is very low, there are other indications, such as abnormal voltages and resistance values.

Determining the output voltage amplitude

If you are wondering what output voltage to expect from an oscillator, check the power-supply or source voltage. The output voltage should be slightly less than the supply (in most cases). For example, in Fig. 5-D, the square-wave and triangular-wave outputs can be no greater than the ±15-V supply. (The amplitude is arbitrarily adjusted for ±12 V.) The same is usually true for the RF oscillator in Fig. 5-E(D). However, in the oscillator of Fig. 5-E(B), the output voltage will be less than 20 V because of the drop across the collector resistor of Q1.

Oscillator/generator circuits

Measuring amplitude with a scope

If you want to measure oscillator signal amplitude with a scope, use the following procedure. Adjust the scope controls so you can measure the amplitude of several cycles (along the vertical scale). For example, as shown in Fig. 5-F, there are two traces. Trace A is a sine wave taken at the input, with an amplitude of two vertical divisions. Each division is 100 mV, so the input is 200 mV. Trace B is a pulse or trigger response taken at the output with an amplitude of about two vertical divisions. Each division is 2 V, so the output is 4 V (adjustable by the 2.5-k pot).

Generator waveshape problems

If a generator produces a signal at the correct frequency and amplitude, but the waveshape is not correct, suspect the leakage. Usually, leaking capacitors and/or a leaking transistor (in discrete circuits) are the culprits. For example, if either or both the triangular-wave and square-wave outputs in Fig. 5-D are of the correct amplitude and frequency but are distorted (square-wave sides not straight, tops not flat, triangular-wave ramps bending in or out, etc.) suspect capacitor CO.

Capacitor quick checks

During the troubleshooting process, suspected capacitors can be removed from the circuit and tested on bridge-type checkers. This check establishes that the capacitor value is correct. With a correct value, it is responsible to assume that the capacitor is not open, shorted, or leaking. From the opposite standpoint, if the capacitor shows no shorts, opens, or leaks, it also is fair to say that the capacitor is good. So, from a practical troubleshooting standpoint, a simple test that checks for shorts, opens, or leaks is usually sufficient.

Checking capacitors with circuit voltages

As shown in Fig. 5-G, using circuit voltages involves disconnecting one lead of the capacitor (the ground or cold end) and connecting a voltmeter between the disconnected lead and ground. In a good capacitor, there should be a momentary voltage indication (or surge) as the capacitor charges up to the voltage at the hot end.

If the voltage indication remains high, the capacitor is probably shorted. If the voltage indication is steady, but not necessarily high, the capacitor is probably leaking. If there is no voltage indication whatsoever, the capacitor is probably open. Notice that this test is good only where one end of the capacitor is connected to a point in the circuit with a measurable voltage above ground (such as the capacitor in Fig. 5-F).

Checking capacitors with an ohmmeter

As shown in Fig. 5-G(B), using an ohmmeter involves disconnecting one lead of the capacitor (usually the hot end) and connecting an ohmmeter across the capacitor. Make certain that all power is removed from the circuit. As a precaution, short across the capacitor (after the power is removed) to make sure that no charge is re-

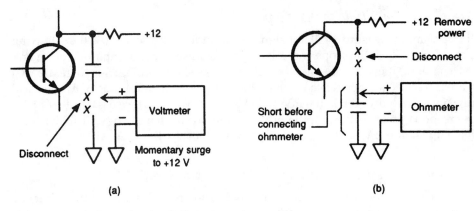

(a)   (b)

Fig. 5-G   In-circuit capacitor tests.

tained. In a good capacitor, there should be a momentary resistance indication (or surge) as the capacitor charges up to the voltage of the ohmmeter battery.

If the resistance indication is near zero and remains so, the capacitor is probably shorted. If the resistance indication is steady at some high value, the capacitor is probably leaking. If there is no resistance indication (or surge) whatsoever, the capacitor is probably open.

Oscillator/generator circuit titles and descriptions

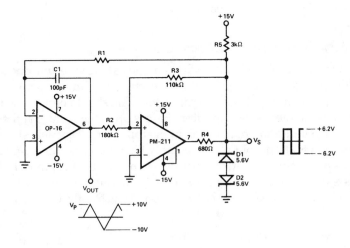

Triangular-wave generator

Fig. 5-1 The op-amp generator shown in the illustration produces triangular waveshapes of ±10 V for 100 Hz to 500 Hz with the values of R1 from 15 MΩ to 3 kΩ, as given by:

Oscillator/generator circuits

$$R1 = \frac{VS}{4\,VP \cdot f_0 \cdot C1}, \quad VS = 6.2\ \text{V}, \quad PV = 10\ \text{V}$$

The amplitude of the triangular wave can be adjusted via R2 as follows:

$$R2 = \frac{VP \cdot 110\ \text{k}}{6.2\ \text{V}}$$

Analog Devices, Applications Reference Manual, 1993, p. 13-52.

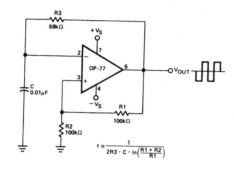

Op-amp free-running square-wave oscillator

Fig. 5-2 The simple oscillator shown in the illustration generates a square-wave output of $\pm(V_S - 2\ \text{V})$ at 1 kHz for the values shown. Analog Devices, Applications Reference Manual, 1993, p. 13-60.

Pinout

HA7210
TOP VIEW

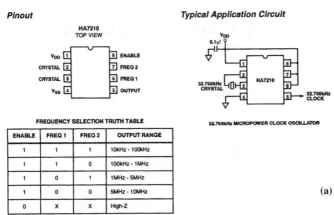

VDD	1		8	ENABLE
CRYSTAL	2		7	FREQ 2
CRYSTAL	3		6	FREQ 1
VSS	4		5	OUTPUT

Typical Application Circuit

32.768kHz MICROPOWER CLOCK OSCILLATOR

(a)

FREQUENCY SELECTION TRUTH TABLE

ENABLE	FREQ 1	FREQ 2	OUTPUT RANGE
1	1	1	10kHz - 100kHz
1	1	0	100kHz - 1MHz
1	0	1	1MHz - 5MHz
1	0	0	5MHz - 10MHz
0	X	X	High-Z

CAUTION: These devices are sensitive to electrostatic discharge. Users should follow proper I.C. Handling Procedures.

Simplified Block Diagram

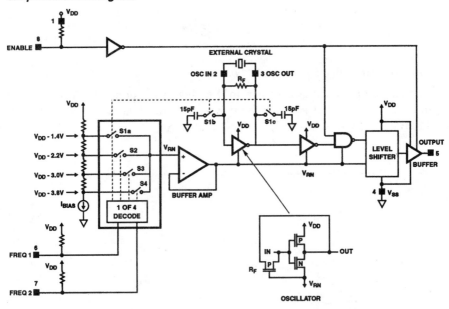

ENABLE	FREQ 1	FREQ2	SWITCH	OUTPUT RANGE
1	1	1	S1a, b, c	10kHz - 100kHz
1	1	0	S2	100kHz - 1MHz
1	0	1	S3	1MHz - 5MHz
1	0	0	S4	5MHz - 10MHz+
0	X	X	X	High Impedance

(b)

Oscillator/generator circuits

Continued

Absolute Maximum Ratings

Supply Voltage 10.0V
Voltage (any pin) . . . . V_{SS}-0.3V to V_{DD}+0.3V
Junction Temperature (Plastic Package) . . . . +150°C
ESD Rating (Note 2) . . . . >4000V
Lead Temperature (Soldering 10 Sec.) . . . . +300°C

Operating Conditions

Operating Temperature (Note 3) . . . . -40°C to +85°C
Storage Temperature Range . . . . -65°C to +150°C

CAUTION: Stresses above those listed in "Absolute Maximum Ratings" may cause permanent damage to the device. This is a stress only rating and operation of the device at these or any other conditions above those indicated in the operational sections of this specification is not implied.

Electrical Specifications V_{DD} = 5V, V_{SS} = GND, T_A = +25°C, Unless Otherwise Specified.

PARAMETER	HA7210I			UNITS
	MIN	TYP	MAX	
V_{DD} Supply Range (f_{OSC} = 32kHz)	2	5	7	V
I_{DD} Supply Current				
$\quad f_{OSC}$ = 32kHz, V_{DD} = 5V, EN = 0 Standby	-	5.0	9.0	µA
$\quad f_{OSC}$ = 32kHz, V_{DD} = 5V, C_L = 10pF (Note 1), EN = 1, Freq1 = 1, Freq2 = 1	-	5.2	10.2	µA
$\quad f_{OSC}$ = 32kHz, V_{DD} = 5V, C_L = 40pF, EN = 1, Freq1 = 1, Freq2 = 1	-	10	15	µA
$\quad f_{OSC}$ = 32kHz, V_{DD} = 3V, C_L = 10pF (Note 1), EN = 1, Freq1 = 1, Freq2 = 1	-	3.6	6.1	µA
$\quad f_{OSC}$ = 32kHz, V_{DD} = 3V, C_L = 40pF, EN = 1, Freq1 = 1, Freq2 = 1	-	6.5	9	µA
$\quad f_{OSC}$ = 1MHz, V_{DD} = 5V, C_L = 10pF (Note 1), EN = 1, Freq1 = 0, Freq2 = 1	-	130	200	µA
$\quad f_{OSC}$ = 1MHz, V_{DD} = 5V, C_L = 40pF, EN = 1, Freq1 = 0, Freq2 = 1	-	270	350	µA
$\quad f_{OSC}$ = 1MHz, V_{DD} = 3V, C_L = 10pF (Note 1), EN = 1, Freq1 = 0, Freq2 = 1	-	90	160	µA
$\quad f_{OSC}$ = 1MHz, V_{DD} = 3V, C_L = 40pF, EN = 1, Freq1 = 0, Freq2 = 1	-	140	250	µA
V_{OH} Output High Voltage (I_{OUT} = -1mA)	4.0	4.9	-	V
V_{OL} Output Low Voltage (I_{OUT} = 1mA)	-	0.07	0.4	V
I_{OH} Output High Current (V_{OUT} ≥ 4V)	-	-10	-5	mA
I_{OL} Output Low Current (V_{OUT} ≤ 0.4V)	5.0	10.0	-	mA
I_{IN} Enable, Freq1, Freq2 Input Current (V_{IN} = V_{SS} to V_{DD})	-	0.4	1.0	µA
V_{IH} Input High Voltage Enable, Freq1, Freq2	2.0	-	-	V
V_{IL} Input Low Voltage Enable, Freq1, Freq2	-	-	0.8	V
t_R Output Rise Time (10% - 90%, f_{OSC} = 32kHz, C_L = 40pF)	-	20	40	ns
t_R Output Rise Time (10% - 90%, f_{OSC} = 1MHz, C_L = 40pF)	-	12	-	ns
t_F Output Fall Time (10% - 90%, f_{OSC} = 32kHz, C_L = 40pF)	-	25	50	ns
t_F Output Fall Time (10% - 90%, f_{OSC} = 1MHz, C_L = 40pF)	-	12	-	ns
Duty Cycle (C_L = 40pF) f_{OSC} = 32kHz, Packaged Part Only (Note 4)	40	50	60	%
Frequency Stability vs. Supply Voltage (f_{OSC} = 32kHz, V_{DD} = 5V, C_L=10pF)	-	1	-	ppm/V
Frequency Stability vs. Temperature (f_{OSC} = 32kHz, V_{DD} = 5V, C_L=10pF)	-	0.1	-	ppm/°C
Frequency Stability vs. Load (f_{OSC} = 32kHz, V_{DD} = 5V, C_L=10pF)	-	0.01	-	ppm/pF

NOTES:
1. Calculated using the equation I_{DD} = I_{DD} (No Load) + (V_{DD}) (f_{OSC})(C_L).
2. Human body model.
3. This product is production tested at +25°C only.
4. Duty cycle will vary with supply voltage, oscillation frequency, and parasitic capacitance on the crystal pins.

(c)

Low-power crystal oscillator

Fig. 5-3 The IC shown is a low-power, crystal-controlled oscillator that can be externally programmed to operate between 10 kHz and 10 MHz. Figures 5-3B and 5-3C show the internal functions and characteristics, respectively. For normal operation, the IC requires only the addition of a crystal and possibly a bypass capacitor for the supply. The IC also features a disable mode that switches the output to a high impedance state. This mode is useful for minimizing power dissipation during standby and when multiple oscillator circuits are used. HARRIS SEMICONDUCTORS, LINEAR & TELECOM ICS, 1994, PP. 7-104, 105, 106.

(a)

Absolute Maximum Ratings

Supply Voltage (V- to V+) 36V
Power Dissipation (Note 1) 750mW
Input Voltage (Any Pin) .... V- to V+
Input Current (Pins 4 and 5) 25mA
Output Sink Current (Pins 3 and 9) 25mA
Lead Temperature (Soldering 10 Sec.) .... +300°C

Operating Conditions

Operating Temperature Range
ICL8038AM, ICL8038BM .... -55°C to +125°C
ICL8038AC, ICL8038BC, ICL8038CC .... 0°C to +70°C
Storage Temperature Range .... -65°C to +150°C

CAUTION: Stresses above those listed in "Absolute Maximum Ratings" may cause permanent damage to the device. This is a stress only rating and operation of the device at these or any other conditions above those indicated in the operational sections of this specification is not implied.

Electrical Specifications

$V_{SUPPLY} = \pm10V$ or +20V, $T_A = +25°C$, $R_L = 10k\Omega$, Test Circuit Unless Otherwise Specified

PARAMETERS	SYMBOL	TEST CONDITIONS	ICL8038CC MIN	TYP	MAX	ICL8038BC(BM) MIN	TYP	MAX	ICL8038AC(AM) MIN	TYP	MAX	UNITS
Supply Voltage Operating Range	V_{SUPPLY}											
Single Supply	V+		+10		+30	+10		30	+10		30	V
Dual Supplies	V+, V-		±5		±15	±5		±15	±5		±15	V
Supply Current	I_{SUPPLY}	$V_{SUPPLY} = \pm10V$ (Note 2)										
8038AM 8038BM							12	15		12	15	mA
8038AC, 8038BC, 8038CC				12	20		12	20		12	20	mA
FREQUENCY CHARACTERISTICS (ALL WAVEFORMS)												
Max. Frequency of Oscillation	f_{MAX}		100			100			100			kHz
Sweep Frequency of FM Input	f_{SWEEP}			10			10			10		kHz
Sweep FM Range (Note 3)				35:1			35:1			35:1		
FM Linearity		10:1 Ratio		0.5			0.2			0.2		%
Frequency Drift with Temperature (Note 5)	$\Delta f/\Delta T$											
8038 AC, BC, CC		0°C to +70°C		250			180			120		ppm/°C
8038 AM, BM		-55°C to +125°C						350			250	ppm/°C
Frequency Drift with Supply Voltage	$\Delta f/\Delta V$	Over Supply Voltage Range		0.05			0.05			0.05		%/V
OUTPUT CHARACTERISTICS												
Square Wave												
Leakage Current	I_{OLK}	$V_9 = 30V$			1			1			1	μA
Saturation Voltage	V_{SAT}	$i_{SINK} = 2mA$		0.2	0.5		0.2	0.4		0.2	0.4	V
Rise Time	t_R	$R_L = 4.7k\Omega$		180			180			180		ns
Fall Time	t_F	$R_L = 4.7k\Omega$		40			40			40		ns
Typical Duty Cycle Adjust (Note 6)	ΔD		2		98	2		98	2		98	%
Triangle/Sawtooth/Ramp												
Amplitude	$V_{TRIANGLE}$	$R_{TRI} = 100k\Omega$	0.30	0.33		0.30	0.33		0.30	0.33		xV_{SUPPLY}
Linearity				0.1			0.05			0.05		%
Output Impedance	Z_{OUT}	$I_{OUT} = 5mA$		200			200			200		Ω
Sine Wave												
Amplitude	V_{SINE}	$R_{SINE} = 100k\Omega$	0.2	0.22		0.2	0.22		0.2	0.22		xV_{SUPPLY}
THD	THD	$R_S = 1M\Omega$ (Note 4)		2.0	5		1.5	3		1.0	1.5	%
THD Adjusted	THD	Use Figure 14		1.5			1.0			0.8		%

NOTES:

1. Derate ceramic package at 12.5mW/°C for ambient temperatures above 100°C.
2. R_A and R_B currents not included.
3. $V_{SUPPLY} = 20V$; R_A and $R_B = 10k\Omega$, $f \cong 10kHz$ nominal; can be extended 1000 to 1. See Figures 15A and 15B.
4. 82kΩ connected between pins 11 and 12, Triangle Duty Cycle set at 50%. (Use R_A and R_B.)
5. Figure 1, pins 7 and 8 connected, $V_{SUPPLY} = \pm10V$. See Typical Curves for T.C. vs V_{SUPPLY}.
6. Not tested, typical value for design purposes only.

(b)

Oscillator/generator circuits

Continued

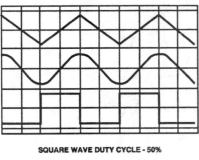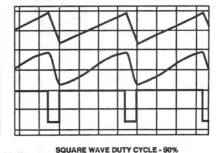

SQUARE WAVE DUTY CYCLE - 50%   SQUARE WAVE DUTY CYCLE - 80%

(c)

Waveform generator and VCO (voltage-controlled oscillator)

Fig. 5-4 The circuit in the illustration uses an ICL8038 to provide simultaneous sine, square, and triangular outputs, which can be sweep and frequency modulated. The frequency range is from 0.001 Hz to 300 kHz and is set by the values of R_A, R_B, and C, using the equation:

$$\text{frequency} = \frac{0.33}{RC} \text{ where } R = R_A = R_B$$

The characteristics are shown in Fig. 5-4B. Figure 5-4C shows the phase relationship of the waveforms. HARRIS SEMICONDUCTORS, LINEAR & TELECOM ICS, 1994, PP. 7-121, 123, 125.

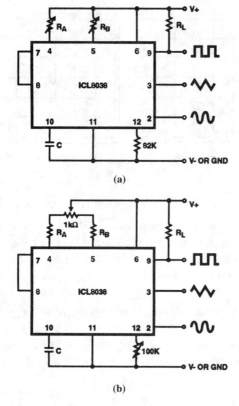

(a)

(b)

Waveform generator with variable duty cycle

Fig. 5-5 The circuit in the illustration is similar to that of Fig. 5-4, but it has alternate timing configurations and a duty-cycle control. For best results, use Fig. 5-5A where the timing resistors are separate. Resistor R_A controls the rising portion of the triangle and sine wave and l-state of the square wave. The duty cycle is 50% when $R_A = R_B$. If the duty cycle is to be varied over a small range about the 50%-point, the connections of Fig. 5-5B are more convenient. Adjust the 100-kΩ pot for minimum sine-wave distortion of about 1%. HARRIS SEMICONDUCTORS, LINEAR & TELECOM ICs, 1994, P. 7-126.

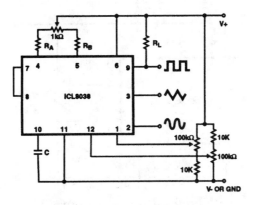

Waveform generator with minimum sine-wave distortion

Fig. 5-6   The circuit in the illustration is similar to that of Fig. 5-4, except that sine-wave distortion can be reduced to less than 0.5%. HARRIS SEMICONDUCTORS, LINEAR & TELECOM ICs, 1994, P. 7-126.

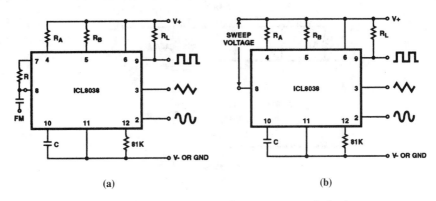

(a)   (b)

Waveform generator with sweep and frequency modulation

Fig. 5-7   These circuits are similar to that of Fig. 5-4 except that the circuits provide for *FM* (frequency modulation) or sweep modulation. Use circuit 5-7A for FM (small deviations, about ±10%). Use circuit 5-7B for a sweep range of 1000:1. The external resistor between pins 7 and 8 is not absolutely necessary, but it can be used to increase input impedance from about 8 kΩ (pins 7 and 8 connected together) to about (R + 8 kΩ). The capacitor at pin 8 provides decoupling to the FM source. As a guideline, the sweep frequency approaches 0 Hz when the voltage at pin 8 equals the supply voltage V+ and reaches maximum at the lower pin-8 voltage limit of $V+ -(\frac{1}{3} V_{SUPPLY} -2)$. Waveform symmetry variations can be minimized when a 10-MΩ resistor is added between pins 5 and 11. HARRIS SEMICONDUCTORS, LINEAR & TELECOM ICs, 1994, P. 7-127.

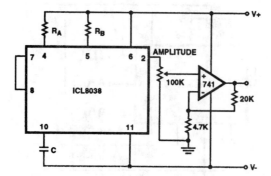

Waveform generator with buffered sine-wave output

Fig. 5-8 The circuit in the illustration is similar to that of Fig. 5-4, except that the sine-wave output is buffered. In addition to buffering the output, this circuit provides for amplification and adjustment of the sine wave. HARRIS SEMICONDUCTORS, LINEAR & TELECOM ICs, 1994, P. 7-127.

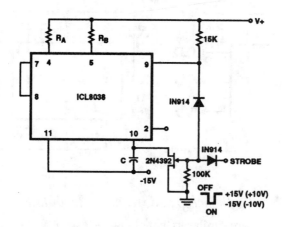

Strobed tone-burst generator

Fig. 5-9 The circuit in the illustration is similar to that of Fig. 5-4, except that the output can be *strobed* (switched on and off) with an external pulse. When the strobe is applied, the timing capacitor C at pins 10/11 is shorted to ground, halting oscillation. With this arrangement, the output always starts on the same slope. HARRIS SEMICONDUCTORS, LINEAR & TELECOM ICs, 1994, P. 7-127.

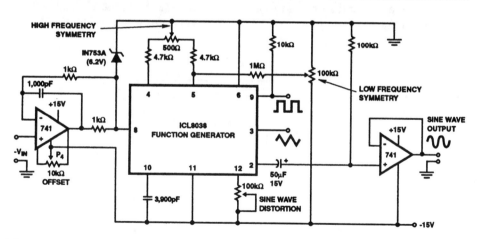

Variable audio oscillator (20 Hz to 20 kHz)

Fig. 5-10 The circuit in the illustration is similar to that of Fig. 5-4, except that the duty cycle, frequency, distortion, and duty-cycle variations are all adjustable. (The 15-MΩ resistor is adjusted to reduce duty-cycle variations with sweep.) The diode lowers the supply voltage so a 1000:1 sweep range is possible. HARRIS SEMICON-DUCTORS, LINEAR & TELECOM ICs, 1994, P. 7-127.

Linear voltage-controlled oscillator

Fig. 5-11 The circuit in the illustration uses the ICL8038 (Fig. 5-4) and two sections of a 741 op amp (or similar device) to provide sine-wave output controlled by an input voltage. Both the high-frequency and low-frequency symmetry are adjustable, as is sine-wave distortion. HARRIS SEMICONDUCTORS, LINEAR & TELECOM ICs, 1994, P. 7-128.

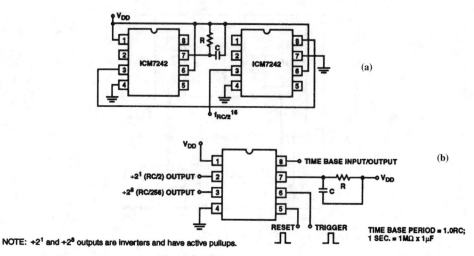

(a)

(b)

NOTE: ÷2¹ and ÷2⁸ outputs are inverters and have active pullups.

Low-frequency reference oscillator (timer)

Fig. 5-12 The circuit in the illustration uses two timers connected in cascade to produce a low-frequency reference signal. Figure 5-12B shows the basic timer connections and equations for operating frequency. Keep in mind that the frequency is the reciprocal of the time-base period. The minimum value for timing capacitor C is 10 pF. The range for the timing resistor R is from 1 kΩ to 22 MΩ. V_{DD} can be from 2 V to 16 V. HARRIS SEMICONDUCTORS, LINEAR & TELECOM ICs, 1994, PP. 7-141, 144.

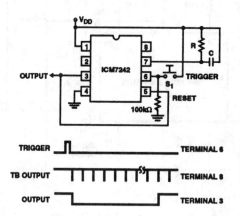

Monostable oscillator (timer)

Fig. 5-13 The circuit in the illustration shows the monostable (one-shot) connections for the ICM7242 timer. The equations for operating frequency and period shown in Fig. 5-12B apply. HARRIS SEMICONDUCTORS, LINEAR & TELECOM ICs, 1994, P. 7-144.

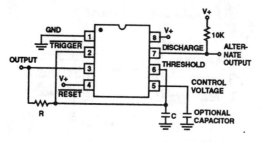

Astable oscillator (timer)

Fig. 5-14 The circuit in the illustration shows the ICM7555/7556 general-purpose timer connected for astable (or free-running) operation as an oscillator. The output swings from rail to rail and is a true 50% duty-cycle square wave. The supply can range from 5 V to 15 V. The oscillator frequency is set by the equation:

$$F = \frac{1}{(1.4\,RC)}$$

HARRIS SEMICONDUCTORS, LINEAR & TELECOM ICS, 1994, P. 7-152.

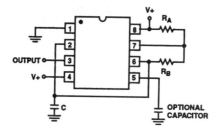

Alternate astable oscillator (timer)

Fig. 5-15 The circuit in the illustration is an alternate to the circuit of Fig. 5-14. Again, the supply can range from 5 V to 15 V, with a rail-to-rail output. However, the duty cycle is set by the ratio of R_A and R_B using the equation:

$$\text{duty cycle } D = \frac{(R_A + R_B)}{(R_A + 2R_B)}$$

The frequency is set by the equation:

$$F = \frac{1.44}{(R_A + 2R_B)}\,C$$

HARRIS SEMICONDUCTORS, LINEAR & TELECOM ICS, 1994, P. 7-152.

Oscillator/generator circuit titles and descriptions

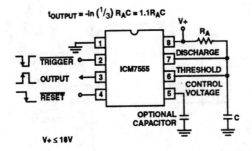

Alternate monostable oscillator (timer)

Fig. 5-16   The circuit in the illustration shows the ICM7555/7556 general-purpose timer connected for monostable (one-shot) operation. The output timing is set by the equation:

$$t_{OUTPUT} = -1n(\tfrac{1}{3})\, R_A C = 1.1\, R_A C$$

HARRIS SEMICONDUCTORS, LINEAR & TELECOM ICS, 1994, P. 7-152.

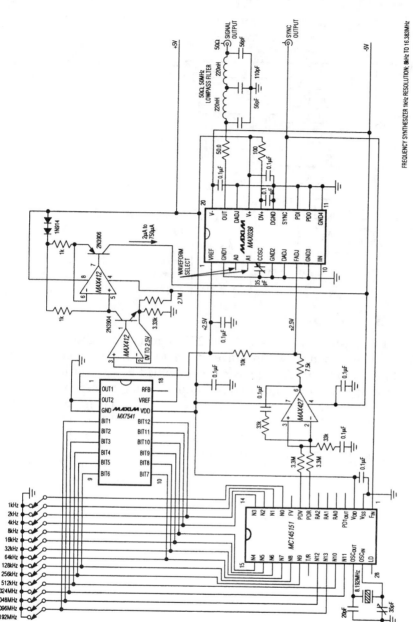

Crystal-controlled digitally programmed frequency synthesizer

Fig. 5-17 The circuit in the illustration shows a frequency synthesizer (based on the MAX038 waveform generator) that produces accurate and stable sine, square, or triangular waves with frequency range of 8 kHz to 16.383 MHz in 1-kHz increments. The manual switches set the output frequency (opening any switch increases the frequency). MAXIM NEW RELEASES DATA BOOK, 1995, P. 10-16.

Oscillator/generator circuit titles and descriptions

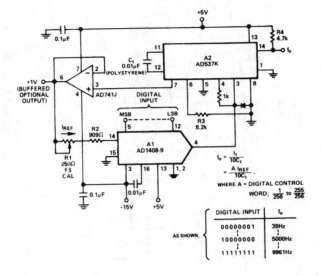

Eight-bit programmable oscillator

Fig. 5-18 The circuit in the illustration shows an 8-bit (255-frequency) programmable oscillator with a TTL-compatible square-wave output. The digital input produces a corresponding current from A1. The current is applied to A2, which produces a square-wave output with a frequency proportional to the numerical value of the digital input word. Using the values shown, the circuit has a nominal full-scale frequency of 10 kHz (9961 Hz for all 1s). To calibrate, apply all 1s to A1 and adjust R1 for output frequency of 9961 Hz. Worst-case nonlinearity is 0.16%. ANALOG DEVICES, APPLICATIONS REFERENCE MANUAL, 1993, P. 8-72.

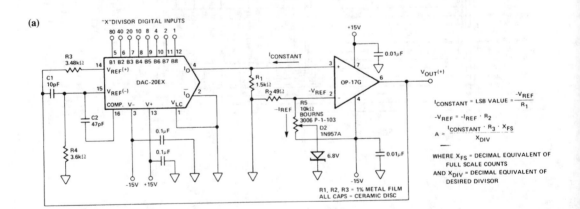

(b)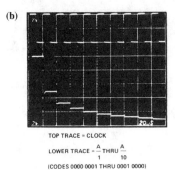

(c)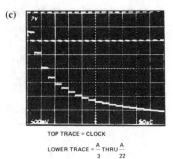

(d)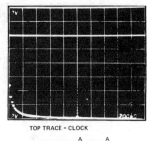

TOP TRACE = CLOCK

LOWER TRACE = $\frac{A}{1}$ THRU $\frac{A}{10}$

(CODES 0000 0001 THRU 0001 0000)

TOP TRACE = CLOCK

LOWER TRACE = $\frac{A}{3}$ THRU $\frac{A}{22}$

(CODES 0000 0011 THRU 0010 0010)

TOP TRACE = CLOCK

LOWER TRACE = $\frac{A}{1}$ THRU $\frac{A}{99}$

(CODES 0000 0001 THRU 1001 1001)

A/X function generator

Fig. 5-19 The circuit in the illustration generates hyperbolic functions of the type A/X, where A indicates an analog constant and X represents a decimally expressed digital advisor. A constant current ($I_{CONSTANT}$) equal to the value of one *LSB* (least significant bit) flows into the DAC output terminal i_O. Simultaneous adjustment of the scale factor and output-amplifier offset voltage is enabled by a multiturn, low-tempo pot R5, which adjusts current $-I_R$, producing voltage $-V_R$ across R2. The LSB value (scale factor) equals $-V_R/R_1$. ANALOG DEVICES, APPLICATIONS REFERENCE MANUAL, 1993, P. 8-77.

(a)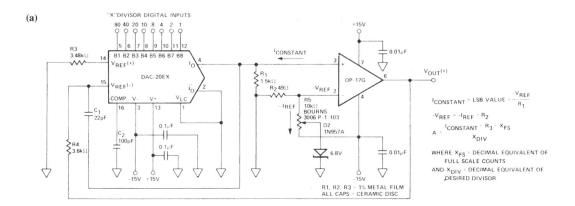

Continued

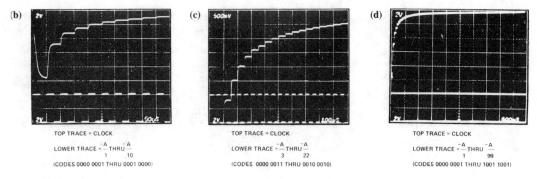

(b) 2v ... 2V ... 50µS

TOP TRACE = CLOCK

LOWER TRACE = $\frac{-A}{1}$ THRU $\frac{-A}{10}$

(CODES 0000 0001 THRU 0001 0000)

(c) 500mV ... 2V ... 100µS

TOP TRACE = CLOCK

LOWER TRACE = $\frac{-A}{3}$ THRU $\frac{-A}{22}$

(CODES 0000 0011 THRU 0010 0010)

(d) 2v ... 2V ... 500µS

TOP TRACE = CLOCK

LOWER TRACE = $\frac{-A}{1}$ THRU $\frac{-A}{99}$

(CODES 0000 0001 THRU 1001 1001)

A/X function generator

Fig. 5-20 The circuit in the illustration is similar to that of Fig. 5-19 except that the output is the $-A/X$ function. The output is accomplished by reversing both the DAC reference amplifier and output-amplifier terminals. Capacitors C1 and C2 provide a phase compensation. ANALOG DEVICES, APPLICATIONS REFERENCE MANUAL, 1993, P. 8-78.

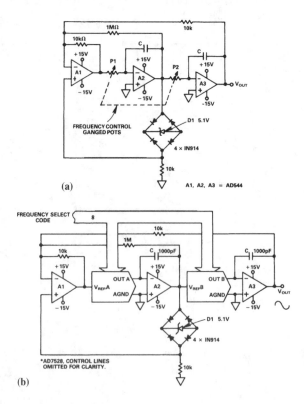

(a)

(b)

Continued

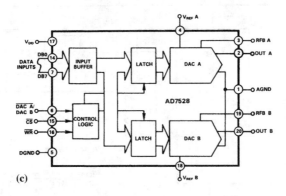

(c)

Programmable sine-wave oscillator

Fig. 5-21 The circuits in the illustration produce a sine-wave oscillator with linear control using a state-variable filter. Figure 5-21A shows the circuit configuration where the frequency is set by ganged potentiometers P1 and P2. Figure 5-21B shows the same circuit with P1 and P2 replaced by a matched pair of AD7528s. Figure 5-21C shows the internal functions and pin configuration for the AD7528. Frequency is expressed as:

$$\text{Frequency in Hz} = \frac{N}{256\,(6.28\,RC)}$$

where R = DAC ladder resistance (V_{REF} input resistance), C is as shown in Fig. 5-21B, N = decimal representation of digital input code. For example, N = 128 for input code 10000000. For the values given in Fig. 5-21B, output frequency is variable from 0 to 15 kHz. Output amplitude is set by D1. THD is –53 dB at 1 kHz and –43 dB at 14 kHz. A cosine output also is available at the output of A2. ANALOG DEVICES, APPLICATIONS REFERENCE MANUAL, 1993, PP. 8-113, 115.

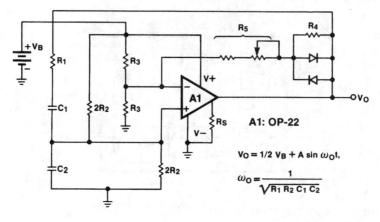

Battery-operated Wien-bridge oscillator

Fig. 5-22 The circuit in the illustration produces a 6-V_{p-p} output when operated from a 9-V battery. The output frequency is 1004 Hz with the following component values: $C_1 = C_2 = 0.01$ μF, $R_1 = 15.8$ kΩ, $2R_2 = 31.8$, $R_3 = 50$ kΩ, $R_4 = 10$ kΩ, $R_5 = 40$ kΩ (nominal), diodes = 1N914 or 1N4148, $R_S = 1$ MΩ. Resistor R5 is adjusted for best amplitude stability. Supply current is about 100 μA. Use an OP-27 instead of the OP-22 if frequencies greater than 1 kHz are required. ANALOG DEVICES, APPLICATIONS REFERENCE MANUAL, 1993, P. 13-81.

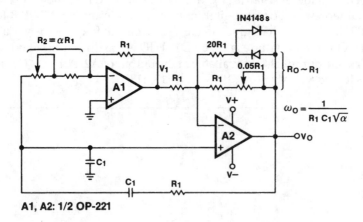

A1, A2: 1/2 OP-221

Wien-bridge oscillator with tuning control

Fig. 5-23 The circuit in the illustration adds tuning capability to the classic Wien-bridge oscillator. Use 10 kΩ for the first trial value of R_1. ANALOG DEVICES, APPLICATIONS REFERENCE MANUAL, 1993, P. 13-83.

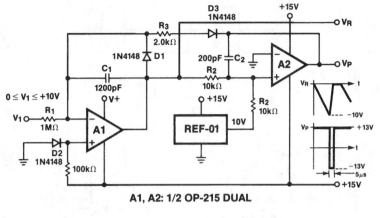

A1, A2: 1/2 OP-215 DUAL

Precision ramp generator

Fig. 5-24 The circuit in the illustration produces both a ramp (V_R) and pulse (V_P) over a range of about 10 Hz to 1 kHz. Repetition rate is controlled by voltage V_1. The reset time interval is about 5 μs. ANALOG DEVICES, APPLICATIONS REFERENCE MANUAL, 1993, P. 13-85.

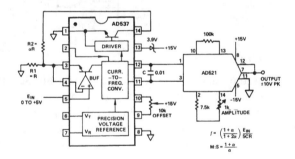

Asymmetric triwave generator

Fig. 5-25 The circuit in the illustration produces a triwave output where the mark-space ratio can be set by adjustment of the factor alpha. For one half-cycle (pin 14 open circuit), the only path for timing current is through R1. When pin 14 is a short circuit, both R1 and R2 are in the circuit, and the half-cycle is thus shortened. The mark-space ratio (M:S) is maintained over the frequency-control range. As shown by the equation, frequency (f) is controlled by C and R1, as well as the voltage E_{IN} at pin 5 of the AD537 . If a symmetrical waveform is needed, R2 is omitted, releasing pin 14 of the AD437 for use as a square-wave output. The mark-space relationship can be inverted by reversing pins 1 and 3 on the AD521. Note that the +15-V supply to pin 13 of the AD537 is reduced by the zener to remain within the common-mode range of the AD521 inputs. This feature also reduces the input signal range. ANALOG DEVICES, APPLICATIONS REFERENCE MANUAL, 1993, P. 23-26.

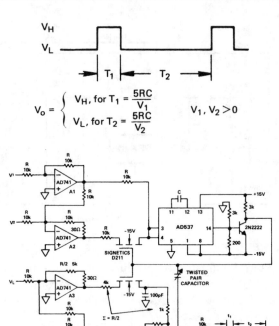

Voltage-programmable pulse generator

Fig. 5-26 The circuit in the illustration produces a pulse-train output with continuously variable output frequency, duty cycle, and output levels. As shown by the waveforms and equations, the output is controlled by voltages V_H, V_L, V_1, and V_2, as well as the values of R and C. The value of timing resistance R is set at 10 kΩ, and the C value is chosen for the maximum square-wave running frequency. The minimum square-wave frequency is set by circuit offsets. Frequency speeds in excess of 1000 to 1 (corresponding to a 10 mV to 10 V range for V_1 and V_2) are possible. Additional range is possible by trimming offsets. Matching of the resistors around A1 and A2 will influence duty-cycle accuracy. If 0.01% resistors are used, a duty-cycle sweep of 1000 to 1 with less than 10% error is possible. The 30-Ω resistor shown in the feedback of A2 and A3 compensate for the FET on-resistance. Output amplifier A5 is compensated slightly with the 100 pF and 1 kΩ at the inputs. This compensation results in a rise time of about 300 ns, with minimal overshoot and noise on the transitions. ANALOG DEVICES, APPLICATIONS REFERENCE MANUAL, 1993, P. 23-26.

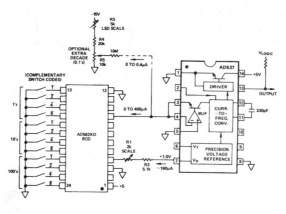

Digitally programmable clock source

Fig. 5-27 The circuit in the illustration combines a DAC (Chapter 9) with a V/F (Chapter 7) to provide digital control of frequency. Using the values shown, the circuit provides three-digit control of frequency from 100 Hz to 99.9 kHz. Either an external binary word or manually set thumb switches can be used. If a straight-binary AD562 is used instead of *BCD* (binary-coded decimal) as shown, the 330-pF capacitor should be changed to 510 pF. To calibrate, set 999 into the thumb switches and adjust R1 for an output frequency of 99.9 kHz. A fourth decade can be added in the LSD position as shown by the dotted lines. ANALOG DEVICES, APPLICATIONS REFERENCE MANUAL, 1993, P. 23-26.

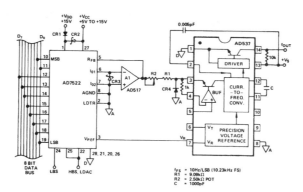

Microprocessor-programmable frequency source

Fig. 5-28 The circuit in the illustration is similar to that of Fig. 5-27, but it features full microprocessor control of frequency. The 10-bit AD7522 is directly microprocessor compatible and can be treated as either two output ports or two memory locations. The low-order eight bits are loaded into the AD7522 holding register on the rising edge of LSB, and the two most-significant bits are loaded on the rising edge of the HBS. The HBS signal also is connected onto the LDAC, which strobes the complete 10-bit word onto the actual DAC control lines simultaneously with loading of the two most significant bits. The reference for the AD7522 is provided by the 1.00-V output of the AD537. The 10-kΩ impedance of the DAC reference input lowers the full-scale frequency of the AD537 because of loading on the 1.00-V output. Resistor R2 provides sufficient trim range to compensate for this effect. The DAC current output is converted to a voltage by the AD517. This voltage has a range of 0 to 1 V full-scale and is applied to the AD537 (connected for a negative input voltage). The output frequency is scaled (by R2) for 10 Hz/LSB. ANALOG DEVICES, APPLICATIONS REFERENCE MANUAL, 1993, P. 23-29.

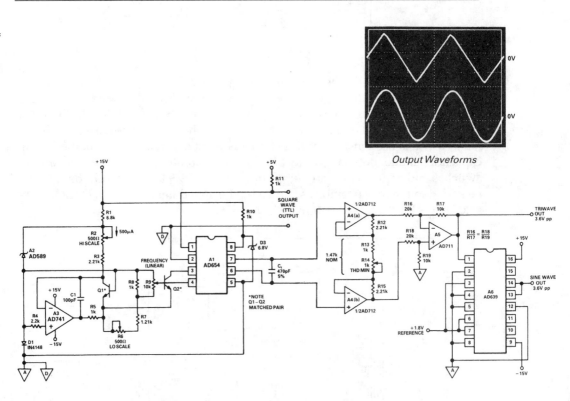

Output Waveforms

Function generator (V/F based)

Fig. 5-29 The circuit in the illustration shows a function generator, based on an AD654 V/F (Chapter 7), with a wide-range, exponentially generated timing current for frequency control. This allows a standard linear pot to be used for a four-decade control range without crowding the low end of the control span. The control-current generator comprises A2, A3, Q1, and Q2. This circuit produces a current from the collector of Q2, which ranges downward under control of FREQUENCY pot R9. R2 calibrates the high end of the scale for a 100-kHz output. Resistor R6 sets the lower end to 10 Hz. Q1 and Q2 are a matched pair (2N5089), etc.).

Oscillator/generator circuits

Resistor R14 serves generally as an amplitude control but is best adjusted for minimum sine-wave distortion (THD). ANALOG DEVICES, APPLICATIONS REFERENCE MANUAL, 1993, P. 23-50.

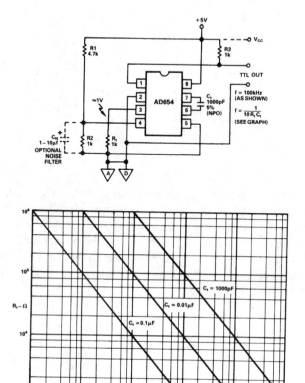

Simple fixed-frequency clock source

Fig. 5-30 The circuit in the illustration is most useful where a single spot frequency is desired and a fixed RC combination can be selected. The nomograph (based on a 1-V input at pin 3) shows various combinations of R and C for the recommended frequency range and can be used with other RC circuits. ANALOG DEVICES, APPLICATIONS REFERENCE MANUAL, 1993, P. 23-51.

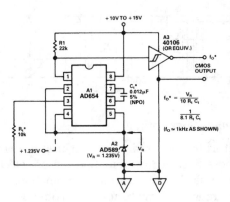

10/1 manually tuned clock generator

Fig. 5-31   The circuit in the illustration is similar to that of Fig. 5-30, but the simple 1-V supply divider is replaced by an LED reference circuit that improves supply stability and adds coarse temperature compensation for the diode drift. Note that R4 and R5 must be combined ($R_4 + R_5 = R_t$) to use the nomograph of Fig. 5-30. The table in Fig. 5-31 shows recommended capacitor values for various full-scale frequencies. The circuit is tuned by linear pot R2 and is calibrated for full scale by R5. ANALOG DEVICES, APPLICATIONS REFERENCE MANUAL, 1993, P. 23-51.

Self-biasing precision clock

Fig. 5-32   The circuit in the illustration improves on both the circuits shown in Figs. 5-30 and 5-31; also, an output buffer is added. The nomograph of Fig. 5-30 can be used to determine frequency. However, the capacitor curves must be multiplied by a factor of 1.2 (1200 pF, 12,000 pF, etc.) when selecting C and R_t for a given frequency. Also, if an exact frequency is required, R_t must be trimmed. ANALOG DEVICES, APPLICATIONS REFERENCE MANUAL, 1993, P. 23-52.

Oscillator/generator circuits

12-bit D/F converter

Fig. 5-33 The circuit in the illustration is a digitally programmed frequency source with a linearity of about 0.1%. Using the values shown, the output frequency is adjusted for 99,976 Hz by R4 when the digital input is 1111 1111 1111. ANALOG DEVICES, APPLICATIONS REFERENCE MANUAL, 1993, P. 23-52.

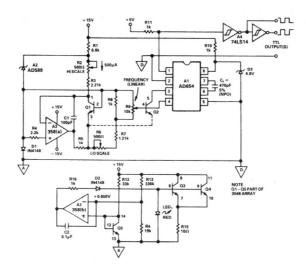

Log-controlled oscillator

Fig. 5-34 The circuit in the illustration is tunable over four decades from 10 Hz to 100 kHz, but it does not require a log-taper pot. Instead, the circuit is tuned by linear pot R9. R2 and R6 provide high and low full-scale calibration, respectively. ANALOG DEVICES, APPLICATIONS REFERENCE MANUAL, 1993, P. 23-54.

Oscillator/generator circuit titles and descriptions

=6=

Digital/
microprocessor
circuits

The discussions in this chapter assume that you are already familiar with digital and microprocessor basics (number systems, logic elements, gates, microprocessors, digital test equipment, etc.). If not, read *Lenk's Digital Handbook*, McGraw-Hill, 1992. The following paragraphs summarize both testing and troubleshooting for digital circuits. This information is included so those readers who are not familiar with electronic procedures can both test the circuits in this chapter and localize problems if the circuits fail to perform.

Digital/microprocessor circuit testing and troubleshooting

Both testing and troubleshooting for the circuits of this chapter can be performed with conventional test equipment (meters, generators, scopes, etc.), covered in other chapters. However, a logic or digital probe and a digital pulser can make life much easier if you must regularly test and troubleshoot digital devices. So, this section begins with a brief description of the probe and pulser. The description is followed by testing and troubleshooting for the various types of circuits in this chapter.

Logic or digital probe

Logic probes are used to monitor in-circuit pulse or logic activity. By means of a simple lamp indicator, a logic probe shows you the logic state of the digital signal and allows brief pulses to be detected (that you might miss with a scope). Logic probes detect and indicate high and low (1 or 0) logic levels, as well as intermediate or "bad" logic levels (indicating an open circuit) at the terminals of a logic element (the inputs and outputs of a gate, a digital-to-analog converter, or a microprocessor, for example).

Not all logic probes have the same functions, and you must learn the operating characteristics of your probe. For example, on the more sophisticated probes, the indicator lamp can give any of four indications: off, dim (about half brilliance), bright (full brilliance), or flashing on and off.

The lamp is normally in the dim state and must be driven to one of the other three states by voltage levels at the probe tip. The lamp is bright for inputs at or above the 1 state and off for inputs at or below 0. The lamp is dim for voltages between the 1 and 0 states and for open circuits. Pulsating inputs cause the lamp to flash at about 10 Hz (regardless of the input pulse rate.) The probe is particularly effective when it is used with the logic pulser.

Logic pulser

The hand-held logic pulser (similar in appearance to the logic probe) is an in-circuit stimulus device that automatically outputs pulses of the required logic polarity, amplitude, current, and width to drive lines and other test points high and low. A typical logic also has several pulse-burst and stream modes available.

Logic pulsers are compatible with most digital devices. Pulse amplitude depends on the equipment supply voltage, which also is the supply voltage for the pulser. Pulse current and width depend on the load being pulsed. The frequency and number of pulses that are generated by the pulser are controlled by operation of a switch. A flashing LED indicator on the pulser tip indicates the output mode.

The logic pulser forces overriding pulses onto lines or test points and can be programmed to output single pulses, pulse streams, or bursts. The pulser can be used to force ICs to be enabled or clocked. Also, you can pulse the circuit inputs while observing the effects on the circuit outputs with a logic probe.

Interface and translator circuits

A circuit such as shown in Fig. 6-A can be tested by applying pulses at the input and monitoring the output. This test can be done with a generator or pulser at the input and a scope or probe at the output. For example, pulses with ECL levels can be applied at the ECL-gate input and TTL pulses can be monitored at the output. If TTL pulses are absent at pin 7 of the 4805 comparator, check for pulses at pins 2 and 3. This procedure will isolate the problem to the gate or the comparator.

Bus and line-driver circuits

Digital bus and line drivers such as shown in Fig. 6-B can be tested by applying pulses at the input and monitoring the output. This test can be done with a generator or pulser at the input and a scope or probe at the output. For example, pulses can be applied at pins 4 and 13 of the RM3182 differential driver, and pulses can be monitored at pins 6 and 11. If the output pulses are absent, suspect the RM3182.

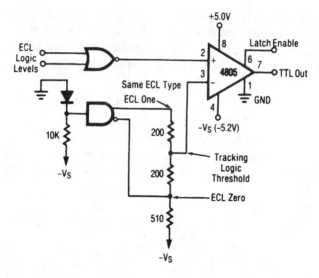

Fig. 6-A ECL-to-TTL translator (tracking).

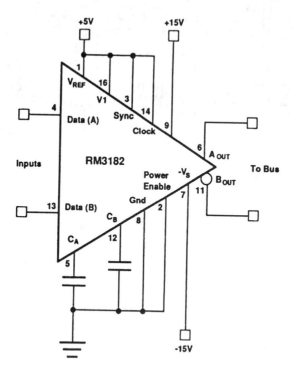

Fig. 6-B Differential line driver.

Digital/microprocessor circuit testing and troubleshooting

Before pulling the IC, make certain that the power sources are applied at all of the pins indicated. Also note that the frequency limitation of the IC is set by the values of the capacitors at pins 5 and 12 (a high frequency requires a larger capacitance). Keep this in mind if a digital circuit appears to perform properly at frequencies below that of the normal system clock.

Digital displays

LED or LCD display circuits such as shown in Fig. 6-C can be tested by applying a specific voltage at the input and noting the display readout. For example, in the circuit of Fig. 6-C, the LCD display should follow voltage applied at pins 10 and 11 of the CA3126E.

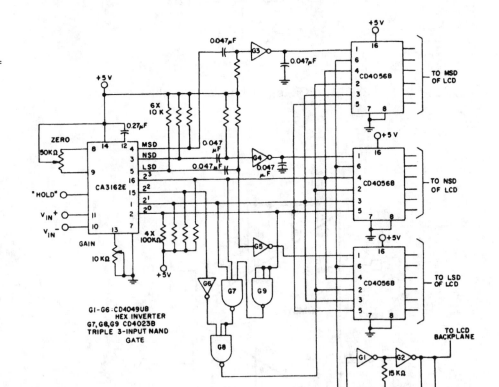

Fig. 6-C   A/D converter for three-digit LCD display.

If the display is slightly off, try correcting the problem with adjustment. Adjust the 50-k pot at pins 8 and 9 of the CA3126E for 000 on the display, with pins 10 and 11 shorted. Then, adjust the 10-k pot at pins 7 and 13 for 900 on the display, with 900 mV applied at pins 10 and 11.

Digital/microprocessor circuits

If all of the display digits are blank, suspect the LCD backplane oscillator (G1 and G2). Notice that an LED display does not require a backplane drive signal.

If only one of the display digits is bad, check power (pin 16) and ground (pins 7 and 8) of the corresponding CD4056B decoder/driver. Also check the corresponding signal from the CA3163E to pin 1 of the CD4056B. For example, the MSD signal from pin 4 of the CA3182E is applied to pin 1 of the MSD CD4056B through a 0.047-µF capacitor and inverter G3.

If certain segments of the display are absent on all three digits, check the corresponding inputs at pins 2 through 6 of the CD4056B from pins 1, 2, 15, and 16 of the CA3163E. Notice that the inputs to pin 2 of the CD4056Bs are made through inverter G6 and gates G7, G8, and G9.

Optoisolators

Circuits involving optoisolators (Figs. 6-D and 6-E) can be tested by applying a voltage or signal to the optoisolator and checking the corresponding response. For example, in the circuit of Fig. 6-D, the 2N6071B triac is triggered by a signal from pin 4 of the MOC3011 when a pulse is applied to the buffer input. With the triac triggered on, 115-V power is applied to the load.

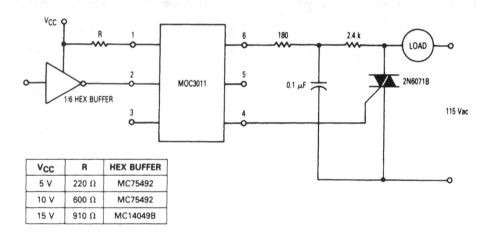

Fig. 6-D   MOS-to-ac load interface.

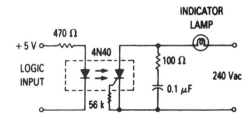

Fig. 6-E   25-W logic-indicator lamp driver.

Digital/microprocessor circuit testing and troubleshooting

If this sequence does not occur, first check that pulses are arriving at pin 2 of the MOC3011. If not, suspect the buffer. Then, check that there is a trigger at pin 4 of MOC3011 when pulses are applied to pin 2. If not, suspect the MOC3011. If you get a trigger at pin 4 of the MOC3011, but power is not applied to the load, suspect the triac.

For an optocoupler that is used as a solid-state relay (Fig. 6-E), apply 5 V to the 4N40 and check that the indicator lamp turns on. If not, suspect the 4N40 (or the lamp). Of course, it is possible that one of the three resistors is open (terminal broken), or that the capacitor is shorted (or leaking badly), but this is not likely.

Single-chip digital devices

Single-chip devices, such as shown in Figs. 6-F and 6-G, can best be tested by checking their function in the circuit. For example, the digital power monitor DS1231 (Fig. 6-F) should produce an RST or $\overline{\text{RST}}$ (reset) and $\overline{\text{NMI}}$ (nonmaskable interrupt) to a microprocessor when voltage is removed from the input. To test this function, apply 10 V to the voltage sense point of the Fig. 6-F circuit and monitor the RST, $\overline{\text{RST}}$, and $\overline{\text{NMI}}$ pins of the DS1231 (make certain that +5 V is applied to V_{CC} and MODE pins). Now remove the voltage from the voltage sense point and check that RST, $\overline{\text{RST}}$, and $\overline{\text{NMI}}$ signals appear at the corresponding pins of the DS1231.

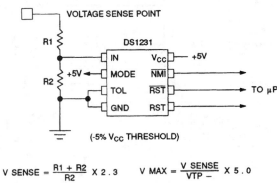

$$V \text{ SENSE} = \frac{R1 + R2}{R2} \times 2.3 \qquad V \text{ MAX} = \frac{V \text{ SENSE}}{VTP -} \times 5.0$$

EXAMPLE: V SENSE = 8 VOLTS AT TRIP POINT AND A
MAXIMUM VOLTAGE OF 17.5V WITH R2 = 10K

$$\text{THEN } 8 = \frac{R1 + 10K}{10K} \times 2.3 \qquad R1 = 25K$$

Fig. 6-F   Digital power monitor.

The full-duplex circuit of Fig. 6-G can be tested by applying pulses at both inputs and monitoring for pulses at both outputs. For example, apply pulses to the DIN pin of the DS1275 and check that pulses appear on pin 3 (RXD) of the PC serial port. There should be no substantial difference between the input and output pulses (same amplitude, shape, etc.). Then apply pulses to pin 2 (TXD) of the PC serial port and check for pulses at the DOUT pin of the other DS1275. If either re-

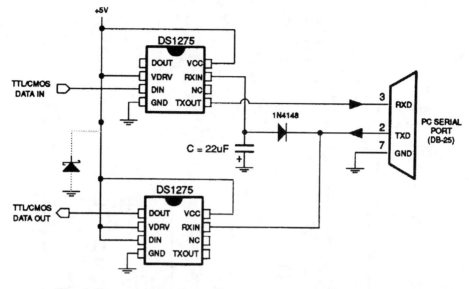

Fig. 6-G Full-duplex RS-232 that uses negative-charge storage.

ceive (signals to the PC port) or transmit (signals from the PC port) functions are abnormal or absent, suspect the corresponding DS1275. Of course, it is possible that the diode or capacitor is leaking or shorted, but this is not likely.

General digital-IC troubleshooting tips

The following troubleshooting tips apply to digital circuits where the majority of components are contained in ICs.

Power and ground connections The first step in tracing problems in a digital circuit with ICs is to check all power and ground connections to the ICs. Many ICs have more than one power and one ground connection. For example, the LTC1043 in Fig. 6-H requires +5 V at pin 4 and –5 V at pin 17. Also, the LTC1090 in Fig. 6-I has both a digital ground (DGND) and an analog ground (AGND). Likewise, the DAC-8565 in Fig. 6-J has both an analog common (pin 5) and a digital common (pin 12).

Reset, chip-select, read, write, and start signals With all power and ground connections confirmed, check that all the ICs receive reset, chip-select, start, and any other function signals, as required. For example, the DAC-4881 in Fig. 6-K requires a chip-select at pin 1, as well as address-decode signals at pins 2 and 28. Likewise, the ADC0808/0809 in Fig. 6-L requires, start, ALE, EOC, output enable signals from the microprocessor or control logic; see Fig. 6-L(B). If any of these signals are absent or abnormal (incorrect amplitude, improper timing, etc.) circuit operation comes to an immediate halt.

In some cases, control signals to digital ICs are pulses (usually timed in a certain sequence) and other control signals are steady (high or low). If any of the lines carrying the signals to the ICs are open, shorted to ground, or to power (typically

Digital/microprocessor circuit testing and troubleshooting

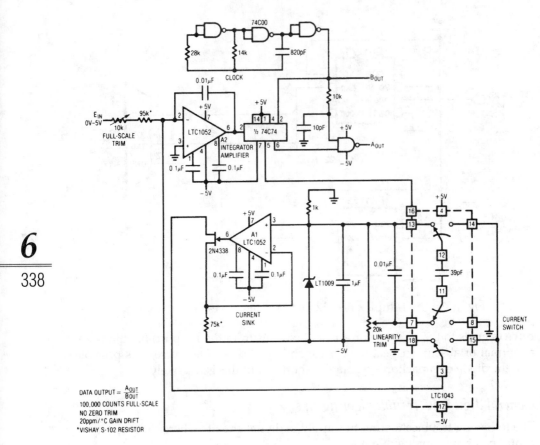

Fig. 6-H   16-bit A/D converter (chopper stabilized).

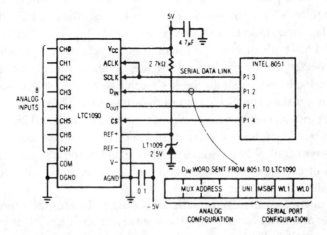

Fig. 6-I   Data-acquisition IC with four-wire microprocessor interface.

Digital/microprocessor circuits

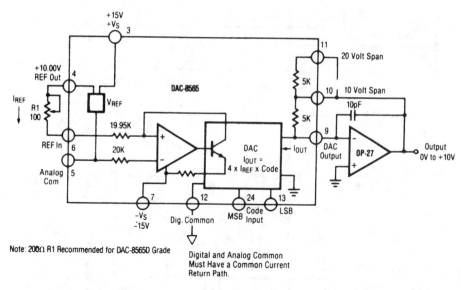

Fig. 6-J D/A converter with 0- to +10-V unipolar voltage output.

+5 V or +12 V, and 3 V or 3.3 V for newer digital circuits), the IC will not function. So if you find an IC control pin that is always high, always low, or apparently is connected to nothing (floating), check the PC traces or other wiring to that pin carefully. This guideline applies to all control pins, unless the circuit calls for the control function to be steady. For example, when the DAC4881 is connected as a 12-bit straight binary D/A (Chapter 9) converter as shown in Fig. 6-M, the chip-select (pin 1) and address-decode (pins 2, 28) are connected to ground. If the DAC4881 is connected as an 8-bit with complementary input D/A (Fig. 6-K), the chip-select must receive a write (WR) signal, and the address-decode pins must receive address bits, from the microprocessor.

Clock signals Many digital ICs require clocks. For example, there is a clock at the CP pin of the SAR2504 in Fig. 6-N and a clock at the CP of the 74C905 SAR in Fig. 6-O. Figure 6-L(B) shows the clock periods for the ADC in Fig. 6-L. In some cases, the clock comes from an external source (Figs. 6-L and 6-N). In other cases, the clock is part of the circuit (Fig. 6-O).

Generally, the presence of pulse activity on any pin of a digital IC indicates the presence of a clock but do not count on it. Check directly at the clock pins (typically, all ICs that require a clock are connected to the same clock source).

It is possible to measure the presence of a clock signal with a scope or logic probe. However, a frequency counter provides the most accurate measurement. Obviously, if any ICs do not receive required clock signals, the IC cannot function. On the other hand, if the clock is off frequency, all of the ICs might appear to have a clock signal, but the IC function can be impaired. Notice that crystal-controlled clocks do not usually drift off frequency but can go into some overtone frequency (typically a third overtone) beyond the capacity of the IC.

Input-output signals Once you are certain that all ICs are good and have proper

6

339

Digital/microprocessor circuit testing and troubleshooting

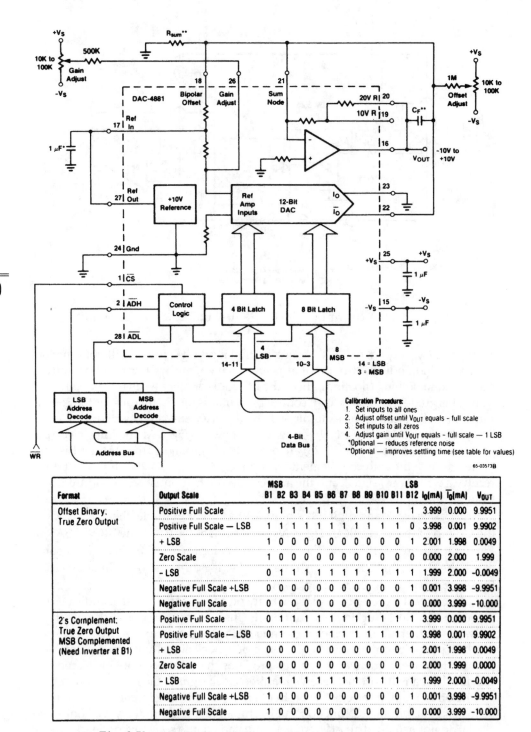

Calibration Procedure:
1. Set inputs to all ones
2. Adjust offset until V_{OUT} equals − full scale
3. Set inputs to all zeros
4. Adjust gain until V_{OUT} equals − full scale — 1 LSB
 *Optional — reduces reference noise
**Optional — improves settling time (see table for values)

65-03573B

Format	Output Scale	MSB B1	B2	B3	B4	B5	B6	B7	B8	B9	B10	B11	LSB B12	I_0(mA)	$\overline{I}_0$(mA)	V_{OUT}
Offset Binary, True Zero Output	Positive Full Scale	1	1	1	1	1	1	1	1	1	1	1	1	3.999	0.000	9.9951
	Positive Full Scale — LSB	1	1	1	1	1	1	1	1	1	1	1	0	3.998	0.001	9.9902
	+ LSB	1	0	0	0	0	0	0	0	0	0	0	1	2.001	1.998	0.0049
	Zero Scale	1	0	0	0	0	0	0	0	0	0	0	0	0.000	2.000	1.999
	− LSB	0	1	1	1	1	1	1	1	1	1	1	1	1.999	2.000	-0.0049
	Negative Full Scale +LSB	0	0	0	0	0	0	0	0	0	0	0	1	0.001	3.998	-9.9951
	Negative Full Scale	0	0	0	0	0	0	0	0	0	0	0	0	0.000	3.999	-10.000
2's Complement; True Zero Output MSB Complemented (Need Inverter at B1)	Positive Full Scale	0	1	1	1	1	1	1	1	1	1	1	1	3.999	0.000	9.9951
	Positive Full Scale — LSB	0	1	1	1	1	1	1	1	1	1	1	0	3.998	0.001	9.9902
	+ LSB	0	0	0	0	0	0	0	0	0	0	0	1	2.001	1.998	0.0049
	Zero Scale	0	0	0	0	0	0	0	0	0	0	0	0	2.000	1.999	0.0000
	− LSB	1	1	1	1	1	1	1	1	1	1	1	1	1.999	2.000	-0.0049
	Negative Full Scale +LSB	1	0	0	0	0	0	0	0	0	0	0	1	0.001	3.998	-9.9951
	Negative Full Scale	1	0	0	0	0	0	0	0	0	0	0	0	0.000	3.999	-10.000

Fig. 6-K   8-bit D/A converter with microprocessor interface.

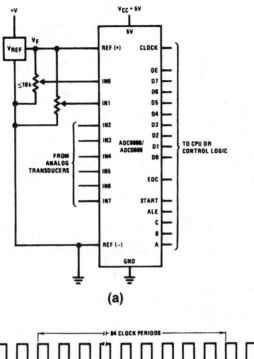

(a)

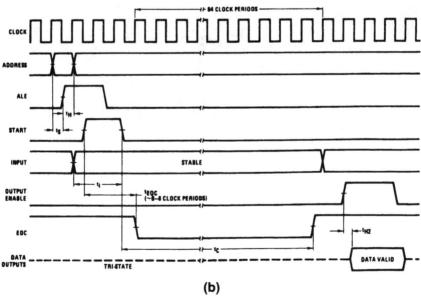

(b)

Fig. 6-L   Ratiometric A/D converter with separate interface.

Digital/microprocessor circuit testing and troubleshooting

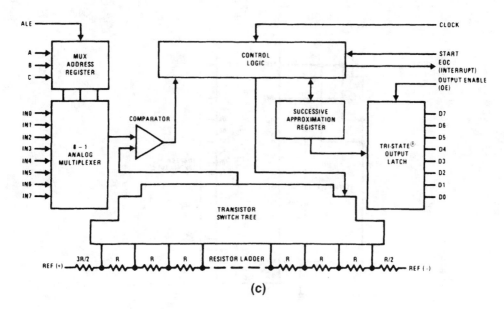

(c)

Fig. 6-L Continued.

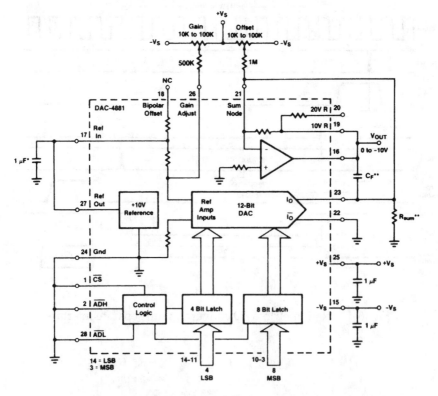

Fig. 6-M Twelve-bit straight-binary D/A converter.

Calibration Procedure:

1. Set inputs to all zeros
2. Adjust offset until V_{OUT} equals 0V
3. Set inputs to all ones

4. Adjust gain until V_{OUT} equals correct full scale value
   *Optional — reduces reference noise
   **Optional — improves settling time (see table for values)

Format	Output Scale	MSB B1	B2	B3	B4	B5	B6	B7	B8	B9	B10	B11	LSB B12	I_0(mA)	$\overline{I_0}$(mA)	V_{OUT}
Straight Binary. Unipolar with True Input Code. True Zero Output	Positive Full Scale	1	1	1	1	1	1	1	1	1	1	1	1	3.999	0.000	9.9976
	Positive Full Scale — LSB	1	1	1	1	1	1	1	1	1	1	1	0	3.998	0.001	9.9951
	LSB	0	0	0	0	0	0	0	0	0	0	0	1	0.0001	3.998	0.0024
	Zero Scale	0	0	0	0	0	0	0	0	0	0	0	0	0.000	3.999	0.0000
Complementary Binary. Unipolar with Complementary Input Code. True Zero Output	Positive Full Scale	0	0	0	0	0	0	0	0	0	0	0	0	0.000	3.999	9.9976
	Positive Full Scale — LSB	0	0	0	0	0	0	0	0	0	0	0	1	0.001	3.998	9.9951
	LSB	1	1	1	1	1	1	1	1	1	1	1	0	3.998	0.001	0.0024
	Zero Scale	1	1	1	1	1	1	1	1	1	1	1	1	3.999	0.000	0.0000

Fig. 6-M Continued.

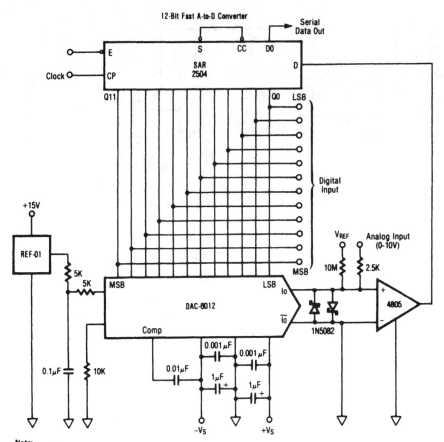

Note:
Device(s) connected to analog input must be capable of sourcing 4.0mA
a buffer may be required

Fig. 6-N Fast 12-bit A/D converter.

Digital/microprocessor circuit testing and troubleshooting

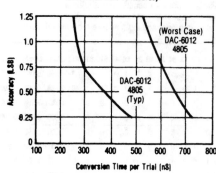

Conversion Time vs Accuracy

Conversion Time (nS)	Typ	Worst Case
SAR	33nS	55nS
4805	92nS	125nS
Total	375nS	680nS
x 13	4.9µS	8.8µS

Fig. 6-N Continued.

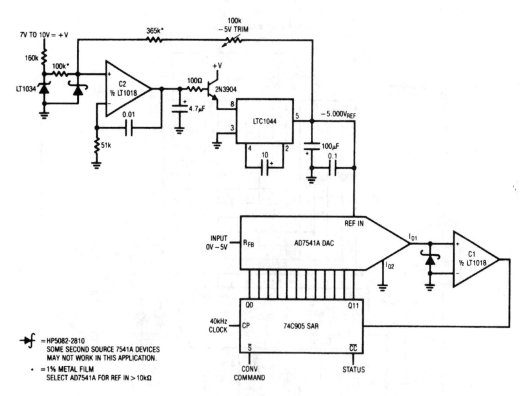

Fig. 6-O Micropower 12-bit A/D converter.

power and ground connections and that all control signals (reset, chip-select, write, start, etc.), and clock signals are available, the next step is to monitor all input and output signals at each IC. This test can be done with either a scope or probe.

Digital/microprocessor circuit titles and descriptions

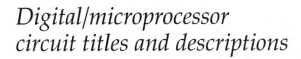

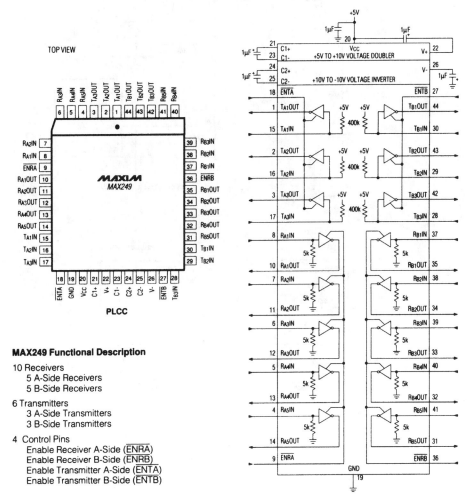

MAX249 Functional Description

10 Receivers
    5 A-Side Receivers
    5 B-Side Receivers

6 Transmitters
    3 A-Side Transmitters
    3 B-Side Transmitters

4 Control Pins
    Enable Receiver A-Side ($\overline{\text{ENRA}}$)
    Enable Receiver B-Side ($\overline{\text{ENRB}}$)
    Enable Transmitter A-Side ($\overline{\text{ENTA}}$)
    Enable Transmitter B-Side ($\overline{\text{ENTB}}$)

Multichannel RS-232 driver/receiver

Fig. 6-1 In the circuit shown, separate transmitter-enable inputs control two sets of three transmitters. Separate receiver-enable inputs control two sets of five receivers. The IC requires four external 1-µF capacitors. V_{CC} should be decoupled to ground with a capacitor of the same value as C1 and C2 and connected as close as possible to the device in applications that are sensitive to power-supply noise. RS-232 receivers and drivers on all devices invert. MAXIM NEW RELEASES DATA BOOK, 1992, P. 2-54.

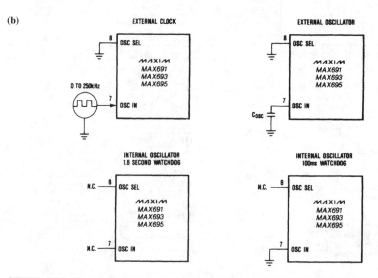

(a)

(b)

OSC SEL	OSC IN	Watchdog Timeout Period		Reset Timeout Period	
		Normal	Immediately After Reset	MAX691/93	MAX695
Low	External Clock Input	1024 clks	4096 clks	512 clks	2048 clks
Low	External Capacitor	$\dfrac{400ms}{47pF} \times C$	$\dfrac{1.6\ sec}{47pF} \times C$	$\dfrac{200ms}{47pF} \times C$	$\dfrac{800ms}{47pF} \times C$
Floating	Low	100ms	1.6 sec	50ms	200ms
Floating	Floating	1.6 sec	1.6 sec	50ms	200ms

Note 1: The MAX690/692/694 watchdog timeout period is fixed at 1.6 seconds nominal, the MAX690/692 Reset pulse width is fixed at 50ms nominal and the MAX694 is 200ms nominal.

Note 2: When the MAX691 OSC SEL pin is low, OSC IN can be driven by an external clock signal, or an external capacitor can be connected between OSC IN and GND. The nominal internal oscillator frequency is 10.24kHz. The nominal oscillator frequency with external capacitor is

$$F_{OSC}(Hz) = \frac{184,000}{C(pF)}$$

Note 3: See Electrical Characteristics Table for minimum and maximum timing values.

Digital/microprocessor circuits

Microprocessor supervisory circuit

Fig. 6-2 In the circuit shown, a CMOS RAM is powered from V_{OUT}, which is internally connected to V_{CC} when 5-V power is present, or to V_{BATT} when V_{CC} is less than the battery voltage. V_{OUT} can supply 50 mA from V_{CC}. However, if more current is required, an external pnp transistor can be added (as shown by the dotted lines). When V_{CC} is higher than V_{BATT}, the BATT ON output goes low, providing 25 mA of base drive for the external transistor. When V_{CC} is lower than V_{BATT}, an internal MOSFET connects the backup battery to V_{OUT}. The quiescent current in the battery backup mode is 1 µA maximum when V_{CC} is between 0 V and V_{BATT} − 700 mV. A voltage detector monitors V_{CC} and generates a RESET output to hold the microprocessor's reset line low when V_{CC} is below 4.65 V (4.4 V for the MAX693). An internal monostable holds RESET low for 50 ms (200 ms for the MAX695) after V_{CC} rises above 4.65 V (4.4 V for the MAX693). This prevents repeated toggling of RESET even if the 5-V power drops out and recovers with each power-line cycle. When the voltage at PFI falls below 1.3 V, the PFO drives the microprocessor NMI input low. If a power-fail threshold of 4.8 V is chosen (by election of resistors at pin 9), the microprocessor will have the time when V_{CC} falls from 4.8 V to 4.65 V to save data into RAM. If V_{CC} falls below the reset threshold, CE OUT (connected to the RAM chip-select) goes high and prevents the microprocessor from writing erroneous data into RAM during power-up, power-down, brownouts, and momentary power interruptions. The watchdog output (WDO) goes low if the watchdog timer is not serviced within its timeout period (see Fig. 6-2B). This action indicates that the microprocessor is not functioning and is not applying toggle pulses to the WDI input. Once WDO goes low, it remains low until a transition occurs at WDI. The watchdog timer feature can be disabled by leaving WDI not connected. OSC IN and OSC SEL also allow other watchdog timing options, as shown in Fig. 6-2B. MAXIM NEW RELEASES DATA BOOK, 1992, PP. 5-23, 29.

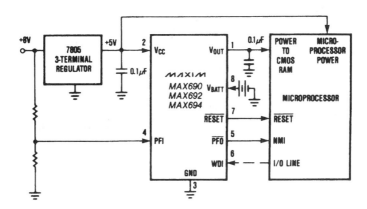

Alternate microprocessor supervisory circuit

Fig. 6-3 Operation of the circuit in the illustration is similar to that of the Fig. 6-2 circuit. However, in the Fig. 6-3 circuit, the PFI monitors the unregulated input to the 7805 regulator. The $\overline{\text{RESET}}$ output goes low when V_{CC} falls below 4.65 V (4.4 V for the MAX692). This circuit does not have a BATT ON output for an external transistor, and current consumption of the battery-backup bus must be less than 50 mA. The circuit does not have chip-enable (CE) input and outputs. However, in many systems, CE gating is not needed because a low input to the microprocessor $\overline{\text{RESET}}$ line prevents the microprocessor from writing to RAM during power-up and power-down transients. The Fig. 6-3 circuit watchdog time has a fixed 1.6-s timeout period. If WDI remains either low or high for more than 1.6 s (indicating a microprocessor failure), a $\overline{\text{RESET}}$ pulse is sent to the microprocessor. The watchdog timer is disabled if WDI is left floating. MAXIM NEW RELEASES DATA BOOK, 1992, P. 5-24.

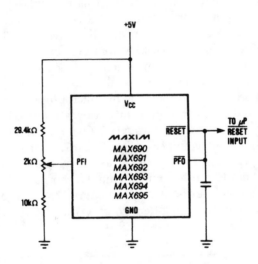

Externally adjustable V_{CC} reset threshold (for microprocessor)

Fig. 6-4 In the circuit shown, the ICs of Fig. 6-2 and 6-3 are connected so the *PFI* (power-fail detector) is used to initiate a system reset when V_{CC} falls to 4.85 V. Because the threshold of the power-fail detector is not as accurate as the on-board reset voltage detector, a trimpot must be used to adjust the voltage-detection threshold. Both $\overline{\text{PFO}}$ and $\overline{\text{RESET}}$ outputs have high sink-current capability and only 10 µA of source-current drive. This allows the two outputs to be connected directly to each other in a wired-OR configuration. MAXIM NEW RELEASES DATA BOOK, 1992, P. 5-30.

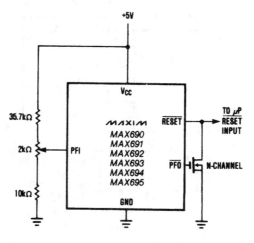

Reset on overvoltage or undervoltage (for microprocessor)

Fig. 6-5 In the circuit shown, the ICs of Figs. 6-2 and 6-3 are connected so the microprocessor resets whenever the nominal 5-V V_{CC} is above 5.5 V. MAXIM NEW RELEASES DATA BOOK, 1992, P. 5-30.

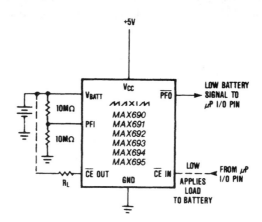

Backup battery monitor with optional test load (for microprocessor)

Fig. 6-6 In the circuit shown, the ICs of Figs. 6-2 and 6-3 are connected to show the status of the memory backup battery. If desired, the CE OUT can be used to apply a test load to the battery. Because CE OUT is forced high during the battery-backup mode, the test load will not be applied to the battery (when the battery is being used) even if the microprocessor is not powered. MAXIM NEW RELEASES DATA BOOK, 1992, P. 5-30.

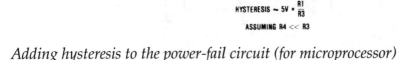

$$V_H = 9.125V$$
$$V_L = 7.9V$$
$$HYSTERESIS = 1.23V$$

$$V_H = 1.3V \left(1 + \frac{R1}{R2} + \frac{R1}{R3} \right)$$

$$V_L = 1.3V \left(1 + \frac{R1}{R2} - \frac{(5V - 1.3V)}{1.3V} \frac{R1}{(R3 + R4)} \right)$$

$$HYSTERESIS \approx 5V \times \frac{R1}{R3}$$

ASSUMING R4 << R3

Adding hysteresis to the power-fail circuit (for microprocessor)

Fig. 6-7　In the circuit shown, the ICs of Figs. 6-2 and 6-3 are connected to add hysteresis to the power-fail voltage comparator. Because the power-fail circuit is noninverting, hysteresis is added by connecting a resistance between the $\overline{PFO}$ output and the PFI input as shown. When $\overline{PFO}$ is low, R3 sinks current from the summing junction at the PFI pin. When $\overline{PFO}$ is high, the series combination of R3 and R4 source current into the PFI summing junction. MAXIM NEW RELEASES DATA BOOK, 1992, P. 5-31.

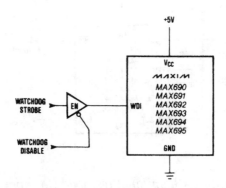

Disabling the watchdog under program control (for microprocessor)

Fig. 6-8　In the circuit shown, the ICs of Figs. 6-2 and 6-3 are connected so that the watchdog feature can be enabled and disabled under program control. Enabling and disabling is done by driving WDI with a three-state buffer. A drawback to this circuit is that a software fault might set the buffer into the three-state (intermediate) condition and prevent the IC from detecting that the microprocessor is no longer working. MAXIM NEW RELEASES DATA BOOK, 1992, P. 5-31.

Digital/microprocessor circuits

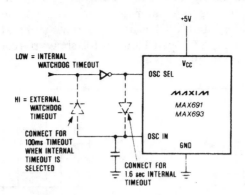

Extending the watchdog timeout (for microprocessor)

Fig. 6-9 The circuit in the illustration is an alternate for the circuit of Fig. 6-8 and extends the watchdog period, rather than disabling the watchdog. When the control input (from the system program) is high, the OSC SEL pin is low and the watchdog timeout is set by the external capacitor. (A 0.01-µF capacitor sets as the watchdog timeout delay of 100 s.) When the control input is low, the OSC CEL pin is driven high, selecting the internal oscillator. The 100-ms or the 1.6-s period is chosen, depending on which diode is used. Note that this circuit can be applied only to the MAX691 or MAX693 (Fig. 6-2). MAXIM NEW RELEASES DATA BOOK, 1992, P. 5-31.

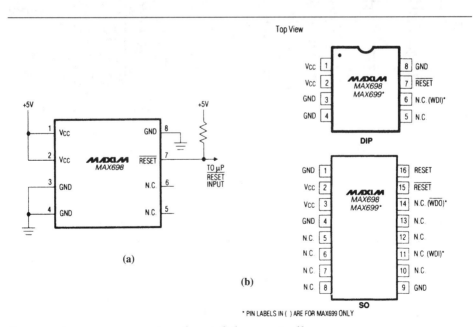

Low-cost power-on reset and watchdog controller

Fig. 6-10 The MAX698 and MAX699 monitor the +5-V supply in microprocessor and digital systems. The ICs apply a $\overline{\text{RESET}}$ pulse of at least 140 ms duration on power-up, power-down, and during low-voltage brownout conditions. The MAX699 also includes a watchdog feature to monitor microprocessor activity. The RESET output goes low if the watchdog input (WDI) is not toggled within 1 s. The watchdog feature can be disabled by leaving WDI open. Figure 6-10B shows pin configurations. MAXIM NEW RELEASES DATA BOOK, 1992, P. 5-45.

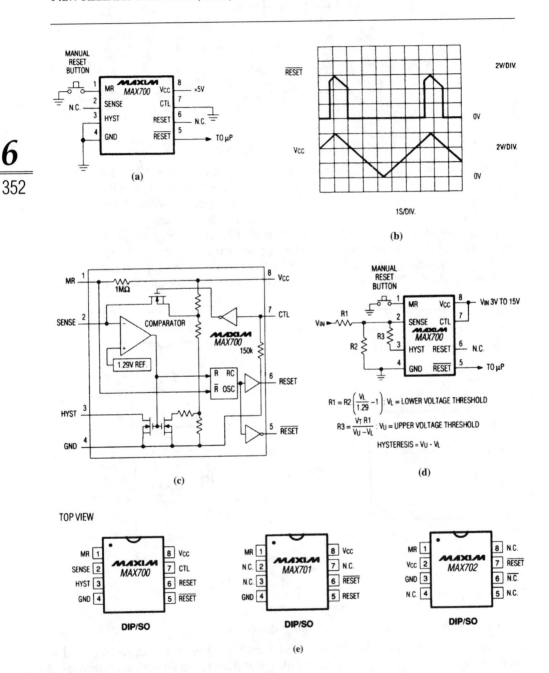

Power-supply monitor with reset

Fig. 6-11 The MAX700/701/702 are used to monitor the power supplies in micro-processor and digital systems. The $\overline{\text{RESET}}$/RESET outputs are guaranteed to be in the correct states for V_{CC} voltages down to +1 V (see Fig. 6-11B). The MAX702 $\overline{\text{RESET}}$ goes low when V_{CC} falls to 4.65 V and includes a debounced manual reset input (Fig. 6-11C). The MAX701 performs the same functions, but also has both $\overline{\text{RESET}}$ and RESET. Both ICs provide an active reset signal for low supply voltages and for at least 200 ms after the supply voltage reaches normal operating value. The MAX700 also provides adjustable hysteresis and preset or adjustable voltage detection so thresholds other than 4.65 V can be selected (see Fig. 6-11D). Figure 6-11E shows pin configurations. MAXIM NEW RELEASES DATA BOOK, 1992, PP. 5-49, 52.

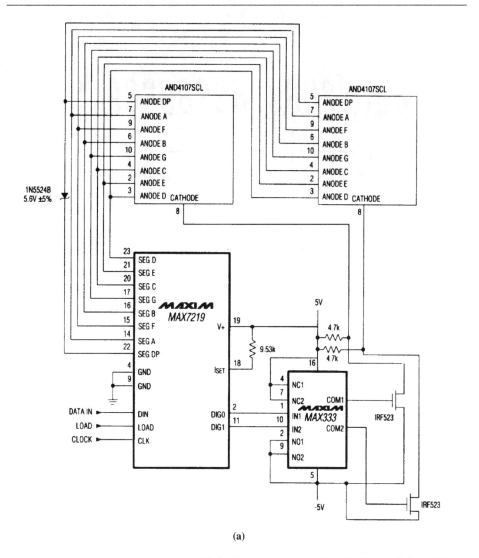

(a)

Digital/microprocessor circuit titles and descriptions

NUMBER OF DIGITS DISPLAYED	MAXIMUM SEGMENT CURRENT	
1	10mA	**(b)**
2	20mA	
3	30mA	

I_{SEG} (mA)	V_{LED} (V)					
	1.5	2	2.5	3	3.5	
40	11.3	10.4	9.8	8.9	7.8	**(c)**
30	16.3	15	14	12.9	11.4	
20	26.2	24.6	22.8	20.9	18.6	
10	60.1	56	51.7	47	41.9	

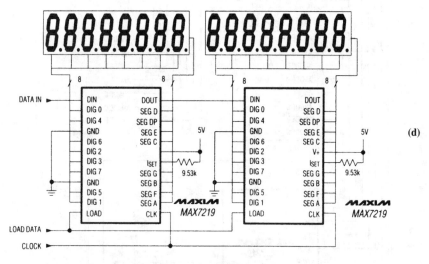

(d)

Serially interfaced, eight-digit LED digital display

Fig. 6-12   The circuit in the illustration uses a MAX7219 segment driver, a MAX333 analog switch, and external drivers to drive AND4107SCL common-cathode displays. The 5.6-V zener is used in series with the decimal-point LED, because the forward voltage for this LED is typically 4.2 V. The forward voltage for the remaining LEDs is typically 8 V. The value of RESET (connected between pins 18 and 19 of the MAX7219) sets the drive current for the LED segments, as shown in Figs. 6-12B and 6-12C. Because external transistors are used to sink current, peak segment currents of 40 mA are allowed, even though only two digits are displayed. Figure 6-12D shows how two MAX7219s can be cascaded to drive 16 seven-segment LEDs. MAXIM NEW RELEASES DATA BOOK, 1992, PP. 10-7, 8, 9.

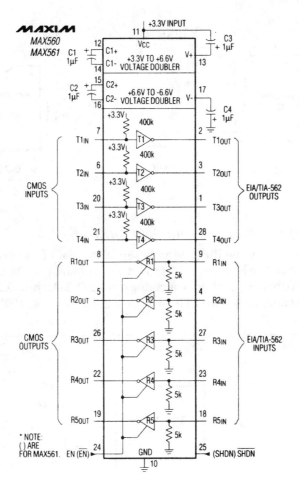

Multichannel RS-232 driver/receiver (3.3 V)

Fig. 6-13 The circuit in the illustration is similar to that of Fig. 6-1, except that the IC is for 3.3-V digital systems. MAXIM NEW RELEASES DATA BOOK, 1993, P. 2-60.

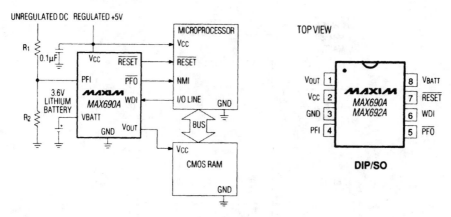

Microprocessor supervisory circuit (four function)

Fig. 6-14 The circuit in the illustration is similar to that of Figs. 6-2 and 6-3 but with less functions. The MAX690/692 provides reset (4.65 V for MAX690A and 4.4 V for MAX692A), battery-backup, watchdog timing (1.6 s), and a 1.25-V threshold low-battery functions only. MAXIM NEW RELEASES DATA BOOK, 1993, P. 5-3.

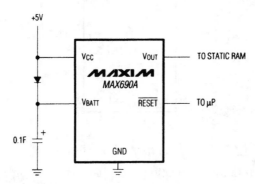

Using a capacitor as a backup power source

Fig. 6-15 In the circuit shown, the IC of Fig. 6-14 is connected to use a SuperCap™ as a backup power source rather than a battery. The 0.1-F SuperCap™ rapidly charges to within a diode drop of V_{CC}. However, after a long time, the diode leakage current will pull the SuperCap voltage up to V_{CC} (which cannot exceed 4.75 V + 0.6 V = 5.35 V). MAXIM NEW RELEASES DATA BOOK, 1993, P. 5-8.

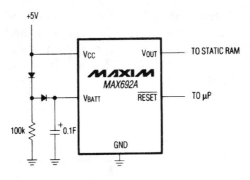

Using a capacitor as a backup power source (alternate)

Fig. 6-16 The circuit in the illustration is similar to that of Fig. 6-15, but it can be used where the power supply is +5 V ±10%, rather than ±5%. The Fig. 6-16 circuit ensures that the SuperCap™ changes to V_{CC} −0.5 V. At the maximum V_{CC} of 5.5 V, the SuperCap™ charges up to 5.0 V. This charge is only 0.5 V above the maximum reset threshold and well within the required 0.6 V. MAXIM NEW RELEASES DATA BOOK, 1993, P. 5-8.

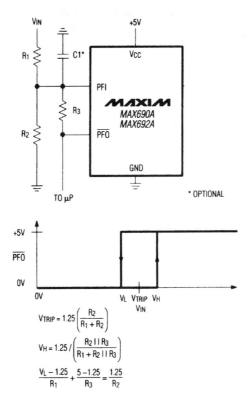

Adding hysteresis to four-function microprocessor supervisor

Fig. 6-17 In the circuit shown, the IC of Fig. 6-14 is connected to add hysteresis to the power-fail voltage comparator. Select the ratio of R1 and R2 so that PFI sees 1.25 V when V_{IN} falls to the trip point (V_{TRIP}) as shown by the equations. Make R3 10 kΩ or larger. The current through R1/R2 should be at least 1 μA. MAXIM NEW RELEASES DATA BOOK, 1993, P. 5-9.

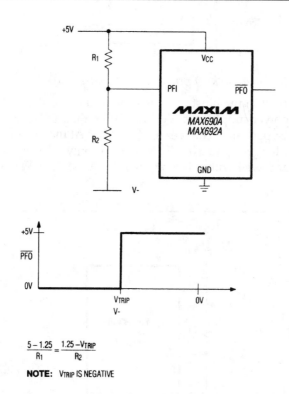

$$\frac{5-1.25}{R_1} = \frac{1.25-V_{TRIP}}{R_2}$$

NOTE: V_{TRIP} IS NEGATIVE

Monitoring a negative voltage

Fig. 6-18 In the circuit shown, the IC of Fig. 6-14 is connected to monitor a negative voltage. When the negative rail is good (the negative voltage is large) $\overline{PFO}$ is low. When the negative voltage is of lesser magnitude, $\overline{PFO}$ goes high. The accuracy of this circuit is affected by the PFI threshold tolerance, the V_{CC} line, and the resistors. MAXIM NEW RELEASES DATA BOOK, 1993, P. 5-9.

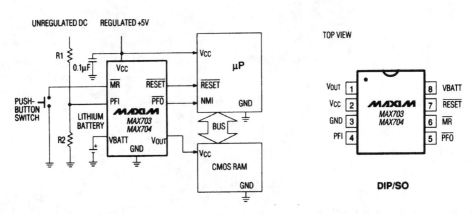

Low-cost microprocessor supervisory circuit

Fig. 6-19 The circuit in the illustration is similar to that of Figs. 6-2 and 6-3 but with less functions. The MAX703/704 provide reset (4.65 V for MAX703 and 4.4 V for MAX704), battery-backup, power-fail or battery monitoring, and active-low manual reset. MAXIM NEW RELEASES DATA BOOK, 1993, P. 5-13.

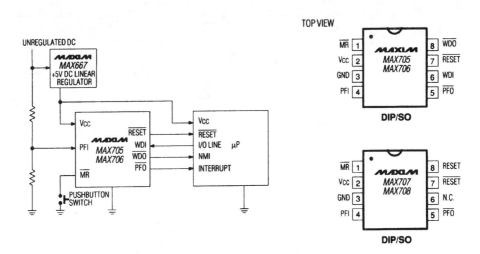

Low-cost microprocessor supervisory circuit (with watchdog)

Fig. 6-20 The circuit in the illustration is similar to that of Figs. 6-2 and 6-3 but with less functions. The MAX705/706 provide reset, watchdog timing (1.6 s), a 1.25-V threshold low-battery or power-fail, and a manual reset. The MAX707/708 are the same as the MAX705/706, except that an active-high reset is substituted for the watchdog timer. Reset for the MAX705/707 is 4.65 V. The MAX706/708 have 4.4-V reset. MAXIM NEW RELEASES DATA BOOK, 1993, P. 5-15.

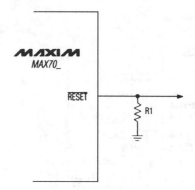

Ensuring a valid reset when $V_{CC} = 0$

Fig. 6-21   In the circuit shown, the IC of Fig. 6-20 is connected to ensure that there will be a valid reset even when V_{CC} drops to zero. When V_{CC} falls below 1 V the reset output no longer sinks current and becomes an open circuit. High-impedance CMOS logic inputs can drift to undetermined voltages if left undriven. By adding Rl (typically 100 kΩ) any stray charge or leakage will be drained to ground, holding RESET low. MAXIM NEW RELEASES DATA BOOK, 1993, P. 5-20.

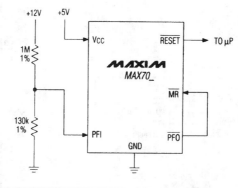

PARAMETER	MIN	TYP	MAX	UNIT
+12V Reset Threshold at +25°C	10.67	10.87	11.50	V

Monitoring both +5 V and +12 V

Fig. 6-22   In the circuit shown, the IC of Fig. 6-20 is connected to monitor both +5 V and +12 V. The IC will assert RESET when the +5-V supply falls below the reset threshold, or when the +12-V supply falls below about 11 V. MAXIM NEW RELEASES DATA BOOK, 1993, P. 5-21.

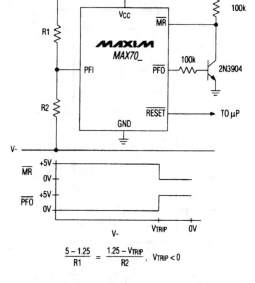

$$\frac{5 - 1.25}{R1} = \frac{1.25 - V_{TRIP}}{R2}, \quad V_{TRIP} < 0$$

Monitoring negative voltages

Fig. 6-23   In the circuit shown, the IC of Fig. 6-20 is connected to monitor negative voltages. When the negative rail is good (the negative voltage is large), $\overline{PFO}$ is low. PFO is high when the negative voltage is of lesser magnitude. The accuracy of this circuit is affected by the $\overline{PFI}$ threshold tolerance, the V_{CC} line, and the resistors. MAXIM NEW RELEASES DATA BOOK, 1993, P. 5-21.

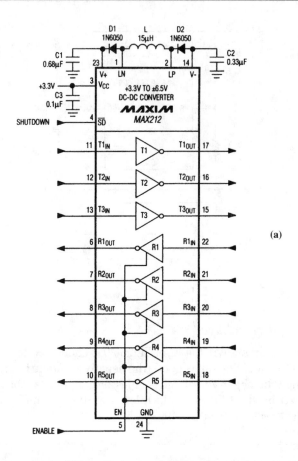

(a)

(b)

MANUFACTURER	PART NUMBER	PHONE NUMBER	FAX NUMBER
Murata	LQH4N150K-TA	USA (404) 831-9172 Japan (075) 951-9111	USA (404) 436-3030 Japan (075) 955-6526
TDK	NLC453232T-150K	USA (708) 803-6100 Japan (03) 3278-5111	USA (708) 803-6296 Japan (03) 3278-5358
Sumida	CD43150	USA (708) 956-0666 Japan (03) 3607-5111	USA (708) 956-0702 Japan (03) 3607-5428

Multichannel RS-232 driver/receiver (3.0 V)

Fig. 6-24   The circuit in the illustration is similar to that of Fig. 6-1 except that the IC is for 3-V digital systems. Figure 6-24B shows typical coil suppliers. (The circuits of Figs. 6-1 and 6-13 use the charge-pump principle. The circuit of Fig. 6-24 uses the switching-regulator principle and thus requires a coil L.) The operating voltage range extends from 3.6 V to 3.0 V and maintains true RS-232 and EIA/TIA-562 voltage levels. MAXIM NEW RELEASES DATA BOOK, 1994, P. 2-39.

Digital/microprocessor circuits

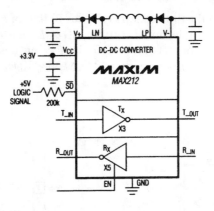

Operating a 3-V RS-232 driver/receiver with 5-V logic

Fig. 6-25 The circuit in the illustration is similar to that of Fig. 6-24, but it will operate with 5-V logic inputs simultaneously with a 3-V supply. MAXIM NEW RELEASES DATA BOOK, 1994, P. 2-40.

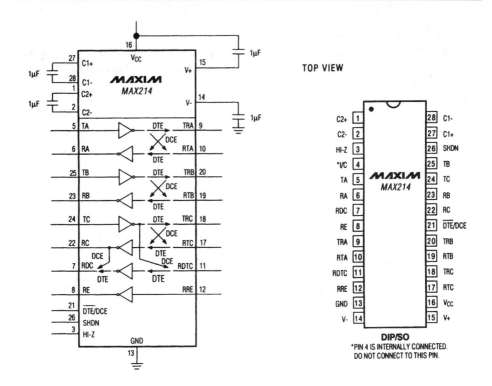

Programmable DTE/DCE RS-232 transceiver

Fig. 6-26   The circuit in the illustration is similar to that of Fig. 6-1, except that the IC functions as a *DTE* (data terminal equipment) or *DCE* (data circuit-terminating equipment) transceiver for AT-compatible laptop or desktop computers and modems, printers, and other peripherals. MAXIM NEW RELEASES DATA BOOK, 1994, P. 2-43.

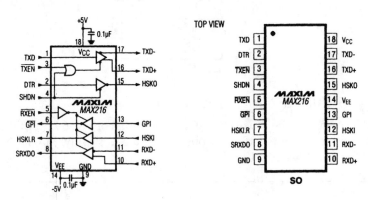

Low-power AppleTalk™ interface transceiver

Fig. 6-27   The circuit in the illustration is similar to that of Fig. 6-1, except that the IC is designed specifically for communication with AppleTalk™ interfaces, including EIA/TIA-232/562 to RS-422 conversion. MAXIM NEW RELEASES DATA BOOK, 1994, P. 2-45.

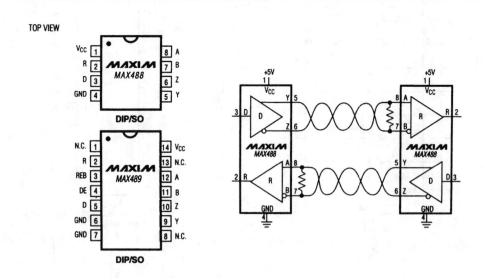

Low-power RS-485/RS-422 transceivers

Fig. 6-28 The circuit in the illustration is similar to that of Fig. 6-1, except that the ICs are designed specifically for RS-485 and RS-422 communication. The ICs allow full-duplex communication with separate receiver inputs and driver outputs and feature a reduced slew rate that minimizes EMI (electromagnetic interference) and reflections caused by improperly terminated cables. MAXIM NEW RELEASES DATA BOOK, 1994, P. 2-103.

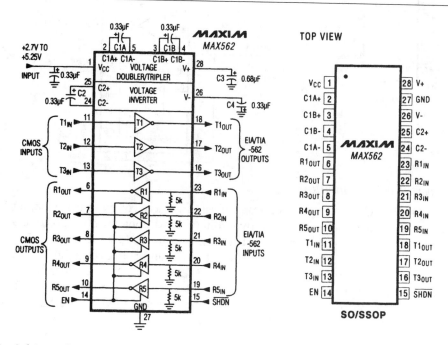

Serial interface for notebook/palmtop computers

Fig. 6-29 The circuit in the illustration is similar to that of Fig. 6-1, except the IC is designed specifically for notebook and palmtop computers that need to transfer data quickly. The IC runs at rates up to 230 Kbps, is LapLink™ compatible, has a guaranteed 4-V/µs slew rate, and meets the EIA/TIA-562 standard that guarantees compatibility with RS-232 interfaces. In the keep-awake mode, the transmitters are shut down, but all receivers are active, allowing unidirectional communication. In the shutdown mode, the entire IC is disabled and all outputs are in a high-imped-ance state. MAXIM NEW RELEASES DATA BOOK, 1994, P. 2-107.

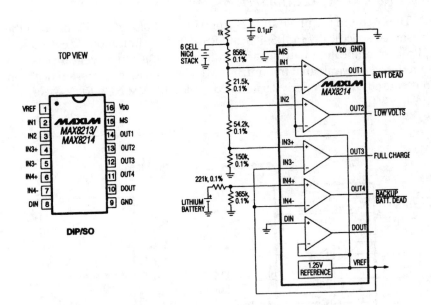

TOP VIEW

DIP/SO

Universal voltage monitors

Fig. 6-30 The ICs shown contain four precision voltage comparators (Chapter 8) that can monitor undervoltage and overvoltage conditions for both positive and negative supplies. Accurate trip-point setting is facilitated by the internal 1.25-V reference. Trip-level accuracy is guaranteed to ±1%, and trip levels of all channels are guaranteed to match each other within ±1%. A fifth comparator channel monitors microprocessor voltages and generates delayed reset signals. The MAX8213 has open-drain outputs. Active pull-up outputs are provided by the MAX8214. MAXIM NEW RELEASES DATA BOOK, 1994, P. 5-107.

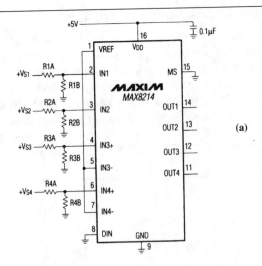

(a)

Continued

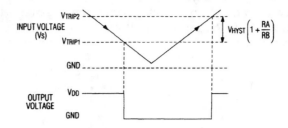

(b)

TO DETERMINE THE TRIP VOLTAGES
FROM PARTICULAR RESISTOR
VALUES:

$$V_{TRIP1} = V_{TH}\left(1 + \frac{RA}{RB}\right)$$

$$V_{TRIP2} = \left(V_{TH} + V_{HYST}\right)\left(1 + \frac{RA}{RB}\right)$$

TO CALCULATE THE REQUIRED
RESISTOR RATIOS FOR PARTICULAR
TRIP VOLTAGES:

$$\frac{RA}{RB} = \frac{V_{TRIP1}}{V_{TH}} - 1$$

$$\frac{RA}{RB} = \frac{V_{TRIP2}}{V_{TH} + V_{HYST}} - 1$$

6

367

NOTE: V$_{TH}$ IS THE VOLTAGE ON THE INVERTING PIN OF EACH COMPARATOR.

Quad undervoltage detector

Fig. 6-31 In the circuit shown, the IC of Fig. 6-30 is connected to detect when a monitored voltage has dropped below a certain level. Figure 6-31B shows the waveforms and calculations for resistor values. MAXIM NEW RELEASES DATA BOOK, 1994, PP. 5-114, 115.

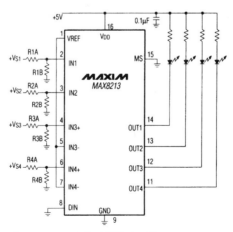

Quad undervoltage detector with LED indicators

Fig. 6-32 In the circuit shown, the IC of Fig. 6-30 is connected to detect and indi-

Digital/microprocessor circuit titles and descriptions

cate when a monitored voltage has dropped below a certain level. A low at a comparator output indicates an undervoltage condition and causes the associated LED to turn on. MAXIM NEW RELEASES DATA BOOK, 1994, P. 5-114.

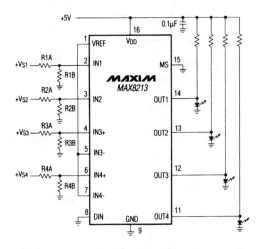

Quad overvoltage detector with LED indicators

Fig. 6-33 In the circuit shown, the IC of Fig. 6-30 is connected to detect and indicate when a monitored voltage has risen above a certain level. A low at a comparator output indicates an overvoltage condition and causes the associated LED to turn on. MAXIM NEW RELEASES DATA BOOK, 1994, P. 5-115.

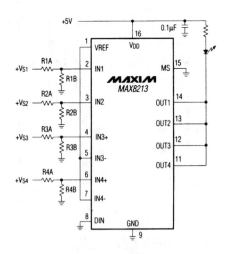

Quad undervoltage detector with single LED indicator

Fig. 6-34 In the circuit shown, the IC of Fig. 6-30 is connected to detect and indi-

Digital/microprocessor circuits

cate when any of four monitored voltages has dropped below a certain level. A low at any comparator output indicates an undervoltage condition on the corresponding voltage and causes the single LED to turn on. MAXIM NEW RELEASES DATA BOOK, 1994, P. 5-115.

(a)

(b)

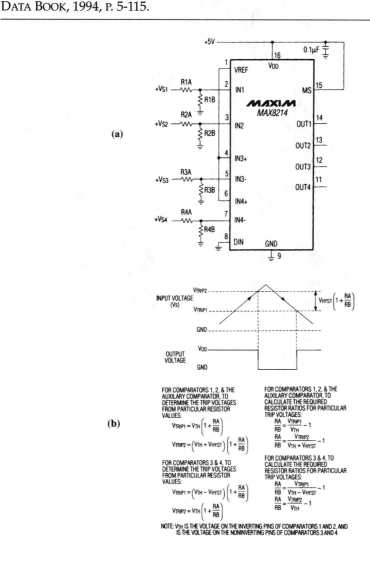

FOR COMPARATORS 1, 2, & THE AUXILARY COMPARATOR, TO DETERMINE THE TRIP VOLTAGES FROM PARTICULAR RESISTOR VALUES:

$$V_{TRIP1} = V_{TH}\left(1 + \frac{RA}{RB}\right)$$

$$V_{TRIP2} = \left(V_{TH} + V_{HYST}\right)\left(1 + \frac{RA}{RB}\right)$$

FOR COMPARATORS 3 & 4, TO DETERMINE THE TRIP VOLTAGES FROM PARTICULAR RESISTOR VALUES:

$$V_{TRIP1} = \left(V_{TH} - V_{HYST}\right)\left(1 + \frac{RA}{RB}\right)$$

$$V_{TRIP2} = V_{TH}\left(1 + \frac{RA}{RB}\right)$$

FOR COMPARATORS 1, 2, & THE AUXILARY COMPARATOR, TO CALCULATE THE REQUIRED RESISTOR RATIOS FOR PARTICULAR TRIP VOLTAGES:

$$\frac{RA}{RB} = \frac{V_{TRIP1}}{V_{TH}} - 1$$

$$\frac{RA}{RB} = \frac{V_{TRIP2}}{V_{TH} + V_{HYST}} - 1$$

FOR COMPARATORS 3 & 4, TO CALCULATE THE REQUIRED RESISTOR RATIOS FOR PARTICULAR TRIP VOLTAGES:

$$\frac{RA}{RB} = \frac{V_{TRIP1}}{V_{TH} - V_{HYST}} - 1$$

$$\frac{RA}{RB} = \frac{V_{TRIP2}}{V_{TH}} - 1$$

NOTE: V_{TH} IS THE VOLTAGE ON THE INVERTING PINS OF COMPARATORS 1 AND 2, AND IS THE VOLTAGE ON THE NONINVERTING PINS OF COMPARATORS 3 AND 4.

Quad overvoltage detector (alternate)

Fig. 6-35 In the circuit shown, when an input voltage rises above a preset trip level, determined by the calculations of Fig. 6-35B, the corresponding comparator output goes low. The LED indicator circuits of Figs. 6-33 and 6-34 can be used if needed. MAXIM NEW RELEASES DATA BOOK, 1994, P. 5-116.

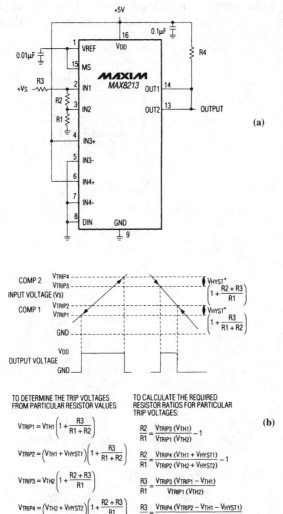

(a)

(b)

TO DETERMINE THE TRIP VOLTAGES
FROM PARTICULAR RESISTOR VALUES:

$$V_{TRIP1} = V_{TH1}\left(1 + \frac{R3}{R1 + R2}\right)$$

$$V_{TRIP2} = \left(V_{TH1} + V_{HYST1}\right)\left(1 + \frac{R3}{R1 + R2}\right)$$

$$V_{TRIP3} = V_{TH2}\left(1 + \frac{R2 + R3}{R1}\right)$$

$$V_{TRIP4} = \left(V_{TH2} + V_{HYST2}\right)\left(1 + \frac{R2 + R3}{R1}\right)$$

TO CALCULATE THE REQUIRED
RESISTOR RATIOS FOR PARTICULAR
TRIP VOLTAGES:

$$\frac{R2}{R1} = \frac{V_{TRIP3}\,(V_{TH1})}{V_{TRIP1}\,(V_{TH2})} - 1$$

$$\frac{R2}{R1} = \frac{V_{TRIP4}\,(V_{TH1} + V_{HYST1})}{V_{TRIP2}\,(V_{TH2} + V_{HYST2})} - 1$$

$$\frac{R3}{R1} = \frac{V_{TRIP3}\,(V_{TRIP1} - V_{TH1})}{V_{TRIP1}\,(V_{TH2})}$$

$$\frac{R3}{R1} = \frac{V_{TRIP4}\,(V_{TRIP2} - V_{TH1} - V_{HYST1})}{V_{TRIP2}\,(V_{TH2} + V_{HYST2})}$$

NOTE: V_{TH1} AND V_{TH2} ARE THE VOLTAGES ON THE INVERTING INPUTS OF COMPARATOR
1 AND COMPARATOR 2, RESPECTIVELY (BOTH ARE EQUAL TO THE REFERENCE
VOLTAGE IN THIS CASE).

Overvoltage/undervoltage detector (window detector)

Fig. 6-36 The circuit in the illustration shows how two comparators can be connected to detect when a voltage level is between two trip voltages. The combination of comparator 1 (overvoltage) and comparator 2 (undervoltage) creates this window-detector function. Figure 6-36B shows the waveforms and calculations for resistor values. Maxim New Releases Data Book, 1994, p. 5-116.

Digital/microprocessor circuits

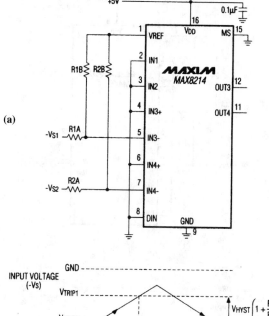

(a)

(b)

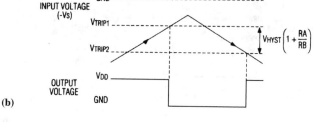

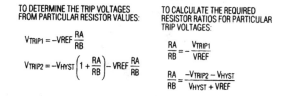

TO DETERMINE THE TRIP VOLTAGES
FROM PARTICULAR RESISTOR VALUES:

$$V_{TRIP1} = -V_{REF} \frac{RA}{RB}$$

$$V_{TRIP2} = -V_{HYST}\left(1 + \frac{RA}{RB}\right) - V_{REF}\frac{RA}{RB}$$

TO CALCULATE THE REQUIRED
RESISTOR RATIOS FOR PARTICULAR
TRIP VOLTAGES:

$$\frac{RA}{RB} = -\frac{V_{TRIP1}}{V_{REF}}$$

$$\frac{RA}{RB} = \frac{-V_{TRIP2} - V_{HYST}}{V_{HYST} + V_{REF}}$$

Dual negative undervoltage detector

Fig. 6-37 In the circuit shown, the IC of Fig. 6-30 is connected to detect when a monitored negative voltage has dropped below a certain level. Figure 6-37B shows the waveforms and calculations for resistor values. MAXIM NEW RELEASES DATA BOOK, 1994, P. 5-117.

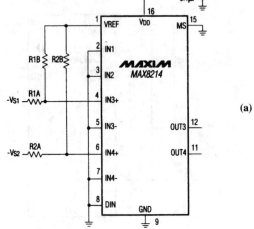

(a)

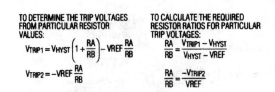

(b)

TO DETERMINE THE TRIP VOLTAGES
FROM PARTICULAR RESISTOR
VALUES:

$$V_{TRIP1} = V_{HYST}\left(1 + \frac{RA}{RB}\right) - VREF\frac{RA}{RB}$$

$$V_{TRIP2} = -VREF\frac{RA}{RB}$$

TO CALCULATE THE REQUIRED
RESISTOR RATIOS FOR PARTICULAR
TRIP VOLTAGES:

$$\frac{RA}{RB} = \frac{V_{TRIP1} - V_{HYST}}{V_{HYST} - VREF}$$

$$\frac{RA}{RB} = \frac{-V_{TRIP2}}{VREF}$$

Dual negative overvoltage detector

Fig. 6-38 In the circuit shown, the IC of Fig. 6-30 is connected to detect when a monitored negative voltage has risen above a certain level. Figure 6-38B shows the waveforms and calculations for resistor values. MAXIM NEW RELEASES DATA BOOK, 1994, P. 5-118.

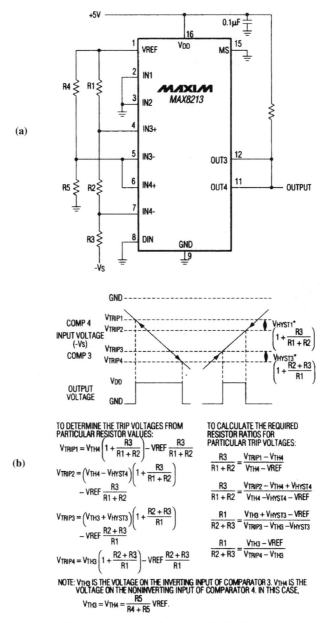

Negative overvoltage/undervoltage detector (window detector)

Fig. 6-39 The circuit in the illustration shows how two comparators can be connected to detect when a negative voltage level is between two trip voltages. The combination of comparator 1 (overvoltage) and comparator 2 (undervoltage) creates this window-detector function. Figure 6-39B shows the waveforms and calculations for resistor values. MAXIM NEW RELEASES DATA BOOK, 1994, P. 5-118.

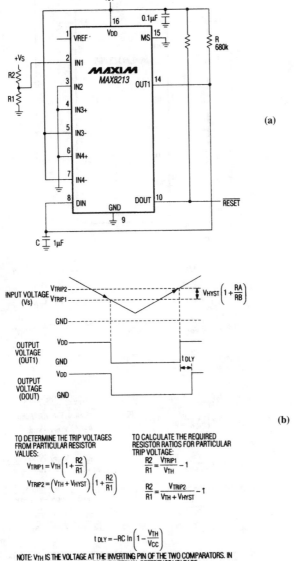

(a)

(b)

TO DETERMINE THE TRIP VOLTAGES FROM PARTICULAR RESISTOR VALUES:

$$V_{TRIP1} = V_{TH}\left(1 + \frac{R2}{R1}\right)$$

$$V_{TRIP2} = (V_{TH} + V_{HYST})\left(1 + \frac{R2}{R1}\right)$$

TO CALCULATE THE REQUIRED RESISTOR RATIOS FOR PARTICULAR TRIP VOLTAGE:

$$\frac{R2}{R1} = \frac{V_{TRIP1}}{V_{TH}} - 1$$

$$\frac{R2}{R1} = \frac{V_{TRIP2}}{V_{TH} + V_{HYST}} - 1$$

$$t_{DLY} = -RC \ln\left(1 - \frac{V_{TH}}{V_{CC}}\right)$$

NOTE: V_{TH} IS THE VOLTAGE AT THE INVERTING PIN OF THE TWO COMPARATORS. IN THIS CASE IT IS EQUAL TO THE INTERNAL REFERENCE VOLTAGE.

Microprocessor reset circuit with time delay

Fig. 6-40 In the circuit shown, the IC of Fig. 6-30 is connected to provide a reset output to a microprocessor when the supply voltage drops below a certain level. The output remains low for 200 ms after the supply voltage goes above the threshold. Figure 6-40B shows the waveforms and calculations for resistor values. MAXIM NEW RELEASES DATA BOOK, 1994, P. 5-119.

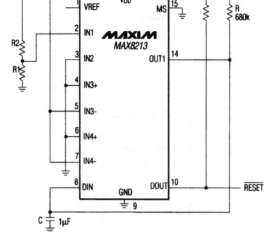

Microprocessor reset circuit with time delay (alternate)

Fig. 6-41 The circuit in the illustration is similar to that of Fig. 6-40 except that the Fig. 6-41 circuit monitors its own supply voltage, rather than the microprocessor supply. The waveforms and calculations of Fig. 6-40B still apply. MAXIM NEW RE-LEASES DATA BOOK, 1994, P. 6-119.

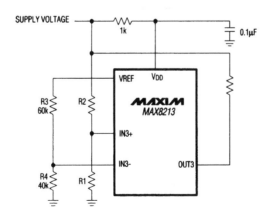

Undervoltage monitoring for low voltages

Fig. 6-42 In the circuit shown, the IC of Fig. 6-30 is connected to provide an output (to an indicator or other load) when the supply drops to 2.25 V (military temperature range) or 2.1 V (commercial temperature range). Resistors R3 and R4 divide the reference to create 0.5 V at IN3. R1 and R2 are used to set the trip level. This current will trip when the supply voltage reduces to:

$$0.5 \text{ V} \left(1 + \frac{R2}{R3} \right)$$

Maxim New Releases Data Book, 1994, p. 5-120.

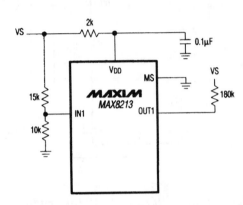

Undervoltage monitoring for 3.3-V digital supplies

Fig. 6-43 In the circuit shown, the IC of Fig. 6-30 is connected to provide an output (to an indicator or other load) when the supply drops to 3.125 V. After OUT1 goes low, the zero-level (0 level) is maintained down to about 0.8 V. Maxim New Releases Data Book, 1994, p. 5-120.

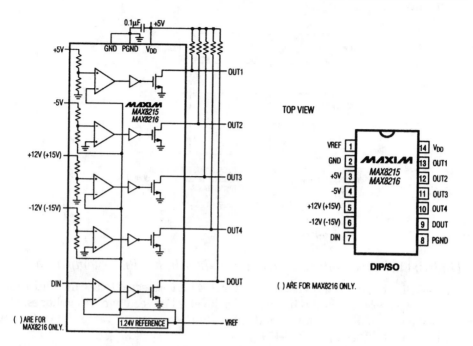

Dedicated microprocessor voltage monitors

Fig. 6-44 The MAX8215 contains five voltage comparators: four are for monitoring +5 V, –5 V, +12 V, and –12 V, and the fifth monitors any desired voltage. The MAX8216 is identical except that it monitors ±15-V supplies instead of ±12 V. The resistors required to monitor these voltages and provide comparator hysteresis are included on chip. All comparators have open-drain outputs. These devices consume 250-μA maximum supply current over temperature. MAXIM NEW RELEASES DATA BOOK, 1994, P. 5-123.

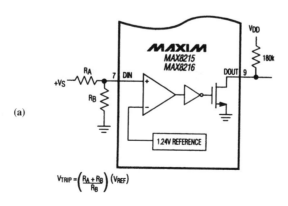

(a)

$$V_{TRIP} = \left(\frac{R_A + R_B}{R_B} \right) (V_{REF})$$

Continued

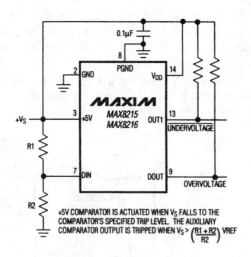

V_{TRIP2}

INPUT VOLTAGE
(V_S)

$V_{HYST} \left(1 + \frac{RA}{RB}\right)$

V_{TRIP1}

GND

OUTPUT
VOLTAGE

V_{DD}

(b)

GND

TO DETERMINE THE TRIP VOLTAGES
FROM PARTICULAR RESISTOR
VALUES:

$$V_{TRIP1} = V_{REF} \left(1 + \frac{RA}{RB}\right)$$

$$V_{TRIP2} = (V_{REF} + V_{HYST}) \left(1 + \frac{RA}{RB}\right)$$

$$V_{HYST} = 16mV\ TYP$$

TO CALCULATE THE REQUIRED
RESISTOR RATIOS FOR PARTICULAR
TRIP VOLTAGES:

$$\frac{RA}{RB} = \frac{V_{TRIP1}}{V_{REF}} - 1$$

$$\frac{RA}{RB} = \frac{V_{TRIP2}}{V_{REF} + V_{HYST}} - 1$$

Undervoltage/overvoltage comparator using the auxiliary comparator

Fig. 6-45 In the circuit shown, the IC of Fig. 6-44 is connected to use the (fifth) auxiliary comparator to detect when a voltage level is between two trip voltages. Figure 6-45B shows the waveforms and calculations for resistor values. MAXIM NEW RELEASES DATA BOOK, 1994, P. 5-129.

$+V_S$

R1

R2

$0.1\mu F$

8

PGND

2 GND

V_{DD} 14

MAXIM

MAX8215
MAX8216

3 +5V

OUT1 13
UNDERVOLTAGE

7 DIN

DOUT 9
OVERVOLTAGE

+5V COMPARATOR IS ACTUATED WHEN V_S FALLS TO THE
COMPARATOR'S SPECIFIED TRIP LEVEL. THE AUXILIARY
COMPARATOR OUTPUT IS TRIPPED WHEN $V_S > \left(\frac{R1 + R2}{R2}\right) V_{REF}$

Monitoring the supply for overvoltage/undervoltage

Fig. 6-46 In the circuit shown, the IC of Fig. 6-44 is connected to monitor the voltages at V_{DD} for both undervoltage and overvoltage. The +5-V comparator checks undervoltage. Overvoltage is monitored by the auxiliary comparator. MAXIM NEW RELEASES DATA BOOK, 1994, P. 5-129.

Digital/microprocessor circuits

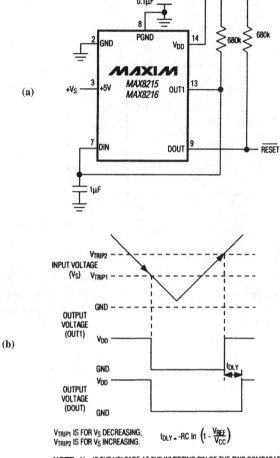

Voltage monitor with microprocessor reset

Fig. 6-47 In the circuit shown, the IC of Fig. 6-44 is connected to provide a reset output to a microprocessor when the supply voltage drops below a certain level. The output remains low for 200 ms after the supply voltage goes above the threshold. Figure 6-47B shows the waveforms and calculations for resistor values. MAXIM NEW RELEASES DATA BOOK, 1994, P. 5-130.

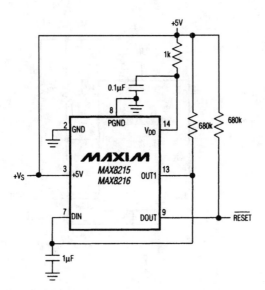

Voltage monitor with microprocessor reset (alternate)

Fig. 6-48 The circuit in the illustration is similar to that of Fig. 6-44 except that the Fig. 6-48 circuit monitors its own supply voltage, rather than the microprocessor supply. The waveforms and calculations of Fig. 6-47B still apply. MAXIM NEW RE-LEASES DATA BOOK, 1994, P. 5-131.

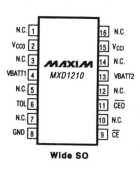

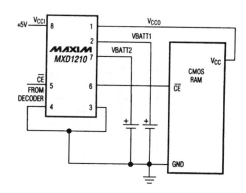

PART	TEMP. RANGE	PIN-PACKAGE
MXD1210CPA	0°C to +70°C	8 Plastic DIP
MXD1210CSA	0°C to +70°C	8 SO
MXD1210CWE	0°C to +70°C	16 Wide SO
MXD1210C/D	0°C to +70°C	Dice*
MXD1210EPA	-40°C to +85°C	8 Plastic DIP
MXD1210ESA	-40°C to +85°C	8 SO
MXD1210EWE	-40°C to +85°C	16 Wide SO
MXD1210MJA	-55°C to +125°C	8 CERDIP

*Contact factory for dice specifications.

6

381

Nonvolatile RAM (random-access memory) controller

Fig. 6-49 The MXD1210 controller is a very low-power CMOS device that converts standard (volatile) CMOS RAM into nonvolatile memory. The IC also continually monitors the power supply to provide RAM write-protection when power to the RAM is in a marginal (out-of-tolerance) condition.

When the power supply begins to fail, the RAM is write protected, and the IC switches to battery-backup mode. The RAM power-supply switch directs power to the RAM from the incoming supply or to the selected battery, whichever is the greater voltage. The write-protection function is enabled when a power failure is detected. The power-failure detection range is set by the condition of the TOL pin. With TOL at ground, the range is from 4.75 V to 4.50 V. With TOL at VCCO, the range is from 4.50 V to 4.25 V. The write-protection function holds the $\overline{CEO}$ output to within 0.2 V of V_{CCI}, or of the selected battery, whichever is greater. If $\overline{CE}$ input is low (active) when power failure is detected, then $\overline{CEO}$ is held low until $\overline{CE}$ is brought high, at which time $\overline{CEO}$ is gated high for the duration of the power failure. This occurs during the RD/WR cycle, preventing data corruption if the RAM access is a WR cycle.

The second battery is optional. When two batteries (typically both 3 V) are used, the stronger battery is selected to provide RAM backup and to power the MXD1210. The battery-status warning notifies the system (by inhibiting the second memory cycle) when the stronger of the two batteries (or the one battery) measures less than 2 V. MAXIM NEW RELEASES DATA BOOK, 1994, P. 5-133.

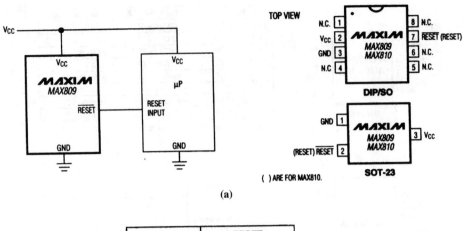

Three-pin microprocessor reset monitor

Fig. 6-50 These simple, single-function ICs assert a reset signal to the microprocessor whenever the V_{CC} supply declines below a preset threshold and keeps the reset asserted for at least 140 ms after V_{CC} has risen above the reset threshold. The MAX809 has an active-low $\overline{\text{RESET}}$ output (which is guaranteed to be in the correct state for V_{CC} down to 1 V). The MAX810 has an active-high RESET output. Figure 6-50B shows the reset threshold for various models of the IC. MAXIM NEW RELEASES DATA BOOK, 1995, P. 5-71.

=7=

Voltage/frequency and frequency/voltage circuits

This chapter is devoted to circuits that convert voltage to frequency (V/F) and frequency to voltage (F/V). Before you get into testing and troubleshooting for these circuits, read these V/F and F/V converter basics.

V/F converter operation (IC)

Most present-day V/F and F/V circuits use a V/F IC as the basic converter element. Figure 7-A shows such an IC connected as a V/F converter. The circuit is essentially a relaxation oscillator with an output frequency proportional to the input voltage. The voltage to be converted is applied to a comparator within the IC at pin 7. The comparator output is applied to a one-shot. Output pulses from the one-shot are applied to pin 3 (the circuit output) through Q1 and to a current switch.

The current switch interrupts current to RL and it thus provides current pulses at pins 1 and 6. Except at zero, the current pulses keep the average voltage across CL (and at pin 6 of the comparator) slightly greater than the input voltage, so the one-shot continues to produce pulses. Resistor RS is made adjustable so the current source can be set to a given scale factor (frequency out for a given voltage in), with R_T and C_T selected for some given frequency range.

F/V converter operation (IC)

The same V/F IC shown in Fig. 7-A also is used as an F/V converter. Figure 7-B shows typical connections. Note that for F/V, the frequency input (pulse or square

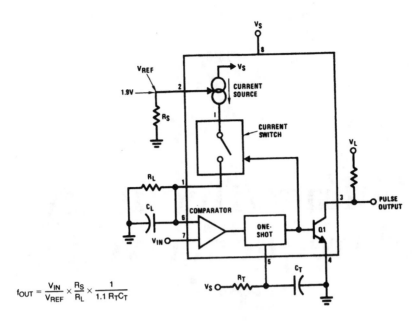

$$f_{OUT} = \frac{V_{IN}}{V_{REF}} \times \frac{R_S}{R_L} \times \frac{1}{1.1\, R_T C_T}$$

Fig. 7-A   V/F IC connected as a V/F converter.

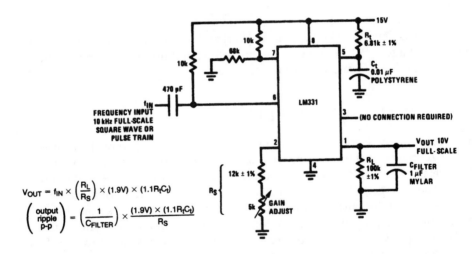

$$V_{OUT} = f_{IN} \times \left(\frac{R_L}{R_S}\right) \times (1.9V) \times (1.1 R_f C_t)$$

$$\left(\begin{array}{c} \text{output} \\ \text{ripple} \\ \text{p-p} \end{array}\right) = \left(\frac{1}{C_{FILTER}}\right) \times \frac{(1.9V) \times (1.1 R_f C_t)}{R_S}$$

Fig. 7-B   Basic F/V converter.

wave) is applied to the comparator at pin 6. The other comparator input (pin 7) is connected to a fixed voltage. This connection produces pulses at the comparator output, which corresponds in frequency to the input. The comparator pulses trigger the one-shot. In turn, the one-shot controls the amount of current at pin 1 (and thus the circuit output voltage) through the current switch. Resistor RS is made adjustable so the current source can be set to produce a given scale factor (voltage out for a given frequency in) with R_T and C_T selected for some given frequency range.

Voltage/frequency and frequency/voltage circuits

In the circuit of Fig. 7-B, R_S is adjusted so the voltage output (pin 1) is 10 V when the input frequency is 10 kHz.

V/F converter operation (discrete component)

V/F ICs cannot be used for all voltage-to-frequency applications, so a number of discrete-component circuits have been developed to meet special needs. The following is a summary of discrete-component V/F circuit techniques.

Ramp-comparator V/F

Figure 7-C shows the basic ramp-comparator V/F concept. The input drives an integrator, and the slope of the integrator ramp varies with the input-derived current. When the ramp crosses V_{REF}, the comparator turns on the switch, discharging the capacitor and restarts the cycle. The frequency of this action directly relates to input voltage. In some designs, one op amp serves as both integrator and comparator.

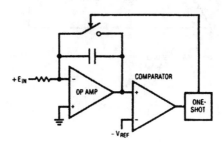

Fig. 7-C   Ramp-comparator V/F.

Charge-pump V/F

In the circuit of Fig. 7-D, the integrator is enclosed in a charge-dispensing loop. Capacitor C1 charges to V_{REV} during the integrator-ramp time. When the comparator trips, C1 is discharged into the op-amp summing point, forcing the op amp high.

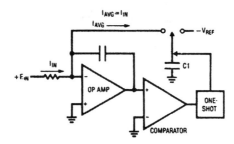

Fig. 7-D   Charge-pump V/F.

After C1 discharges, the op amp begins to ramp and the cycle repeats (frequency is related to input voltage).

Current-balance V/F

In the circuit of Fig. 7-E, the current sink pulls current from the summing point each time the op-amp output trips the comparator. Current is pulled from the summing point for the timing-reference duration, forcing the integrator positive. At the end of the current-sink period, the integrator output again goes negative. The frequency of this action is input-related.

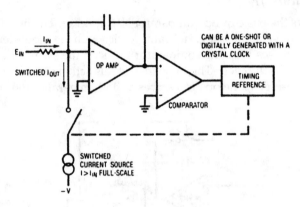

Fig. 7-E   Current-balance V/F.

Loop charge-pump V/F

In the circuit of Fig. 7-F, the circuit output switches a charge pump. The output of the charge pump, integrated to dc, is compared to the input voltage. The amplifier (an op amp operating as a dc amplifier) forces the V/F operating frequency to be a direct function of input voltage. The frequency-compensating capacitor, required because of loop delays, limits loop response time.

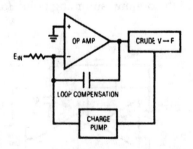

Fig. 7-F   Loop charge-pump V/F.

Loop-DAC V/F

In the circuit of Fig. 7-G, the charge pump is replaced by digital counters, a quartz time base and a DAC (Chapter 9). The loop forces the DAC LSB to oscillate around the ideal value. These oscillations are integrated to dc in the loop-compensation capacitor so the circuit tracks input shifts much smaller than a DAC LSB. Typically, a 12-bit DAC (4096 steps) yields 1 part in 50,000 resolutions. Circuit linearity is set by the DAC specifications.

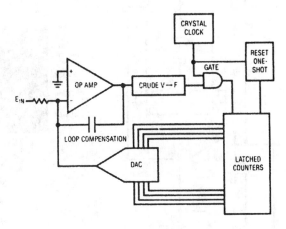

Fig. 7-G Loop DAC V/F.

V/F-converter tests

The obvious test for any V/F converter is to vary the input voltage over the range and check that the output frequency varies accordingly (use a digital meter at the input and a frequency counter at the output). Using the discrete-component circuit of Fig. 7-H as an example, the output frequency should vary between 0 and 30 kHz when the input voltage is varied between 0 and 3 V (1-V input produces 10-kHz output, 2-V in for 20-kHz out, etc.). If practical, the circuit also can be subjected to temperature changes and the output frequency monitored for drift. With the circuit of Fig. 7-H, the drift is supposed to be about 20 ppm/°C.

Using the circuit of Fig. 7-I as an example, two inputs are required for testing, and the output should be the ratio of the two inputs. That is, if V_1 is –10 V and V_2 is –8 V, the ratio is 10/8 = 1.25, and the output should be 12.5 kHz. Note that *FS* (full-scale) for the circuit of Fig. 7-I is 15 kHz, so ratios beyond 1.5 cannot be measured.

F/V converter tests

The test for an F/V is the reverse of that for a V/F. That is, you vary the input frequency over the range and check that output voltages vary accordingly. For most

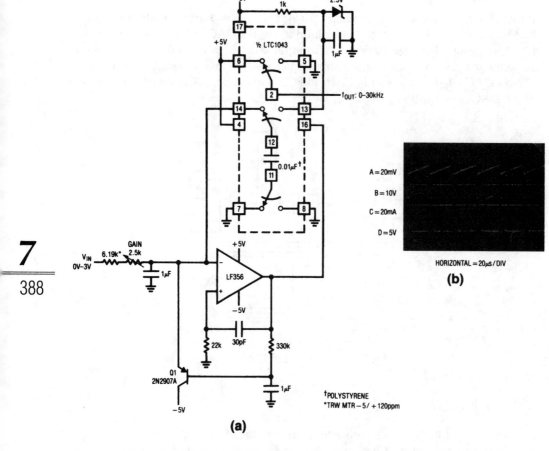

Fig. 7-H Charge-pump V/F converter.

of the circuits in this chapter, use a pulse or square-wave generator and a frequency counter at the input with a digital meter at the output. In some cases, you can use a simple RC differentiator to convert from sine waves to pulses at the inputs, but this might disturb operation of certain circuits (but a sine-wave input should only be used for testing when a pulse or square-wave generator is not available).

V/F-converter troubleshooting

The first step in troubleshooting V/F-converter circuits involves checking that the desired output frequency is produced by a given input voltage. If the output is not correct, try correcting the problem with adjustment.

If the problem cannot be corrected by adjustment, trace signals using a meter or scope from input (typically a dc voltage) to output (typically pulses). From that

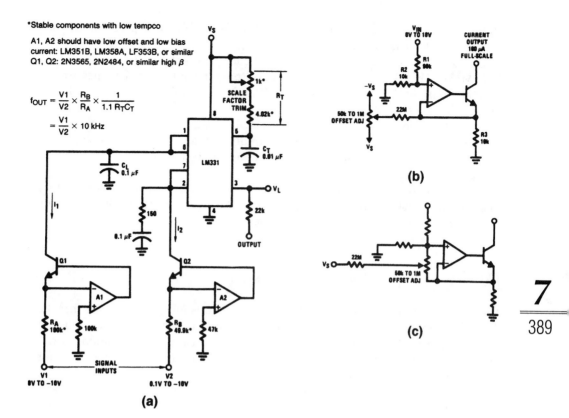

Fig. 7-I   Ratio V/F converter.

point on, it is a matter of voltage measurements and/or point-to-point resistance measurements. The following are typical examples.

In the circuit of Fig. 7-H, begin by checking for pulses at pin 2 of the LTC1043. Pin 2 switches between pins 5 (ground) and 6 (+5 V), at a frequency determined by the signal at clock pin 16. If there is no signal at pin 16, the LTC switches at a fixed rate near 200 kHz. In this circuit, the LTC1043 clock is synchronized with the signal at pin 16.

If pin 2 is not switching at any frequency, suspect the LTC1043. If pin 2 is switching at a fixed frequency, with a variable V_{IN}, suspect the LF356 and associated parts. The output of the LF356, trace B in Fig. 7-H(B) should be a series of positive ramps (trace A). Current flowing from the LF356 summing point into the 0.01-μF capacitor at the end of the ramp should produce a series of negative spikes, trace C in Fig. 7-H(B). Simultaneously, there should be a series of pulses (trace D) at the noninverting input of the LF356.

Note that Q1 prevents the LF356 from going to the negative rail (and staying there) by pulling the summing point negative if the A1 output stays low long enough to charge the 1-μF/330-kΩ RC during startup. Also note that if the circuit shows excessive drift, or nonlinearity in output frequency (for a given input voltage), suspect the 0.01-μF capacitor.

In the circuit of Fig. 7-I, begin by checking for pulses at the 22-kΩ resistor (pin 3 of LM331). There should be pulses at this output no matter what voltages are applied at the input (including zero input). With both V_1 and V_2 at zero, scale-factor trim is adjusted so the output is 10 kHz.

If there are no pulses, suspect the LM331, or possibly the timing capacitor CT. The same is true if there are pulses, but the frequency cannot be brought within the desired range or if the frequency does not change with changing voltage at the input (pins 1/6 and 2/7).

If there are pulses, the pulse frequency is not controlled by voltages V_1 and V_2, suspect Q1, Q2, A1, and the associated parts. Make certain that the voltages at pins 1/6 and 2/7 vary when V1 and V2 are varied.

7 *F/V-converter troubleshooting*

The first step in troubleshooting F/V converter circuits involves checking that the desired output voltage is produced by a given input frequency. If the output is not correct, try correcting the problem with adjustment, using the calibration or trim procedures described for the circuit. Note that calibration (or trim) is described for all circuits in this chapter that require adjustment.

If the problem cannot be corrected by adjustment, trace signals using a meter or scope from input (typically pulse or square-wave signals) to the output (typically a dc voltage). After tracing, make voltage measurements and/or a point-to-point resistance measurement. The following are typical examples.

In the circuit of Fig. 7-J, begin by checking for pulses at pin 14 of the LRC1043. Pin 12 switches between pins 13 and 14 at a frequency that is determined by the signal at clock pin 16 (trace A, Fig. 7-J), which is the F/V circuit input in this case. If pin 14 is not switching at any frequency, suspect the LTC1043 (of course, check for proper voltages at pins 4, 13, and 17, as well as ground at pins 7 and 8). If the LTC1043 and voltages are good, but there are no pulses at pin 14, the 1000-pF capacitor at pins 11 and 12 is the prime suspect (trace B shows the capacitor signal). The capacitor might be shorted or badly leaking.

If there are pulses at the inverting input of the LF356, but there is no output voltage, or if the output voltage does not vary with changes in pulse frequency at the input, suspect the LF356 or associated parts. The feedback resistors determine the LF356 gain, and the 1-μF feedback capacitor averages the pulse input to a dc output (trace C shows the negative and positive swing of the LF356 output).

Note that if the circuit shows excessive drift, or nonlinearity in output voltage (for a given input frequency), suspect the 100-pF capacitor at pins 11 and 12.

In the circuit of Fig. 7-B, begin by checking for a dc voltage at pin 1 of the LM331, with a signal at the input (if the input to the LM331 is zero, the voltage at pin 1 is zero). If the voltage does not vary at pin 1, when the frequency of the sig-

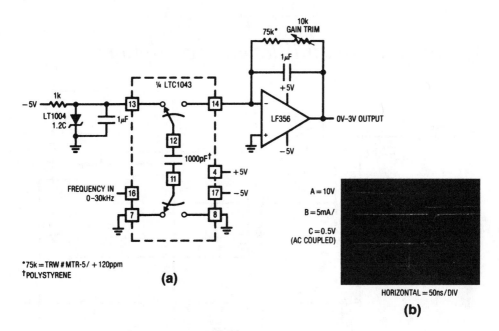

Fig. 7-J   Charge-pump F/V converter.

nal at pin 6 is varied (between 0 and 10 kHz), suspect the LM331. It also is possible that the timing capacitor is shortened or badly leaking.

Try monitoring the output voltage (with the input frequency steady) while varying the gain-adjust control. If there is no change in output voltage, check voltage and resistance from pin 2 of the LM331 to ground. If the connection from pin 2 is good, suspect the LM331.

F/V-converter troubleshooting

V/F and F/V circuit titles and descriptions

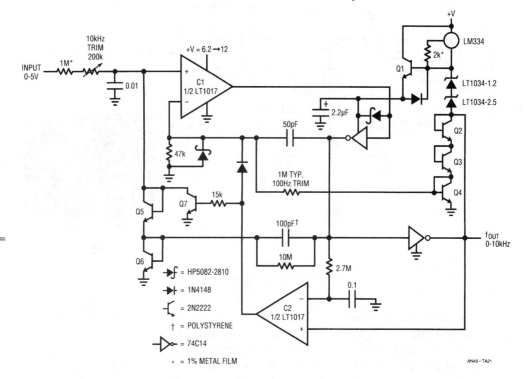

Micropower V/F converter

Fig. 7-1   The circuit in the illustration has a 0.05% linearity but consumes only 90 μA of supply current. To calibrate, select a value for the fixed resistor at the input that produces a 100-Hz output for a 50-mV input. Then, apply 5 V at the input and trim the input pot for a 10-kHz output. LINEAR TECHNOLOGY, LINEAR APPLICATIONS HANDBOOK, 1993, P. AN45-15.

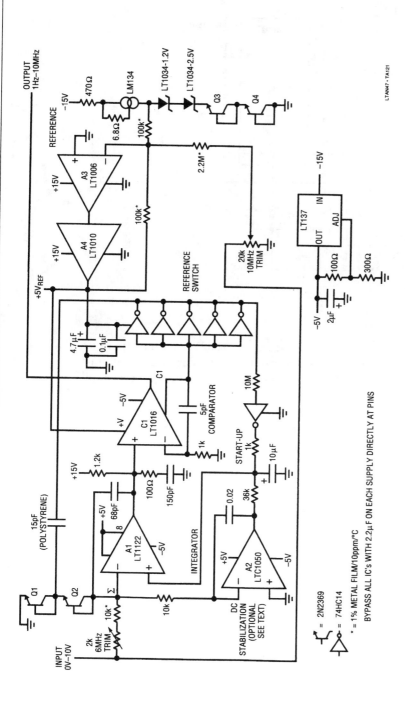

Wide-range V/F converter

Fig. 7-2 The circuit in the illustration converts inputs from 0 V to 10 V into an output from 1 Hz to 10 MHz, with a linearity of 0.03% and 50 ppm/°C. To calibrate, apply 6.000 V at the input and adjust the 2-kΩ pot for 6.000-MHz output. Then apply 10.000 V and trim the 20-kΩ pot for 10.000-MHz output. Repeat these adjustments until both points are fixed. The low-drift of A2 eliminates a zero adjustment. If operation below 600 Hz is not required, A2 and the associated parts can be deleted. LINEAR TECHNOLOGY, LINEAR APPLICATIONS HANDBOOK, 1993, P. AN47-54.

V/F and F/V circuit titles and descriptions

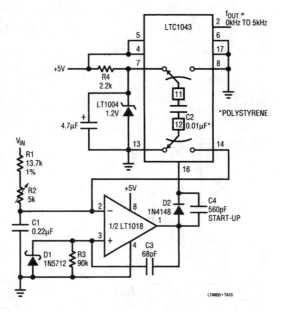

Single-supply V/F converter

Fig. 7-3 The circuit in the illustration has a linearity of 0.0025%, with temperature coefficient of 20C/ppm°C and consumes 3 mA of quiescent current. To calibrate, apply 5 V at the input and adjust R2 for a 5-kHz output. You can get other full-scale frequencies by selecting R_1 and C_2, using:

$$F_{OUT} = \frac{V_{IN}}{(R_{IN} \times V_{REF} \times C_T)}$$

where $R_{IN} = R_1 + R_2$, and $C_T = C_2$. LINEAR TECHNOLOGY, LINEAR APPLICATIONS HANDBOOK, 1993, P. AN50-7.

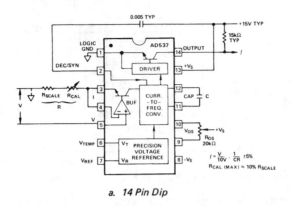

a. 14 Pin Dip

b. 10 Pin Metal Can

Basic V/V converter

Fig. 7-4 The circuit in the illustration shows the basic connections for an AD537 V/F converter. Note that the logic ground is strapped to the supply ground. The value of R is chosen so the FS (full-scale) input voltage produces a current of 1 mA. For example, for a 0 to +10-V input, R is 10 kΩ. The value of C is set by the equation shown in Fig. 7-4A. For a 10-V FS input, the equation is simplified to:

$$C = \frac{1}{CR}$$

Typically, $C = 0.01$ μF for 10-kHz FS (1 Hz/mV), or 1000 pF for 100-kHz FS (10 Hz/mV), assuming an R of 10 k. To calibrate, begin with 0 V in and adjust the optional offset pot ROS (pins 9 and 10) for zero out. (A scope at the output will show when the IC stops oscillating and produces zero out.) Then apply the FS voltage input and adjust R_{CAL} for the full-scale output. For example, with R at 10 kΩ and C at 1000 pF, apply 10 V and adjust R_{CAL} for an output of 100 kHz. When FS is small, adjustment of R_{CAL} might affect offset, so it might be necessary to readjust ROS. Typical input-voltage drift (after offset nulling) is 1 μV/°C. ANALOG DEVICES, APPLICATIONS REFERENCE MANUAL, 1993, P. 23-15.

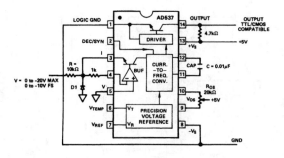

V/F converter for negative-input operation

Fig. 7-5 The circuit in the illustration is similar to that of Fig. 7-4, except for the negative-input capability. ANALOG DEVICES, APPLICATIONS REFERENCE MANUAL, 1993, P. 23-16.

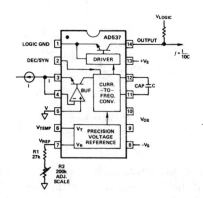

V/F converter with scale adjustment for current inputs

Fig. 7-6 The circuit in the illustration is similar to that of Fig. 7-4, except that the input signal is in the form of a negative current. Capacitor C is selected to be 5% below the normal value. With R2 in the mid-position, the output frequency is given by:

$$F = \frac{I}{10.5 \times C}$$

where *F* is in kHz, *I* is in mA and *C* is in µF. For example, for an FS frequency of 10 kHz at an FS input of 1 mA, *C* = 9500 pF. To calibrate, apply FS input and adjust R2 and FS output. ANALOG DEVICES, APPLICATIONS REFERENCE MANUAL, 1993, P. 23-16.

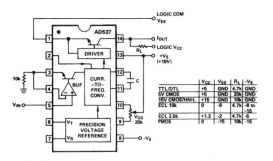

V/F converter with digital interfacing

Fig. 7-7 The circuit in the illustration shows the connections for interfacing the V/F converter of Fig. 7-4 with various logic families and components. The required logic common voltage, logic supply voltage, pull-up resistor, and $-V_S$ supply are shown in the table. In the TTL mode, up to 12 standard gates (20 mA) can be driven at a maximum low voltage of 0.4 V. ANALOG DEVICES, APPLICATIONS REFERENCE MANUAL, 1993, P. 23-16.

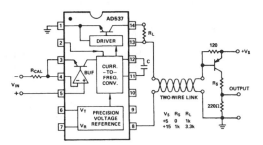

V/F converter with two-wire data transmission

Fig. 7-8 The circuit in the illustration shows the connections for operating the V/F converter of Fig. 7-4 at the remote end of a single wire-pair. The pnp converts current modulation into a voltage signal suitable for driving digital logic. The wire-pair line supplies power to the V/F. Using the values shown, the supply current through the line is:

	Output off	**Output on**
Zero signal	1.2 mA	5.2 mA
Full-scale	3.5 mA	7.5 mA
(1 mA)		

Approximately 500 mV of variation appears on the remote end of the supply line but does not affect operation of the V/F. ANALOG DEVICES, APPLICATIONS REFERENCE MANUAL, 1993, P. 23-17.

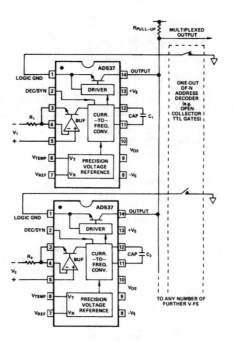

V/F converter with signal multiplexing

Fig. 7-9 The circuit in the illustration shows the connections for multiplexing the outputs of several V/F converters (Fig. 7-4). In this circuit, all V/F converters are operating continuously, but only the device having the LOGIC COMMON pin grounded (through the open-collector decoder or other digital-switching element) transmits an output. ANALOG DEVICES, APPLICATIONS REFERENCE MANUAL, 1993, P. 23-17.

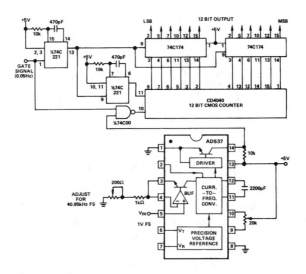

V/F converter as an A/D converter

Fig. 7-10 The circuit in the illustration shows the V/F converter of Fig. 7-4 connected as a 12-bit analog-to-digital (A/D) converter (Chapter 9). Using the values shown, the circuit generates a binary output of 111111111111 (decimal 4095) and the first bit occurs for an input of 244 μV. The offset pot is adjusted as described for Fig. 7-4. Then, the input pot is adjusted for an output frequency (at pin 14) of 40.95 kHz as shown. ANALOG DEVICES, APPLICATIONS REFERENCE MANUAL, 1993, P. 23-19.

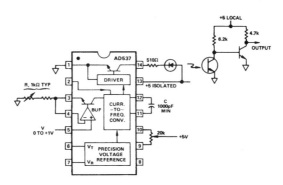

V/F converter with opto-coupling

Fig. 7-11 The circuit in the illustration shows the V/F converter of Fig. 7-4 connected for opto-coupling to a transmission line. The output pulses at pin 14 are amplified by the circuit to a level suitable for use with a fiberoptic transmitter or directly to a metal line. ANALOG DEVICES, APPLICATIONS REFERENCE MANUAL, 1993, P. 23-20.

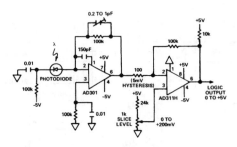

Photodiode preamplifier for V/F-converter output

Fig. 7-12   The circuit in the illustration shows a photodiode preamplifier suitable for converting pulses from a V/F converter and transmitted over a fiberoptic cable into a logic output. The circuit has sufficient bandwidth to accept optical inputs up to 20 kHz. ANALOG DEVICES, APPLICATIONS REFERENCE MANUAL, 1993, P. 23-20.

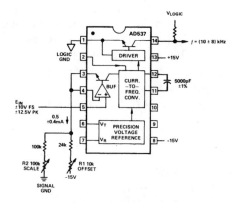

V/F converter with bipolar input

Fig. 7-13   The circuit in the illustration shows the V/F converter of Fig. 7-4 connected to accept bipolar inputs. To calibrate, set the input to zero and adjust R1 for an output frequency of 10 kHz. Then apply an input of +10 V and adjust R2 for an output of 18 kHz. ANALOG DEVICES, APPLICATIONS REFERENCE MANUAL, 1993, P. 23-20.

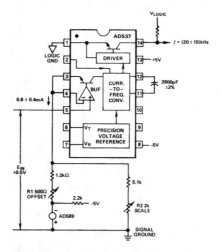

V/F converter with bipolar input (stable reference)

Fig. 7-14 The circuit in the illustration is similar to that of Fig. 7-13, but with an AD589 voltage-reference added. ANALOG DEVICES, APPLICATIONS REFERENCE MANUAL, 1993, P. 23-20.

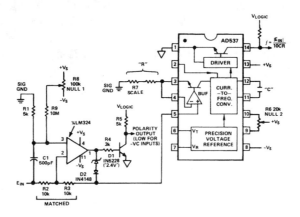

V/F converter with bipolar input (absolute value)

Fig. 7-15 The circuit in the illustration accepts bipolar inputs but has double the range of linear operation (compared to the circuits of Figs. 7-13 and 7-14). Selection of values for *R* and *C* is the same as for the Fig. 7-4 circuit, as is calibration. However, the Fig. 7-15 circuit also requires that the op-amp offset be nulled by R8. The Fig. 7-15 circuit is scaled for an input of ±10 V full-scale. You can get best accuracy by nulling the input to the AD537 (with a digital voltmeter at pin 5). Note that if

V/F and F/V circuit titles and descriptions

R2 and R3 are not properly matched, there might be a reversal at the input. ANALOG DEVICES, APPLICATIONS REFERENCE MANUAL, 1993, P. 23-21.

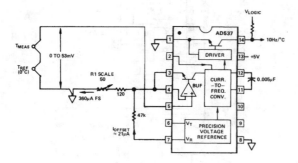

V/F converter with thermocouple interface

Fig. 7-16 The circuit in the illustration is used to indicate temperature in °C on a counter/display with a 100 ms gate width. The V/F converter must deliver 7 kHz for an input of 53.14 mV. (The output of a Chromel-Constantan, Type C, thermocouple, with a reference junction of 0°C, varies from 0 to 53.14 mV over the temperature range of 0 to +700°C with a slope of 80.678 μV/degree over most of the range and some nonlinearity from 0 to +200°C.) The circuit of Fig. 7-16 provides the greatest accuracy from +300 to +700°C. To calibrate, raise the thermocouple to a known reference temperature (preferably near +500°C) and adjust R1 for the correct readout on the counter/display. (See Chapter 10 for thermocouple discussion.) The error should be within ±0.2% over the range 400–700°C. ANALOG DEVICES, APPLICATIONS REFERENCE MANUAL, 1993, P. 23-21.

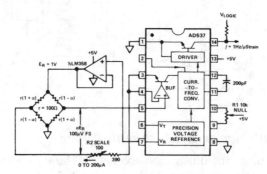

V/F converter with strain-gauge input

Fig. 7-17 The circuit in the illustration shows the V/F converter of Fig. 7-4 connected to accept a strain-gauge input and is calibrated to generate a scale of Hz per

Voltage/frequency and frequency/voltage circuits

microstrain (100 kHz at the assumed FS value). To calibrate, adjust R1 for zero off-set, then apply an FS input (100 mV) and adjust R2 for 100 kHz at the output. ANA-LOG DEVICES, APPLICATIONS REFERENCE MANUAL, 1993, P. 23-22.

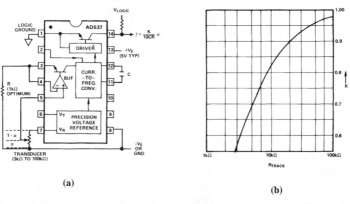

(a)

(b)

V/F converter with resistive-transducer interface

Fig. 7-18 The circuit in the illustration shows the V/F converter of Fig. 7-4 connected to accept resistive-transducer inputs. (Such transducers include: linear displacement, rotary servo pot, level, light-comparator with photo-resistors, etc.) As shown by the equation, the output frequency depends on transducer resistance R, capacitance C, and a constant K (see Fig. 7-18B). ANALOG DEVICES, APPLICATIONS REFERENCE MANUAL, 1993, P. 23-22.

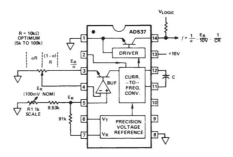

V/F converter with resistive-transducer interface (linear-period)

Fig. 7-19 The circuit in the illustration is similar to that of Fig. 7-18 except that the circuit of Fig. 7-19 operates under linear-period control. That is, linear motion of the transducer-pot slider results in linear control of the period, rather than the frequency, of the output (frequency is inversely proportional to time period). Unlike the circuit of Fig. 7-18, the circuit of Fig. 7-19 can be calibrated for a given output frequency. ANALOG DEVICES, APPLICATIONS REFERENCE MANUAL, 1993, P. 23-23.

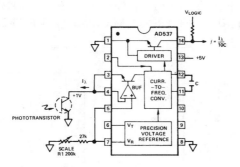

V/F converter with phototransistor interface

Fig. 7-20 The circuit in the illustration shows the V/F of Fig. 7-4 connected to accept phototransistor inputs. As shown by the equation, the output frequency depends on capacitor C and the current generated by the phototransistor. The scale is calibrated to the desired output frequency (for a given phototransistor current) by R1. A<small>NALOG</small> D<small>EVICES</small>, A<small>PPLICATIONS</small> R<small>EFERENCE</small> M<small>ANUAL</small>, 1993, P. 23-22.

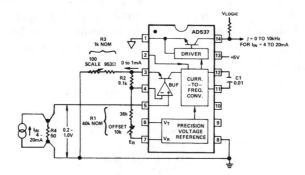

V/F converter for a 4- to 20-mA loop operation

Fig. 7-21 The circuit in the illustration uses the V/F of Fig. 7-4 to convert instrumentation signals in the 4- to 20-mA format to a frequency format. (The output frequency is zero when the input current is 4 mA.) Resistor R1 sets the offset for zero output (input 4 mA), and R3 sets the FS frequency of 10 kHz (input 20 mA). A<small>NA-</small>LOG D<small>EVICES</small>, A<small>PPLICATIONS</small> R<small>EFERENCE</small> M<small>ANUAL</small>, 1993, P. 23-23.

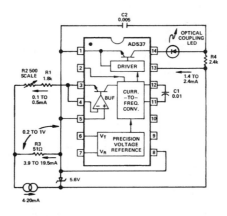

V/F converter for 4- to 20-mA loop operation (self-powered)

Fig. 7-22 The circuit in the illustration is similar to that of Fig. 7-21, except that the IC is powered by the 4- to 20-mA current, and the output is optically coupled to provide isolation. Using the values shown, R2 is adjusted for a 5-kHz output when the input current is 20 mA. ANALOG DEVICES, APPLICATIONS REFERENCE MANUAL, 1993, P. 23-24.

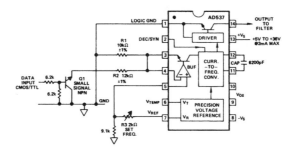

V/F converter as a Bell System data encoder

Fig. 7-23 The circuit in the illustration shows the V/F of Fig. 7-4 connected as an *FSK* (frequency-shift keying) encoder suitable for Bell System modem communication (mark frequency of 1200 Hz and space frequency of 2200 Hz). Resistor R3 sets the frequency of both mark and space outputs. The square-wave output must be filtered before transmission over a public-telephone line. ANALOG DEVICES, APPLICATIONS REFERENCE MANUAL, 1993, P. 23-24.

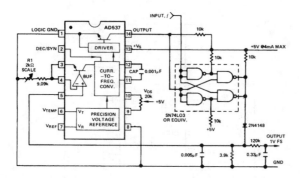

V/F converter as an F/V converter

Fig. 7-24 The circuit in the illustration shows the V/F converter of Fig. 7-24 connected as an F/V converter. The circuit can lock onto any frequency from x to full-scale (10 kHz in this example) within four or five cycles. The dc output (taken from a filter outside the loop) is +1 V for a full-scale input. To calibrate, set the V_{OS} pot to midscale, apply a frequency input of 10 kHz and adjust R1 for +1 V. Then, apply a 10-Hz input signal and trim V_{OS} for a 1-mV output. Retrim R1, as necessary, at 10 kHz. ANALOG DEVICES, APPLICATIONS REFERENCE MANUAL, 1993, P. 23-25.

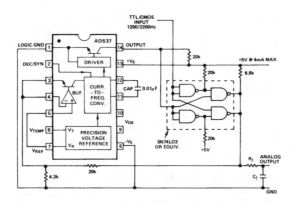

V/F converter as a Bell System data decoder

Fig. 7-25 The circuit in the illustration shows the V/F of Fig. 7-4 connected as an FSK decoder for the Bell System modem communication (mark frequency of 1200 Hz and space frequency of 2200 Hz). No calibration is required. The operating range of this circuit is 800 to 2600 Hz. ANALOG DEVICES, APPLICATIONS REFERENCE MANUAL, 1993, P. 23-25.

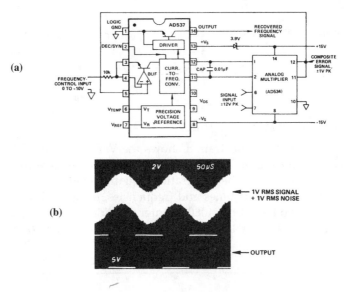

(a)

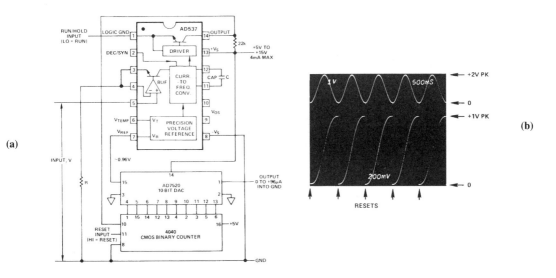

(b)

V/F converter as a PLL (phase-locked loop)

Fig. 7-26 The circuit in the illustration shows the V/F converter of Fig. 7-14 connected as a PLL (phase-locked loop). Figure 7-26B shows the waveforms. The output at pin 14 is a noise-free square wave having exactly the same frequency as the input signal at pins 6/7 of the analog multiplier. Frequency control is applied at pin 4, and no adjustments are required. ANALOG DEVICES, APPLICATIONS REFERENCE MANUAL, 1993, P. 23-25.

V/F and F/V circuit titles and descriptions

Continued

$$N = \frac{1}{T}\int_0^t V(t)dt \quad , \quad T = 10CR$$

T	C	R	f(FS)
1 second	0.01μF	1.8K + 500Ω adj	4096Hz
1 minute	0.5μF	2.2K + 500Ω adj	68.27Hz
1 hour	10.0μF	6.8k + 1kΩ adj	1.138Hz

(c)

(d)

V/F converter as an analog integrator

Fig. 7-27 The circuit in the illustration shows the V/F converter of Fig. 7-4 connected as an analog integrator. Figure 7-27B shows the waveforms (output of integrator). Figure 7-27C shows the integration equation. Figure 7-27D shows the values of C and R for various times and frequencies, using a +1-V analog input. Note that the first two bits of the 4080 counter provide a prescaler, allowing the V/F to run four times faster. ANALOG DEVICES, APPLICATIONS REFERENCE MANUAL, 1993, P. 23-27.

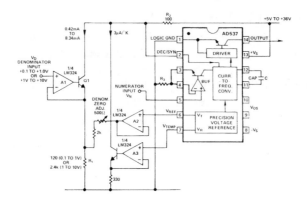

V/F converter as an analog divider

Fig. 7-28 The circuit in the illustration shows the V/F converter of Fig. 7-4 connected as an analog divider (with frequency output). That is, the output frequency is proportional to the ratio of the two input voltages V_D and V_N representing the denominator and numerator. The output frequency depends on the values of R1, R3, and C. For an R_1 of 2.4 kΩ, the output frequency is V_N/VCR_3. To adjust the denominator offset, connect the V_N and V_D inputs together and trim to maintain frequency independent of input voltage. Linearity of division is typically ±0.1%. ANALOG DEVICES, APPLICATIONS REFERENCE MANUAL, 1993, P. 23-27.

Voltage/frequency and frequency/voltage circuits

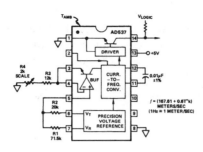

V/F converter as a sound-velocity monitor

Fig. 7-29 The circuit in the illustration shows the V/F converter of Fig. 7-4 connected as a sound-velocity monitor. The circuit uses the temperature and reference outputs of the AD537 because the velocity of sound is related to temperature. The relationship is: $V_S = (331.5 + 0.6T_C) = (167.6 + 0.6T_K)$, where V_S is sound velocity in m/s (meters per second), T_C is Celsius temperature, and T_K is Kelvin temperature. Using the values shown, the voltage on pin 5 is 452.8 mV at 300K (degrees Kelvin), which is scaled by R4 to an output frequency of 347.6 Hz (corresponding to the velocity of sound at 300K). As shown by the equation of Fig. 7-29, R4 is adjusted so the output is 1 Hz when the V/F monitors a sound velocity of 1 m/s. ANALOG DEVICES, APPLICATIONS REFERENCE MANUAL, 1993, P. 23-28.

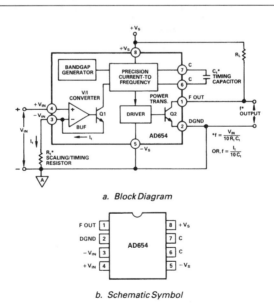

a. Block Diagram

b. Schematic Symbol

Basic V/F converter with square-wave output

Fig. 7-30 The circuit in the illustration shows the basic connections for an AD654 V/F converter. The circuit is similar in function to that of Fig. 7-4, but with some

Continued

differences (such as a square-wave output for the V/F of Fig. 7-30). As shown by the equations, the output frequency is related to input voltage V_{IN} (or the value of input current I) and the values of R_t and C_t. In practical circuits, R_t is made adjustable to provide a specific scale of input voltage to output frequency. ANALOG DEVICES, APPLICATIONS REFERENCE MANUAL, 1993, P. 23-31.

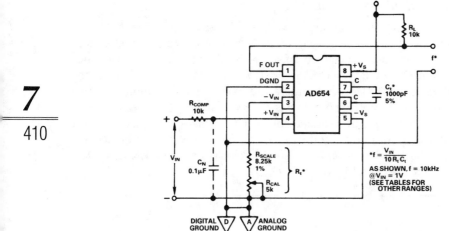

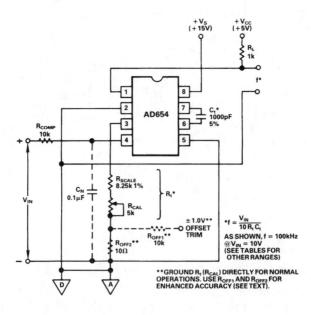

(a)

Voltage/frequency and frequency/voltage circuits

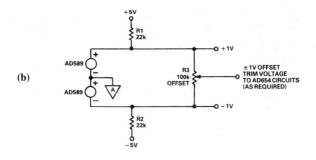

(b)

(c)

FS V$_{IN}$	R$_t$ FS I$_t$ = 100µA	R$_t$ FS I$_t$ = 1mA
100V*	1 meg	100k
10V	100k	10k
1V	10k	1k
100mV	1k	100Ω

NOTE
*Applies *only* to Figure 3.

(d)

FS f	C$_t$ FS I$_t$ = 100µA	C$_t$ FS I$_t$ = 1mA
≥1MHz**	*	≤100pF**
500kHz	*	200pF
250kHz	*	390pF
100kHz	*	1000pF
10kHz	1000pF	10000pF

Notes
*Not recommended, see text.
**"Exalted" operation, see text.

(e)

FS V$_{IN}$	FS Frequency 10kHz	100kHz	500kHz
10V	10V → 10kHz 1mV → 1Hz	10V → 100kHz 1mV → 10Hz	10V → 500kHz** 1mV → 50Hz*
1V	1V → 10kHz 1mV → 10Hz	1V → 100kHz 1mV → 100Hz	1V → 500kHz** 1mV → 500Hz*
100mV	100mV → 10kHz 1mV → 100Hz	100mV → 100kHz 1mV → 1kHz	100mV → 500kHz** 1mV → 5kHz*

Notes
*Adjust OFFSET (if used) as noted.
**Adjust FS cal as noted.

V/F converter for positive-input operation

Fig. 7-31 The circuits in the illustration show the V/F of Fig. 7-30 connected to convert positive inputs to a corresponding output frequency. Figure 7-31B shows an offset-trim bias network (if required). Figures 7-31C and 7-31D show the values of R_t and C_t for various full-scale voltage inputs and output frequencies. To cali-

V/F and F/V circuit titles and descriptions

Continued

brate, apply a known full-scale input voltage and adjust R_{CAL} for the desired FS output, using the values of Fig. 7-31E. If the offset circuit, Fig. 7-31B is used, begin calibration by applying a zero input, (short the inputs together) and adjust the offset pot (R3) for zero output. (A scope at the output pin will show when the IC stops oscillating and produces zero out.) It might be necessary to work between the FS and offset adjustments because they are interactive. ANALOG DEVICES, APPLICATIONS REFERENCE MANUAL, 1993, PP. 23-33, 34, 35.

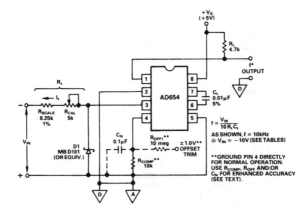

V/F converter for negative-input operation (square-wave output)

Fig. 7-32 The circuit in the illustration is similar to that of Fig. 7-30, except for the negative-input capability. ANALOG DEVICES, APPLICATIONS REFERENCE MANUAL, 1993, P. 23-36.

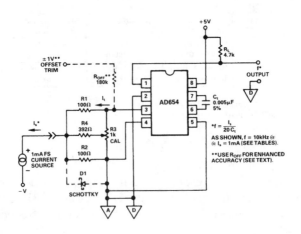

Voltage/frequency and frequency/voltage circuits

V/F converter with negative current input

Fig. 7-33 The circuit in the illustration is similar to that of Fig. 7-30, except that the input signal is in the form of a negative current. ANALOG DEVICES, APPLICATIONS REFERENCE MANUAL, 1993, P. 23-37.

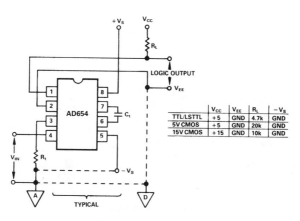

	V_{CC}	V_{EE}	R_L	$-V_S$
TTL/LSTTL	+5	GND	4.7k	GND
5V CMOS	+5	GND	20k	GND
15V CMOS	+15	GND	10k	GND

a. Standard Interfacing

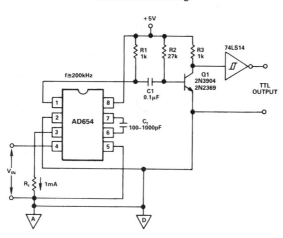

b. High-Speed TTL Output Buffer

V/F converter with standard-logic interfacing

Fig. 7-34 The circuit in the illustration shows the connections for interfacing the V/F converter of Fig. 7-30 with various logic families and components. The required logic common voltage, logic supply, pull-up resistor, and $-V_S$ supply are shown in the table. In the TTL mode, up to 12 standard gates (20 mA) can be driven at a maximum low voltage of 0.4 V. With pin 2 grounded, pin 1 shorted to +15 V, and the oscillator running, the average power in the output stage is about 265 mW. The power is reduced to about one-third when the supply is +5 V. However, if +15 V is used and the output is in the on-state for long periods (low-frequency/long-period), the peak dissipation is about 1260 mW. Also, the dissipation can cause heating (a special problem for the plastic package). ANALOG DEVICES, APPLICATIONS REFERENCE MANUAL, 1993, P. 23-39.

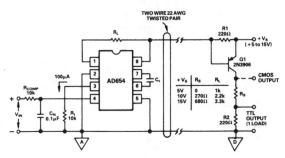

a. Simple Phantom Power Driver

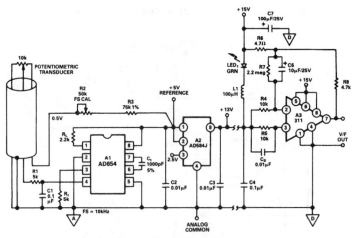

b. Phantom Power Driver/Regulator

V/F converter with phantom power

Fig. 7-35 These circuits show the V/F of Fig. 7-30 connected to receive power from the system or component to which the V/F output is applied. ANALOG DEVICES, APPLICATIONS REFERENCE MANUAL, 1993, P. 23-40.

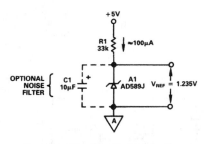

a. Basic Positive Bias Reference Source

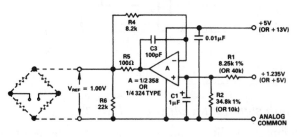

b. Buffered Low Power Reference

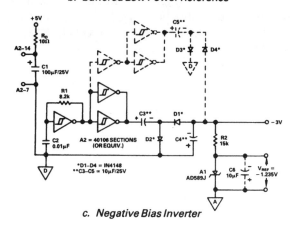

c. Negative Bias Inverter

References for V/F converters

Fig. 7-36 These circuits show several different reference configurations for the V/F converters described in this chapter. Such reference circuits can be used for bias, offset, or the establishing of scaling in a V/F application. ANALOG DEVICES, APPLICATIONS REFERENCE MANUAL, 1993, P. 23-41.

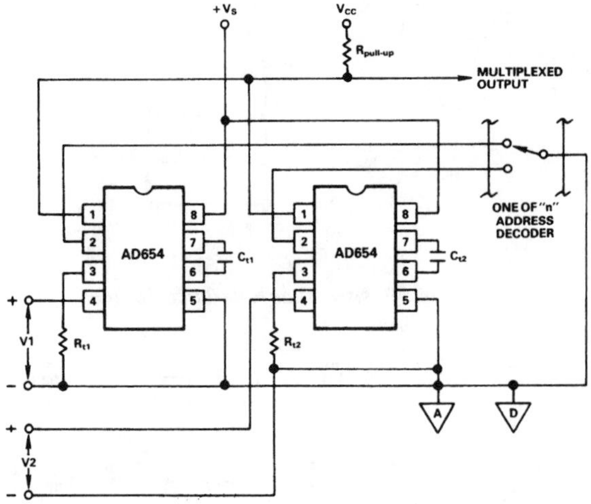

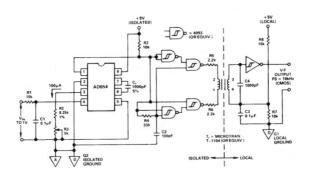

V/F converter with signal multiplexing (alternate)

Fig. 7-37 The circuit in the illustration shows the connections for multiplexing the outputs of several V/F converters (Fig. 7-30). In this circuit, all V/F converters are operating continuously, but only the device having the D_{GND} (pin 2, digital ground) pin grounded transmits an output. ANALOG DEVICES, APPLICATIONS REFERENCE MANUAL, 1993, P. 23-41.

V/F converter with transformer signal isolation

Fig. 7-38 The circuit in the illustration shows the V/F of Fig. 7-30 connected to provide an output signal to a system with a separate digital ground. ANALOG DEVICES, APPLICATIONS REFERENCE MANUAL, 1993, P. 23-42.

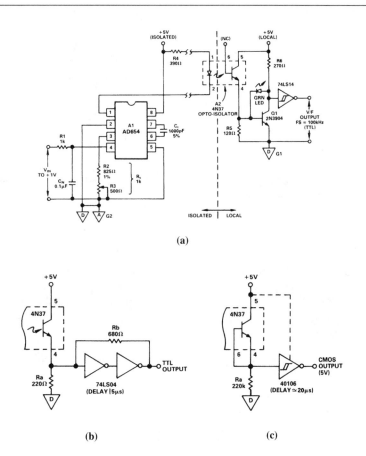

(a)

(b)   (c)

V/F converter with medium-speed opto-coupling (100 kHz)

Fig. 7-39   The circuit in the illustration shows the V/F of Fig. 7-30 connected for opto-coupling to a transmission line. Figures 7-39B and 7-39C show alternate TTL and CMOS output buffers, respectively. ANALOG DEVICES, APPLICATIONS REFERENCE MANUAL, 1993, P. 23-42.

V/F and F/V circuit titles and descriptions

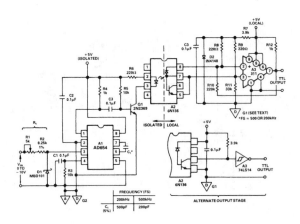

V/F converter with high-speed opto-coupling (200–500 kHz)

Fig. 7-40 The circuit in the illustration shows the V/F of Fig. 7-30 connected for opto-coupling to a transmission line, where the V/F output frequency is substantially higher than 100 kHz. The circuit is useful up to about 1 MHz, but the greatest linearity is at about 200 kHz. ANALOG DEVICES, APPLICATIONS REFERENCE MANUAL, 1993, P. 23-42.

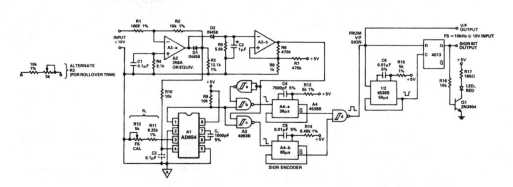

Single-supply bipolar V/F

Fig. 7-41 The circuit in the illustration shows the V/F converter of Fig. 7-30 connected to accept bipolar inputs. To calibrate, apply a –10-V input and adjust R12 for an output of 10 kHz. If the alternate R3 is used to compensate for rollover error (a gain difference between equal-magnitude positive and negative inputs), trim R3 for a positive gain equal to negative gain. (A +10-V input should produce a 10-kHz output, as does a –10-V input.) The signal-bit output should be high, and the LED should be on, for a positive input. ANALOG DEVICES, APPLICATIONS REFERENCE MANUAL, 1993, P. 23-43.

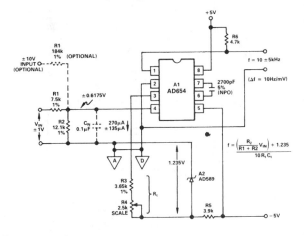

Dual-supply bipolar V/F

Fig. 7-42   The circuit in the illustration shows the V/F of Fig. 7-30 connected to accept bipolar inputs but requires a dual supply (±5 V in this example). ANALOG DEVICES, APPLICATIONS REFERENCE MANUAL, 1993, P. 23-44.

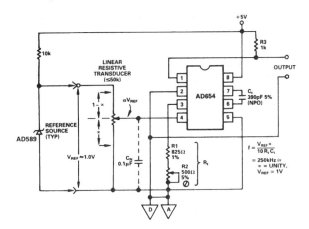

V/F converter with linear resistive-transducer interface

Fig. 7-43   The circuit in the illustration shows the V/F of Fig. 7-30 connected to accept linear resistive-transducer inputs. With the values shown, R2 is adjusted so that the output frequency is 250 kHz (full-scale) when alpha is at unity. ANALOG DEVICES, APPLICATIONS REFERENCE MANUAL, 1993, P. 23-45.

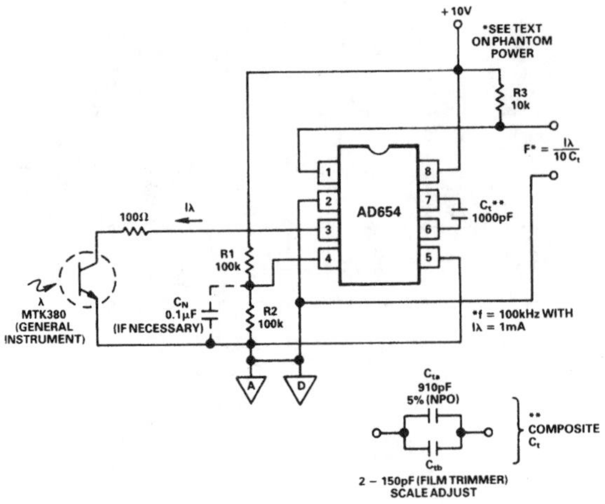

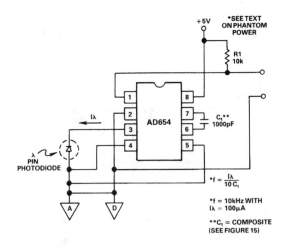

V/F converter with phototransistor input

Fig. 7-44   The circuit in the illustration shows the V/F of Fig. 7-30 connected to accept phototransistor inputs. If necessary, trim C_T (film trimmer) to produce the desired output frequency. ANALOG DEVICES, APPLICATIONS REFERENCE MANUAL, 1993, P. 23-45.

V/F converter with photodiode input

Fig. 7-45   The circuit in the illustration shows the V/F of Fig. 7-30 connected to accept photodiode inputs. If necessary, trim C_T (film trimmer) to produce the desired output frequency. The Figure 15 shown in the illustration refers to Fig. 7-44 in this book. ANALOG DEVICES, APPLICATIONS REFERENCE MANUAL, 1993, P. 23-45.

Voltage/frequency and frequency/voltage circuits

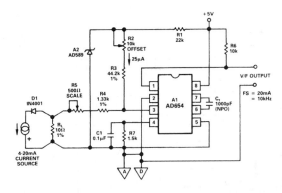

V/F converter for 4- to 20-mA loop operation (alternate)

Fig. 7-46   The circuit in the illustration uses the V/F of Fig. 7-30 to convert instrumentation signals in the 4- to 20-mA format to a frequency format. Resistor R2 sets the offset for zero output (when input is 4 mA), and R5 sets the FS frequency of 10 kHz (when the input is 20 mA). These adjustments are interactive. ANALOG DEVICES, APPLICATIONS REFERENCE MANUAL, 1993, P. 23-47.

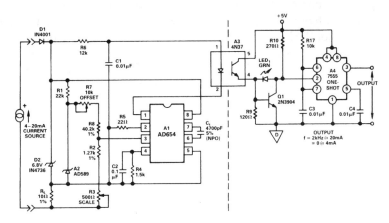

Self-powered V/F converter for 4- to 20-mA loop

Fig. 7-47   The circuit in the illustration is similar to that of Fig. 7-46, except that the V/F is powered by the 4- to 20-mA current, and the output is optically coupled to provide isolation. Resistor R7 sets the offset for zero output (input 4 mA), and R3 sets the FS frequency of 2 kHz (input 20 mA). These adjustments are interactive. ANALOG DEVICES, APPLICATIONS REFERENCE MANUAL, 1993, P. 23-47.

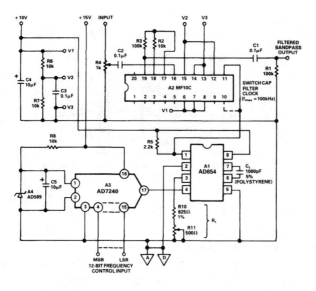

Digitally tuned switched-capacitor filter

Fig. 7-48   The circuit in the illustration uses the V/F of Fig. 7-30 as a single-supply, digitally programmed clock source for a switched-capacitor filter. Using the values shown, the V/F output is a 10-V square wave (f_{CLK}), which satisfies the clocking requirements of the MF10C filter. In this circuit, the MF10C is programmed for bandpass operation. Resistor R11 is adjusted so the maximum 12-bit input produces a clock of 100 kHz. Resistor R4 sets the level of the signal to be filtered. ANALOG DEVICES, APPLICATIONS REFERENCE MANUAL, 1993, P. 23-48.

Voltage/frequency and frequency/voltage circuits

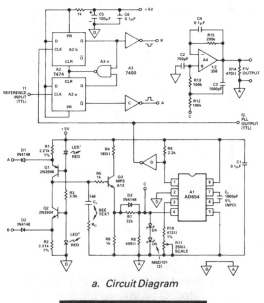

a. Circuit Diagram

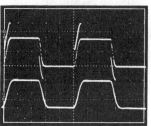

b. AD654 PLL Performance (Fast Response Mode)
Upper Trace: Phase Detector Output (Point "C")
Lower Trace: F/V Filtered Output

Scales: 0.2V/div, 500μs/div
Source: 50/100kHz   FSK @ 400Hz Rate
Condition: $C_L = 0.1\mu F$, $R_d = 500\Omega$

V/F converter as a PLL (alternate)

Fig. 7-49 The circuit in the illustration shows the V/F converter of Fig. 7-30 connected as a *PPL* (phase-locked loop). Figure 7-26B shows the waveforms. The F/V output is a noise-free square wave having exactly the same frequency as the input signal. Resistor R11 is adjusted for exactly 0.5 V at the V/F input (for full-scale). Resistance R_D and capacitance C_L set the dynamic range and/or speed. The waveforms are for a C_L of 0.1 μF and an R_D of 500 Ω. With the loop adjusted for a wide dynamic range (C_L = 1 μF, R_D = 160), the circuit will maintain a frequency and phase lock from 100 kHz down to about 500 Hz. ANALOG DEVICES, APPLICATIONS REFERENCE MANUAL, 1993, P. 23-49.

V/F and F/V circuit titles and descriptions

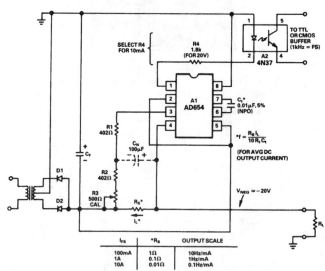

a. −5V Buss Monitor

V/F converter as a bus monitor

Fig. 7-50 The circuit in the illustration shows the V/F of Fig. 7-50 connected as a monitor for a −5-V buss. Resistor R4 is adjusted to provide a scale of 1 Hz per mV. (The output frequency should be 5000 Hz when the buss is at 5 V.) Analog Devices, Applications Reference Manual, 1993, p. 23-53.

b. Negative Supply Current Monitor

V/F converter as a negative supply-current monitor

Fig. 7-51 The circuit in the illustration shows the V/F of Fig. 7-30 connected as a monitor for a negative supply current (–20 V in this example). As shown by the table, the current limit to be monitored is set by the value of R_S. Resistor R3 sets the corresponding scale. For example, to monitor a maximum current of 1 A, use 0.1 for R_S and adjust R3 for a scale of 1 Hz per mA. (The output frequency should be 1000 Hz when the buss current is 1 A.) ANALOG DEVICES, APPLICATIONS REFERENCE MANUAL, 1993, P. 23-53.

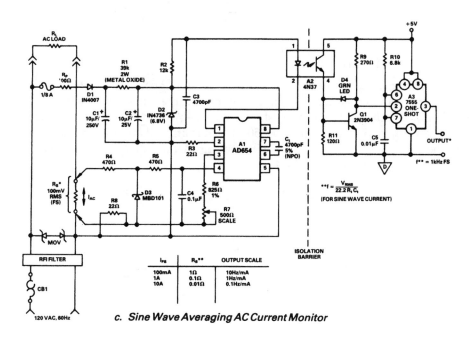

c. Sine Wave Averaging AC Current Monitor

V/F converter as a sine-wave-averaging alternating-current monitor

Fig. 7-52 The circuit in the illustration shows the V/F of Fig. 7-30 connected as a monitor for an alternating current (120-V line current in this case). As shown by the table, the current limit to be monitored is set by the value of R_S. Resistor R7 sets the corresponding scale. For example, to monitor a maximum current of 10 A, use 0.01 for R_S and adjust R_7 for a scale of 0.1 Hz per mA. (The output frequency should be 1000 Hz when the line current is 10 A.) ANALOG DEVICES, APPLICATIONS REFERENCE MANUAL, 1993, P. 23-53.

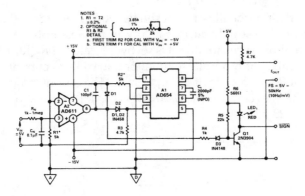

Bipolar V/F with ultra-high input impedance

Fig. 7-53 The circuit in the illustration shows the V/F converter of Fig. 7-30 connected to accept bipolar inputs. The circuit input impedance is set by the value of R_{IN}. For the mechanical calibration shown, apply –5 V and trim R2 for 50-kHz output. Then apply +5 V and trim R1 for 50-kHz output. SIGN output should be high, and the LED should be on, for a positive input. ANALOG DEVICES, APPLICATIONS REFERENCE MANUAL, 1993, P. 23-54.

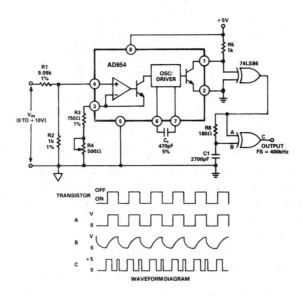

V/F converter with output doubling

Fig. 7-54 The circuit in the illustration shows the V/F of Fig. 7-30 connected to convert the signal normal square-wave output of the AD654 into a pulse train (effectively doubling the output frequency) but still preserving the low-frequency

Voltage/frequency and frequency/voltage circuits

linearity of the V/F. In this circuit, R1/R2 scale the 0 to +10-V input down to 0 to +1 mV at pin 4. R4 is adjusted to produce a 400-kHz FS output for a +10-V input. ANALOG DEVICES, APPLICATIONS REFERENCE MANUAL, 1993, P. 23-56.

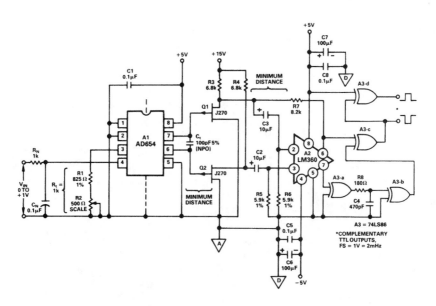

V/F converter with output doubling (2 MHz)

Fig. 7-55 The circuit in the illustration is similar to that of Fig. 7-54, except with a much greater output-frequency capability. In this circuit, R2 is adjusted to produce a 2-MHz FS output for a +1-V input. ANALOG DEVICES, APPLICATIONS REFERENCE MANUAL, 1993, P. 23-56.

=8=

Comparator circuits

This chapter is devoted to comparator circuits. Because the present trend is to use comparator ICs as the basic elements, this chapter concentrates on such circuits. However, circuits that use a different approach also are included. Before you get into testing and troubleshooting, here is a review of some comparator basics.

Comparator operation

Comparator ICs are essentially high-gain op amps (Chapter 1) designed for open-loop operation. Typically, a comparator produces an output when the input goes above or below a certain level or if it crosses zero. For example, the LM111 comparator IC shown in Figs. 8-A and 8-B produces a logic-1 output at pin 7 with a positive signal between the two inputs. A logic-0 output is produced with a negative input.

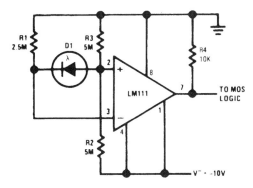

Fig. 8-A   Level detector for photodiode.

Parameter	Limits			Units
	Min	Typ	Max	
Input Offset Voltage		0.7	3	mV
Input Offset Current		4	10	nA
Input Bias Current		60	100	nA
Voltage Gain		100		V/mV
Response Time		200		ns
Common Mode Range	0.3		3.8	V
Output Voltage Swing			50	V
Output Current			50	mA
Fan Out (DTL/TTL)	8			
Supply Current		3	5	mA

Fig. 8-B   Characteristics of an LM111 comparator.

Threshold or level detection is accomplished by putting a reference voltage on one input and the signal to be compared on the other input. The output then changes states when the signal input goes above or below the reference-input level. An op amp can be used as a comparator, except that op-amp response time is typically in the tens of microseconds (often too slow for many applications). Figure 8-B shows the characteristics of a classic comparator IC (the LM111 operated with a 5-V supply at 25°C).

IC comparator application tips

One of the problems with any IC comparator is the tendency to oscillate. Here are some tips to keep oscillation and other comparator problems to a minimum.

Keep output and balance leads (such as pins 5, 6, and 7 of Fig. 8-C) apart, if possible, to avoid stray coupling between output and balance. If the balance is not used (Fig. 8-A), tie the balance pins together. When balance is required, try connecting a 0.1-μF capacitor between the balance leads to minimize oscillation. Normally, bypass capacitors are not required for IC comparators. If required to eliminate large voltage spikes into the supplies when the comparator changes states, keep the leads as short as possible between the IC and bypass capacitors.

When source resistances between 1 and 10 kΩ are used, the impedance (both capacitive and resistive) on both inputs should be equal. The equal impedance tends to reject the feedback signal (and oscillation). Use positive feedback to increase hysteresis, as shown in Figs. 8-D and 8-E.

When driving the inputs from a low-impedance source, use a limiting resistor in series with the input lead. The resistor limits peak current and is especially important when the inputs go where they can accidentally be connected to a high-voltage source. Low-impedance sources do not cause a problem unless the output voltage exceeds the negative supply. However, because the supplies go to zero when turned off, the isolation might be needed.

Comparator circuits

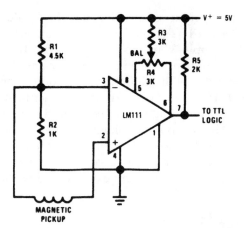

Fig. 8-C   Zero-crossing detector for magnetic transducer.

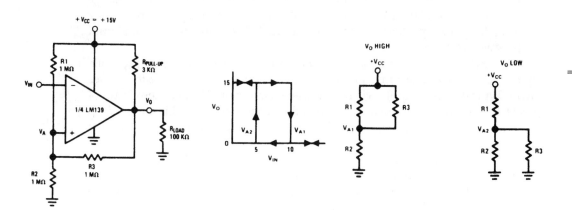

Fig. 8-D   Inverting comparator with hysteresis.

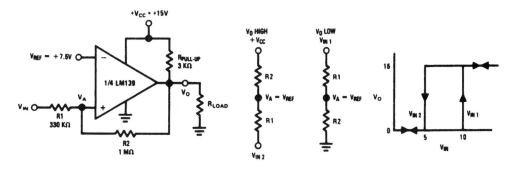

Fig. 8-E   Noninverting comparator with hysteresis.

IC comparator application tips

Large capacitors at the input (greater than 0.1 µF) should be treated as a low source impedance and should be isolated with a resistor. Such capacitors can hold a charge larger than the supply when the supply is abruptly shut off.

Avoid reversing the supplies on comparators (or any IC). Typically, reverse voltage in excess of 1 V can melt the aluminum interconnections if current is high. Use a clamp diode with adequate peak-current rating across the supply bus.

Do not operate an IC comparator with the ground terminal at a voltage that exceeds either supply. Also, the output voltage (such as the 50 V in Fig. 8-B) generally applies to the potential between the output and the V-terminal. Therefore, if the comparator is operated from a negative supply (Fig. 8-A), the maximum output voltage must be reduced by an amount equal to (or less than) the V-voltage.

Comparator tests

The test for any comparator circuit is to change the input and check that the output changes. For example, in the circuit of Fig. 8-A, the output should change states when the diode D1 current reaches 1 µA. Check by covering D1, measuring the voltage at pin 7, and then exposing D1 to light. Typically, the output should switch between 0 V and near −10 V.

In the circuits of Figs. 8-D and 8-E, monitor the output across R_{LOAD} while varying the input voltage above and below the V_{REF} point (7.5 V using the values shown). Typically, the output switches between 0 and near 15 V. For the circuit of Fig. 8-D, the output should go to 15 V when the input is at 5 V, and it should drop to 0 V when the input is increased to 10 V. The opposite should occur for the circuit of Fig. 8-E.

In the circuit of Fig. 8-F, apply voltage to V_{IN} and check that the lamp L1 turns on and off. If V_{IN} is greater than V_A, or less than V_B, the lamp should be off. Lamp L1 goes on only when V_{IN} is less than V_A but greater than V_B. The voltage range where L1 is on depends on the values of R_1, R_2, and R_3.

In the circuit of Fig. 8-G, apply a sine-wave signal at the input and check that the output produces a square wave of the same frequency. The square-wave amplitude depends on the value of the pull-up resistance in relation to R5/R6.

Comparator troubleshooting

The first step in troubleshooting comparator circuits is to check that the desired output-voltage change is produced for a given change at the input. If not, try correcting the problem with adjustment (as described for the circuit).

If the problem cannot be corrected by adjustment (or there is no adjustment), trace signals using a meter or scope from input (typically, a voltage level) to output (typically, a rapid voltage-level change, or possibly a square-wave/pulse-signal output). Follow this with voltage and/or point-to-point resistance measurement. The following are some typical examples.

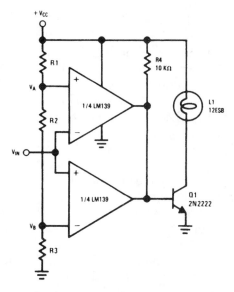

Fig. 8-F   Limit comparator with lamp driver.

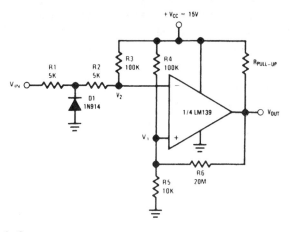

Fig. 8-G   Zero-crossing detector for squaring a sine wave.

   In the circuit of Fig. 8-A, if the output does not change states when D1 is alternately exposed to light and dark, check for any change at pin 2 of the LM111, with D1 covered and uncovered. Although the voltage change will be small (D1 produces about 1 μA of current), the change should be measurable. If there is no change at pin 2 of the LM111, suspect D. If there is a measurable change at pin 2 but not at pin 7, suspect the LM111.

Comparator troubleshooting

In the circuits of Figs. 8-D and 8-E, check that the output switches between zero and near 15 V when the input is varied between 5 and 10 V. Of course, the circuits produce opposite results or outputs for the same input, as shown by the hysteresis graphs (Fig. 8-D is inverting, and Fig. 8-E is noninverting). If there is no change in output for a given input change, suspect the LM139 (unless, of course, the circuit is not properly wired, in which case you shall have no pie!).

In the circuit of Fig. 8-F, if lamp L1 stays on or off when V_{IN} is varied above and below the limits of V_A and V_B, check for a change at the base of Q1 when V_{IN} is varied. If there is a change at Q1, but the lamp does not respond, suspect Q1.

For example, if the base of Q1 goes negative, the lamp should turn off, and vice versa. Of course, if the lamp is always off, the problem could be the lamp (which you should have checked first!). Remember that the point at which the lamp turns on is set by the values of R1, R2, and R3. Assuming a V_{CC} of 10 V, and that R1, R2, and R3 are all the same value, L1 should be off if V_{IN} is greater than 6.6 V, or less than 3.3 V. Lamp L1 stays on when V_{IN} is greater than 3.3 V but less than 6.6 V.

In the circuit of Fig. 8-G, if there is no square-wave output for a sine-wave input, suspect the LM139 (again assuming good wiring and proper resistor values). Of course, if D1 is leaking badly (shorted), the input signal might be prevented from reaching the LM139. A level of about −700 mV at the junction of R1 and R2 indicates that D1 is probably good.

Remember that the zero reference is set by the values of R4 and R5. With $R_1 + R_2$ equal to R_5, V_1 should equal V_2 when V_{IN} is zero, and the square-wave output should switch states each time the sine-wave input crosses zero. With the values shown, the no-signal voltage at both inputs of the LM139 is about 1.5 V.

Comparator response-time problems

One difficult troubleshooting problem for comparators is when the circuit operates, but the response time is not correct (too slow). It is difficult to tell if the problem is one of circuit components or the comparator IC. The test connections and corresponding response-time graphs shown in Fig. 8-H can help pinpoint the problem. Note that there are no external components (except for the 5.1-kΩ load). The output is monitored on a scope, and a pulse generator output is applied at the input. (The pulse generator must have a rise time faster than the anticipated comparator response time.) The graphs of Fig. 8-H show the response to both positive and negative pulses. Note that response time increases for lower input voltages, so the tests should be made with the same pulse voltage as is used with the circuit. If the response time of the comparator IC is well within the required tolerance (when tested without external components) the problem is localized. However, if response time for the IC is too slow, use a faster IC such as the RM4805 comparator shown in Fig. 8-I.

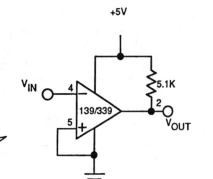

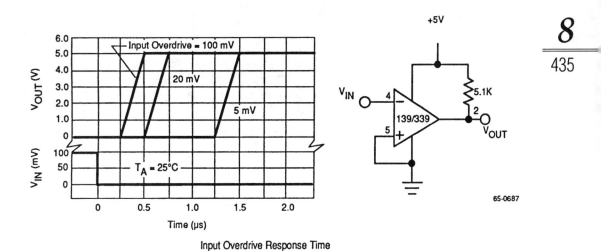

Input Overdrive Response Time

Fig. 8-H   Response-time test circuit and graphs (low speed).

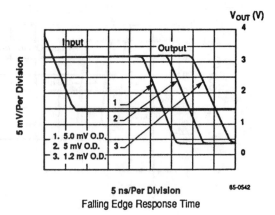

1. 5.0 mV O.D.
2. 5 mV O.D.
3. 1.2 mV O.D.

V_{OUT} (V)

5 mV/Per Division

Input

Output

5 ns/Per Division   65-0543

Rising Edge Response Time

V_{OUT} (V)

Input   Output

5 mV/Per Division

1. 5.0 mV O.D.
2. 5 mV O.D.
3. 1.2 mV O.D.

5 ns/Per Division   65-0542

Falling Edge Response Time

Internal to Generator

FET Probe

50

4.5 pF

5K

2

+

FET Probe

100:1 Divider

3   4805   7

V_{OUT}

50   50

—

4.5pF

65-0541

Response Time Test Setup

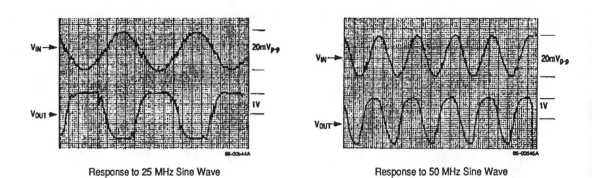

V_{IN}

20mV$_{p-p}$

V_{OUT}

1V

65-0544A

Response to 25 MHz Sine Wave

V_{IN}

20mV$_{p-p}$

V_{OUT}

1V

65-0545A

Response to 50 MHz Sine Wave

Fig. 8-1   Response-time test circuit and graphs (high speed).

Comparator circuits

Comparator circuit titles and descriptions

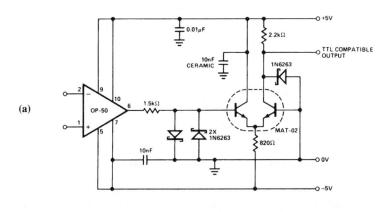

(a)

(b)

INPUT OVERDRIVE	100mV	10mV	1mV	100μV	10μV
Positive Output Delay	3.2μs	5μs	40μs	340μs	2.4ms
Negative Output Delay	1.8μs	5μs	50μs	380μs	4.5ms

High-sensitivity voltage comparator

Fig. 8-1 The op-amp comparator shown can resolve a sub-microvolt difference signal. The OP-50, operating without feedback, drives a second gain stage that generates a TTL-compatible output signal. Schottky-clamp diodes prevent over-driving of the transistor pair and stop saturation of the output transistor. The supply voltage is set at ±5 V to lower the quiescent power dissipation and minimize thermal feedback because of output-stage dissipation. Operating from ±5 V also reduces the OP-50 rise and fall times, thus reducing output response time. Although the comparator is not fast, the circuit is sensitive and can detect signal differences as low as 0.3 μV. With large input overdrives, the circuit responds in about 3 μs. If sharp transitions are needed, use a TTL Schmitt-trigger input. Figure 8-1B shows the response time versus input overdrive. ANALOG DEVICES, APPLICATIONS REFERENCE MANUAL, 1993, P. 13-51.

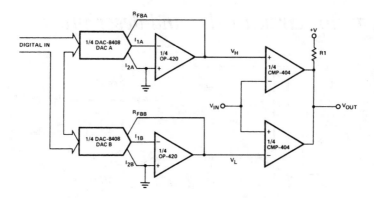

Op-amp dual programmable window comparator

Fig. 8-2 Only three ICs are required to fully implement two independent programmable window comparators. The quad, latched, 8-bit CMOS DAC-8408 (Chapter 9), together with the quad micropower OP-420 provide digitally programmable HIGH and LOW thresholds to the CMP-404, quad low-power comparator. The outputs of the threshold comparators are wired ORed with a common pull-up resistor producing $V_{OUT} = +V$ only when $V_L < V_{IN} < V_H$. Total supply current for the full circuit is less than 2 mA. ANALOG DEVICES, APPLICATIONS REFERENCE MANUAL, 1993, P. 13-55.

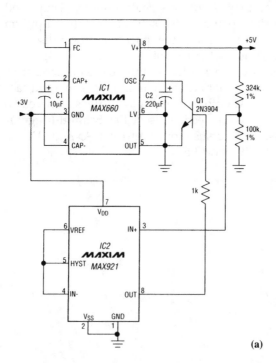

(a)

Comparator circuits

Continued

LOAD RESISTANCE (Ω)	OUTPUT VOLTAGE (V)	OUTPUT RIPPLE (mV$_{p-p}$)
∞	5.00	30
10k	5.00	35
1k	5.00	100
100	4.96	100
50	4.59	150

a. SUPPLY = +3.0V

(b)

LOAD RESISTANCE (Ω)	OUTPUT VOLTAGE (V)	OUTPUT RIPPLE (mV$_{p-p}$)
∞	5.01	55
10k	5.01	55
1k	5.01	55
100	4.98	170
50	4.90	170

b. SUPPLY = +3.3V

LOAD RESISTANCE (Ω)	OUTPUT VOLTAGE (V)	OUTPUT RIPPLE (mV$_{p-p}$)
∞	4.98	10
10k	4.98	25
1k	4.98	25
100	4.64	70
50	4.29	90

c. SUPPLY = +2.7V

Combined comparator and charge pump to convert 3 V to 5 V

Fig. 8-3 The circuit in the illustration combines charge-pump IC1 with comparator IC2 to convert 3 to 5 V. Figure 8-3B shows the output ripple for various load resistances and output voltages. The internal 45-kHz oscillator of IC1 transfers charge from C1 to C2, causing the regulated output to rise. When the feedback voltage at pin 3 of IC2 exceeds 1.18 V, IC2 output goes high and turns off the IC1 oscillator (through Q1). The resistance values across the +5-V output determine the threshold. MAXIM ENGINEERING JOURNAL, VOLUME 11, P. 12.

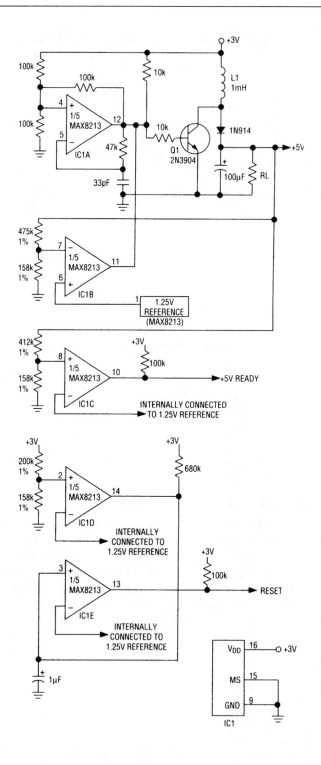

Comparator circuits

3-V to 5-V regulator with microprocessor reset

Fig. 8-4 The circuit in the illustration uses a single five-comparator IC to produce 5 V from 3 V, as well as power-on (+5 V ready) and reset signals to system microprocessor. IC1A is connected as an oscillator with a square-wave output to drive Q1 that in turn is connected as a dc/dc converter (Chapter 1) with L1, D2, and C2. When V_{OUT} exceeds 5 V, IC1B pulls the oscillator signal low, resulting in a 5-V regulated output. Coil L1 should have a series resistance of about 25 Ω. Output ripple is about 50 mV. IC1C provides an active-high +5 V ready signal when the Q1 output reaches 4.5 V. IC1D and IC1E provide a reset for the microprocessor when the 3-V supply is too low (below 2.83 V). RESET remains low for 200 ms after the +3-V supply goes above 2.87 V. MAXIM ENGINEERING JOURNAL, VOLUME 11, P. 13.

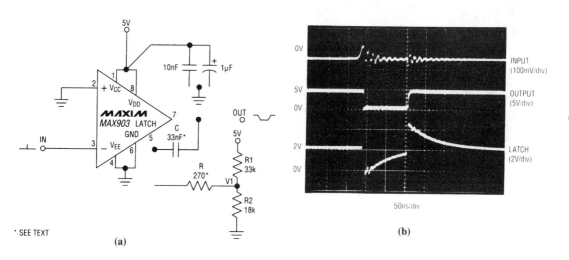

* SEE TEXT

(a)

(b)

Comparator as a pulse stretcher

Fig. 8-5 The circuit in the illustration shows a comparator connected as a pulse stretcher. Figure 8-5B shows the waveforms. When an input pulse as narrow as 15 ns is applied, the circuit output (pin 7) goes low, and capacitor C pulls the TTL-compatible latch input (pin 5) low, latching the output. Capacitor discharges through R until the latch input-voltage crosses the 1.4-V threshold, releasing the latch. With the values shown for R and C, the output pulse is about 100 ns (see Fig. 8-5B). To assure stable operation, the values of R1/R2 are selected so that voltage V_1 is about 2 V—a substantial portion of the discharge curve for capacitor C, as shown by the negative swing of the latch waveform (Fig. 8-5B). Output pulse widths range between 50 ns and 500 ns when R is kept between 0.27 kΩ and 1.0 kΩ, and C is between 10 pF and 100 pF. MAXIM ENGINEERING JOURNAL, VOLUME 11, P. 15.

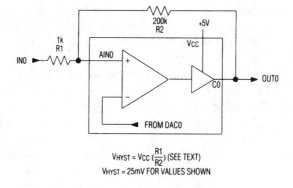

$$V_{HYST} = V_{CC} \left(\frac{R1}{R2}\right) \text{ (SEE TEXT)}$$
$$V_{HYST} = 25mV \text{ FOR VALUES SHOWN}$$

Basic circuit for adding hysteresis to a comparator

Fig. 8-6 The circuit in the illustration shows the basic connections for adding hysteresis to a comparator. The relationship of V_{HYST} to V_{CC} and R_1/R_2 is:

$$V_{TH} = V_T \left(\frac{R_1}{R_2} + 1 \right)$$

$$V_{TL} = \left(V_T \frac{R_1}{R_2} + 1 \right) - V_{CC} \left(\frac{R_1}{R_2} \right)$$

$$V_{HYST} = V_{TH} - V_{TL}$$

$$V_{HYST} = V_{CC} \left(\frac{R_1}{R_2} \right)$$

where V_t is the threshold voltage set by the internal circuits of the comparator with no hysteresis, V_{TL} is the shifted high-going threshold with hysteresis added, V_{TL} is the shifted low-going threshold with hysteresis, and V_{HYST} is the total hysteresis ($V_{TH} - V_{TL}$). Note that V_{HYST} and V_{TL} change with V_{CC} and that R_1 can add offset error, depending on the input bias current. MAXIM NEW RELEASES DATA BOOK, 1993, P. 3-40.

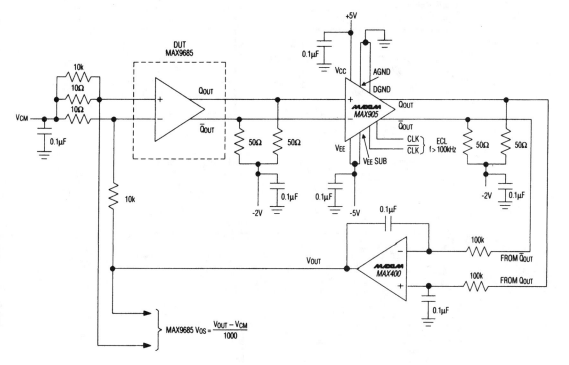

$$\text{MAX9685 } V_{OS} = \frac{V_{OUT} - V_{CM}}{1000}$$

Measuring VOS of ultra high-speed comparators

Fig. 8-7 The circuit in the illustration shows the connections for measuring V_{OS} (input offset voltage) for a comparator or other high-speed component. In this example, the circuit is used to measure V_{OS} of the MAX9685, an ultra high-speed comparator that tends to oscillate when tested in a standard op-amp circuit (such as described in Chapter 1). In this circuit, the loop is broken by the MAX905 comparator and MAX400 connected as a differential integrator, thus preventing oscillation. The test loop forces the MAX9685 to switch with a precise 50% duty cycle. MAXIM NEW RELEASES DATA BOOK, 1993, P. 3-50.

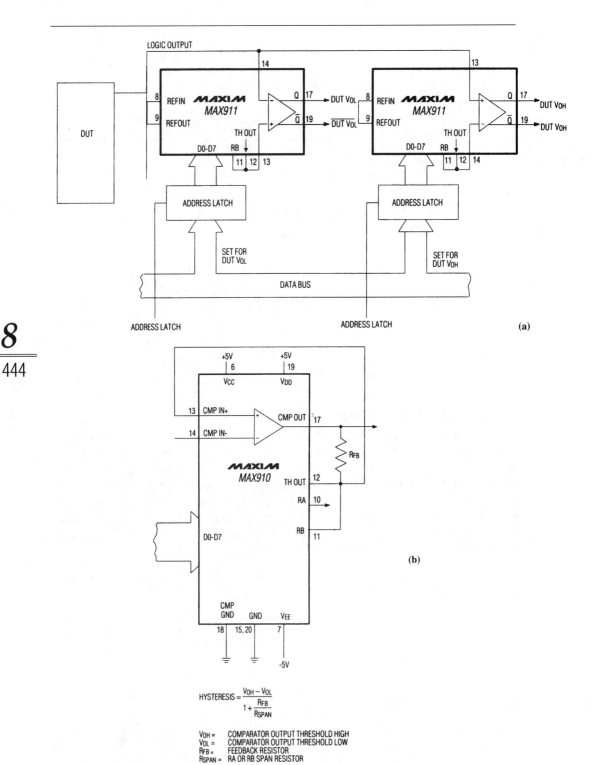

Comparator circuits

Logic threshold detector for automatic test equipment

Fig. 8-8 The circuit in the illustration shows two MAX911 comparators connected to detect the upper and lower threshold limits of a logic output from a device under test (DUT). Such a circuit can be used in automatic test equipment (ATE). One MAX911 is programmed for the upper threshold limit of the DUT by logic inputs from a data bus. The lower threshold limit is covered by the other MAX911. The threshold range is –2.4 V to +2.56 V, with 1 LSB = 20 mV. The MAX911's switch output states when the threshold limits are crossed. Figure 8-8B shows a circuit for adding hysteresis, if needed. MAXIM NEW RELEASES DATA BOOK, 1993, P. 3-70, 71.

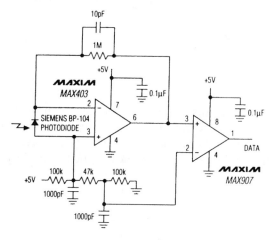

Battery-operated infrared data link

Fig. 8-9 The circuit in the illustration provides for reception of infrared data at speeds exceeding the standard 1-Mbps (million bits per second) data rate. The MAX403 op amp converts the photo diode current to a voltage, and the MAX907 determines if the amplifier output is high enough to be called a logic-1. The current consumption of the MAX403 and MAX907 are 250 μA and 700 μA, respectively. MAXIM NEW RELEASES DATA BOOK, 1994, P. 3-100.

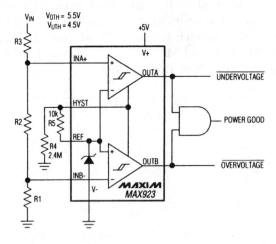

Auto-off power source

Fig. 8-10 The circuit in the illustration shows a power supply with a timed, automatic power-off function. The comparator output is the power-supply output. With a 10-mA load, the circuit provides a voltage of (V_{BATT} −0.12 V) but draws only 3.5 µA quiescent current. With the values shown, the hysteresis is ±50 mV and the IN+ voltage is set at 100 mV. This gives an IN+ trip threshold of about 50 mV for IN+ falling. The maximum power-on time of the OUT pin (before power-down occurs) is set by: $R \times C \times 4.6$ s. For example, 2 MΩ × 10 µF × 4.6 = 92 s. The actual time will vary with both the leakage current of C and the battery voltage applied. MAXIM NEW RELEASES DATA BOOK, 1994, P. 3-125.

Comparator circuits

Ultra low-power window detector

Fig. 8-11 The circuit in the illustration is a window detector with a 4.5-V undervoltage threshold and a 5.5-V overvoltage threshold, when $R_1 = 294$ kΩ, $R_2 = 61.9$ kΩ, and $R_3 = 1$ MΩ (all 1% standard values). Other thresholds can be selected with the following equations: R_1 should be between 100 kΩ and 1 MΩ to provide at least 100 nA through R_1 (for accurate thresholds):

$$R_2 + R_3 = R_1 \times \left(\frac{V_{\text{OTH}}}{V_{\text{REF}} + V_{\text{H}}} - 1 \right)$$

$$R_2 = (R_1 + R_2 + R_3) \times \frac{(V_{\text{REF}} - V_{\text{H}})}{V_{\text{UTH}}} - R_1$$

$$R_3 = (R_2 + R_3) - R_2$$

where $V_{\text{REF}} = 1182$, $V_{\text{H}} = 0.005$. MAXIM NEW RELEASES DATA BOOK, 1994, P. 3-126.

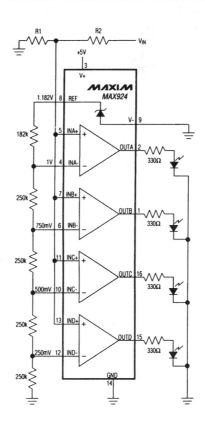

Bar-graph level gauge

Fig. 8-12 The circuit in the illustration is essentially a four-stage level detector. The full-scale threshold (all LEDs on) is

$$V_{IN} = \frac{(R_1 + R_2)}{R_1} \text{ V}$$

The other thresholds are at ¾ full scale, ½ full scale, and ¼ full scale. The output resistors limit current into the LEDs. MAXIM NEW RELEASES DATA BOOK, 1994, P. 3-127.

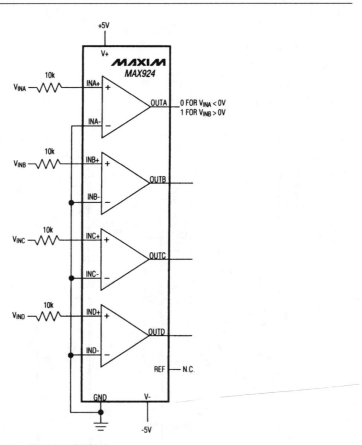

Level shifter (±5 V to CMOS/TTL)

Fig. 8-13 The circuit in the illustration shifts bipolar ±5 V inputs to logic outputs. See Fig. 8-12 for pin numbers. The 10-kΩ resistors protect the comparator inputs but do not materially affect circuit operation. MAXIM NEW RELEASES DATA BOOK, 1994, P. 3-127.

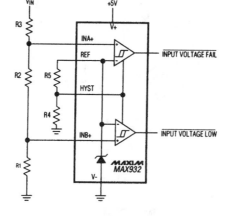

Two-stage low-voltage detector

Fig. 8-14 The circuit in the illustration monitors an input voltage in two steps. When V_{IN} is higher than the LOW and FAIL thresholds, the outputs are high. Threshold calculations are similar to those described for the window detector (Fig. 8-11), where LOW equals the higher threshold and FAIL equals the lower threshold. Use 10 kΩ for R_5 and 2.4 MΩ for R_4. MAXIM NEW RELEASES DATA BOOK, 1995, P. 3-59.

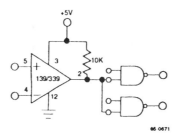

TTL driver

Fig. 8-15 The circuit in the illustration shows an LM139/339 connected as a basic driver for TTL circuits. The input can be applied at either pin 4 or 5, with the other input pin grounded. RAYTHEON SEMICONDUCTOR DATA BOOK, 1994, P. 3-680.

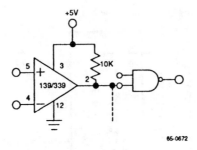

65-0672

CMOS driver

Fig. 8-16 The circuit in the illustration shows an LM139/339 connected as a basic driver CMOS circuit. The input can be applied at either pin 4 or 5, with the other input pin grounded. RAYTHEON SEMICONDUCTOR DATA BOOK, 1994, P. 3-680.

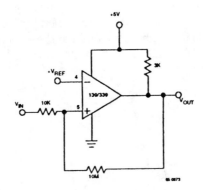

Comparator with hysteresis

Fig. 8-17 The circuit in the illustration shows an LM139/339 connected to provide normal hysteresis for most applications. RAYTHEON SEMICONDUCTOR DATA BOOK, 1994, P. 3-680.

Comparator circuits

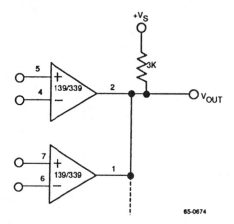

65-0674

ORing the output of a comparator

Fig. 8-18 The circuit in the illustration shows the connections for ORing the outputs of two or more comparators. RAYTHEON SEMICONDUCTOR DATA BOOK, 1994, P. 3-680.

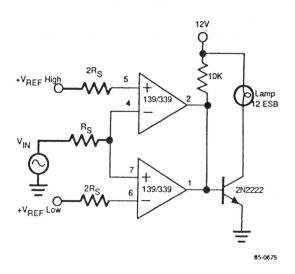

65-0675

Limit comparator

Fig. 8-19 The circuit in the illustration shows an LM139/339 connected to indicate when an input voltage goes above and below fixed thresholds. RAYTHEON SEMICONDUCTOR DATA BOOK, 1994, P. 3-680.

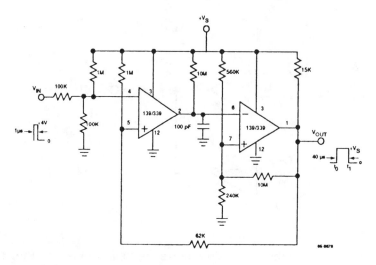

One-shot multivibrator with input lock-out

Fig. 8-20 With this circuit, pulse width is independent of supply voltage and a 99% duty cycle is possible. When triggered, the output is high, causing the voltage at the noninverting input of comparator 1 to go to $+V_S$. This prevents any additional input pulses from disturbing the circuit until the output pulse has been completed. RAYTHEON SEMICONDUCTOR DATA BOOK, 1994, P. 3-680.

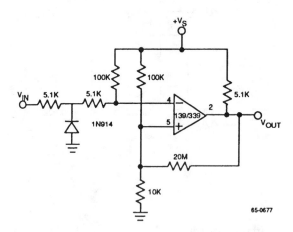

Comparator circuits

Zero-crossing detector (single supply)

Fig. 8-21 The circuit in the illustration symmetrically squares a sine wave (centered about 10 V, if V_S is 15 V) by introducing a small amount of positive feedback. The diode ensures that the inverting input never goes below –100 mV. RAYTHEON SEMICONDUCTOR DATA BOOK, 1994, P. 3-681.

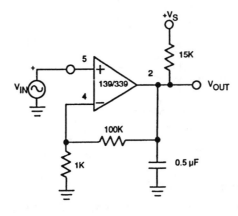

Low-frequency op amp

Fig. 8-22 The circuit in the illustration functions as an op amp (Chapter 1), but at low frequencies. RAYTHEON SEMICONDUCTOR DATA BOOK, 1994, P. 3-681.

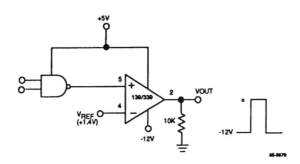

TTL to MOS logic converter

Fig. 8-23 The circuit in the illustration shows an LM139/339 connected to provide a 12-V (MOS level) output for a 5-V (TTL level) input. RAYTHEON SEMICONDUCTOR DATA BOOK, 1994, P. 3-681.

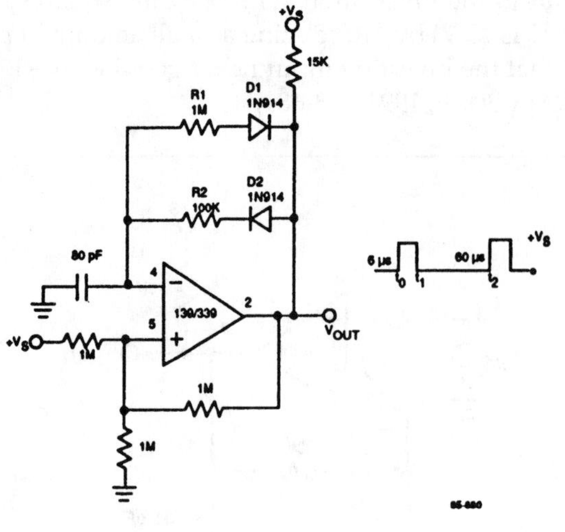

Pulse generator using a comparator

Fig. 8-24 In the circuit in the illustration, pulse width is set by R_2, and pulse frequency is set by R_1. RAYTHEON SEMICONDUCTOR DATA BOOK, 1994, P. 3-681.

=9=

Analog/digital and digital/analog circuits

This chapter is devoted to circuits that convert analog voltages or signals to digital form (A/D or ADC, whichever you prefer) and convert digital signals to analog form (D/A or DAC). Because these circuits are essentially digital, all the test and troubleshooting information of Chapter 6 applies to A/D and D/A circuits, whether IC or a combination of IC and discrete component. For example, power, ground, reset, chip-select, clock, and input-output signals must be checked as with any IC.

A/D or ADC converters

Analog-to-digital converters can be tested by applying precision voltages at the input and monitoring the output for corresponding digital values. For example, in the A/D converter of Fig. 9-A, a fixed voltage between 0 and +10 V can be applied to pin 2 of A1, and the corresponding digital value can be read out at pins 5 through 14 of the DAC-10, or at the lines between the DAC-10 and the 2504 *SAR* (successive approximation register). The lines should go to +5 V for a logic-1 and to ground or 0 V for a logic-0.

Notice that there is a serial digital output at pin 2 of the 2504 SAR. This output is best monitored on a scope. The rate at which conversions are done is controlled by the 1- to 2-MHz clock input at pin 15 of the 2504. Pin 14 of the 2504 must receive a start-conversion input signal (typically from the system microprocessor) to initiate each conversion cycle. At the end of the conversion cycle, the conversion-complete, pin 3 produces an output to the microprocessor (indicating status, conversion-complete, or conversion-not-complete).

If the output readings of the circuit are slightly off, try correcting the problem by adjusting R7 at the low end (all logic-0 when the analog input is 0 V) or try

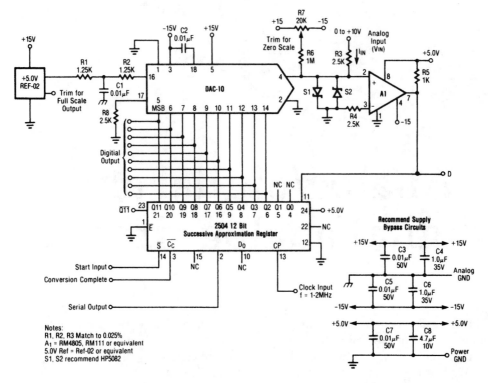

Fig. 9-A Typical A/D converter.

trimming the +5-V reference at the high end (all logic-1 when the analog input is +10 V). If the outputs are absent or are way off, suspect the DAC-10, comparator A1, or the 2504. Also note that the accuracy of this circuit depends on the precision of R1, R2, and R3.

D/A or DAC converters

Digital-to-analog converters can be tested by applying digital inputs and monitoring the output for corresponding voltages. For example, in the D/A converter of Fig. 9-B, the inputs at pins 4 through 11 of the RM/RC4888 can be connected to ground (for a 0) or to 5 V (for a 1), and the output can be monitored with a precision voltmeter at pin 13.

If the output voltage is slightly off, try correcting the problem using the calibration procedure. For example, with all of the digital inputs at 0 (pins 4 through 11 grounded), adjust the offset pot until V_{OUT} is 0.0000. Then, with all digital inputs at 5 V (logic 1), adjust the gain pot until V_{OUT} is 9.9609. It might be necessary to work between these two adjustments until all output voltages are within tolerance.

Analog/digital and digital/analog circuits

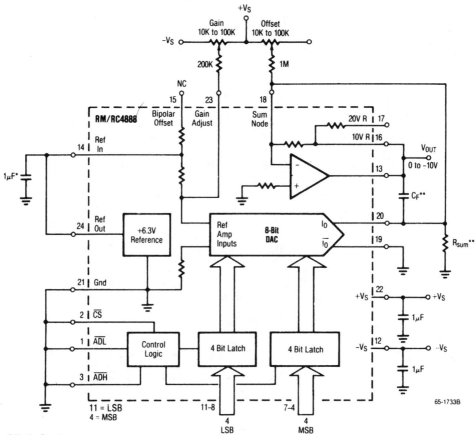

65-1733B

Calibration Procedure:
1. Set inputs to all zeros
2. Adjust offset until V_{OUT} equals 0V
3. Set inputs to all ones

4. Adjust gain until V_{OUT} equals correct full scale value
*Optional — reduces reference noise
**Optional — improves settling time (see table for values)

Format	Output Scale	MSB DB7	DB6	DB5	DB4	DB3	DB2	DB1	LSB DB0	I_0 (mA)	$\overline{I_0}$ (mA)	V_{OUT}
Straight Binary: Unipolar With True Input Code. True Zero Output	Positive Full Scale	1	1	1	1	1	1	1	1	3.999	0.000	9.9609
	Positive Full Scale – LSB	1	1	1	1	1	1	1	0	3.984	0.001	9.9219
	LSB	0	0	0	0	0	0	0	1	0.0001	3.984	0.0391
	Zero Scale	0	0	0	0	0	0	0	0	0.000	3.999	0.0000
Complementary Binary: Unipolar With Complementary Input Code. True Zero Output	Positive Full Scale	0	0	0	0	0	0	0	0	0.000	3.999	9.9609
	Positive Full Scale – LSB	0	0	0	0	0	0	0	1	0.001	3.984	9.9219
	LSB	1	1	1	1	1	1	1	0	3.984	0.001	0.0391
	Zero Scale	1	1	1	1	1	1	1	1	3.999	0.000	0.0000

Fig. 9-B   Eight-bit straight-binary D/A converter.

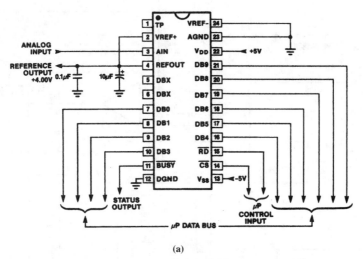

(a)

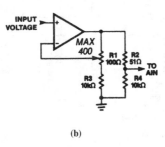

(b)

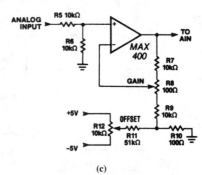

(c)

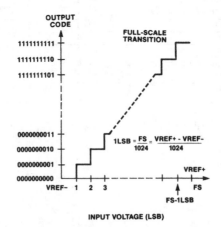

(d)

Continued

PIN	NAME	FUNCTION
1	TP	Test Pin, leave open.
2	VREF+	Positive Ladder Input, upper limit of reference span. Set the full-scale input voltage. Range: 1V to V_{DD}.
3	AIN	(Sampling) Analog Input
4	REFOUT	+4V Reference Output, usually connected to VREF+.
5-6	DBX	(Reserved for DB0-1, future 12-bit version, = LOW.)
7-10	DB0-DB3	Three-State Data Output, Bits 0-3
11	$\overline{BUSY}$	Busy Status Output, low when conversion is in progress.
12	DGND	Digital Ground.

PIN	NAME	FUNCTION
13	V_{SS}	Negative Supply, –5V.
14	$\overline{CS}$	Chip Select Input, must be low for the ADC to recognize $\overline{RD}$.
15	$\overline{RD}$	Active Low Read Input, starts conversion when $\overline{CS}$ is low. $\overline{RD}$ also enables the output drivers when $\overline{CS}$ is low.
16-21	DB4-DB9	Three-State Data Output, Bits 4-9
22	V_{DD}	Positive Supply, +5V
23	AGND	Analog Ground
24	VREF–	Negative Voltage Ladder Input, lower limit of reference span. Set the zero-code voltage. Range: AGND ±0.1V.

(e)

300-kHz 10-bit A/D converter

Fig. 9-1 Figure 9-1A shows the basic operational diagram for the MAX151 A/D converter with internal reference and track/hold. Figures 9-1B, C, D, and E show the full-scale trim circuit, offset and gain trim circuit, transfer function, and pin functions, respectively. The circuit of Fig. 9-1B is used when only the full-scale output need be trimmed. Use the preferred circuit of Fig. 9-1C for both offset trim (±20 mV) and gain trim (±0.5%) as follows. Apply 1/2 LSB (2 mV) at the analog input and adjust R12 so the output changes between 00000 00000 and 00000 00001. Then, apply FS-2/2 LSB (3.994 mV) and adjust R8 until the output code changes between 11111 11110 and 11111 11111. There might be interaction between these adjustments. MAXIM NEW RELEASES DATA BOOK, 1992, PP. 7-25, 26, 28, 29.

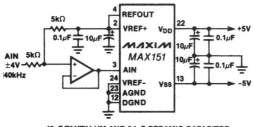

10μF TANTALUM AND 0.1μF CERAMIC CAPACITOR
SHOULD BE AS CLOSE TO MAX151 AS POSSIBLE.

Ten-bit A/D converter with ±4-V bipolar input

Fig. 9-2 The circuit in the illustration shows the A/D of Fig. 9-1 connected to accept ±4-V bipolar inputs. A MAX400 is recommended for the driving amplifier at pin 3. MAXIM NEW RELEASES DATA BOOK, 1992, P. 7-27.

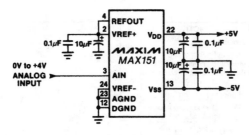

Ten-bit A/D converter with 0-V to +4-V input

Fig. 9-3 The circuit in the illustration shows the A/D of Fig. 9-1 connected to accept 0-V to 4-V inputs, using the internal 4-V reference. MAXIM NEW RELEASES DATA BOOK, 1992, P. 7-27.

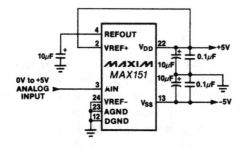

Ten-bit A/D converter with 0-V to +5-V input

Fig. 9-4 The circuit in the illustration shows the A/D of Fig. 9-1 connected to accept 0-V to +5-V power supply as a reference. MAXIM NEW RELEASES DATA BOOK, 1992, P. 7-27.

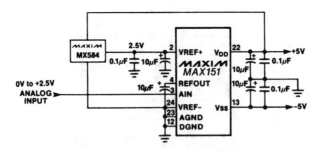

Ten-bit A/D converter with 0-V to +2.5-V input

Fig. 9-5 The circuit in the illustration shows the A/D of Fig. 9-1 connected to accept 0-V to +2.5-V inputs using an external MX584 voltage reference. MAXIM NEW RELEASES DATA BOOK, 1992, 7-28.

Analog/digital and digital/analog circuits

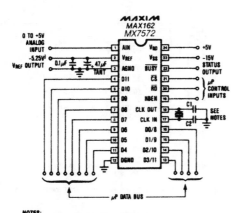

NOTES:
MAX162 - 4MHZ CRYSTAL/CERAMIC RESONATOR.
MX7572XX05 - 2.5MHz CRYSTAL/CERAMIC RESONATOR.
MX7572XX12 - 1.0MHz CRYSTAL/CERAMIC RESONATOR.
C1 AND C2 CAPACITANCE VALUES DEPEND ON CRYSTAL/CERAMIC
RESONATOR MANUFACTURER. TYPICAL VALUES ARE FROM 0 TO 100pF.

(a)

(b)

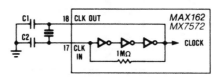

NOTES:
MAX162 - 4MHZ CRYSTAL/CERAMIC RESONATOR.
MX7572XX05 - 2.5MHz CRYSTAL/CERAMIC RESONATOR.
MX7572XX12 - 1.0MHz CRYSTAL/CERAMIC RESONATOR.
C1 AND C2 CAPACITANCE VALUES DEPEND ON CRYSTAL/CERAMIC
RESONATOR MANUFACTURER. TYPICAL VALUES ARE FROM 0 TO 100pF.

(c)

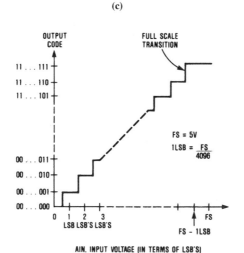

(d)

Analog/digital and digital/analog circuit titles and descriptions

Continued

PIN	NAME	FUNCTION
1	AIN	Analog Input, 0 to +5V unipolar input
2	V_{REF}	-5.25V Reference Output
3	AGND	Analog Ground
4-11	D11-D4	Three-State Data Outputs
12	DGND	Digital Ground
13-16	D3/11-D0/8	Three-State Data Outputs
17	CLKIN	Clock Input. An external TTL/CMOS compatible clock may be applied to this pin or a crystal can be connected between CLKIN and CLKOUT.
18	CLKOUT	Clock Output. An inverted CLKIN signal appears at this pin.

PIN	NAME	FUNCTION
19	HBEN	High Byte Enable Input. This pin is used to multiplex the internal 12-bit conversion result into the lower bit outputs (D7-D0/8). HBEN also disables conversion starts when HIGH.
20	$\overline{RD}$	READ Input. This active low signal starts a conversion when $\overline{CS}$ and HBEN are low. $\overline{RD}$ also enables the output drivers when $\overline{CS}$ is low.
21	$\overline{CS}$	The CHIP SELECT Input must be low for the ADC to recognize $\overline{RD}$ and HBEN inputs.
22	$\overline{BUSY}$	The $\overline{BUSY}$ Output is low when a conversion is in progress.
23	V_{SS}	Negative Supply, -15V for MX7572 and -15V or -12V for MAX162.
24	V_{DD}	Positive Supply, +5V.

Data Bus Output, $\overline{CS}$ & $\overline{RD}$ = LOW

	Pin 4	Pin 5	Pin 6	Pin 7	Pin 8	Pin 9	Pin 10	Pin 11	Pin 13	Pin 14	Pin 15	Pin 16
MNEMONIC*	D11	D10	D9	D8	D7	D6	D5	D4	D3/11	D2/10	D1/9	D0/8
HBEN = LOW	DB11	DB10	DB9	DB8	DB7	DB6	DB5	DB4	DB3	DB2	DB1	DB0
HBEN = HIGH	DB11	DB10	DB9	DB8	LOW	LOW	LOW	LOW	DB11	DB10	DB9	DB8

Note:
* D11 . . . D0/8 are the ADC data output pins.
  DB11 . . . DB0 are the 12-bit conversion results, DB11 is the MSB.

(e)

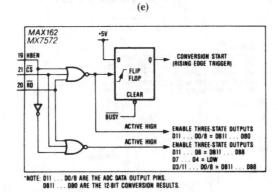

(f)

High-speed 12-bit CMOS A/D converter

Fig. 9-6 Figure 9-6A shows the basic operational diagram for the MAX162/7572 A/D converter. Conversion times are 3 µs (MAX162) and 5/12 µs (MAX7572). Figures 9-6B, C, D, E, and F show the analog-equivalent circuit internal clock circuit, transfer function, pin functions, and logic equivalents, respectively. MAXIM NEW RELEASES DATA BOOK, 1992, PP. 7-41, 42, 44, 50.

Analog/digital and digital/analog circuits

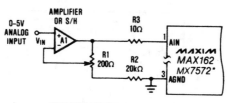

*ADDITIONAL PINS OMITTED FOR CLARITY

Twelve-bit A/D converter with 0-V to 5-V input

Fig. 9-7 The circuit in the illustration shows the A/D of Fig. 9-6 connected to accept 0-V to 5-V inputs. To adjust the full-scale range, apply FS-3/2 LSB (4.99817 V) at the analog input and adjust R1 until the output code changes between 1111 1111 1110 and 1111 1111 1111. MAXIM NEW RELEASES DATA BOOK, 1992, P. 7-50.

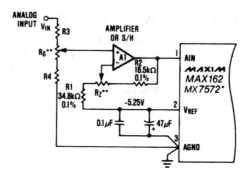

*ADDITIONAL PINS OMITTED FOR CLARITY
**OMIT IF ERROR ADJUST IS NOT REQUIRED

(a)

V_{IN} Range (Volts)	R3* (kΩ)	R4* (kΩ)	R_Z (Ω)	R_G (Ω)	1/2LSB (mV)	FS/2-3/2LSBs (Volts)
±2.5	3.83	8.25	500	500	0.61	2.49817
±5.0	33.2	16.9	500	1000	1.22	4.99634
±10.0	47.5	9.53	500	500	2.44	9.99268

Notes:
* R3 and R4 have a 0.1% tolerance.
  All resistors are standard EIA/MIL decade values.

(b)

Analog/digital and digital/analog circuit titles and descriptions

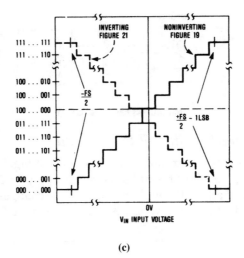

(c)

Twelve-bit A/D converter with noninverting bipolar inputs

Fig. 9-8 The circuit in the illustration shows the A/D of Fig. 9-6 connected to accept noninverting bipolar inputs. Figures 9-8B and C show the input voltage ranges and transfer function, respectively. Adjust the offset first, by applying +1/2 LSB (Fig. 9-8B) to the analog input and adjust RZ until the output code flickers between 1000 0000 0000 and 1000 0000 0001. Then apply FS-3/2 LSB, Fig. 9-8B, and adjust RG until the output code flickers between 1111 1111 1110 and 1111 1111 1111. MAXIM NEW RELEASES DATA BOOK, 1992, P. 7-50.

9

464

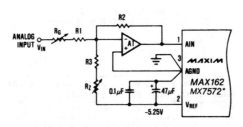

*ADDITIONAL PINS OMITTED FOR CLARITY

(a)

V_{IN} Range (Volts)	R1* (kΩ)	R2* (kΩ)	R3* (kΩ)	R_Z (Ω)	R_G (Ω)	1/2LSB (mV)	FS/2-3/2LSBs (Volts)
±2.5	20	20.5	42.2	2000	1000	0.61	2.49817
±5.0	20	10.2	21	1000	1000	1.22	4.99634
±10.0	20	5.11	10.5	500	1000	2.44	9.99268

Notes:
* R1, R2, and R3 have a 0.1% tolerance.
  All resistors are standard EIA/MIL decade values.

(b)

Analog/digital and digital/analog circuits

Twelve-bit A/D converter with inverting bipolar inputs

Fig. 9-9 The circuit in the illustration shows the A/D of Fig. 9-6 connected to accept inverting bipolar inputs. Figure 9-9B shows the input voltage ranges. Adjust the offset first, by applying +1/2 LSB (Fig. 9-9B) to the analog input and adjust R_Z until the output code flickers between 0111 1111 1111 and 0111 1111 1110. Then apply FS-3/2 LSB, Fig. 9-9B, and adjust RG until the output code flickers between 0000 0000 0001 and 0000 0000 0000. MAXIM NEW RELEASES DATA BOOK, 1992, P. 7-51.

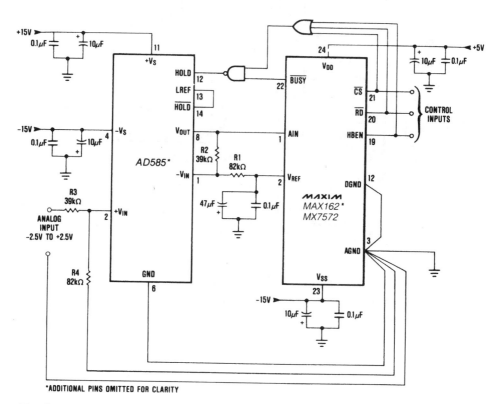

*ADDITIONAL PINS OMITTED FOR CLARITY

Twelve-bit A/D converter with low-speed sample/hold interface

Fig. 9-10 The circuit in the illustration shows the A/D of Fig. 9-6 connected to accept –2.5-V to +2.5-V inputs, with a sampling rate of 64.5 kHz and a 1-MHz clock. MAXIM NEW RELEASES DATA BOOK, 1992, P. 7-48.

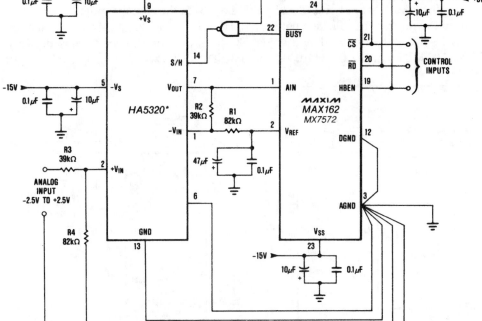

*ADDITIONAL PINS OMITTED FOR CLARITY

Twelve-bit A/D converter with high-speed sample/hold interface

Fig. 9-11 The circuit in the illustration shows the A/D of Fig. 9-6 connected to accept –2.5-V to +2.5-V inputs, with a sampling rate of 125 kHz and a 2-MHz clock. MAXIM NEW RELEASES DATA BOOK, 1992, P. 7-49.

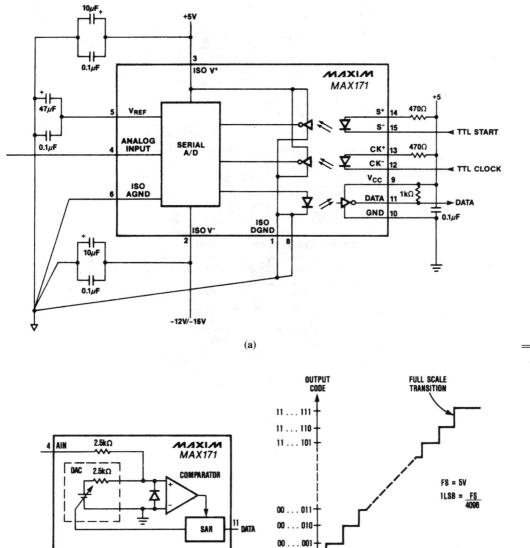

(a)

(b)

(c)

Analog/digital and digital/analog circuit titles and descriptions

PIN DIP	NAME	FUNCTION
1	ISO DGND	Isolated Digital Ground
2	ISO V⁻	Analog Negative Supply, −12V or −15V
3	ISO V⁺	Analog Positive Supply, +5V
4	AIN	Analog Input, 0V to +5V Unipolar
5	REF	Reference Voltage Output, −5.25V
6	ISO AGND	Isolated Analog Ground. Normally tied to ISO DGND
7	TP	Test Pin. Leave unconnected
8	ISO DGND	Isolated Digital Ground
ELECTRICAL ISOLATION BARRIER		
9	V_{CC}	Digital Positive Supply. +5V
10	GND	Digital Ground
11	DATA	Serial Data Output
12	CK⁻	Clock⁻ Input
13	CK⁺	Clock⁺ Input
14	S⁺	Conversion Start⁺ Input
15	S⁻	Conversion Start⁻ Input
16	N.C.	No Connect

(d)

Opto-isolated serial-output 12-bit A/D converter

Fig. 9-12   Figure 9-12A shows the basic operational diagram for the MAX171 A/D converter with serial output and full opto-isolation (1500 V). Conversion time is 5.8 µs. Figures 9-12B, C, and D show the analog equivalent circuit, transfer function, and pin functions, respectively. MAXIM NEW RELEASES DATA BOOK, 1992, PP. 7-70, 71, 73.

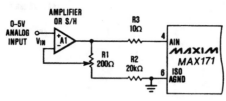

*ADDITIONAL PINS OMITTED FOR CLARITY

Analog/digital and digital/analog circuits

Opto-isolated serial A/D converter with 0-V to 5-V input

Fig. 9-13   The circuit in the illustration shows the A/D of Fig. 9-12 connected to accept 0-V to 5-V inputs. To adjust the full-scale range, apply FS-3/2 LSB (4.9817 V) at the analog input and adjust R1 until the output code changes between 1111 1111 1110 and 1111 1111 1111. MAXIM NEW RELEASES DATA BOOK, 1992, P. 7-73.

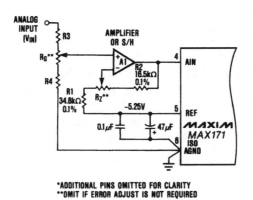

(a)

V_{IN} Range (Volts)	R3* (kΩ)	R4* (kΩ)	R_Z (Ω)	R_G (Ω)	1/2LSB (mV)	FS/2-3/2LSBs (Volts)
±2.5	3.83	8.25	500	500	0.61	2.49817
±5.0	33.2	16.9	500	1000	1.22	4.99634
±10.0	47.5	9.53	500	500	2.44	9.99268

*R3 and R4 have a 0.1% tolerance. All resistors are standard EIA/MIL decade values.

(b)

Opto-isolated serial A/D converter with noninverting bipolar inputs

Fig. 9-14   The circuit in the illustration shows the A/D of Fig. 9-12 connected to accept noninverting bipolar inputs. Figure 9-14B shows the input voltage ranges. Adjust the offset first, by applying +1/2 LSB (Fig. 9-14B) to the analog input and adjust R_Z until the output code flickers between 1000 0000 0000 and 1000 0000 0001. Then apply FS-3/2 LSB (Fig. 9-14B) and adjust RG until the output code flickers between 1111 1111 1110 and 1111 1111 1111. MAXIM NEW RELEASES DATA BOOK, 1992, P. 7-74.

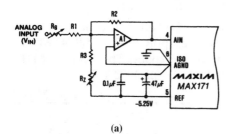

V$_{IN}$ Range (Volts)	R1* (kΩ)	R2* (kΩ)	R3* (kΩ)	R$_Z$ (Ω)	R$_G$ (Ω)	1/2LSB (mV)	FS/2 -3/2 LSBs (Volts)
±2.5	20	20.5	42.2	2000	1000	0.61	2.49817
±5.0	20	10.2	21	1000	1000	1.22	4.99634
±10.0	20	5.11	10.5	500	1000	2.44	9.99268

*R1, R2 and R3 have a 0.1% tolerance. All resistors are standard EIA/MIL decade values.

(a)   (b)

Opto-isolated serial A/D converter with inverting bipolar inputs

Fig. 9-15   The circuit in the illustration shows the A/D of Fig. 9-12 connected to accept inverting bipolar inputs. Figure 9-15B shows the input voltage ranges. Adjust the offset first, by applying +1/2 LSB, Fig. 9-15B, to the analog input and adjust R_Z until the output code flickers between 0111 1111 1111 and 0111 1111 1110. Then apply FS-3/2 LSB (Fig. 9-15B) and adjust R_G until the output code flickers between 0000 0000 0001 and 0000 0000 0000. MAXIM NEW RELEASES DATA BOOK, 1992, P. 7-74.

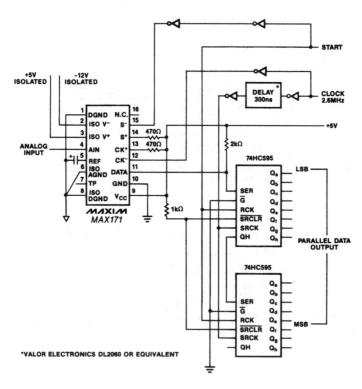

Opto-isolated serial A/D converter with parallel output

Fig. 9-16   The circuit in the illustration shows the A/D of Fig. 9-12 connected to provide a parallel-data output. MAXIM NEW RELEASES DATA BOOK, 1992, P. 7-72.

Analog/digital and digital/analog circuits

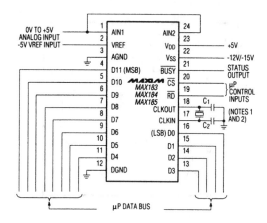

NOTE 1: MAX183— 4.0MHz CRYSTAL/CERAMIC RESONATOR
MAX184— 2.5MHz CRYSTAL/CERAMIC RESONATOR
MAX185—1.25MHz CRYSTAL/CERAMIC RESONATOR
NOTE 2: C1 AND C2 CAPACITANCE VALUES DEPEND ON CRYSTAL/CERAMIC RESONATOR MANUFACTURER. TYPICAL VALUES ARE FROM 0pF TO 100pF.

(a)

PIN	NAME	FUNCTION
1	AIN1	Analog Input
2	VREF	Voltage-Reference Input
3	AGND	Analog Ground
4-11	D11-D4	Three-State Data Outputs. They are active when $\overline{CS}$ and $\overline{RD}$ are low. DB11 is the most significant bit.
12	DGND	Digital Ground
13-16	D3-D0	Three-State Data Outputs
17	CLKIN	Clock Input. Connect an external TTL-compatible clock to CLKIN. Alternatively, insert a crystal or ceramic resonator between CLKIN and CLKOUT.
18	CLKOUT	Clock Output. When using an external clock, an inverted CLKIN signal appears on CLKOUT. See CLKIN description.
19	$\overline{RD}$	READ Input. Along with $\overline{CS}$, this active low signal enables the three-state drivers and starts a conversion.
20	$\overline{CS}$	CHIP SELECT. Along with $\overline{RD}$, this active low signal enables the three-state drivers and starts a conversion.
21	$\overline{BUSY}$	BUSY. Low while a conversion is in progress. $\overline{BUSY}$ indicates converter status.
22	Vss	Negative Supply, -12V to -15V
23	VDD	Positive Supply, +5V
24	AIN2	Analog Input

(b)

Analog/digital and digital/analog circuit titles and descriptions

Continued

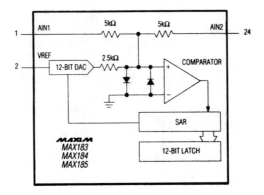

(c)

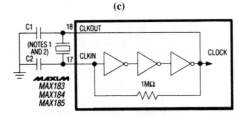

NOTE 1: MAX183— 4.0MHz CRYSTAL/CERAMIC RESONATOR
MAX184— 2.5MHz CRYSTAL/CERAMIC RESONATOR
MAX185—1.25MHz CRYSTAL/CERAMIC RESONATOR
NOTE 2: C1 AND C2 CAPACITANCE VALUES DEPEND ON CRYSTAL/CERAMIC
RESONATOR MANUFACTURER. TYPICAL VALUES ARE FROM 30pF TO 100pF.

(d)

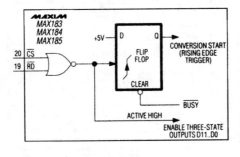

(e)

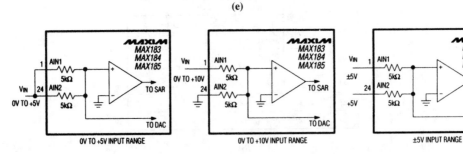

(f)

Analog/digital and digital/analog circuits

Continued

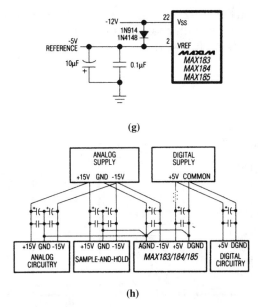

(g)

(h)

High-speed 12-bit A/D converter with external reference

Fig. 9-17 Figure 9-17A shows the basic operational diagram for the MAX183/ 184/185 A/D converter. Conversion times are 3 μs (MAX183), 5 μs (MAX184) and 10 μs (MAX185). Figures 9-17B, C, D, E, and F show the pin functions, analog inputs, internal clock, logic inputs, and analog-input range configurations, respectively. The external reference-input range is –5.1 V to –4.9 V, and the reference should be bypassed as shown in Fig. 9-17G. (The diode ensures that V_{SS} is applied before V_{REF}, when these voltages are from different sources.) Figure 9-17H shows recommended power-source connections. MAXIM NEW RELEASES DATA BOOK, 1992, PP. 7-146, 147, 148, 150, 152.

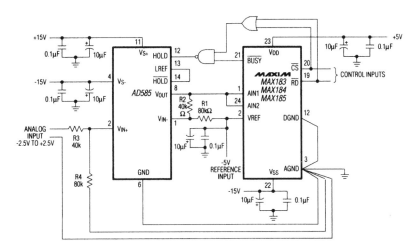

Analog/digital and digital/analog circuit titles and descriptions

A/D converter with sample/hold (external reference)

Fig. 9-18   The circuit in the illustration shows the A/D of Fig. 9-17 connected to accept –2.5-V to +2.5-V inputs, with a sampling rate of 64.5 kHz and a 1-MHz clock. MAXIM NEW RELEASES DATA BOOK, 1992, P. 7-151.

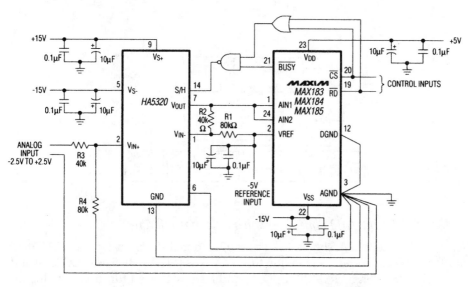

A/D converter with high-speed sample/hold (external reference)

Fig. 9-19   The circuit in the illustration shows the A/D of Fig. 9-17 connected to accept –2.5-V to +2.5-V inputs, with a sampling rate of 125 kHz and a 2-MHz clock. Note that if the MAX183 is used with this circuit, the sampling rate is increased to 210 kHz, with a 4-MHz clock. MAXIM NEW RELEASES DATA BOOK, 1992, P. 7-151.

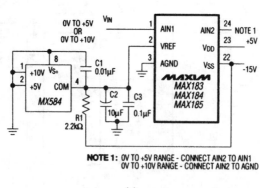

(a)

Continued

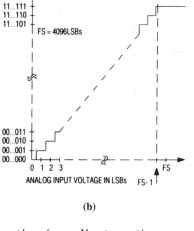

(b)

A/D with unipolar operation (no adjustment)

Fig. 9-20   The circuit in the illustration shows the A/D of Fig. 9-17 connected to accept either 0-V to +5-V, or 0-V to +10-V, inputs. Figure 9-20B shows the ideal transfer function. MAXIM NEW RELEASES DATA BOOK, 1992, P. 7-152.

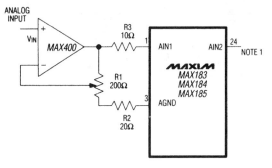

NOTE 1: 0V TO +5V RANGE - CONNECT AIN2 TO AIN1
0V TO +10V RANGE - CONNECT AIN2 TO AGND

A/D with unipolar operation (gain adjustment)

Fig. 9-21   The circuit in the illustration shows the A/D of Fig. 9-17 connected to accept either 0-V to +5-V, or 0-V to +10-V, inputs, with gain adjustment. To adjust the full-scale range, apply FS-3/2 LSB (last code transition) at the analog input, and adjust R1 until the output code switches between 1111 1111 1110 and 1111 1111 1111. For the 0-V to +5-V configuration, FS-3/2 LSB = 4.99817 V. For the 0-V to +10-V configuration, FS-3/2 LSB = 9.99634 V. MAXIM NEW RELEASES DATA BOOK, 1992, P. 7-153.

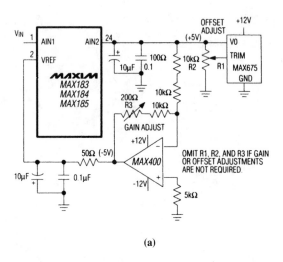

(a)

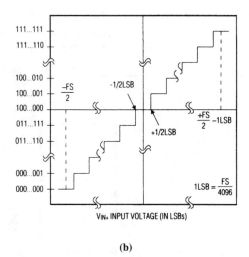

(b)

A/D with bipolar operation (external reference)

Fig. 9-22   The circuit in the illustration shows the A/D of Fig. 9-17 connected to accept −5-V to +5-V inputs. Figure 9-22B shows the transfer functions. To calibrate, apply −4.99878 V and adjust R1 until the output code switches between 0000 0000 0000 and 0000 0000 0001. Then apply +4.99634 V and adjust R3 until the output code switches between 1111 1111 1110 and 1111 1111 1111. MAXIM NEW RELEASES DATA BOOK, 1992, P. 7-153.

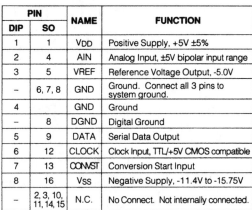

(a)

PIN		NAME	FUNCTION
DIP	SO		
1	1	VDD	Positive Supply, +5V ±5%
2	4	AIN	Analog Input, ±5V bipolar input range
3	5	VREF	Reference Voltage Output, -5.0V
–	6, 7, 8	GND	Ground. Connect all 3 pins to system ground.
4		GND	Ground
–	8	DGND	Digital Ground
5	9	DATA	Serial Data Output
6	12	CLOCK	Clock Input, TTL/+5V CMOS compatible
7	13	CONVST	Conversion Start Input
8	16	VSS	Negative Supply, -11.4V to -15.75V
–	2, 3, 10, 11, 14, 15	N.C.	No Connect. Not internally connected.

(b)

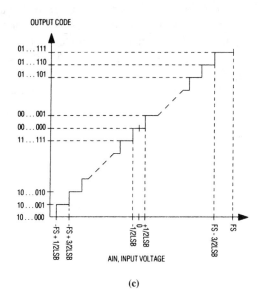

(c)

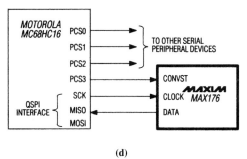

(d)

Analog/digital and digital/analog circuit titles and descriptions

Continued

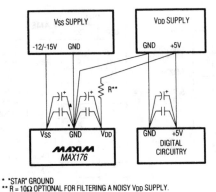

* "STAR" GROUND
** R = 10Ω OPTIONAL FOR FILTERING A NOISY V$_{DD}$ SUPPLY.

(e)

Twelve-bit A/D converter with 250-ksps serial output

Fig. 9-23 Figure 9-23A shows the basic operation diagram for the MAX176 A/D converter. The sampling rate is 250 ksps (thousands of samples per second). Figures 9-23B, C, D, and E show the pin functions, transfer functions, microprocessor interface, and recommended grounding, respectively. MAXIM NEW RELEASES DATA BOOK, 1994, PP. 7-76, 79, 80, 84.

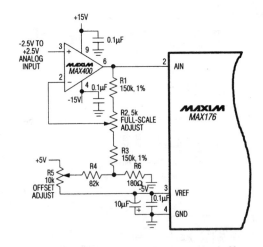

A/D converter (250-ksps) with noninverting gain/offset

Fig. 9-24 The circuit in the illustration shows the A/D of Fig. 9-23 connected to accept –2.5-V to +2.5-V inputs. To calibrate, apply 1/2 LSB (1.22 mV) at the analog input and adjust R5 until the output code changes between 0000 0000 0000 and 0000 0000 0001. Then apply –FS + 1/2 LSB (–2.49939 V) and adjust R2 until the output code changes between 1000 0000 0000 and 1000 0000 0001. MAXIM NEW RELEASES DATA BOOK, 1994, P. 7-80.

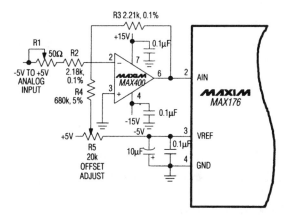

A/D converter (250-ksps) with inverting gain/offset

Fig. 9-25 The circuit in the illustration is similar to that of Fig. 9-24, except that –5-V to +5-V inputs can be accepted, and the input is inverted. MAXIM NEW RELEASES DATA BOOK, 1994, P. 7-80.

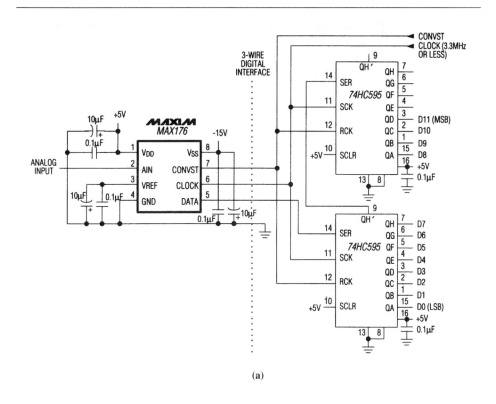

(a)

Analog/digital and digital/analog circuit titles and descriptions

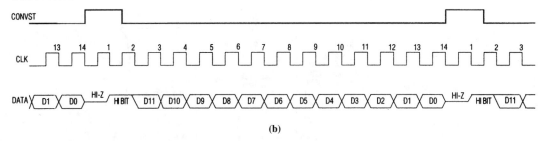

(b)

A/D converter (250-ksps) with three-wire interface to parallel port

Fig. 9-26 The circuit in the illustration shows the A/D of Fig. 9-23 connected for a three-wire interface to a parallel port. Figure 9-26B shows the timing. MAXIM NEW RELEASES DATA BOOK, 1994, P. 7-81.

480

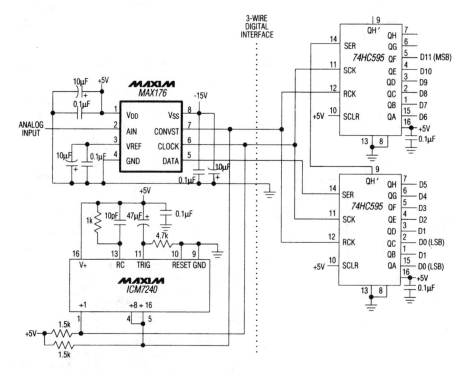

A/D (250-ksps) converter with three-wire interface and timer chip

Fig. 9-27 The circuit in the illustration is similar to that of Fig. 9-26, except that the ICM7240 timer permits stand-alone operation (no external clock required). MAXIM NEW RELEASES DATA BOOK, 1994, P. 7-82.

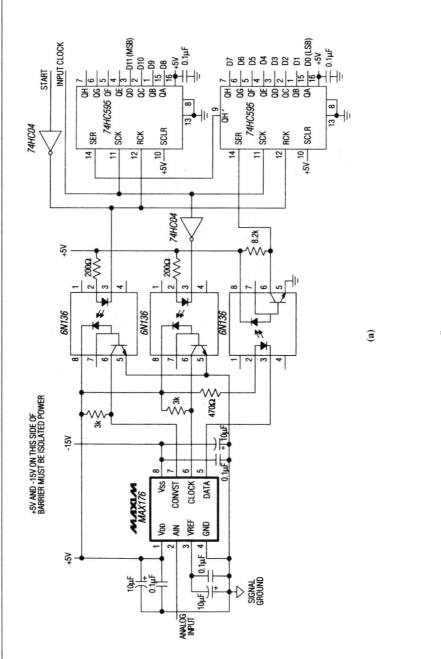

(a)

9

481

Analog/digital and digital/analog circuit titles and descriptions

Continued

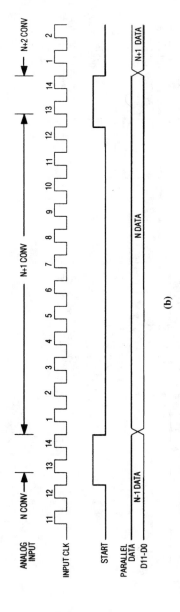

(b)

A/D (250-ksps) converter with isolated interface to parallel port

Fig. 9-28 The circuit in the illustration is similar to that of Fig. 9-26 except that the A/D is isolated from the parallel-port chips. Figure 9-28B shows the timing diagram. Conversion time is 100 μs. MAXIM NEW RELEASES DATA BOOK, 1994, P. 7-83.

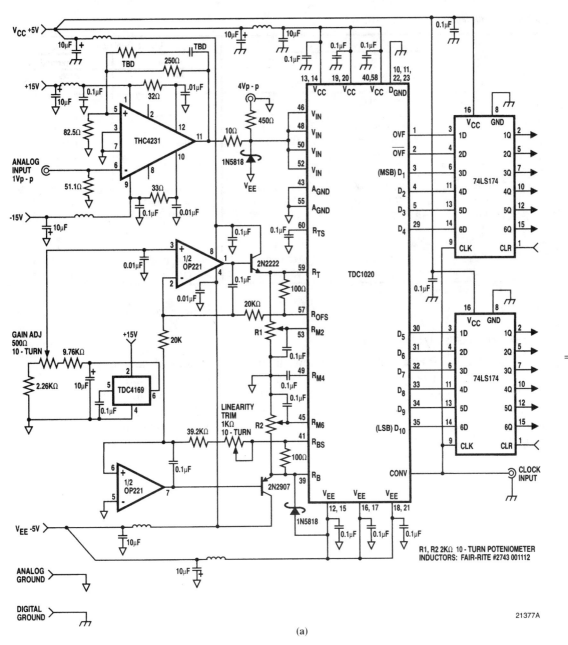

Analog/digital and digital/analog circuit titles and descriptions

Continued

Output Coding Table

Input	Binary		Offset Two's Complement	
	True NMINV = 1 NLINV = 1	Inverted NMINV = 0 NLINV = 0	True NMINV = 0 NLINV = 1	Inverted NMINV = 1 NLINV = 0
	MSB – LSB (OVF)			
>2.000V	0000000000(1)	1111111111(1)	1000000000(1)	0111111111(1)
2.000V	0000000000(0)	1111111111(0)	1000000000(0)	0111111111(0)
1.996V	0000000001(0)	1111111110(0)	1000000001(0)	0111111110(0)
⋮	⋮	⋮	⋮	⋮
0.004V	0111111111(0)	1000000000(0)	1111111111(0)	0000000000(0)
0.000V	1000000000(0)	0111111111(0)	0000000000(0)	1111111111(0)
−0.004V	1000000001(0)	0111111110(0)	0000000001(0)	1111111110(0)
⋮	⋮	⋮	⋮	⋮
−1.996V	1111111110(0)	0000000001(0)	0111111110(0)	1000000001(0)
−2.000V	1111111111(0)	0000000000(0)	0111111111(0)	1000000000(0)

Note: Input voltages are at code centers.

(b)

Ten-bit 20-Msps A/D converter

Fig. 9-29 The circuit in the illustration shows a 20-Msps (millions of samples per second) A/D converter, complete with interface. To calibrate, center R1 and R2, apply 1.998 V at the analog input and adjust the GAIN pot so the output toggles between full-scale and one LSB below full-scale (1111111111 and 1111111110 (Fig. 9-29B) for binary conversions). Then apply –1.998 V and adjust the LINEARITY TRIM pot so that the output toggles between 0 and 1 (000000000 and 0000000001 (Fig. 9-29B) for binary conversions). RAYTHEON SEMICONDUCTOR DATA BOOK, 1994, PP. 3-8, 15.

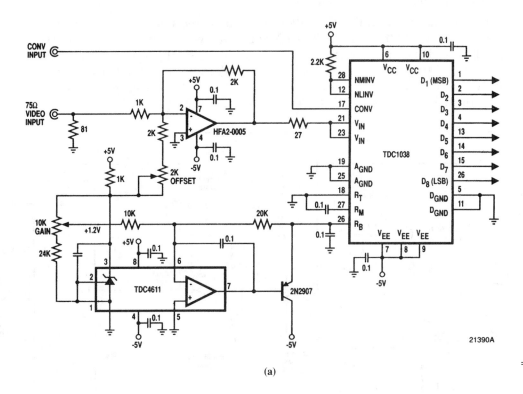

(a)

Input Voltage	Binary		Offset Two's Complement	
	True	Inverted	True	Inverted
	NMINV=HIGH NLINV=HIGH	NMINV=LOW NLINV=LOW	NMINV=LOW NLINV=HIGH	NMINV=HIGH NLINV=LOW
0.0000V	0000 0000	1111 1111	1000 0000	0111 1111
-0.0078V	0000 0001	1111 1110	1000 0001	0111 1110
.	.	.	.	.
.	.	.	.	.
.	.	.	.	.
-0.9922V	0111 1111	1000 0000	1111 1111	0000 0000
-1.0000V	1000 0000	0111 1111	0000 0000	1111 1111
-1.0078V	1000 0001	0111 1110	0000 0001	1111 1110
.	.	.	.	.
.	.	.	.	.
.	.	.	.	.
-1.9844V	1111 1110	0000 0001	0111 1110	1000 0001
-1.9922V	1111 1111	0000 0000	0111 1111	1000 0000

Notes: 1. NMINV and NLINV are to be considered DC controls. They may be tied to +5V for a logic 1 or tied to ground for a logic 0.
2. Voltages are code midpoints.

(b)

Analog/digital and digital/analog circuit titles and descriptions

Video A/D converter (low power)

Fig. 9-30 The circuit in the illustration shows a 20-Msps, low-power A/D converter complete with interface. The 7-MHz full-power bandwidth and 30-MHz small-signal bandwidth make the A/D suitable for video applications. Power dissipation is 700 mW. To calibrate, apply the voltages shown in Fig. 9-30B and adjust the OFFSET (low end) and GAIN (high end) pots for the correct output codes. RAYTHEON SEMICONDUCTOR DATA BOOK, 1994, PP. 3-29, 33.

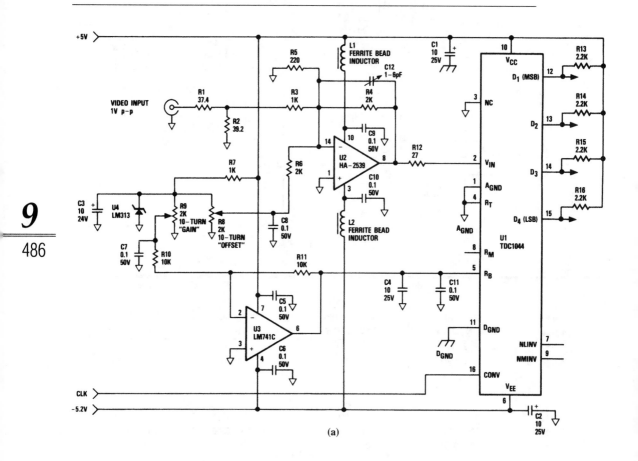

(a)

9

486

Continued

Output Coding Table [1]

Range	Binary		Offset Two's Complement	
	True	Inverted	True	Inverted
−1.00V FS	NMINV = 1	0	0	1
	NLINV = 1	0	1	0
0.000V	0000	1111	1000	0111
−0.067V	0001	1110	1001	0110
−0.133V	0010	1101	1010	0101
−0.200V	0011	1100	1011	0100
−0.267V	0100	1011	1100	0011
−0.333V	0101	1010	1101	0010
−0.400V	0110	1001	1110	0001
−0.467V	0111	1000	1111	0000
−0.533V	1000	0111	0000	1111
−0.600V	1001	0110	0001	1110
−0.667V	1010	0101	0010	1101
−0.733V	1011	0100	0011	1100
−0.800V	1100	0011	0100	1011
−0.867V	1101	0010	0101	1010
−0.933V	1110	0001	0110	1001
−1.000V	1111	0000	0111	1000

Note:

1. Input voltages are at code centers.

(b)

Video A/D converter (4-bit, 25 Msps)

Fig. 9-31 The circuit in the illustration shows a 25-Msps A/D converter complete with interface. Figure 9-31B shows the output codes for various input voltages, using the values of R1 and R2 shown. Other input-voltage ranges and input impedances can be selected with different values for R_1/R_2, as follows:

$$R_1 = \frac{1}{\left(\frac{2V_R}{Z_{IN}}\right) - \frac{1}{1000}} \qquad R_2 = Z_{IN} - \left(\frac{1000\,R_1}{1000 + R_1}\right)$$

where V_R is the input-voltage range, Z_{IN} is the input impedance, and the constant 1000 is the value of R3. As shown, the circuit is set up for 1 V_{p-p} and a typical 75-Ω video input. To calibrate, assuming a 0-V to −1-V range shown in Fig. 9-31B, continuously strobe the converter with −0.0033 V (1/2 LSB from 0.000 V) on the analog input and adjust R8 (offset) until the output toggles between 0000 and 0001. Then apply −0.967 V (1/2 LSB from −1.000 V) and adjust R9 (gain) so the output toggles between 1110 and 1111. If necessary, adjust C12 for optimum pulse/frequency response of U2. RAYTHEON SEMICONDUCTOR DATA BOOK, 1994, PP. 3-41, 43.

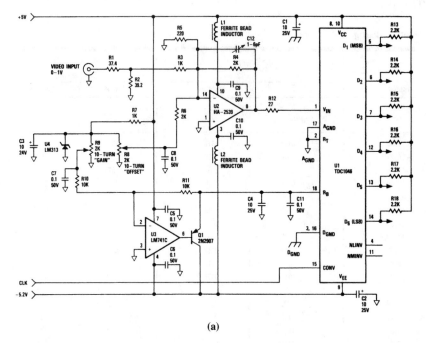

(a)

Output Coding Table [1]

Range	Binary		Two's Complement	
	True	**Inverted**	**True**	**Inverted**
	NMINV = 1	0	0	1
15.8730mV Step	NLINV = 1	0	1	0
0.0000V	000000	111111	100000	011111
−0.0159V	000001	111110	100001	011110
•	•	•	•	•
•	•	•	•	•
•	•	•	•	•
−0.4921V	011111	100000	111111	000000
−0.5079V	100000	011111	000000	111111
−0.5238V	100001	011110	000001	111110
•	•	•	•	•
•	•	•	•	•
•	•	•	•	•
−0.9841V	111110	000001	011110	100001
−1.0000V	111111	000000	011111	100000

Note: 1. Voltages are code midpoints when calibrated (see *Calibration* section).

(b)

Video A/D converter (6-bit, 25 Msps)

Fig. 9-32   The circuit in the illustration is similar to that of Fig. 9-31, except that the output is 6-bit. Figure 9-32B shows the output codes and input voltage range. Use −0.0079 V for offset and −0.9921 V for gain adjustments. RAYTHEON SEMICONDUCTOR DATA BOOK, 1994, PP. 3-47, 52.

VIDEO INPUT

+5V

CLK

-5.2V

Notes:

1. Unless otherwise specified, all resistors are 1/4W, 2%.

2. $R1 = Z_{IN} - \left(\dfrac{1000\,R2}{1000 + R2} \right)$

3. $R2 = \dfrac{1}{\left(\dfrac{2V_{Range}}{V_{REF}\,Z_{IN}} \right) - 0.001}$

(a)

Continued

Output Coding

Step	Range	Binary		Offset Two's Complement	
		True	Inverted	True	Inverted
	−1.0000V FS	NMINV - 1	0	0	1
	7.874mV STEP	NLINV - 1	0	1	0
000	0.0000V	0000000	1111111	1000000	0111111
001	−0.0078V	0000001	1111110	1000001	0111110
.	.	.	.	.	.
.	.	.	.	.	.
.	.	.	.	.	.
063	−0.4960V	0111111	1000000	1111111	0000000
064	−0.5039V	1000000	0111111	0000000	1111111
.	.	.	.	.	.
.	.	.	.	.	.
.	.	.	.	.	.
126	−1.9921V	1111110	0000001	0111110	1000001
127	−1.0000V	1111111	0000000	0111111	1000000

Note:

1. Voltages are code midpoints when calibrated (see Calibration Section).

(b)

Video A/D converter (7-bit, 20 Msps)

Fig. 9-33 The circuit in the illustration is similar to that of Fig. 9-31, except that the output is 7 bit. Fig. 9-31B shows the output codes and input voltage range. The A/D compares the input signal with 127 reference voltages to produce an *N*-of-127 code (so-called *thermometer code*) and then converts the *N*-of-127 code into binary or offset two's-complement coding. Use −0.0039 V for offset (toggling between codes 00 and 01) and −0.9961 V for gain (toggling between codes 126 and 127) adjustments. RAYTHEON SEMICONDUCTOR DATA BOOK, 1994, PP. 3-59, 60.

(a)

Output Coding Table

Step	Range		Binary		Offset Two's Complement	
			True	**Inverted**	**True**	**Inverted**
	−2.0000V FS	−2.0480V FS	NMINV=1	0	0	1
	7.8431mV Step	8.000mV Step	NLINV=1	0	1	0
000	0.0000V	0.0000V	00000000	11111111	10000000	01111111
001	−0.0078V	−0.0080V	00000001	11111110	10000001	01111110
•	•	•	•	•	•	•
•	•	•	•	•	•	•
•	•	•	•	•	•	•
127	−0.9961V	−1.0160V	01111111	10000000	11111111	00000000
128	−1.0039V	−1.0240V	10000000	01111111	00000000	11111111
129	−1.0118V	−1.0320V	10000001	01111110	00000001	11111110
•	•	•	•	•	•	•
•	•	•	•	•	•	•
•	•	•	•	•	•	•
254	−1.9921V	−2.0320V	11111110	00000001	01111110	10000001
255	−2.0000V	−2.0400V	11111111	00000000	01111111	10000000

Notes: 1. NMINV and NLINV are to be considered DC controls. They may be tied to +5V for a logical "1" and tied to ground for a logical "0."
2. Voltages are code midpoints when calibrated by the procedure given below.

(b)

Analog/digital and digital/analog circuit titles and descriptions

Continued

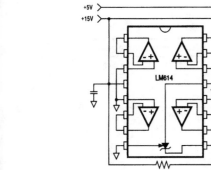

(c)

Video A/D converter (8-bit, 20 Msps)

Fig. 9-34 The circuit in the illustration shows a 20-Msps A/D converter complete with interface. Figures 9-34B and C show the output codes and a recommended circuit for midpoint adjustment, respectively. To calibrate (assuming a 0-V to –2-V input range), apply –0.0039 V and adjust the OFFSET pot so the output toggles between codes 00 and 01. Then apply –1.996 V and adjust the GAIN pot so the output toggles between codes 62 and 63. RAYTHEON SEMICONDUCTOR DATA BOOK, 1994, PP. 3- 67, 71.

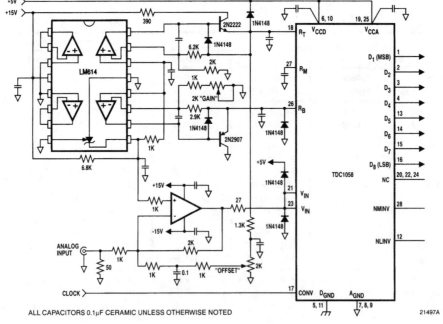

ALL CAPACITORS 0.1µF CERAMIC UNLESS OTHERWISE NOTED

(a)

Analog/digital and digital/analog circuits

Output Coding Table

| | Binary | | Offset Two's Complement | |
| | True | Inverted | True | Inverted |
Input Voltage	NMINV = HIGH NLINV = HIGH	NMINV = LOW NLINV = LOW	NMINV = LOW NLINV = HIGH	NMINV = HIGH NLINV = LOW
5.0000V	0000 0000	1111 1111	1000 0000	0111 1111
4.9922V	0000 0001	11111110	1000 0001	0111 1110
⋮	⋮	⋮	⋮	⋮
4.0078V	0111 1111	1000 0000	1111 1111	0000 0000
4.0000V	1000 0000	0111 1111	0000 0000	1111 1111
3.9922V	1000 0001	0111 1110	0000 0001	1111 1110
⋮	⋮	⋮	⋮	⋮
3.0156V	1111 1110	0000 0001	0111 1110	1000 0001
3.0078V	1111 1111	0000 0000	0111 1111	1000 0000

Notes: 1. NMINV and NLINV are to be considered DC controls. They may be tied to +5V through a 4.7 kOhm resistor for a logic HIGH or tied to ground for a logic LOW.

2. Voltages are code midpoints.

(b)

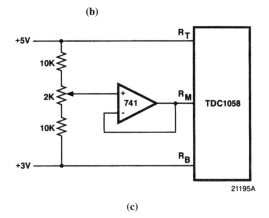

21195A

(c)

Video A/D converter (8-bit, 20 Msps, low power)

Fig. 9-35 The circuit in the illustration shows a video A/D, complete with interface. The 7-MHz full-power bandwidth and 60-MHz small-signal bandwidth make the A/D suitable for video applications. Power dissipation is 600 mW. Figures 7-35B and C show the output codes and recommended circuit for midscale linearity adjustment, respectively. Bipolar inputs to the amplifier can be accommodated by adjusting the OFFSET pot. To calibrate, apply the voltages shown in Fig. 9-35B and adjust the OFFSET (low end) and GAIN (high end) pots for the correct output codes. RAYTHEON SEMICONDUCTOR DATA BOOK, 1994, PP. 3-87, 91, 92.

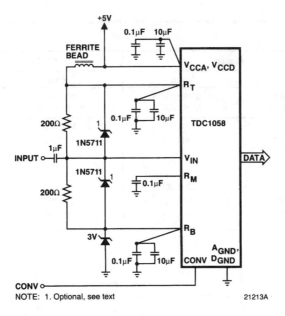

NOTE: 1. Optional, see text   21213A

Video A/D converter (inexpensive interface)

Fig. 9-36 The circuit in the illustration provides the same basic functions as that of Fig. 9-35, but at reduced cost and complexity, where dc response is not required and loss of some power-supply rejection is tolerable. RAYTHEON SEMICONDUCTOR DATA BOOK, 1994, P. 3-92.

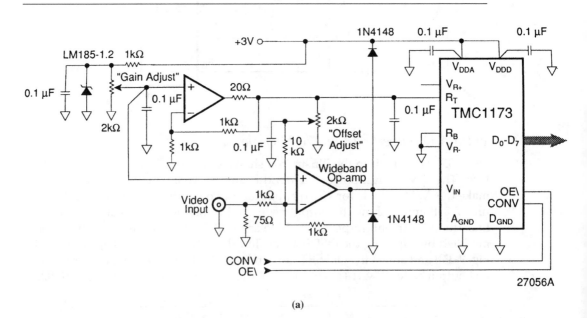

27056A

(a)

Analog/digital and digital/analog circuits

Continued

Input Voltage	Output Code MSB LSB
$>R_T$	11111111
R_T	11111111
R_T - 1 LSB	11111110
•	•
•	•
$(R_T + R_B)/2 + 1$ LSB	10000000
$(R_T + R_B)/2$	01111111
•	•
•	•
R_B + 1 LSB	00000001
R_B	00000000
$<R_B$	00000000

Note: 1 LSB = $(R_T - R_B)/255$

(b)

A/D converter for 2.7-V to 3.6-V operation (10 Msps)

Fig. 9-37 The circuit in the illustration shows an A/D converter operated from a single +3-V supply. Figure 9-37B shows the output codes. Power dissipation is 90 mW. The circuit uses a band-gap reference to generate a variable R_T reference voltage for TMC1173, as well as a bias voltage to offset the wideband amplifier to midrange. An offset adjust also is shown for varying the midrange voltage level. The voltage reference is variable from 0 V to 2.4 V on R_T, with R_B grounded. RAYTHEON SEMICONDUCTOR DATA BOOK, 1994, PP. 3-107, 116.

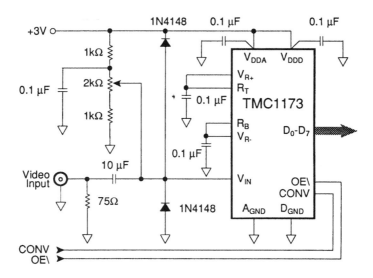

Analog/digital and digital/analog circuit titles and descriptions

A/D converter with input voltage range from 0.4 V to 1.5 V

Fig. 9-38 The circuit in the illustration shows the same A/D as in Fig. 9-37, but with self-bias for R_T and R_B (by connecting V_R+ and V_R). This connection sets up a 0.4-V to 1.5-V input range for V_{IN}. The 2-kΩ pot varies the input offset. RAYTHEON SEMICONDUCTOR DATA BOOK, 1994, P. 3-117.

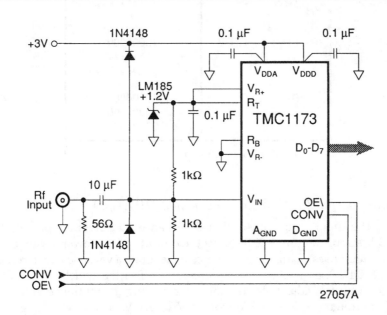

27057A

A/D converter with input voltage range from 0 V to 1.2 V

Fig. 9-39 The circuit in the illustration shows the same A/D as in Fig. 9-37, but with an external LM185 reference to provide a 0-V to 1.2-V input range. The input impedance at Rf is about 50 Ω, and the A/D converter input is biased at the midpoint of the input range. RAYTHEON SEMICONDUCTOR DATA BOOK, 1994, P. 3-117.

Analog/digital and digital/analog circuits

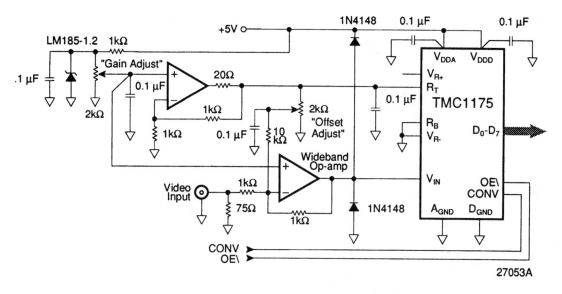

27053A

A/D converter for 5-V operation (40 Msps)

Fig. 9-40 The circuit in the illustration shows an A/D operated from a single +5-V supply. Power dissipation is 150 mW. The circuit is similar in operation to that of Fig. 9-37 except for the +5-V supply. RAYTHEON SEMICONDUCTOR DATA BOOK, 1994, PP. 3-123, 132.

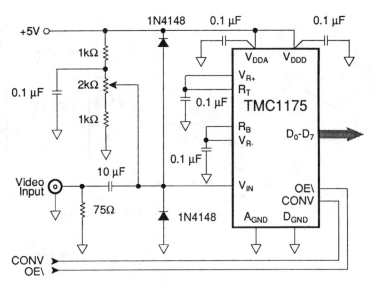

A/D converter with input voltage range from 0.6 V to 2.6 V

Fig. 9-41 The circuit in the illustration shows the same A/D as in Fig. 9-40, but with self-bias for R_T and R_B grounded. This arrangement sets up a 0.6-V to 2.6-V input range for V_{IN}. The 2-kΩ pot varies the input offset. RAYTHEON SEMICONDUCTOR DATA BOOK, 1994, P. 3-133.

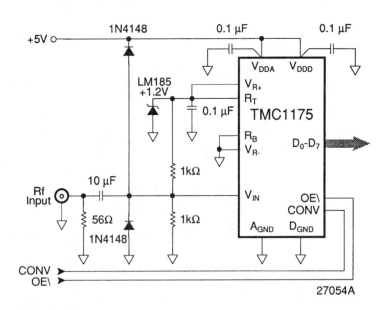

27054A

A/D converter with input voltage range from 0 V to 1.2 V

Fig. 9-42 The circuit in the illustration shows the same A/D as in Fig. 9-40 but with an external LM185 reference to provide a 0-V to 1.2-V input range. The input impedance at Rf is about 50 Ω, and the A/D converter input is biased at the midpoint of the input range. RAYTHEON SEMICONDUCTOR DATA BOOK, 1994, P. 3-133.

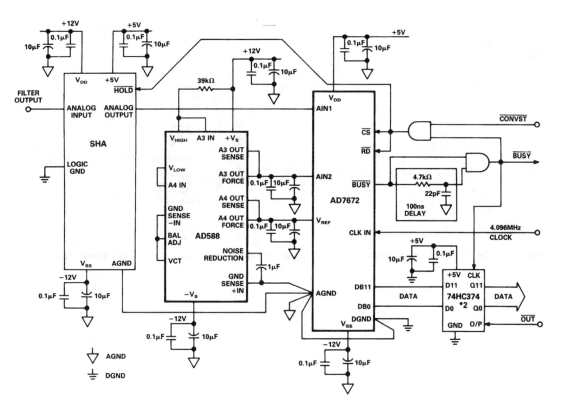

Data-acquisition interface for an A/D converter

Fig. 9-43 The circuit in the illustration is a data-acquisition interface designed for a 200-kHz sampling frequency. The sample and hold can be either an AD346 or AD781. The AD588 reference sets the analog input for a range of ±5 V. Data-access time is under 30 ns. The data format can be a complete parallel load for 16-bit microprocessors, or a two-byte load for 8-bit processors. ANALOG DEVICES, APPLICATIONS REFERENCE MANUAL, 1993, P. 3-30.

Parts List

Resistors			
R1	5K	1/4W	10-turn
R2	1K	1/4W	10-turn
R3	1K	1/4W	5%
R4	43	1/4W	5%
R5	33	1/4W	5%
R6	330	1/4W	5%
R7	750	1/4W	5%
R8,R9	10	1/4W	5%
R10	75	1/4W	2%
R11,R12	10K	1/4W	5%
R13	220	1/4W	5%
R14,R15	100	1/4W	5%
R16,R22	390	1/4W	5%
R17,R18	2K	1/4W	10-turn
R19	1K	1/4W	5%
R20,R21	1K	1/4W	5%

Capacitors		
C1	0.01µF	50V
C2	1.0µF	10V
C3	1.0µF	10V
C4	2.2µF	25V
C5	0.1µF	50V
C6	2–5 pF	50V
C7	0.1µF	50V
C8	0.1µF	50V
C9	0.1µF	50V
C10	0.1µF	50V

RF Chokes	
L1,L2	Ferrite beads

Diodes	
CR1	1N4001

Transistors	
Q1	2N2907
Q2	2N2907
Q3	2N2907
Q4	2N6660
Q5	2N6660

Integrated Circuits	
U1	TRW TDC1016
U2	LM113
U3	HA2539
U4	SN7404

(a)

Continued

Input Coding Table

NDIS	N2C	NFH	NFL	Data	Output	Description
0	x	x	x	xxxxxxxxxx	0.0	Output Disabled
1	1	1	1	1111111111	0.0	Binary (Default State for TTL
1	1	1	1	0000000000	−1.0	Mode Control) Inputs Open
1	1	0	0	1111111111	−1.0	Inverse Binary
1	1	0	0	0000000000	0.0	
1	0	1	1	0111111111	0.0	Two's Complement
1	0	1	1	1000000000	−1.0	
1	0	0	0	0111111111	−1.0	Inverse Two's Complement
1	0	0	0	1000000000	0.0	
1	x	0	1	xxxxxxxxxx	0.0	Force HIGH
1	x	1	0	xxxxxxxxxx	−1.0	Force LOW

Notes:
1. For TTL, $0.0 < V_{IL} < +0.8V$ is logic "0".
2. For TTL, $+2.0 < V_{IH} < +5.0V$ is logic "1".
3. For ECL, $-1.85 < V_{IL} < -1.67V$ is logic "0".
4. For ECL, $-1.0 < V_{IH} < -0.8V$ is logic "1".
5. x = "don't care".

(b)

D/A converter with TTL/ECL compatible inputs (TV video)

Fig. 9-44 The circuit in the illustration shows the schematic and interface for a 10-bit 20-Msps D/A used to reconstruct video signals from digital data. TV timing signals (SYNC and BLANKING) are added at the D/A output and are adjusted by R17/R18. The combined timing signals and D/A output are amplified by U3 as a composite video output. Capacitor C6 adjusts U3 for optimum video signal, and R2 sets the level of the video signal. As shown in Fig. 9-44B, the D/A can be programmed for TTL or ECL logic, through the NDIS, N2C, NFH, and NFL inputs, which are not shown in Fig. 9-44A. To calibrate the D/A, apply a full-scale input and adjust R1 for the correct output voltage. (Full scale can be selected by loading all of the input switches or by making the NFH input a logic-0.) RAYTHEON SEMICONDUCTOR DATA BOOK, 1994, PP. 3-180, 181.

9

501

Figure 7. Typical Interface Circuit

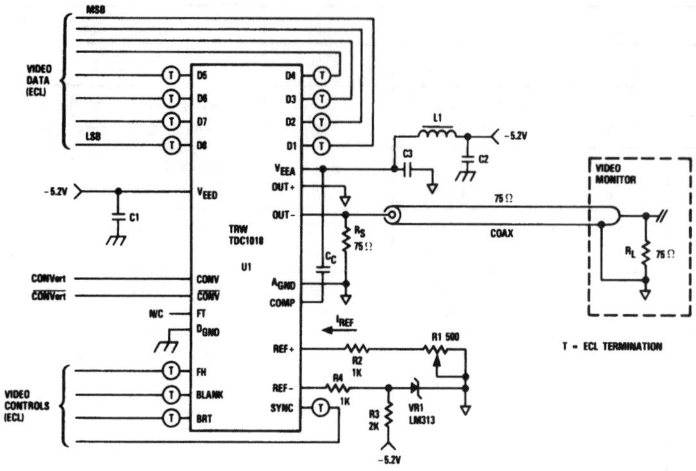

Parts List

Integrated Circuits

U1 TDC1018 D/A Converter

Voltage References

VR1 LM113 or LM313 Bandgap Reference

Inductors

L1 Ferrite Bead Shield Inductor
    Fair-Rite P/N 2743001112 or Similar

Resistors

R1	1KΩ	Pot	10 Turn
R2	1.00KΩ	1/8W	1% Metal Film
R3	2.00KΩ	1/8W	1% Metal Film
R4	1.00KΩ	1/8W	1% Metal Film

Capacitors

C1–C3	0.1μF	50V	Ceramic Disc
C_C	0.01μF	50V	Ceramic Disc

Ordering Information

Product Number	Temperature Range	Screening	Package	Package Marking
TDC1018B7C	STD–T_A=0°C to 70°C	Commercial	24 Pin CERDIP	1018B7C
TDC1018B7C1	STD–T_A=0°C to 70°C	Commercial	24 Pin CERDIP	1018B7C1
TDC1018C3C	STD–T_A=0°C to 70°C	Commercial	28 Contact Chip Carrier	1018C3C
TDC1018C3C1	STD–T_A=0°C to 70°C	Commercial	28 Contact Chip Carrier	1018C3C1

(a)

Sync	Blank	Force High	Bright	Data Input	Out– (mA)[1]	Out– (V)[2]	Out– (IRE)[3]	Description[4]
1	X	X	X	X	28.57	–1.071	–40	Sync Level
0	1	X	X	X	20.83	–0.781	0	Blank Level
0	0	1	1	X	0.00	0.00	110	Enhanced High Level
0	0	1	0	X	1.95	–0.073	100	Normal High Level
0	0	0	0	000...	19.40	–0.728	7.5	Normal Low Level
0	0	0	0	111...	1.95	–0.073	100	Normal High Level
0	0	0	1	000...	17.44	–0.654	17.5	Enhanced Low Level
0	0	0	1	111...	0.00	0.00	110	Enhanced High Level

Notes:

1. Out+ is complementary to Out–. Current is specified as conventional current when flowing into the device.

2. Voltage produced when driving the standard load configuration (37.5 Ohms). See Figure 5.

3. 140 IRE units = 1.00V.

4. RS-343-A tolerance on all control values is assumed.

(b)

Analog/digital and digital/analog circuits

Continued

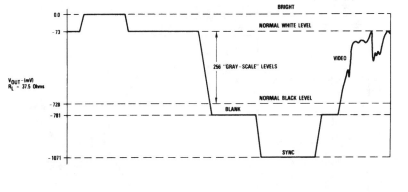

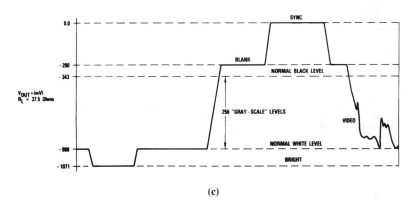

(c)

D/A converter with ECL-compatible inputs (TV video)

Fig. 9-45 The circuit in the illustration shows the schematic and interface for an 8-bit 200-MHz D/A used to reconstruct video signals from digital data. TV signals (SYNC, BLANKING, BRIGHTNESS, etc.) are added in the D/A. To calibrate the D/A, apply a full-scale input and adjust R1 for the correct output voltages (see Fig. 9-45B). Figure 9-45C shows typical video-output waveforms. RAYTHEON SEMICON-DUCTOR DATA BOOK, 1994, PP. 3-192, 193.

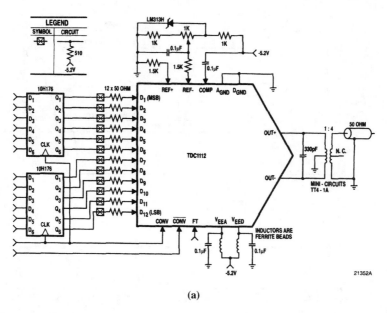

(a)

Output Coding Table [1]

Input Data MSB	D_{1-12} LSB		OUT+ (mA)	V_{OUT+} (mV)	OUT− (mA)	$V_{OUT−}$ (mV)
0000	0000	0000	0.000	0.00	40.000	−1000.00
0000	0000	0001	0.009	−0.24	39.990	−999.75
0000	0000	0010	0.019	−0.49	39.980	−999.52
⋮			⋮	⋮	⋮	⋮
0111	1111	1111	19.995	−499.88	20.005	−500.12
1000	0000	0000	20.005	−500.12	19.995	−499.88
⋮			⋮	⋮	⋮	⋮
1111	1111	1101	39.980	−999.52	0.019	−0.49
1111	1111	1110	39.990	−999.75	0.009	−0.24
1111	1111	1111	40.000	−1000.00	0.000	0.00

Note: 1. $I_{REF} = 625\mu A$, $R_{LOAD} = 25\Omega$.

(b)

Analog/digital and digital/analog circuits

Signal Type	Signal Name	Function	Value	J7, N7 Package Pins	C3, R3 Package Pins
Power	V_{EEA}	Analog Supply Voltage	−5.2V	18	1
	V_{EED}	Digital Supply Voltage	−5.2V	22	5
	A_{GND}	Analog Ground	0.0V	5	13
	D_{GND}	Digital Ground	0.0V	8	17
Reference	REF −	Reference Voltage Input	−1.0V	19	2
	REF +	Reference Current Output	0.625mA	20	3
	COMP	Compensation Capacitor	0.1μF, See Text	21	4
Data Input	D_1 (MSB)	Most Significant Bit Input	ECL	14	24
	D_2		ECL	13	23
	D_3		ECL	12	22
	D_4		ECL	11	21
	D_5		ECL	10	20
	D_6		ECL	9	19
	D_7		ECL	23	6
	D_8		ECL	24	7
	D_9		ECL	1	8
	D_{10}		ECL	2	9
	D_{11}		ECL	3	10
	D_{12} (LSB)	Least Significant Bit Input	ECL	4	11
Feedthrough	FT	Feedthrough Mode Control	ECL	17	28
Convert (Clock)	CONV	Convert (Clock) Input	ECL	16	27
	$\overline{CONV}$	Convert (Clock) Input	ECL	15	26
Analog Output	OUT +	Analog Output	0 to −40mA	6	14
	OUT −	Analog Output	−40 to 0mA	7	15

(c)

D/A converter (10-bit, 50-Msps, ECL, 12-ns settling time) for DDS

Fig. 9-46   The circuit in the illustration shows the schematic and interface for a 10-bit 50-Msps D/A connected to provide DDS (direct digital synthesis) of ECL digital data with a balun output. Figures 9-46B and C show the output coding and package interconnections, respectively. To calibrate, apply a full-scale digital input and adjust the 1-kΩ pot for correct output voltage/current (see Fig. 9-46B). Raytheon Semiconductor Data Book, 1994, pp. 3-211, 212, 221.

9

505

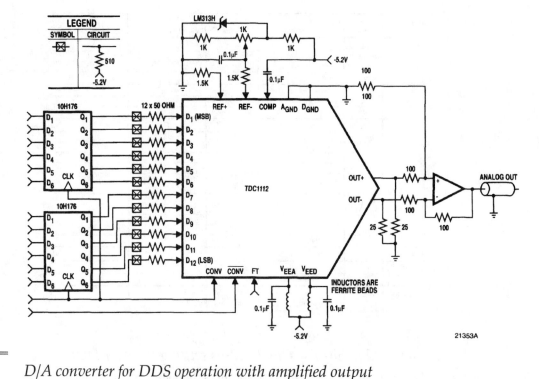

D/A converter for DDS operation with amplified output

Fig. 9-47   The circuit in the illustration is similar to that of Fig. 9-46, except that the output is amplified. RAYTHEON SEMICONDUCTOR DATA BOOK, 1994, P. 3-222.

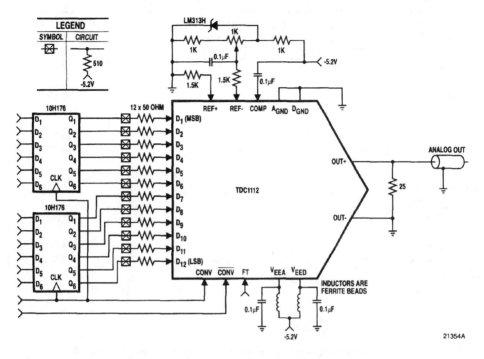

D/A converter for DDS operation with a resistive output

Fig. 9-48 The circuit in the illustration is similar to that of Fig. 9-46, except that the output is a resistive load. RAYTHEON SEMICONDUCTOR DATA BOOK, 1994, P. 3-222.

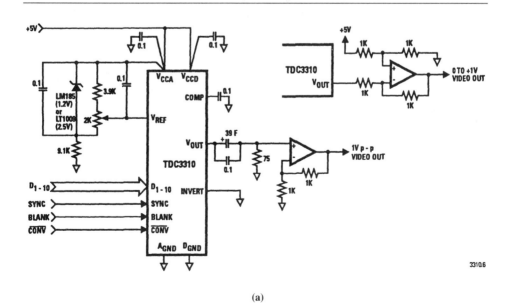

(a)

Analog/digital and digital/analog circuit titles and descriptions

Continued

D1.........D10 (MSB LSB)	BLANK	SYNC	INVERT = LOW V_{OUT} w/r V_{CCA}	V_{OUT} w/r A_{GND}	INVERT = HIGH V_{OUT} w/r V_{CCA}	V_{OUT} w/r A_{GND}
11 1111 1111	0	0	0.000	5.000	-1.000	4.000
11 1111 1110	0	0	-0.001	4.999	-0.999	4.001
11 1111 1101	0	0	-0.002	4.998	-0.998	4.002
·	·	·	·	·	·	·
10 0000 0000	0	0	-0.500	4.500	-0.501	4.499
01 1111 1111	0	0	-0.501	4.499	-0.500	4.500
·	·	·	·	·	·	·
00 0000 0010	0	0	-0.998	4.002	-0.002	4.998
00 0000 0001	0	0	-0.999	4.001	-0.001	4.999
00 0000 0000	0	0	-1.000	4.000	0.000	5.000
xx xxxx xxxx	1	0	-1.081	3.919	-1.081	3.919
xx xxxx xxxx	x	1	-1.514	3.486	-1.514	3.486

Note: $V_{REF} = V_{CCA}$ - 1.000 volts, V_{CCA} = 5.0 volts, no external

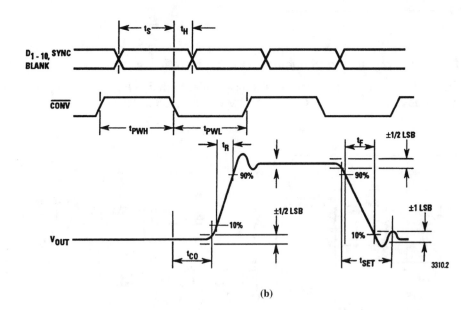

(b)

Continued

Signal Type	Name	Function	Value	N6 Pin	R6 Pin
Power	V_{CCA}	Analog supply voltage	5.0V	10	10
	V_{CCD}	Digital supply voltage	5.0V	17	20
Ground	A_{GND}	Analog ground	0.0V	9	9
	D_{GND}	Digital ground	0.0V	12, 15	1, 13, 17, 19
Reference	V_{REF}	Reference voltage input	V_{CCA}-1	6	6
	COMP	Compensation capacitor	0.1 μF	7	7
Data Input	D_1 (MSB)	Most Significant Bit Input	TTL	19	23
	D_2		TTL	20	24
	D_3		TTL	21	25
	D_4		TTL	22	26
	D_5		TTL	23	27
	D_6		TTL	24	28
	D_7		TTL	25	29
	D_8		TTL	26	30
	D_9		TTL	27	31
	D_{10} (LSB)	Least Significant Bit Input	TTL	28	32
	INVERT	Invert D1-D10	TTL	5	5
	SYNC	SYNC input	TTL	4	3
	BLANK	BLANK input	TTL	3	2
Clock	CONV\	Clock input	TTL	18	22
Output	V_{OUT}	Analog output	+4 to +5	8	8
Not Used	NC	Not connected	Open	1, 2, 11 13, 14, 16	4, 11, 12, 14 15, 16, 18, 21

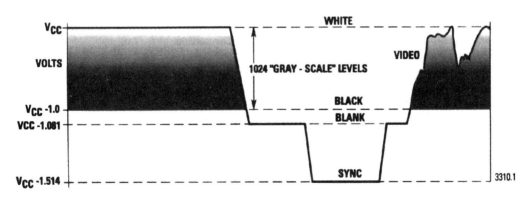

(c)

Analog/digital and digital/analog circuit titles and descriptions

D/A converter with TTL-compatible inputs (TV video)

Fig. 9-49 The circuit in the illustration shows the schematic and interface for a 10-bit 40-Msps D/A used to reconstruct video signals from TTL digital data. TV signals (SYNC, BLANKING, etc.) are added in the D/A. To calibrate the D/A, apply a full-scale input and adjust the 2-kΩ pot for the correct output voltages (see Fig. 9-49B). Figure 9-49C shows package interconnections. RAYTHEON SEMICONDUCTOR DATA BOOK, 1994, PP. 3-242, 243, 248.

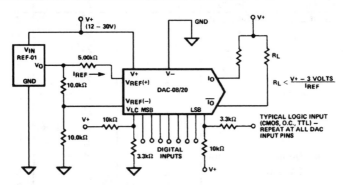

D/A converter operated from a single supply

Fig. 9-50 The circuit in the illustration shows how the classic DAC-08/20 can be operated from a single supply. The resistive voltage-divider inputs to V_{LC} and logic inputs provide the necessary voltage levels for operation from CMOS and open-collector TTL. ANALOG DEVICES, APPLICATIONS REFERENCE MANUAL, 1993, P. 8-33.

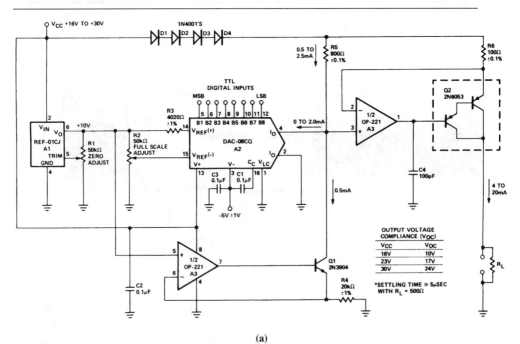

(a)

Analog/digital and digital/analog circuits

Continued

PARTS LIST

Circuit Symbol(s)	Description
A1	+10V Reference, PMI REF-01CJ
A2	8-Bit DAC, PMI DAC-08CQ
A3	Dual Op Amp, PMI OP-221
C1-C3	0.1μF +80%/−20% 50V, Type CK-104
C4	100pF ±5% Mica, DM100ED101J03
D1-D4	Power Diode, 1N4001
Q1	NPN Transistor, 2N3904
Q2	PNP Power Darlington, Motorola 2N6053
R1-R2	50kΩ Potentiometer, Bourns #3006P-1-503
R3	4020Ω ±1%, RN55C4021F
R4	20kΩ ±1#, RN55C2002F
R5	800Ω ±0.1%, GR#8E16D800
R6	100Ω ±0.1%, GR#8E16D100

(b)

4- to 20-mA digital-to-process-current transmitter

Fig. 9-51 The circuit in the illustration shows a classic DAC-08 connected to provide a standard 4- to 20-mA current under control of 8-bit TTL digital signals. Figure 9-51B shows the parts list. To calibrate, apply +23±7 V and −5±1 V to A2 (pins 3/13), with a current-measuring meter connected between the output (pin 4) and ground. Set the digital inputs to all zeros (less than +0.8 V) and adjust R1 until the output current is 4 mA. Change the digital inputs to all ones (greater than +2.0 V) and adjust R2 until the output current is 20 mA. ANALOG DEVICES, APPLICATIONS REFERENCE MANUAL, 1993, PP. 8-75, 76.

9

511

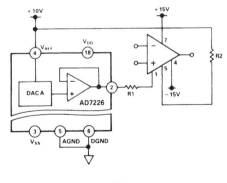

(a)

Analog/digital and digital/analog circuit titles and descriptions

Continued

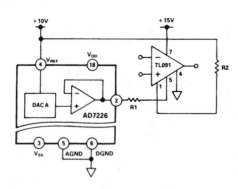

(b)

Op Amp	R2 (kΩ)	R1 (kΩ)	Range (mV)	Final Offset (μV)
AD741	1200	1000	±6.75	−14.5
AD544	620	500	±2.75	3.3
AD542	470	360	±2.4	4.6
TL091*	1000	500	+9.0	2.5

*Operating in Single Supply $V_{SS} = 0V$ $V_{DD} = +15V$.
All other op amps dual supply ±15V.

(c)

Programmable offset-adjust for AD-type op amps

Fig. 9-52 The circuit in the illustration shows a D/A converter connected to provide a digitally programmed offset adjust for op amps (Chapter 1). Figures 9-52B and C show the D/A functional diagram and the op-amp offset values, respectively. The op-amp offset goes more positive when the digital code applied to the D/A increases. ANALOG DEVICES, APPLICATIONS REFERENCE MANUAL, 1993, PP. 8-107, 108.

Analog/digital and digital/analog circuits

Programmable offset-adjust for TL-type op amps

Fig. 9-53 The circuit in the illustration is similar to that of Fig. 9-52, except that the trim terminals are interchanged for TL-type op amps (TL061, 071, etc.). ANALOG DEVICES, APPLICATIONS REFERENCE MANUAL, 1993, P. 8-108.

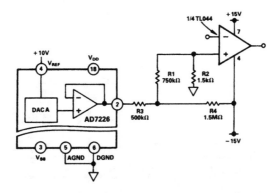

Programmable offset-adjust for op amps without trim terminals

Fig. 9-54 The circuit in the illustration is similar to that of Fig. 9-52 and is used for op amps that do not have trim terminals (such as the TL044 quad op amp). ANALOG DEVICES, APPLICATIONS REFERENCE MANUAL, 1993, P. 8-108.

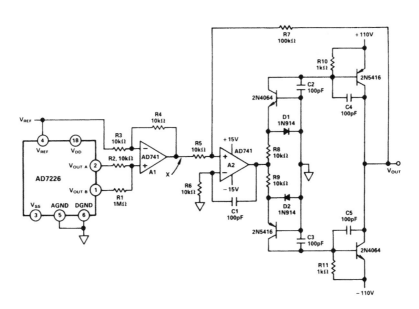

Set-point controller

Fig. 9-55 The circuit in the illustration shows the D/A of Fig. 9-52B connected as a set-point controller. The circuit provides programmable fine and coarse adjust of the output voltage over a 200-V range, into a 2.7-kΩ load (such as might be required in a programmable power supply). The fine-adjustment resolution is 8 mV, which means that an output voltage can be set in the range –100 V to +100 V to within ±4 mV. $V_{OUT\ A}$ (pin 2) provides the coarse adjustment for the –100-V to +100-V output with an LSB of 800 mV, $V_{OUT\ B}$ (pin 1) has a range of 2 V and provides the fine adjust (LSB of 8 mV). ANALOG DEVICES, APPLICATIONS REFERENCE MANUAL, 1993, P. 8-109.

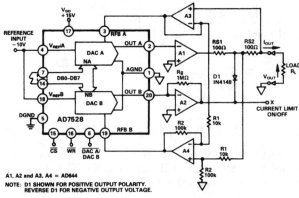

Programmable voltage/current source (unipolar)

Fig. 9-56 The circuit in the illustration shows a D/A connected as a programmable voltage/current (0 to +10 V, 0 to +10 mA) source.

$$V_{OUT} = +V_{REF\ A}\frac{N_A}{256}$$

where N_A is decimal code (0–255).

$$I_{OUT(max)} = V_{REF\ B} \cdot \frac{R_1}{R_2} \cdot \frac{1}{R_{S2}} \cdot \frac{N_B}{256}$$

where N_B = is decimal code (0–255).

When N_A is set to 255 (maximum output voltage capability) and R_L is limited such that:

$$I_{OUT(max)} \cdot R_L < \frac{256}{256} V_{REF\ A}$$

The circuit provides a constant current. The set-current limit corresponds to the load current. Point X changes state when the current limit is reached. ANALOG DEVICES, APPLICATIONS REFERENCE MANUAL, 1993, P. 8-117.

Analog/digital and digital/analog circuits

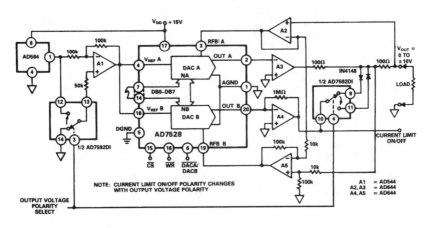

Programmable voltage/current source (bipolar)

Fig. 9-57 The circuit in the illustration is similar to that of Fig. 9-56, except that the output is bipolar (0 to +10 V, 0 to ±10 mA). ANALOG DEVICES, APPLICATIONS REFERENCE MANUAL, 1993, P. 8-117.

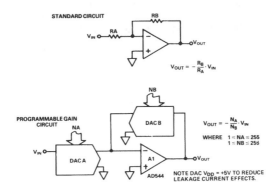

Programmable-gain amplifier

Fig. 9-58 The circuit in the illustration shows a D/A connected to provide programmable gain or attenuation for an amplifier. The equivalent resistance of each DAC from its reference input to the DAC output is used to replace the input and feedback resistors in the standard inverting-amplifier circuit. By loading DAC A and DAC B with suitable codes, programmable gain/attenuation over the range −48 dB to +48 dB is possible. As shown by the equations, V_{OUT} depends on V_{IN} and the ratio of N_A/N_B, where N_A and N_B are the DAC codes in decimal (1–255). ANALOG DEVICES, APPLICATIONS REFERENCE MANUAL, 1993, P. 8-118.

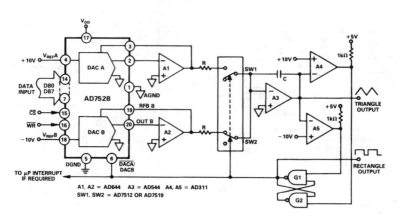

Digitally programmable waveform generator

Fig. 9-59 The circuit in the illustration shows a D/A connected to provide both triangular and rectangular waves under digital control. Such a circuit could be used for vector scan of *CRT* displays. The period (*t*) of the wave is:

$$t = 512 \, RC \left(\frac{1}{N_A} \quad \frac{1}{N_B} \right)$$

where N_A and N_B are the DAC codes in decimal (1–255). If DAC A and DAC B latches contain the same codes, the expression simplifies to:

$$t = \frac{1024 \, RC}{N_A}$$

and output frequency:

$$f = \frac{N_A}{1024 \, RC} \text{ Hz}$$

The mark-to-space ratio of the rectangular wave depends on the ratio of N_A/N_B. ANALOG DEVICES, APPLICATIONS REFERENCE MANUAL, 1993, P. 8-118.

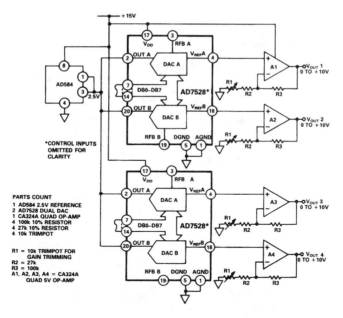

Digitally programmable voltage source (four channel)

Fig. 9-60 The circuit in the illustration shows a D/A connected to provide four output voltages under digital control. The R1 resistors trim individual outputs. ANALOG DEVICES, APPLICATIONS REFERENCE MANUAL, 1993, P. 8-119.

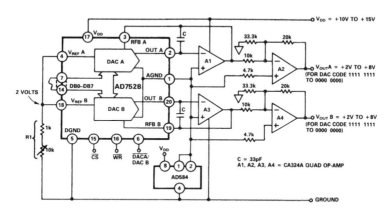

Digitally programmable voltage source (two channel)

Fig. 9-61 The circuit in the illustration shows a D/A connected to provide two output voltages under digital control. Resistor R1 is adjusted until $V_{REF\ A}$ and $V_{REF\ B}$ (pins 4 and 18) are at +2 V. This adjustment is independent of either DAC code. ANALOG DEVICES, APPLICATIONS REFERENCE MANUAL, 1993, P. 8-120.

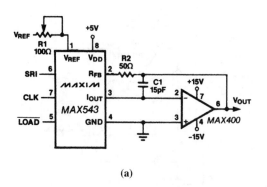

(a)

| DIGITAL INPUT | | ANALOG OUTPUT |
MSB	LSB	
1 1 1 1 1 1 1 1 1 1 1 1		$-V_{REF}\left(\dfrac{4095}{4096}\right)$
1 0 0 0 0 0 0 0 0 0 0 0		$-V_{REF}\left(\dfrac{2048}{4096}\right) = -\dfrac{V_{REF}}{2}$
0 0 0 0 0 0 0 0 0 0 0 1		$-V_{REF}\left(\dfrac{1}{4096}\right)$
0 0 0 0 0 0 0 0 0 0 0 0		0

(b)

Multiplying D/A converter (serial input, unipolar mode)

Fig. 9-62   The circuit in the illustration shows a MAX543 connected for unipolar operation (or two-quadrant multiplication). Figure 9-62B shows the unipolar binary code. To calibrate, apply the 12-bit serial input code and adjust R1 for the correct output voltage. Because of the accuracy of the IC, R1 and R2 can be omitted (in some cases), particularly where the circuit is trimmed for correct output by adjustment of V_{REF}. For wide temperature ranges, use resistors with 30 ppm/°C or less for R1 and R2. MAXIM NEW RELEASES DATA BOOK, 1992, P. 9-36.

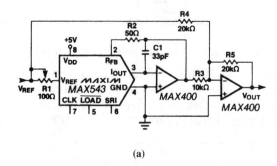

(a)

Analog/digital and digital/analog circuits

Continued

DIGITAL INPUT		ANALOG OUTPUT
MSB	**LSB**	
1 1 1 1 1 1 1 1 1 1 1 1		$+V_{REF}\left(\dfrac{2047}{2048}\right)$
1 0 0 0 0 0 0 0 0 0 0 1		$+V_{REF}\left(\dfrac{1}{2048}\right)$
1 0 0 0 0 0 0 0 0 0 0 0		0
0 1 1 1 1 1 1 1 1 1 1 1		$-V_{REF}\left(\dfrac{1}{2048}\right)$
0 0 0 0 0 0 0 0 0 0 0 0		$-V_{REF}\left(\dfrac{2048}{2048}\right)$

(b)

DIGITAL INPUT		ANALOG OUTPUT
MSB	**LSB**	
0 1 1 1 1 1 1 1 1 1 1 1		$+V_{REF}\left(\dfrac{2047}{2048}\right)$
0 0 0 0 0 0 0 0 0 0 0 1		$+V_{REF}\left(\dfrac{1}{2048}\right)$
0 0 0 0 0 0 0 0 0 0 0 0		0
1 1 1 1 1 1 1 1 1 1 1 1		$-V_{REF}\left(\dfrac{1}{2048}\right)$
1 0 0 0 0 0 0 0 0 0 0 0		$-V_{REF}\left(\dfrac{2048}{2048}\right)$

(c)

9

519

Multiplying D/A converter (serial input, bipolar mode)

Fig. 9-63 The circuit in the illustration is similar to that of Fig. 9-62, except that bipolar operation is obtained by the addition of a second op amp. Figures 9-63B and 9-63C show the offset-binary code and 2's-complement code, respectively. To calibrate, apply 1000 0000 0000 and adjust R1 for 0 V output. MAXIM NEW RE-LEASES DATA BOOK, 1992, P. 9-36.

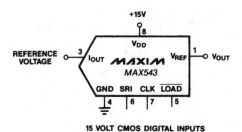

15 VOLT CMOS DIGITAL INPUTS

D/A operating from a single supply (CMOS)

Fig. 9-64 The circuit in the illustration shows the connections for operating the MAX543 (Figs. 9-62 and 9-63) from a single +15-V supply for use with CMOS circuits. Note that the reference voltage is applied to I_{OUT} (pin 3), the output voltage is taken from V_{REF} (pin 1), and R_{FB} (pin 2) is not used (eliminating the need for an R2 and C1). MAXIM NEW RELEASES DATA BOOK, 1992, P. 9-37.

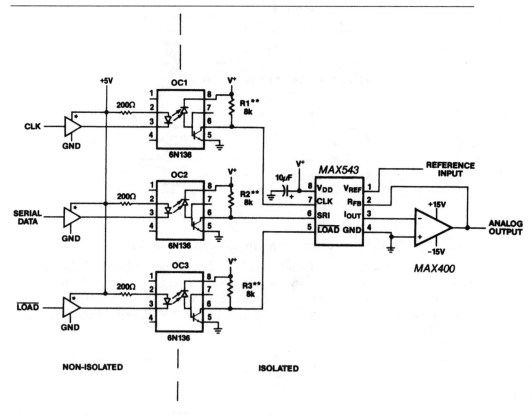

* TTL OR CMOS LOGIC DEVICE
** V⁺ = +15V. FOR V⁺ = +5V USE 3kΩ FOR R1, R2, R3

(a)

Analog/digital and digital/analog circuits

Continued

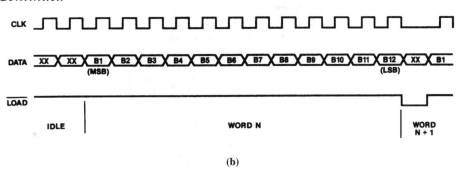

(b)

Multiplying D/A converter with opto-coupling

Fig. 9-65 The circuit in the illustration shows the connections for operating the MAX543 (Figs. 9-62 and 9-63), fully isolated from the serial-data, clock, and load-command signals. Figure 9-65B shows the timing for the opto-isolated circuit. MAXIM NEW RELEASES DATA BOOK, 1992, P. 9-38.

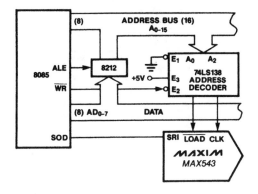

*ANALOG CIRCUITRY OMITTED FOR SIMPLICITY

D/A converter with 8085 interface

Fig. 9-66 The circuit in the illustration shows the connections for interfacing the MAX543 (Figs. 9-62 and 9-63) with an 8085 microprocessor. MAXIM NEW RELEASES DATA BOOK, 1992, P. 9-39.

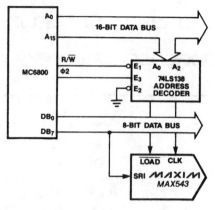

*ANALOG CIRCUITRY OMITTED FOR SIMPLICITY

D/A converter with MC6800 interface

Fig. 9-67 The circuit in the illustration shows the connections for interfacing the MAX543 (Figs. 9-62 and 9-63) with an MC6800 microprocessor. MAXIM NEW RELEASES DATA BOOK, 1992, P. 9-39.

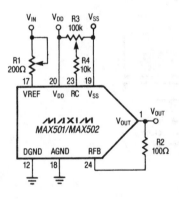

DIGITAL INPUT			ANALOG OUTPUT
1111	1111	1111	$(-V_{IN}) \dfrac{4095}{4096}$
1000	0000	0000	$(-V_{IN}) \dfrac{2048}{4096} = -\dfrac{1}{2} V_{IN}$
0000	0000	0001	$(-V_{IN}) \dfrac{1}{4096}$
0000	0000	0000	0V

(b)

(a)

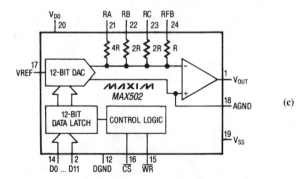

(c)

Analog/digital and digital/analog circuits

Multiplying D/A converter (parallel input, unipolar mode)

Fig. 9-68 The circuit in the illustration shows a MAX501/502 connected for unipolar operation (or two-quadrant multiplication). Figures 9-68B and C show the unipolar binary code and functional diagram, respectively. Resistor R1 adjusts gain, and R3 adjusts zero offset (0 V output). Resistors R1 and R2 can be omitted (in some cases), particularly where the circuit is trimmed for correct output by adjustment of V_{REF}. For wide temperature ranges, use resistors with 300 ppm/°C or less for R1 and R2. MAXIM NEW RELEASES DATA BOOK, 1993, PP. 9-5, 9-14.

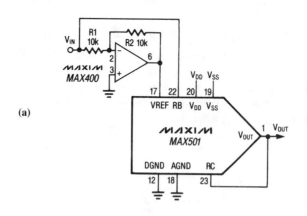

(a)

(b)

DIGITAL INPUT			ANALOG OUTPUT
1111	1111	1111	$(+V_{IN}) \dfrac{2047}{2048}$
1000	0000	0001	$(+V_{IN}) \dfrac{1}{2048}$
1000	0000	0000	0V
0111	1111	1111	$(-V_{IN}) \dfrac{1}{2048}$
0000	0000	0000	$(-V_{IN}) \dfrac{2048}{2048} = V_{IN-}$

Multiplying D/A converter (parallel input, bipolar mode)

Fig. 9-69 The circuit in the illustration is similar to that of Fig. 9-63, except that bipolar operation is obtained by the addition of an op amp. Figure 9-69B shows the bipolar-binary code. Gain error (if any) can be adjusted by changing the R_1 and R_2 ratio. These resistors should be ratio-matched to 0.01% to stay within gain-error specifications and to eliminate trimming. (Internal resistors R_B and R_C are matched to define the offset value.) MAXIM NEW RELEASES DATA BOOK, 1993, P. 9-14.

Analog/digital and digital/analog circuit titles and descriptions

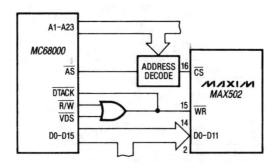

D/A converter with interface for minimum digital feedthrough

Fig. 9-70 The circuit in the illustration shows the connections for interfacing the MAX502 (Figs. 9-68 and 9-69) with a microprocessor to minimize digital feed-through (by means of latches). (High-frequency logic on the data bus can feed through the D/A-package capacitance as noise on the D/A output.) MAXIM NEW RELEASES DATA BOOK, 1993, P. 9-15.

D/A converter with MC68000 interface

Fig. 9-71 The circuit in the illustration shows the connections for interfacing the MAX502 (Figs. 9-68 and 9-69) with a 16-bit MC68000 microprocessor. The MAX502 appears as a memory-mapped peripheral to the MC68000. A write instruction (MOV) loads the MAX502 with the appropriate data. MAXIM NEW RELEASES DATA BOOK, 1993, P. 9-15.

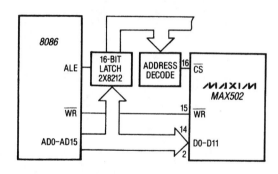

D/A converter with 8086 interface

Fig. 9-72 The circuit in the illustration shows the connections for interfacing the MAX502 (Figs. 9-68 and 9-69) with a 16-bit 8086 microprocessor. The MAX502 appears as a memory-mapped peripheral to the 8086. A write instruction (MOV) loads the MAX502 with the appropriate data. MAXIM NEW RELEASES DATA BOOK, 1993, P. 9-15.

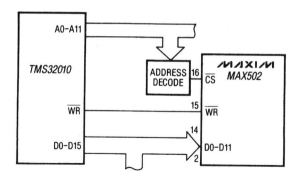

D/A converter with TMS32010 interface

Fig. 9-73 The circuit in the illustration shows the connections for interfacing the MAX502 (Figs. 9-68 and 9-69) with a 16-bit TMS32010 microprocessor. The MAX-502 appears as a memory-mapped peripheral to the TMS32010. A write instruction (OUT) loads the MAX502 with the appropriate data. MAXIM NEW RELEASES DATA BOOK, 1993, P. 9-16.

=10=

Temperature indicator/controller circuits

This chapter is devoted to circuits that indicate or control temperature. Before you read about testing and troubleshooting these circuits, here is a review of some temperature basics.

Temperature terms and scales

Most of the circuits in this chapter are calibrated in degrees Fahrenheit (°F) or Celsius (°C), which is also known as centigrade. Both °F and °C are based on the freezing and boiling points of water at a pressure equal to atmospheric pressure at sea level. Water freezes at 0°C (or 32°F) and boils at 100°C (or 212°F).

To convert from °F to °C, subtract 32 from °F, multiply by 5, and divide by 9. As an example for 77°F: 77 − 32 = 45; 45 × 5 = 225; 225/9 = 25°C.

To convert from °C to °F, multiply °C by 9, divide by 5, and subtract 32. As an example for 25°C: 25 × 9 = 225; 225/5 = 45; 45 + 32 = 77°F.

Some of the circuits involve the Kelvin (K) scale that uses the same increments as °C but is an absolute temperature, where 0K (zero Kelvin) equals 273.18°C. For example, if the Celsius temperature is 25°C, the absolute temperature is 298.18K.

Relative humidity is the ratio of water vapor in the air to the amount possible in saturated air at the same temperature.

Temperature circuit tests

The obvious test for any temperature indicator is to subject the sensor to a given temperature environment and check for corresponding output. For example, in

+ V_S
(+ 5V TO + 20V)

LM34 V_{OUT} = + 10.0 mV/°F

Fig. 10-A   Basic Fahrenheit temperature sensor.

the basic circuit of Fig. 10-A, the output from the LM34 should be 10 mV/°F. If the ambient temperature is 77°F, the output should be 770 mV.

Next, if the sensor is heated and/or cooled (with hair dryer, cooling spray, etc.) the output should vary. Ideally, the sensor should be checked in a precision, temperature-controlled oven. At the opposite test extreme, the sensor temperature can be compared with a simple mercury thermometer.

A crude, but reasonably accurate test is to subject the sensor to an ice bath and check for an output that corresponds to 0°C or 32°F. Then, check the output with a sensor in boiling water (100°C or 212°F). Make certain that the sensor and whatever you hold the sensor with will withstand these temperature extremes! Also remember that the precise boiling point of water depends on altitude (atmospheric pressure).

This simple comparison of sensor temperature against output can be applied to all of the circuits in this chapter, including those with thermocouples, thermistors, or temperature sensor/transducers. The same comparison also can be applied to temperature-controller circuits. Using Fig. 10-B as an example, the 50-kΩ temperature-adjust pot can be set for a temperature increase and the LM395 check for a positive signal at the base. Or the LM34 can be subjected to temperature changes and the output checked at the LM395 base.

Temperature circuit troubleshooting

The first step in troubleshooting temperature-sensor circuits involves testing to see if the desired output is produced by a given temperature. With temperature controllers, test to see if the circuit holds at a selected temperature. In either case, try correcting any malfunctions with adjustment. Note that calibration is described for all circuits in this chapter that require adjustment. If the problem cannot be corrected by adjustment, trace the signals using a meter or scope from input (typically the temperature sensor) to output (typically a voltage or frequency output). From that point on, it is a matter of voltage measurements and/or point-to-point resistance measurements. The following are typical examples.

In the circuit of Fig. 10-B (temperature controller), the output (voltage applied to the base of the LM395 heater from pin 6 of the LM10) depends on the signals at pins 2 and 3 of the LM10. The signal at pin 2 depends on the output of the LM34 (10 mV/°F, as shown). If the LM34 output is incorrect (does not vary with temperature changes or produces the wrong voltage), suspect the LM34. If the LM34 output is good, but the signal at pin 2 of the LM10 is bad, suspect C1, R1, or R2.

Temperature indicator/controller circuits

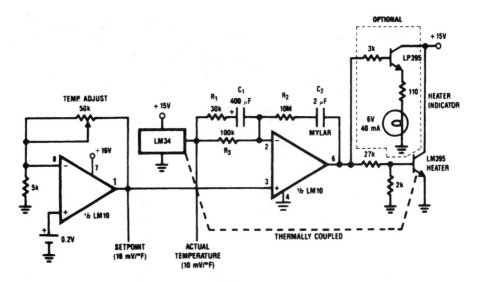

Fig. 10-B   Proportional temperature controller.

The signal at pin 2 of the LM10 depends on the 50-kΩ temperature-adjust setting and on the signal at pin 1 of the LM10 (also 10 mV°F). If this signal is incorrect (wrong voltage or does not vary when different temperatures are selected) suspect the LM10, the 50-kΩ pot, or possibly the 0.2-V battery at the noninverting input.

In the circuit of Fig. 10-C (temperature sensor or converter), the output (5-V pulses that vary in frequency with changes in temperature) depends on the length of the ramp generated by A1 (a longer ramp produces a lower frequency and vice versa). In turn, the ramp length (and output frequency) is set by the 0°C and 100°C pots, and varies when the LM334-3 sensor is subjected to varying temperatures.

If the circuit does not produce pulses at the output, check for a ramp at the A1 output. If the ramp is missing or is abnormal, suspect A1 and/or the associated parts. If the ramp is good but there are no pulses, suspect the output transistor.

If there are output pulses, but not at the correct frequency, calibrate the circuit as follows:

1. Place the LM334 in a 0°C environment (ice bath) and set the 0°C adjustment for 0 Hz at the TTL output.

2. Then put the LM334 in a 100°C environment and set the 100°C adjust control for a 1-kHz output.

3. Repeat this procedure until both points are fixed.

This circuit has a stable 0.1°C resolution with ±1.0°C accuracy. Of course, if the output pulses do not vary when the LM334-3 is subjected to temperature changes, suspect the LM334-3. Note that if the circuit works (produces pulses that vary in frequency) but it is impossible to calibrate the circuit, first try a different LM334-3. If changing the component does not cure the problem, suspect test capacitors are leaking.

Temperature circuit troubleshooting

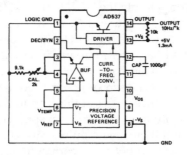

Fig. 10-C Temperature-to-frequency converter.

Temperature indicator/controller circuit titles and descriptions

Absolute temperature transducer using a V/F

Fig. 10-1 The circuit in the illustration uses the V/F of Fig. 7-4 as an absolute temperature transducer. The 1-mV/K output of the AN537 is exactly proportioned to absolute temperature. Figure 10-1 shows the AD537S scaled for 10 Hz/K, corresponding to an output-frequency range of 2180 Hz to 4980 Hz over the specified temperature range of –55°C to +125°C. To calibrate, measure the IC package temperature and adjust the output to the corresponding frequency with the CAL pot at pin 4. ANALOG DEVICES, APPLICATIONS REFERENCE MANUAL, 1993, P. 23-28.

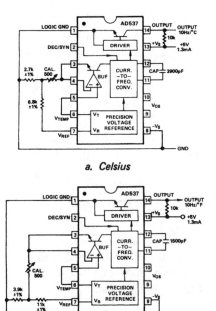

a. Celsius

b. Fahrenheit

Celsius and Fahrenheit temperature transducer with a V/F

Fig. 10-2 These circuits use the V/F of Fig. 7-4 as a Celsius or Fahrenheit temperature transducer. The circuits are similar to that of Fig. 10-1, but they also use the 1.00-V reference output of the AD537 to get the required scale offset. For the Celsius scale, the lower end of the timing resistor is offset by +273.15 mV. This offset reduces the output frequency range to 0 to 1250 Hz, corresponding to 0 to +125°C. The Fahrenheit scale requires an offset of +255.37 mV. This offset reduces the output frequency range to 0 to 2570 Hz, corresponding to 0 to 257°F (–17.78°C to +125°C). Note that the AD537S will perform to +125°C. Where maximum temperatures do not exceed +70°C (+158°F), the J or K grade AD537 can be used with no change in circuit values. ANALOG DEVICES, APPLICATIONS REFERENCE MANUAL, 1993, P. 23-28.

Temperature indicator/controller circuit titles and descriptions

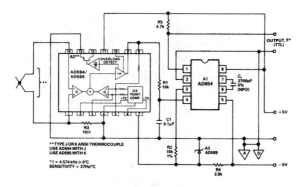

Temperature transducer with thermocouple input

Fig. 10-3 The circuit in the illustration uses the V/F of Fig. 7-30 as a temperature transducer. The signal source to the V/F is an AD594 (type J) or AD595 (type K) thermocouple signal conditioner and corresponds to a signal output of ±1 V for a basic ±100°C span. Using the values shown for R_2 and C_t, the output frequency is 4574 Hz for a temperature of 0°C. ANALOG DEVICES, APPLICATIONS REFERENCE MANUAL, 1993, P. 23-44.

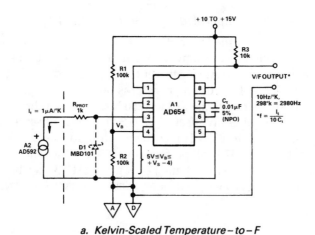

a. Kelvin-Scaled Temperature – to – F

V/F as a Kelvin-scaled temperature-to-frequency converter

Fig. 10-4 The circuit in the illustration uses the V/F of Fig. 7-30 as a Kelvin-scaled temperature transducer with a frequency output. No calibration should be required. However, the circuit can be trimmed with a film trimmer for C_t. ANALOG DEVICES, APPLICATIONS REFERENCE MANUAL, 1993, P. 23-46.

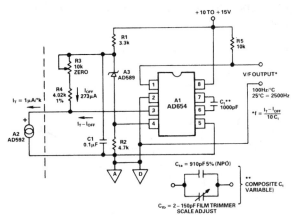

b. **Centigrade-Scaled Temperature – to – F**

V/F as a Celsius-scaled temperature-to-frequency converter

Fig. 10-5 The circuit in the illustration uses the V/F of Fig. 7-30 as a Celsius-scaled temperature transducer with a frequency output. To calibrate, set A2 to zero (ice bath) and adjust R3 for zero (A1 stops oscillating). Then set A2 to 100°C (boiling) and trim C_t (film trimmer) for 10-kHz output. ANALOG DEVICES, APPLICATIONS REFERENCE MANUAL, 1993, P. 23-46.

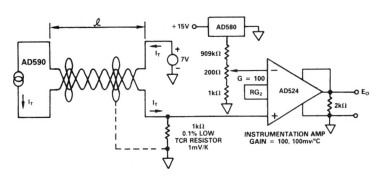

Electronic thermometer

Fig. 10-6 The circuit in the illustration measures temperature from –55°C to +100°C, with an output of 100 mV/°C. Because the AD590 measures absolute temperature (the normal output is 1 μA/K) the output must be offset by 273.2 μA to read out in °C. The AD590 output current flows through the 1-kΩ resistance, developing 1 mV/K. The AD580 2.5-V reference output is divided down to 273.2-mV offset and subtracted from the voltage across the 1-kΩ resistor by the AD524 (connected for a gain of 100). As a result, the circuit output range corresponds to –55°C to +100°C (–5.5 V to +10 V), or 100 mV/°C. The circuit is used with up to 1000 feet of twisted-pair cable (Belden 9461, style 2092). ANALOG DEVICES, APPLICATIONS REFERENCE MANUAL, 1993, P. 19-7.

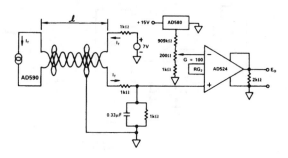

Electronic thermometer with reduced noise

Fig. 10-7   The circuit in the illustration is similar to that of Fig. 10-6 except that the noise is reduced by a factor of 2000. ANALOG DEVICES, APPLICATIONS REFERENCE MANUAL, 1993, P. 19-10.

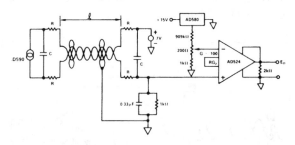

Electronic thermometer with reduced RF noise

Fig. 10-8   The circuit in the illustration is similar to those of Figs. 10-6 and 10-7, except that RF noise is reduced. The values of *R* and *C* are arbitrary, as long as the voltage across the 1-kΩ resistor is sufficient to produce a 1 mV/K. ANALOG DEVICES, APPLICATIONS REFERENCE MANUAL, 1993, P. 19-10.

Temperature indicator/controller circuits

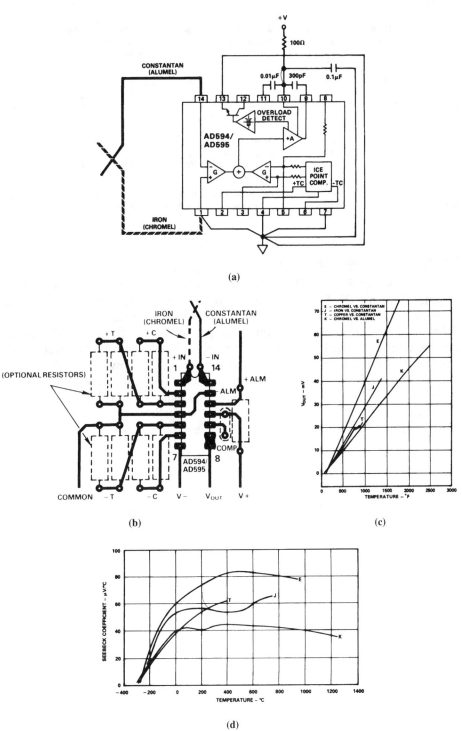

(a)

(b)

(c)

(d)

Temperature indicator/controller circuit titles and descriptions

Continued

Ambient Temp. °C	AD594C Temp. Rej. Error °C	AD594C Total Error °C	AD594A Temp. Rej. Error °C	AD594A Total Error °C	AD595C Temp. Rej. Error °C	AD595C Total Error °C	AD595A Temp. Rej. Error °C	AD595A Total Error °C
−55	4.83	5.83	6.83	9.83	5.28	6.28	7.28	10.28
−25	1.98	2.98	3.23	6.23	2.04	3.04	3.29	6.29
0	0.62	1.62	1.25	4.25	0.62	1.62	1.25	4.25
+25	0.00	1.00	0.00	3.00	0.00	1.00	0.00	3.00
+50	0.62	1.62	1.25	4.25	0.62	1.62	1.25	4.25
+70	1.46	2.46	2.59	5.59	1.38	2.38	2.50	5.50
+85	2.25	3.25	3.75	6.75	1.99	2.99	3.49	6.49
+125	4.90	5.90	7.40	10.40	3.38	4.38	5.88	8.88

NOTE:
Temp. Rej. Error has two components. (a) Difference between actual reference junction and ice point compensation voltage times the gain
(b) Offset and gain TCs extrapolated from 0 to 50°C limits. Total error is temp. rej. plus initial calibration error.

(e)

Basic thermocouple measurement

Fig. 10-9 The circuit in the illustration shows a basic thermocouple-measurement system using an AD594/595. Figures 10-9B, C, D, and E show the recommended PC-board layout, thermocouple-versus-temperature characteristics, Seebeck coefficient, and maximum calculated errors, respectively. Note that the AD594 is factory calibrated to condition a J-type thermocouple (iron-constantan) but can be adjusted to condition an E-type (chromel-constantan) as shown in the data sheet. The AD595 is calibrated for K-type (chromel-alumel), but it can be connected directly to a T-type (copper-constantan) with less than 0.2°C additional error. The nominal circuit output (at pin 9, with feedback to pin 8) is 10 mV/°C. ANALOG DEVICES, APPLICATIONS REFERENCE MANUAL, 1993, PP. 21-4, 5, 6.

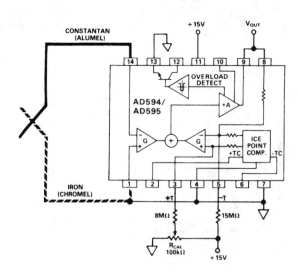

Thermocouple measurement with calibration-error adjustment

Fig. 10-10 The circuit in the illustration is similar to that of Fig. 10-9, except that calibration error can be nulled with a single unidirectional trim. ANALOG DEVICES, APPLICATIONS REFERENCE MANUAL, 1993, P. 21-6.

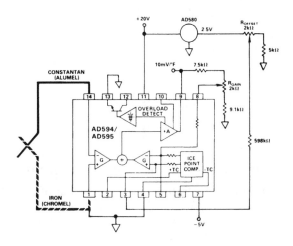

Thermocouple measurement with Fahrenheit output

Fig. 10-11 The circuit in the illustration is similar to that of Fig. 10-9, except that the output is 10 mV/°F. To calibrate, remove the thermocouple and apply a 10-mV 100 Hz signal at pins 1/14. Adjust R_{GAIN} for an output of 3.481 V_{p-p} (AD594) or 4.451 V_{p-p} (AD595). Reconnect the thermocouple (in an ice bath or an ice-point cell set to 0°C) to pins 1/14. Adjust R_{OFFSET} until the output reads 320 mV. The ideal transfer function for a Fahrenheit output AD594/595 when trimmed with a thermocouple at 0°C is: AD594 output = (Type J voltage + 919 µV) × 348.12; AD595 output = (Type K voltage + 719 µV) × 445.14. ANALOG DEVICES, APPLICATIONS REFERENCE MANUAL, 1993, P. 21-6.

10

537

Temperature indicator/controller circuit titles and descriptions

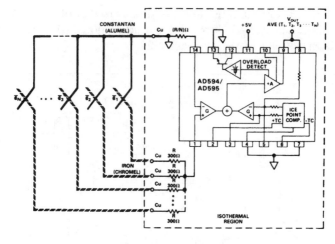

Thermocouple measurement with direct mean temperature

Fig. 10-12 The circuit in the illustration is similar to that of Fig. 10-9, except that average temperature can be measured directly. The circuit is especially useful for monitoring an object with significant thermal gradients. The circuit output equals

$$\frac{(T_1 + T_2 + T_3 + ... \, T_N)}{N}$$

(in °C) times a nominal 10 mV/°C. The correct cold-junction compensation is provided with any number of thermocouple-resistor pairs. ANALOG DEVICES, APPLICATIONS REFERENCE MANUAL, 1993, P. 21-7.

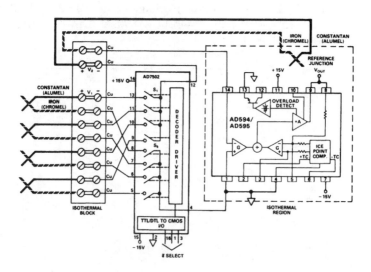

Temperature indicator/controller circuits

Multiplex thermocouple-measurements

Fig. 10-13 The circuit in the illustration is similar to that of Fig. 10-9 except that the outputs of four thermocouples can be multiplexed (Chapter 12) by a single AD594/595. The circuit is useful for large temperature-measuring data-acquisition systems and transforms the usual multiple reference-junction connections from a terminal block to a single junction on the AD594/595. This change is accomplished by placing a thermocouple in thermal contact (beneath) the IC and returning the output to cancel the reference-junction voltages generated at the isothermal block. For a given position of the AD7502 multiplexer, the junctions contribute equal but opposite voltages. As a result, the IC compensates for the reference thermocouple beneath it. An AD7507 (instead of the AD7502 shown) allows you to measure eight temperatures. ANALOG DEVICES, APPLICATIONS REFERENCE MANUAL, 1993, P. 21-7.

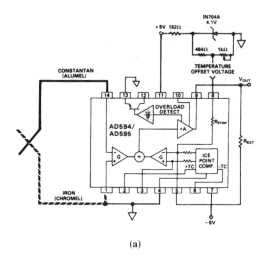

(a)

1. Determine the desired output sensitivity (in mV/°C).
2. Decide on a temperature range T1 to T2.
3. Calculate the average thermocouple sensitivity over that temperature range; $(V_{\tau1} - V_{\tau2})/(T_1 - T_2)$.
4. Divide the desired sensitivity by the average thermocouple sensitivity: result of (1) ÷ calculated value in (3). This value is the new gain (G_{NEW}) of the AD594/AD595. If the calculations are done correctly this result will be dimensionless.
5. Measure the actual feedback resistance (pin 8 to pin 5), R_{FDBK}.
6. $R_{INTERNAL} = \dfrac{R_{FDBK}}{193.4 - 1} = \dfrac{\text{Result of (5)}}{193.4 - 1}$

   NOTE: Use 247.3 for an AD595 instead of 193.4.

7. The new feedback resistance, $R_{EXTERNAL} = (G_{NEW} - 1)(R_{INTERNAL}) = (\text{result from } (4) - 1)(\text{result from } (6))$

(b)

Temperature indicator/controller circuit titles and descriptions

Thermocouple measurements
with zero suppression and output-sensitivity change

Fig. 10-14 The circuit in the illustration is similar to that of Fig. 10-9, except that the zero can be suppressed (the output reads 0 V at any selected temperature) and the output-sensitivity can be changed. Using the values shown, it is possible to measure a temperature range from 300°C to 330°C (with a 5-V supply) and an output of 100 mV/°C beginning with a 0 V output at 300°C. Use the calculations shown in Fig. 10-14B to calculate the value of R_{EXT}. ANALOG DEVICES, APPLICATIONS REFERENCE MANUAL, 1993, P. 21-8.

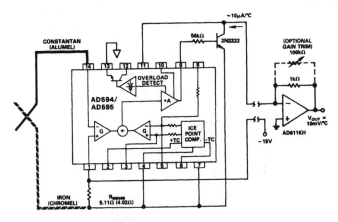

Thermocouple measurements with current-mode transmission

Fig. 10-15 The circuit in the illustration is similar to that of Fig. 10-9, except that the AD594/595 output can be transmitted as a current and then converted back to a voltage at the control point. This transmission method is convenient when sending a temperature signal through a noisy environment. The feedback voltage at pin 9 forces the voltage across R_{SENSE} to equal the thermocouple voltage. Using an R_{SENSE} of 5.11 Ω (AD594) or 4.02 Ω (AD595) generates 10 µA/°C current. Because the voltage across R_{SENSE} equals the thermocouple voltage, the reference-junction voltage appears across the IC input. The minimum temperature that can be measured is 16°C because of the 160 µA quiescent current of the IC. Circuit accuracy is set by the initial IC calibration error and the match between R_{SENSE} and 1-kΩ current-to-voltage conversion resistor at the measurement point. ANALOG DEVICES, APPLICATIONS REFERENCE MANUAL, 1993, P. 21-8.

=11=

Filter circuits

The discussions in this chapter assume that you are already familiar with filter basics, such as filter responses (lowpass, highpass, bandpass, and notch) and common filter types (Butterworth, Chebyshev, Elliptic, and Bessel). The following paragraphs summarize this information, as well as information on testing and troubleshooting for filters.

Filter-circuit testing and troubleshooting

The primary purpose of a filter is to discriminate against the passing of certain groups of frequencies while passing other groups or portions of the frequency spectrum. Although filter circuits range from the very simple to the very complex, there are only two basic types of filters: *active* and *passive*. The active filters in this chapter are essentially frequency-selective amplifiers, but no gain is involved. Passive filters are either *RC* (resistance-capacitance), used primarily for low- or audio-frequency applications, or *LC* (inductance-capacitance) for use at higher frequencies.

Filter frequency response

The four basic filter frequency responses are shown in Fig. 11-A. A *lowpass filter* passes all frequencies lower than a selected value and attenuates higher frequencies. The lowpass filter also is known as a *high-cut filter*. A *highpass filter* passes all frequencies higher than a selected value and attenuates lower frequencies. The highpass filter also is known as a *low-cut filter*. A *bandpass filter* passes a selected band of frequencies. A *notch filter* suppresses a selected band of frequencies while passing all lower and higher frequencies. The notch filter also is known as a *band-elimination, band-stop, band-rejection*, or *band-suppression filter*.

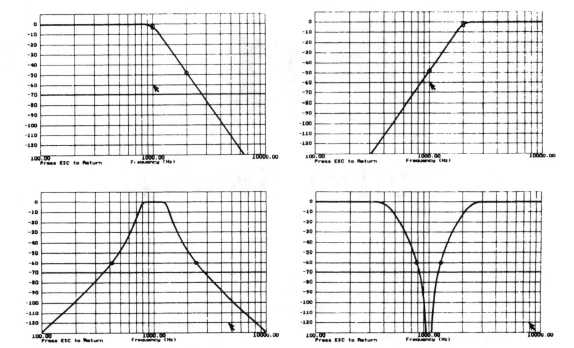

Fig. 11-A   Basic filter frequency responses.

Filter types

Figures 11-B and 11-C show the most common types of filters. Notice the ideal lowpass response in Fig. 11-B (top left). This ideal response is not possible with any of the filter types.

The *Butterworth filter* (top right) has the optimum flatness in the passband, but it has a slope that rolls off more gradually after the cutoff frequency than the other types. The *Chebyshev filter* (bottom left) can have a steeper initial rolloff than Butterworth, but at the expense of more ripple in the passband. The *Elliptic filter* (bottom right) has the steepest rolloff of all but shows ripple in both the passband and the stopband (after the cutoff point). The *Bessel filter* (Fig. 11-C) has a sloping rolloff but much steeper than a Butterworth filter.

Basic filter tests

The obvious test for the filter circuits in this chapter is frequency response. Use the procedure for finding the frequency response of amplifiers from Chapter 1. Of course, there will be no gain, but the response curve will show filter attenuation over the frequency range.

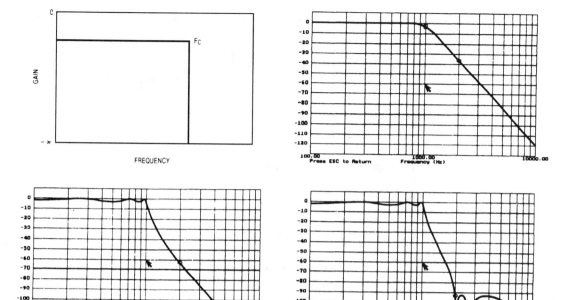

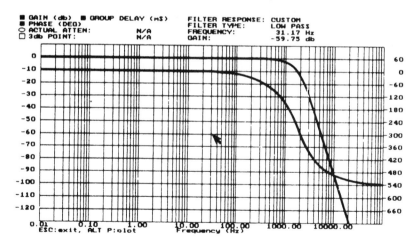

Fig. 11-B   Basic Butterworth, Chebyshev, and Elliptic filter responses.

Fig. 11-C   Basic Bessel filter response.

For example, when testing the filter of Fig. 11-D, the response should be flat (zero attenuation) at frequencies up to about 375 Hz. Then there should be a sharp dropoff (about 45 dB) at 400 Hz. As the frequency is increased, the response should return to 0 dB at about 450 Hz and remain flat at higher frequencies.

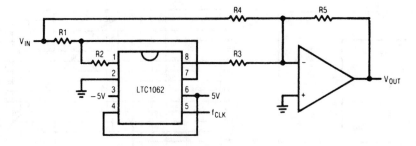

For simplicity use R3 = R4 = R5 = 10k;

$$\frac{R1}{R2} = 1.234, \quad \frac{f_{CLK}}{f_{notch}} = \frac{79.3}{1}$$

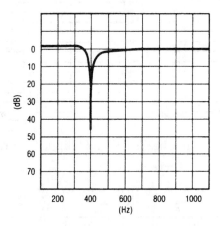

Fig. 11-D Clock-tunable notch filter.

Sometimes it might be necessary to check the phase shift between input and output of a filter as well as noise. If required, use the phase-shift and noise test procedures for amplifiers, as described in Chapter 1.

Basic filter troubleshooting

If a filter circuit fails to produce the desired frequency response, the problem is usually one of incorrect component values, especially components that are out of tolerance. This guideline is given on the assumption that there are no defective parts (especially leaking capacitors) and that the circuit wiring is correct (as it always is with your circuits!).

So, if you get no response, check the wiring and look for bad parts. If you get a response but not the desired response, try correcting the problem with changes in component values. Even slight changes in filter-circuit values can produce substantial changes in frequency response.

Notice that filters have a Q factor, as do all tuned circuits. A high Q produces a sharp frequency response, but a low Q produces a broad response. For example, the circuit of Fig. 11-E has a Q of 5 and should produce a broad response with a notch at 3 kHz. The circuit of Fig. 11-F has a Q of 25 and produces a sharp response at 1 kHz.

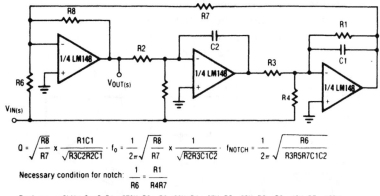

$$Q = \sqrt{\frac{R8}{R7}} \times \frac{R1C1}{\sqrt{R3C2R2C1}} \qquad f_0 = \frac{1}{2\pi}\sqrt{\frac{R8}{R7}} \times \frac{1}{\sqrt{R2R3C1C2}} \qquad f_{NOTCH} = \frac{1}{2\pi}\sqrt{\frac{R6}{R3R5R7C1C2}}$$

Necessary condition for notch: $\dfrac{1}{R6} = \dfrac{R1}{R4R7}$

Ex: f_{NOTCH} = 3kHz, Q = 5, R1 = 270K, R2 = R3 =20K, R4 = 27K, R5 = 20K, R6 = R8 = 10K, R7 = 100K. C1 = C2 = 0.001μF

Better noise performance than the state-space approach

Fig. 11-E   Bi-quad notch filter.

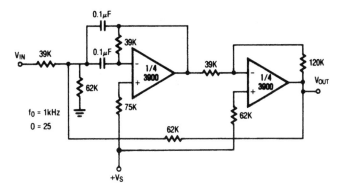

Fig. 11-F   1-kHz bandpass filter with high Q.

Basic filter troubleshooting

Switched-capacitor and instrumentation filters (clocked)

Some of the circuits in this chapter require a clock signal (the switched-capacitor filter of Fig. 11-G, for example). The frequency response of the filter circuit is directly affected by the clock frequency. Therefore, any deviation of the clock from the desired frequency can change the filter frequency response. So, always check clock frequency before you change components to get a desired response. Even a small shift in clock frequency can change the filter-circuit frequency response.

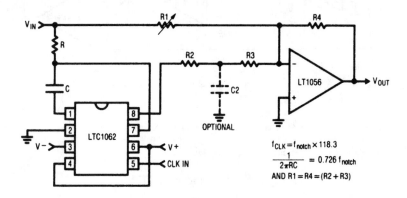

$f_{CLK} = f_{notch} \times 118.3$

$\dfrac{1}{2\pi RC} = 0.726\, f_{notch}$

AND $R1 = R4 = (R2 + R3)$

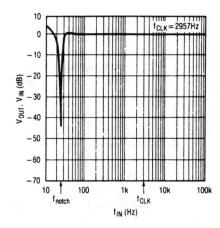

Fig. 11-G   IC notch filter (clock tunable).

Filter circuit titles and descriptions

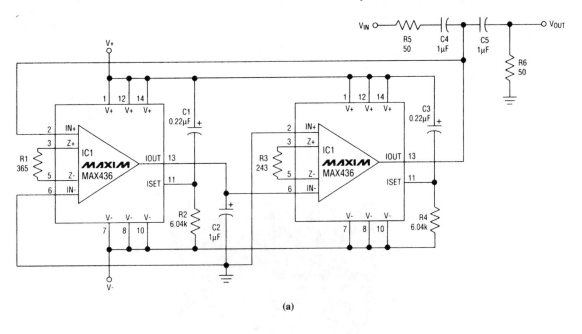

(a)

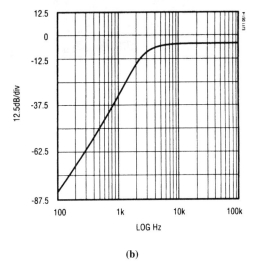

(b)

Third-order highpass filter with synthetic inductor

Fig. 11-1 The circuit in the illustration shows two OTAs (Chapter 1) connected to form a Butterworth highpass filter. Figure 11-1B shows the characteristics (3.2-kHz corner, –6 dB loss). MAXIM ENGINEERING JOURNAL, VOLUME 11, P. 11.

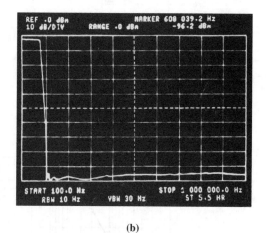

(a)

(b)

Ninth-order Elliptic lowpass filter

Fig. 11-2 The circuit in the illustration shows a discrete-component (ladder type) elliptic lowpass filter. Figure 11-2B shows the characteristics. ANALOG DEVICES, APPLICATIONS REFERENCE MANUAL, 1993, P. 3-33.

Filter circuits

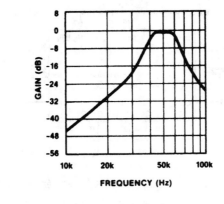

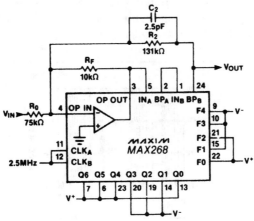

Fourth-order 50-kHz Chebyshev bandpass filter (single clock)

Fig. 11-3 The circuit in the illustration shows a MAX268 pin-programmable filter connected to provide a 10-kHz bandpass, centered at 50 kHz. The maximum passband ripple is 0.1 dB, with a gain of 1 V/V (center frequency) and a Q of 4.2. MAXIM NEW RELEASES DATA BOOK, 1992, P. 6-56.

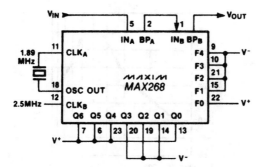

Fourth-order 50-kHz Chebyshev bandpass filter (two clocks)

Fig. 11-4 The circuit in the illustration is similar to Fig. 11-3, except that two clocks are required, but the only external component required is a 1.89-MHz crystal. MAXIM NEW RELEASES DATA BOOK, 1992, P. 6-56.

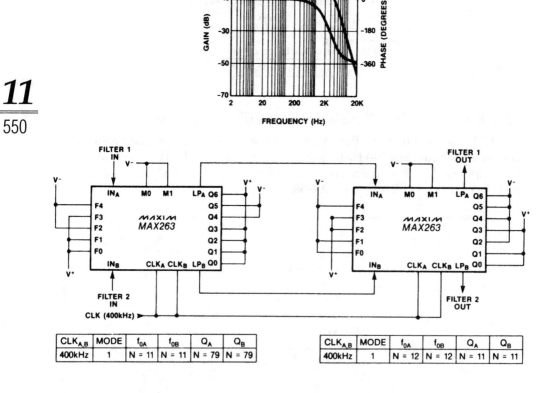

CLK$_{A,B}$	MODE	f$_{0A}$	f$_{0B}$	Q$_A$	Q$_B$
400kHz	1	N = 11	N = 11	N = 79	N = 79

CLK$_{A,B}$	MODE	f$_{0A}$	f$_{0B}$	Q$_A$	Q$_B$
400kHz	1	N = 12	N = 12	N = 11	N = 11

Filter circuits

Dual-tracking 3-kHz 4th-order Butterworth lowpass filter

Fig. 11-5 The circuit in the illustration shows two MAX263 pin-programmable filters connected to track each other and provide a 3-kHz lowpass cutoff, with a *Q* of 1.3 and 0.5. MAXIM NEW RELEASES DATA BOOK, 1992, P. 6-57.

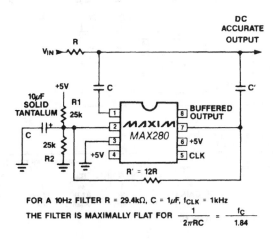

FOR A 10Hz FILTER R = 29.4kΩ, C = 1μF, f_{CLK} = 1kHz

THE FILTER IS MAXIMALLY FLAT FOR $\dfrac{1}{2\pi RC} = \dfrac{f_C}{1.84}$

Single-supply 5th-order lowpass filter (10 kHz)

Fig. 11-6 The circuit in the illustration shows a MAX280 switched-capacitor filter connected to provide lowpass cutoff at 10 kHz. Other cutoff frequencies can be selected by alternating the values of *R* and *C*, as shown by the equations. MAXIM NEW RELEASES DATA BOOK, 1992, P. 6-85.

THE MAX430 IS CONNECTED AS A 2nd ORDER
SALLEN AND KEY LOWPASS FILTER WITH A
CUTOFF FREQUENCY EQUAL TO THE MAX280.
THE ADDITIONAL FILTERING ELIMINATES ANY
10kHz CLOCK FEED THROUGH PLUS DECREASES
THE WIDEBAND NOISE OF THE FILTER.

DC OUTPUT OFFSET (REFERRED TO A DC GAIN OF
UNITY) = 5μV Max.

WIDEBAND NOISE (REFERRED TO A DC GAIN OF
UNIT) = 60μ/V$_{RMS}$

OUTPUT FILTER COMPONENT VALUES						
DC GAIN	R3	R4	R1	R2	C1	C2
1	∞	0	14.3k	53.6k	0.1μF	0.033μF
10	3.57k	32.4k	4.6k	27.4k	0.1μF	0.2μF
101	0.324	32.4k	0.31k	16.9k	0.47μF	1μF

Seventh-order 100-Hz lowpass filter (Sallen and Key)

Fig. 11-7 The circuit in the illustration shows a MAX280 filter and MAX430 amplifier connected to provide lowpass cutoff at 100 Hz. The output gain is set by the values of R1–R4 and C1–C2, as shown by the table. MAXIM NEW RELEASES DATA BOOK, 1992, P. 6-85.

11

552

10Hz, 10th ORDER DC ACCURATE LOW PASS FILTER
60dB/OCTAVE ROLLOFF
0.5dB PASSBAND ERROR, 0dB DC GAIN
MAXIMUM ATTENUATION 110dB (f$_{CLK}$ = 10kHz)
                     100dB (f$_{CLK}$ = 1kHz)
                      95dB (f$_{CLK}$ = 1MHz)

Filter circuits

Cascade 10th-order 10-Hz dc-accurate lowpass filter

Fig. 11-8 The circuit in the illustration shows two MAX280 filters connected in cascade to provide a 10-Hz lowpass filter with 60-dB/octave rolloff (Chapter 1). Use the following for cutoff (f_c) at other frequencies:

$$\frac{1}{(6.28RC)} = \frac{f_c}{1.57}$$

and

$$\frac{1}{(6.28R'C')} = \frac{f_c}{1.6}$$

For example, for f_c = 4.16 kHz, make f_{CLK} = 416 kHz, R = 909 Ω, R' = 107 kΩ, C = 0.066 µF, C' = 574 pF. MAXIM NEW RELEASES DATA BOOK, 1992, P. 6-86.

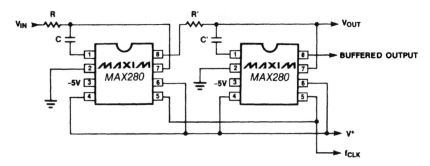

Cascade 10th-order 10-Hz lowpass filter

Fig. 11-9 The circuit in the illustration is similar to that of Fig. 11-8, except that the second stage is driven by the buffered output of the first stage. This introduces a maximum dc error of 2 mV, over temperature, but now the values of R and R' can be similar in value, and the passband-gain error is reduced typically –0.15 dB. Use the following for cutoff frequencies (f_c):

$$\frac{1}{(6.28RC)} = \frac{f_c}{1.59}$$

and

$$\frac{1}{(6.28R'C')} = \frac{f_c}{1.64}$$

MAXIM NEW RELEASES DATA BOOK, 1992, P. 6-86.

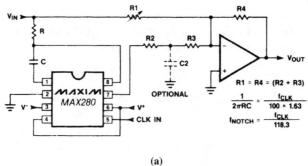

$R1 = R4 = (R2 + R3)$

$$\frac{1}{2\pi RC} = \frac{f_{CLK}}{100 \times 1.63}$$

$$f_{NOTCH} = \frac{f_{CLK}}{118.3}$$

(a)

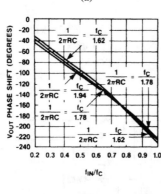

(b)

Notch filter (created from a lowpass filter)

11

554

Fig. 11-10 The circuit in the illustration shows a MAX280 filter and LTC1062 amplifier connected to form a notch filter. Figure 11-10B shows the phase response for various values of input R and C. Use the equations to calculate the desired notch frequency. For example, to get a notch at 60 Hz, the clock frequency should be 7098 Hz, and the input

$$\frac{1}{(6.28RC)}$$

should be about

$$\frac{70.98 \text{ Hz}}{1.63}$$

The optional R_2C_2 at the output is used to filter any clock feedthrough. The

$$\frac{1}{(6.28R_2C_2)}$$

should be about 12 to 15 times the notch frequency. MAXIM NEW RELEASES DATA BOOK, 1992, P. 6-87.

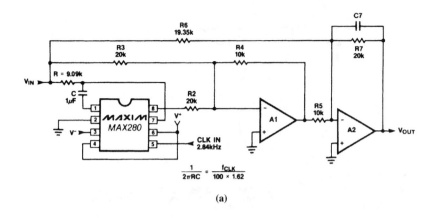

$$\frac{1}{2\pi RC} = \frac{f_{CLK}}{100 \times 1.62}$$

(a)

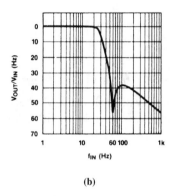

(b)

Lowpass filter with a notch

Fig. 11-11 The circuit in the illustration is similar to that of Fig. 11-10 but with an additional amplifier to provide both lowpass and notch functions. Using the values shown, the notch is at 60 Hz, as shown in Fig. 11-11B. The frequency of the notch is at

$$\frac{f_{CLK}}{47.3}$$

and the value of the ratio

$$\frac{R_6}{R_5}$$

is equal to 1.935. $R_2 = R_3 = R_7$ and $R_4 = R_5 = 0.5R_7$. MAXIM NEW RELEASES DATA BOOK, 1992, P. 6-87.

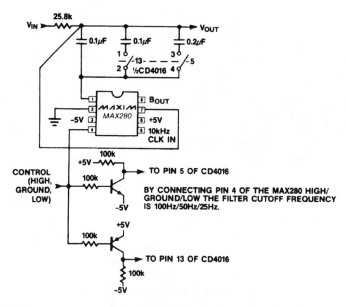

First-order dc-accurate lowpass filter (100, 50, and 25 Hz)

Fig. 11-12 The circuit in the illustration shows a MAX280 filter and a CD4061 connected to provide a dc-accurate lowpass filter with selectable cutoff frequencies. MAXIM NEW RELEASES DATA BOOK, 1992, P. 6-88.

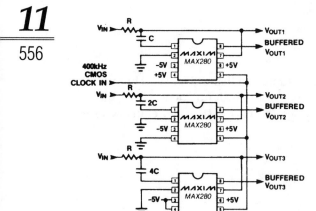

Octave Tuning with a Single Input Clock

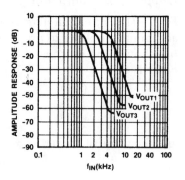

Amplitude Response for the Octave Tuning Circuit

Filter circuits

Lowpass filter with octave tuning

Fig. 11-13 The circuit in the illustration shows three MAX280 filters connected to provide octave tuning with a single input clock. Use the equations of Fig. 11-11 to find the values of *R* and *C*. MAXIM NEW RELEASES DATA BOOK, 1992, P. 6-88.

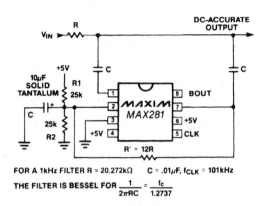

FOR A 1kHz FILTER R = 20.272kΩ   C = .01μF, f_{CLK} = 101kHz

THE FILTER IS BESSEL FOR $\dfrac{1}{2\pi RC} = \dfrac{f_c}{1.2737}$

A. MAX281 WITH A SINGLE +5V SUPPLY

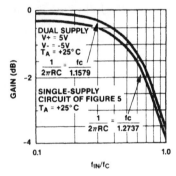

B. SINGLE- AND DUAL-SUPPLY PASSBAND FREQUENCY RESPONSE

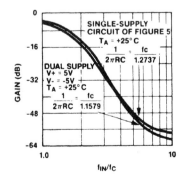

C. SINGLE- AND DUAL-SUPPLY STOPBAND FREQUENCY RESPONSE

Filter circuit titles and descriptions

Fifth-order dc-accurate Bessel lowpass filter

Fig. 11-14 The circuit in the illustration shows a MAX281 switched-capacitor filter connected to provide lowpass cutoff at 1 kHz. Other cutoff frequencies can be selected by alternating the values of R and C, as shown by the equations. MAXIM NEW RELEASES DATA BOOK, 1992, P. 6-97.

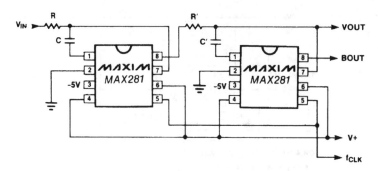

Cascaded lowpass filter with buffered output

Fig. 11-15 The circuit in the illustration is similar to that of Fig. 11-9, except that Bessel MAX281 filters are used. Use the following for cutoff frequency (f_c):

$$\frac{1}{(6.28RC)} = \frac{f_c}{1.1579}$$

MAXIM NEW RELEASES DATA BOOK, 1992, P. 6-98.

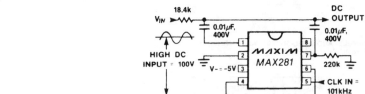

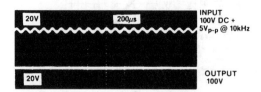

Filter circuits

Filtering ac signals from high dc signals

Fig. 11-16 The circuit in the illustration shows a MAX281 connected to remove undesired ac components from a dc voltage much higher than the operating voltage of the filter IC. This feature is possible because of the shunt architecture of these filters and is not generally possible with conventional active filter structures. MAXIM NEW RELEASES DATA BOOK, 1992, P. 6-98.

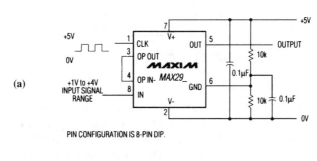

(a)

PIN CONFIGURATION IS 8-PIN DIP.

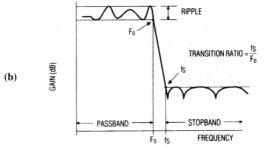

(b)

PIN		NAME	FUNCTION
8-PIN DIP	**16-PIN SO**		
	1,2,7,8,9, 10,15,16	N.C.	No Connect—not internally connected
1	3	CLK	Clock Input—use internal or external clock.
2	4	V-	Negative Supply pin. Dual supplies: -2.375V to -5.5V. Single supply: V- = 0V.
3	5	OP OUT	Uncommitted Op-Amp Output
4	6	OP IN-	Inverting Input to the uncommitted op amp. The noninverting op amp is internally tied to GND.
5	11	OUT	Filter Output
6	12	GND	Ground. In single-supply operation, GND must be biased to the mid-supply voltage level.
7	13	V+	Positive Supply pin. Dual supplies: +2.375V to +5.5V. Single supply: +4.75V to +11.0V.
8	14	IN	Filter Input

(c)

Filter circuit titles and descriptions

Continued

PART	RESPONSE SHAPE	CLOCK: CORNER-FREQUENCY RATIO
MAX291	Butterwoth	100:1
MAX292	Bessel	100:1
MAX293	Elliptic (1.5 transition ratio)	100:1
MAX294	Elliptic (1.5 transition ratio)	100:1
MAX295	Butterworth	50:1
MAX296	Bessel	50:1
MAX297	Elliptic (1.5 transition ratio)	50:1

(d)

Switched-capacitor Elliptic filter

Fig. 11-17   The circuit in the illustration shows a MAX291-297 connected for single-supply operation. Figures 11-17B and C show the Elliptic filter response and pin descriptions, respectively. These filters are either 100:1 or 50:1 clock-to-corner frequency ratio, as shown in Fig. 11-17D. That is, the corner frequency depends directly on the clock frequency, with a range from 0.1 Hz to 50 kHz. 100:1 is recommended for corner frequencies up to 25 kHz. Between 25 kHz and 50 kHz, the 50:1 ratio is recommended. For example, for a 20-kHz corner frequency, with a Bessel response, use the MAX292 and a clock of 2 MHz. The clock can be internal or external, with a maximum of 2.5 MHz. For an internal clock, connect a capacitor between pin 1 and ground, using the equation:

$$\text{clock (kHz)} = \frac{10^5}{(3 \times \text{capacitor in pF})}$$

For example, for a clock of 20 kHz, use a 166-pF capacitor. This capacitance will produce a 200-Hz corner. MAXIM NEW RELEASES DATA BOOK, 1992, PP. 6-44, 6-45.

Filter circuits

22k R2
330pF   C1
22k   22k
R1   R3   4 [OP IN-]
INPUT ─/\/\/─   ─/\/\/─
                              ──── OUTPUT
C2   3
1500pF   OP OUT

PIN CONFIGURATION IS 8-PIN DIP.

MAXIM
MAX29_

(a)

Corner Freq. (Hz)	R1 (kΩ)	R2 (kΩ)	R3 (kΩ)	C1 (F)	C2 (F)
100k	10	10	10	68p	330p
50k	20	20	20	68p	330p
25k	20	20	20	150p	680p
10k	22	22	22	330p	1.5n
1k	22	22	22	3.3p	15n
100	22	22	22	33n	150n
10	22	22	22	330n	1.5µ

NOTE: Some approximations have been made in selecting preferred component values.

(b)

Second-order Butterworth lowpass filter (10 kHz)

Fig. 11-18 The circuit in the illustration uses the uncommitted op amp in the MAX291-297 (of Fig. 11-17) connected as a lowpass filter with a frequency of 10 kHz. Figure 11-18B shows component values for various corner frequencies. This filter is intended for anti-aliasing applications preceding the switch-capacitor filter, but it can be used as a post-filter to reduce clock noise. MAXIM NEW RELEASES DATA BOOK, 1992, P. 6-45.

12

Video circuits

This chapter is devoted to video-frequency circuits. Most of the circuits are a form of amplifiers, buffers, or comparators, so the test/troubleshooting procedures described in Chapters 1 and 8 apply to the circuits in this chapter. Where there are differences in test/troubleshooting, the recommended test circuits (and procedures where it is not obvious) are given for the ICs involved.

Video circuits titles and descriptions

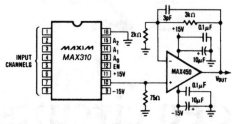

No Insertion Loss, 8 Channel Mux

(a)

Top View

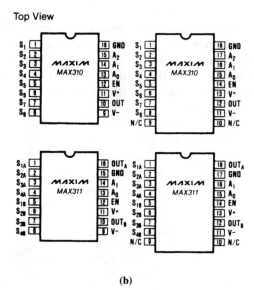

(b)

Continued

ABSOLUTE MAXIMUM RATINGS

Voltage referenced to V⁻
- V⁺ .... +36V
- GND .... +24V
- Digital Inputs .... V⁻ to V⁺

Input Current
- S and COMMON OUT .... ±50mA
- All pins except S and COM. OUT .... ±30mA

Lead Temperature .... +300°C
Storage Temperature .... –65°C to +150°C

Operating Temperature Range
- MAX310C, MAX311C .... 0°C to +70°C
- MAX310E, MAX311E .... –40°C to +85°C
- MAX310M, MAX311M .... –55°C to +125°C

Power Dissipation (16-Pin packages)
- CERDIP (derate 10mW/°C above +75°C) 750mW
- Plastic DIP (derate 7.35mW/°C above +75°C) 550mW
- Small Outline (derate 9mW/°C above +75°C) 550mW

Stresses listed under "Absolute Maximum Ratings" may be applied (one at a time) to devices without resulting in permanent damage. These are stress ratings only, and functional operation of the device at these or any other conditions above those indicated in the operational sections of the specifications is not implied. Exposure to absolute maximum ratings conditions for extended periods may affect device reliability.

ELECTRICAL CHARACTERISTICS
(Over Temperature, V⁺ = +15V, V⁻ = –15V, GND = 0V unless otherwise indicated)

PARAMETER	SYMBOL	CONDITIONS	MIN.	TYP.	MAX.	UNITS
Analog Signal Range		V⁺, V⁻ = ±15V V⁺, V⁻ = ±5V	-15 -5		+12 +2	V
Channel ON Resistance	R_{ON}	V_{IN} = ±5V, I_{OUT} = 10mA T_A = +25°C Over Temp.		150	250 300	Ω
ON Resistance Match	ΔR_{ON}	V_{IN} = ±5V, I_{OUT} = 10mA		6		%
OFF Input Leakage Current	$I_{S(OFF)}$	Figure 10, T_A = +25°C Over Temp		0.4 3	10 100	nA
OFF Output Leakage Current	$I_{D(OFF)}$	Figure 11, T_A = +25°C MAX310 Over Temp. MAX311 Over Temp.		0.8 20 10	10 100 50	nA
ON Channel Leakage Current	$I_{D(ON)}$	Figure 12, T_A = +25°C MAX310 Over Temp. MAX311 Over Temp.		1 30 15	10 200 100	nA
Input Low Threshold	V_{AL}	V⁺/V⁻ = ±15V, ±5V			0.8	V
Input High Threshold	V_{AH}	V⁺/V⁻ = ±15V, ±5V	2.4			V
Input Current (Logic)	I_A	V_A = 0V or 5V			±10	μA
Access Time	t_{ACC}	Figure 7; T_A = +25°C Over Temp.		0.6	1.5 2.0	μs
Enable Delay ON or OFF	$t_{EN(ON/OFF)}$	Figure 8; T_A = +25°C Over Temp.		0.3	1.0 2.0	μs
Break-Before-Make Delay	t_{ON}–t_{OFF}	Figure 9	30	100		ns
OFF Isolation, Single Channel to OUT	ISO_{SC}	Figure 3; T_A = +25°C	-66	-76		dB
OFF Isolation, All Channels to OUT	ISO_{AC}	Figure 4, 5, T_A = +25°C MUX Disabled, EN = +0.8V MUX Enabled, EN = +2.4V		-63 -58		dB
Adjacent Channel Crosstalk	ISO_X	Figure 6, T_A = +25°C		-72		dB
Channel Input Capacitance OFF State ON State	$C_{S(OFF)}$ $C_{S(ON)}$	T_A = +25°C, V_{IN} = 10mV$_{RMS}$ 10 MHz		5 45		pF
Channel Output Capacitance OFF State ON State	$C_{D(OFF)}$ $C_{D(OFF)}$	T_A = +25°C; EN = +0.8V, MAX310 　　　　　　 MAX311 EN = +2.4V, MAX310 　　　　　　 MAX311		38 20 57 40		pF
Charge Injection	Q	Figure 13, T_A = +25°C		110		pC
Supply Current; V⁺ 　　　　　　　V⁻	I⁺ I⁻	EN, A0, A1, A2 = 0V or +5V		75 0.1	200 100	μA
Supply Voltage Range		T_A = +25°C	±4.5		±16.5	V

(c)

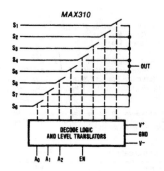

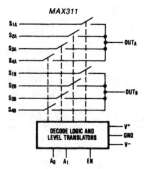

(d)

MAX310					MAX311			
A₂	A₁	A₀	EN	ON Channel	A₁	A₀	EN	ON Channel
0	0	0	1	1	0	0	1	1A + 1B
0	0	1	1	2	0	1	1	2A + 2B
0	1	0	1	3	1	0	1	3A + 3B
0	1	1	1	4	1	1	1	4A + 4B
1	0	0	1	5	X	X	0	ALL OFF
1	0	1	1	6				
1	1	0	1	7				
1	1	1	1	8				
X	X	X	0	ALL OFF				

(e)

INPUT CHANNEL MAX311	OUTPUT – INPUT PHASE SHIFT	
	R_L = 10kΩ	R_L = 75Ω
S_1	–22°	–12°
S_2	–21°	–11.5°
S_3	–20°	–11.5°
S_4	–20°	–11.2°
S_5	–20°	–11.2°
S_6	–20.5°	–11.4°
S_7	–20.7°	–11.5°
S_8	–20.4°	–11.5°

Test Conditions: V^+ = +15V, V^- = –15V, V_{IN} = 1.25V_{RMS} at 10MHz,
OFF inputs terminated with 75Ω.

(f)

SUPPLY VOLTAGE		SIGNAL RANGE	TYPICAL R_{ON} AT V_{IN}	
V⁻	V⁺		NEGATIVE	POSITIVE
–15	+15V	–15V to +12V	104Ω at –10V	265Ω at +10V
		–5V to +5V	115Ω at –5V	150Ω at +5V
GND	+15V	0V to +12V	120Ω at 0V	150Ω at +5V
GND	+30V	0V to +27V	90Ω at 0V	100Ω at +5V
–5V	+5V	–5V to +2V	240Ω at –2V	480Ω at +2V
–10V	+10V	–10V to +7V	140Ω at –5V	220Ω at +5V
–5V	+15V	–5V to +12V	115Ω at –5V	150Ω at +5V

(g)

Video circuits

CMOS RF/Video multiplexers

Fig. 12-1 The circuit in the illustration shows a MAX310 *mux* (multiplexer) combined with a MAX450 amplifier to produce an eight-channel mux with no insertion loss. Figures 12-1B, C, D, E, F, and G show the pin configurations, maximum ratings and characteristics, functional diagram, channel-selection input codes, phase shift, and signal-range and R_{ON} versus supply voltage, respectively. The test circuits for the characteristics given in Fig. 12-1C are shown in Figs. 12-2 through 12-12. MAXIM NEW RELEASES DATA BOOK, 1992, PP. 8-17, 18, 19, 20.

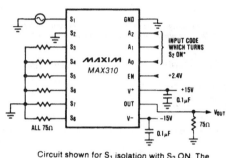

Circuit shown for S₁ isolation with S₂ ON. The ON channel is shorted to GND. Other channels are measured in a similar fashion.

Signal-channel OFF isolation test circuit

Fig. 12-2 The typical values for this test are –66 dB minimum and –76 dB typical (Fig. 12-1C). MAXIM NEW RELEASES DATA BOOK, 1992, P. 8-21.

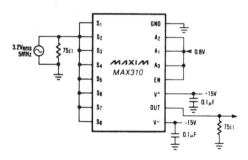

All-channel OFF isolation test circuit (mux disabled)

Fig. 12-3 The typical value for this test is –63 dB (Fig. 12-1C). MAXIM NEW RELEASES DATA BOOK, 1992, P. 8-21.

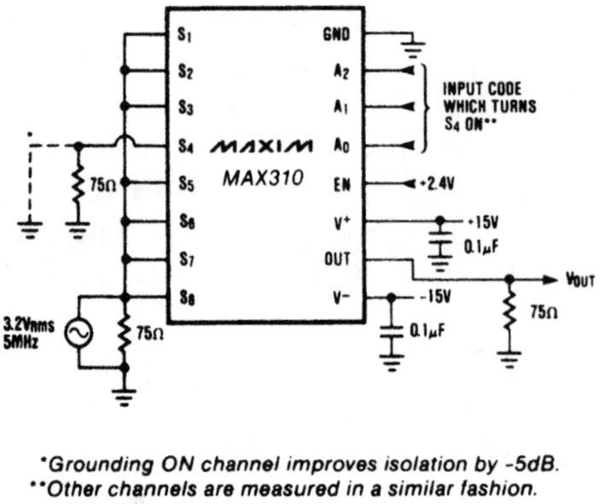

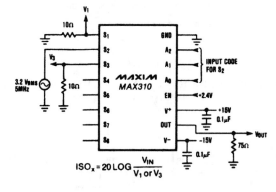

*Grounding ON channel improves isolation by –5dB.
**Other channels are measured in a similar fashion.*

All-channel OFF isolation test circuit (mux enabled)

Fig. 12-4 The typical value for this test is –58 dB (Fig. 12-1C). MAXIM NEW RE-
LEASES DATA BOOK, 1992, P. 8-21.

$$ISO_x = 20 \, LOG \frac{V_{IN}}{V_1 \, or \, V_3}$$

Adjacent-channel crosstalk test circuit

Fig. 12-5 The typical value for this test is –72 dB (Fig. 12-1C). MAXIM NEW RE-
LEASES DATA BOOK, 1992, P. 8-22.

Video circuits

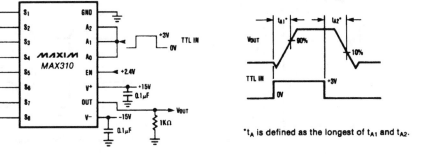

Access-time test circuit

Fig. 12-6 The typical values for this test are 0.6 μs typical and 1.5 μs maximum, at 25°C, and 2 μs over temperature (Fig. 12-1C). Maxim New Releases Data Book, 1992, P. 8-22.

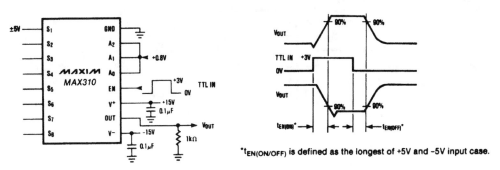

Enable-delay test circuit

Fig. 12-7 The typical values for this test are 0.2 μs typical and 1.0 μs maximum, at 25°C, and 2 μs over temperature (Fig. 12-1C). Maxim New Releases Data Book, 1992, P. 8-22.

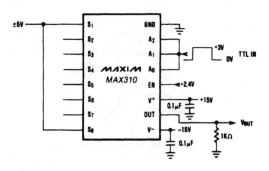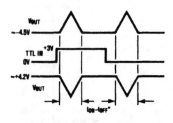

$t_{ON} - t_{OFF}$ is defined as the shortest of +5V and –5V input case.

Break-before-make delay test circuit

Fig. 12-8 The typical values for this test are 30 ns minimum and 100 ns typical (Fig. 12-1C). MAXIM NEW RELEASES DATA BOOK, 1992, P. 8-23.

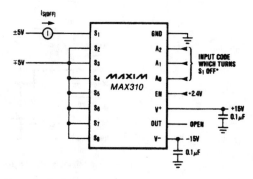

*Shown for S_1. Other channels are measured in a similar fashion.

OFF input-leakage current test circuit

Fig. 12-9 The typical values for this test are 0.4 nA typical and 10 nA maximum at 25°C; 3 nA typical and 100 nA maximum over temperature (Fig. 12-1C). MAXIM NEW RELEASES DATA BOOK, 1992, P. 8-23.

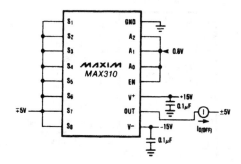

OFF output-leakage current test circuit

Fig. 12-10 The typical values for this test are 0.8 nA typical and 10 nA maximum (Fig. 12-1C). MAXIM NEW RELEASES DATA BOOK, 1992, P. 8-23.

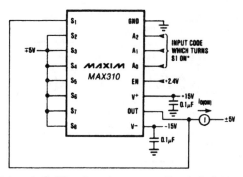

* shown for S₁. Other channels are measured in a similar fashion.

ON output-leakage current test circuit

Fig. 12-11 The typical values for this test are 1 nA typical and 10 nA maximum (Fig. 12-1C). MAXIM NEW RELEASES DATA BOOK, 1992, P. 8-23.

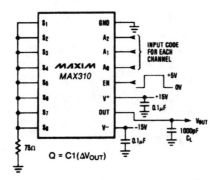

Charge-injection (Q) test circuit

Fig. 12-12 The typical value for this test is 110 pC (picocoulombs) (Fig. 12-1C). MAXIM NEW RELEASES DATA BOOK, 1992, P. 8-23.

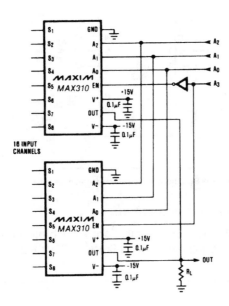

RF/video multiplexer (1-of-16)

Fig. 12-13 The circuit in the illustration shows two MAX310 multiplexers connected in cascade to form a 1-of-16 multiplexer. MAXIM NEW RELEASES DATA BOOK, 1992, P. 8-24.

Video circuits

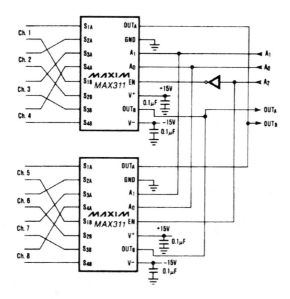

RF/video differential multiplexer (1-of-8)

Fig. 12-14 The circuit in the illustration shows two MAX311 multiplexers connected to form a 1-of-8 differential multiplexer. MAXIM NEW RELEASES DATA BOOK, 1992, P. 8-24.

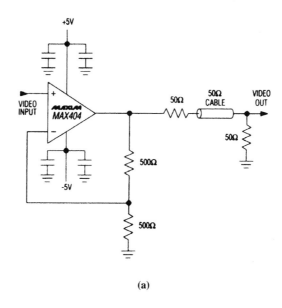

(a)

Continued

TOP VIEW

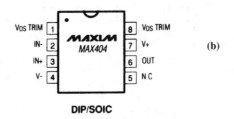

(b)

DIP/SOIC

ELECTRICAL CHARACTERISTICS
(V+ = +5V, V- = -5V, -3V < V$_{IN}$ < +3V, R$_L$ = 100Ω, C$_L$ = 15pF, unless otherwise noted.)

PARAMETER	SYMBOL	CONDITIONS	T$_A$ = +25°C			T$_A$ = T$_{AMIN}$ to T$_{AMAX}$			UNITS
			MIN	TYP	MAX	MIN	TYP	MAX	
Input Offset Voltage	V$_{OS}$	T$_A$ = +25°C		±1	+8				mV
		0°C to +70°C						±10	
		-40°C to +85°C						±12	
Input Offset Voltage Tempco	Δ V$_{OS}$/Δ T			±20					mV/°C
Input Bias Current (Note 1)	I$_{IN+}$, I$_{IN-}$	T$_A$ = +25°C		±1	±3				μA
		0°C to +70°C						æ5	
		-40°C to +85°C						±8	
Input Resistance (Common Mode)	R$_{IN}$	Gain = 1V/V DC		1					MΩ
Input Capacitance	C$_{IN}$			3					pF
Input Common Mode Voltage Range	V$_{IN}$		±3	±3.5		±3			V
Open Loop Voltage Gain	A$_{VOL}$	V$_{OUT}$ = 6V$_{P-P}$	54	66		54			dB
Min. Output Voltage Swing	V$_{OUT}$	R$_{LOAD}$ = 100Ω	±3.0	±3.5		±3.0			V
		R$_{LOAD}$ = 50Ω	æ2.5	±3.0		±2.5			
Output Current Continuous	I$_{OUT}$	R$_{LOAD}$ = 50Ω	æ50			±50			mA
Short Circuit		R$_{LOAD}$ = 0Ω		æ90					
Output Resistance	R$_{OUT}$	Gain = 2V/V		30					mΩ
Power Supply Rejection Ratio	PSRR	|V+, V-| = 4.75 to 5.25V	40	50		40			dB
Common Mode Rejection Ratio	CMRR	V$_{IN}$ = -3V to +3V	60	70		60			dB
Supply Current	I+, I-	T$_A$ = +25°C	25	30	35				mA
		0°C to +70°C				20		45	
		-40°C to +85°C				15		50	

PARAMETER	SYMBOL	CONDITIONS	T$_A$ = +25°C			T$_A$ = T$_{AMIN}$ to T$_{AMAX}$			UNITS
			MIN	TYP	MAX	MIN	TYP	MAX	
AC SPECIFICATIONS									
Gain Bandwidth	GBW	Gain = 10V/V		80					MHz
Slew-Rate	SR	3V step		500					V/μs
Full-Power Bandwidth		V$_{OUT}$ = 6V$_{P-P}$		13					MHz
Closed Loop Bandwidth	BW	Gain = 2V/V		60					MHz
Differential Phase (Note 2)	DP	Gain = 2V/V		0.01					deg
Differential Gain (Note 2)	DG	Gain = 2V/V		0.05					%
Settling Time to 0.1%	ts	Gain = -1V/V, 3V step		70					ns

Note 1: V$_{IN}$ = 0V DC
Note 2: Input test signal: 3.58MHz sine wave of amplitude superimposed on a linear ramp (0 to 100 IRE). 140 IRE = 1.0V.

(c)

Video circuits

Video operational amplifier (80-MHz gain bandwidth)

Fig. 12-15 The circuit in the illustration shows a MAX404 connected as a video cable driver. Figures 12-15B and C show the pin configuration and characteristics, respectively. Use 0.1-µF capacitors for the supply bypass. MAXIM NEW RELEASES DATA BOOK, 1992, PP. 8-25, 26, 27.

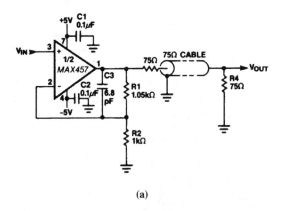

(a)

GAIN (V/V)	f-3dB (MHz)	R1 (Ω)	R2 (Ω)	R_load (Ω)
1	70	39	1000	75
2	50	1050	1000	150
5	40	4170	1000	390
10	25	9420	1000	750

(b)

Top View

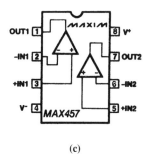

(c)

ELECTRICAL CHARACTERISTICS

(V⁺ = +5V, V⁻ = -5V, -2V ≤ V$_{IN}$ ≤ +2V, Output Load Resistor = 150Ω, T$_A$ = +25°C, unless otherwise noted.)

PARAMETER	SYMBOL	CONDITIONS	MIN	TYP	MAX	UNITS
Input Voltage Range	V$_{IN}$	Over Temperature Range	-2		+2	V
Input Offset Voltage	V$_{OS}$		-5	±2	+5	mV
Offset Voltage Drift	dV$_{OS}$/dT			20	100	μV/°C
Input Bias Current	I$_B$	T$_A$ = +25°C T$_A$ = +70°C T$_A$ = +85°C		0.1 5 15	1 40 100	nA
Input Resistance	R$_{IN}$	T$_A$ = +25°C		10		GΩ
Input Capacitance	C$_{IN}$	Plastic Package		4		pF
Open Loop Voltage Gain	A$_{VOL}$	R$_L$ = 1000Ω R$_L$ = 150Ω R$_L$ = 75Ω	200 45 25	300 65 35		V/V
Open Loop Gain Drift Temperature Coefficient	dA$_{VOL}$/dT	R$_L$ = 150Ω		-0.6		%/°C
Common Mode Rejection Ratio	CMRR	-2V ≤ V$_{IN}$ ≤ +2V	54	66		dB
Power Supply Rejection Ratio	PSRR	±4.5V to ±5.5V	54	66		dB
Slew Rate	SR	(Note 1)	150	300		V/μs
-3dB Bandwidth	GBW1	A$_V$ = 0dB, R$_L$ = 75Ω (Note 1)	50	70		MHz
-3dB Bandwidth	GBW2	A$_V$ = 6dB, R$_L$ = 150Ω (Note 1)	35	50		MHz
Differential Phase Error	DP	(Notes 1, 2)		0.2		deg
Differential Gain Error	DG	(Notes 1, 2)		0.5		%
Settling Time to 1%	t$_S$	R$_L$ = 150Ω, A$_V$ = 6dB		50		ns
Output Impedance	R$_{OUT}$	f = 100kHz, A$_V$ = 0dB		2		Ω
Full Scale Output Current	I$_{OUT}$	R$_L$ = 150Ω	±15	±20		mA
Output Voltage Swing	V$_{OUT}$	R$_L$ = 150Ω	±2.1	±2.5		V
Input Noise, DC to 50MHz	V$_N$	(Note 1)		0.15	0.5	mV$_{RMS}$
Isolation Between Amplifiers	ISOL	f = 5MHz (Note 1)	60	72		dB
Operating Supply Voltage	V⁺, V⁻		±4.5		±5.5	V
Supply Current	I$_S$	T$_A$ = +25°C T$_A$ = +85°C Both Amplifiers	30 34	35 39	42 50	mA

Note 1: Guaranteed by design.
Note 2: Input test signal: 3.58MHz sine wave of amplitude 40 IRE superimposed on a linear ramp (0 to 100 IRE). The amplifier is operated at a gain of 2V/V while driving a 150Ω load. 140 IRE = 1.0V.

(d)

Dual CMOS video amplifier

Fig. 12-16   The circuit in the illustration shows one section of a MAX457 connected as a video cable driver. Figures 12-16B, C, and D show resistor selection for various gains, pin configurations, and characteristics, respectively. MAXIM NEW RELEASES DATA BOOK, 1992, PP. 8-47, 48, 49.

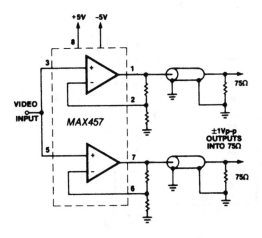

MAX457 Dual Video 70MHz Distribution Amplifier

Dual video 70-MHz distribution amplifier

Fig. 12-17 The circuit in the illustration shows both sections of a MAX457 connected as a video distribution system. See Fig. 12-16 for component values. MAXIM NEW RELEASES DATA BOOK, 1992, P. 8-47.

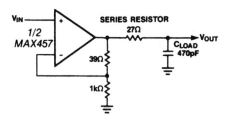

Isolating a capacitive load

Fig. 12-18 The circuit in the illustration shows how a capacitive load can be isolated, using the amplifier of Figs. 12-16 and 12-17. Such isolation is generally not required if the load is less than 10 pF. MAXIM NEW RELEASES DATA BOOK, 1992, P. 8-50.

$$\frac{12}{577}$$

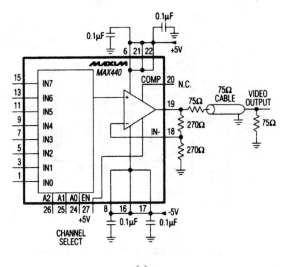

(a)

	MAX440				MAX441		
EN	A2	A1	A0	SELECTED CHANNEL	A1	A0	SELECTED CHANNEL
0	X	X	X	High-Z-Output	0	0	0
1	0	0	0	0	0	1	1
1	0	0	1	1	1	0	2
1	0	1	0	2	1	1	3
1	0	1	1	3			
1	1	0	0	4			
1	1	0	1	5			
1	1	1	0	6			
1	1	1	1	7			

(b)

Video circuits

ELECTRICAL CHARACTERISTICS

(V+ = 5V, V- = -5V, V$_{NS}$ = -5V, R$_L$ = 150Ω, T$_A$ = +25°C, unless otherwise noted.)

PARAMETER	SYMBOL	CONDITIONS		MIN	TYP	MAX	UNITS
DC PERFORMANCE							
Input Voltage Range	V$_{IN}$	T$_A$ = T$_{MIN}$ to T$_{MAX}$		-2		2	V
Input Offset Voltage (All Channels)	V$_{OS}$	T$_A$ = +25°C			±2.5	±10	mV
		0°C to +70°C				±10	
		-40°C to +85°C				±15	
		-55°C to +125°C				±20	
Input Bias Current (Channel On)	I$_B$	V$_{IN}$ = 0V	T$_A$ = +25°C		±1	±2	µA
			0°C to +70°C			±5	
			-40°C to +85°C			±5	
			-55°C to +125°C			±20	
Input Leakage Current (Channel Off)	I$_{LKG}$	V$_{IN}$ = 0V	T$_A$ = +25°C		±0.5	±50	nA
			T$_A$ = T$_{MIN}$ to T$_{MAX}$			±1	µA
Input Resistance (Channel On)	R$_{IN}$	-2V ≤ V$_{CM}$ ≤ 2V	T$_A$ = +25°C	0.5	2		MΩ
			T$_A$ = T$_{MIN}$ to T$_{MAX}$	0.2			
Input Capacitance	C$_{IN}$	Channel on or off			4		pF
DC Output Resistance	R$_{OUT}$	A$_V$ = 0dB			25		mΩ
Disabled Output Resistance	R$_{OUTdis}$	MAX440 only, EN = 0V			130		kΩ
Disabled Output Capacitance	C$_{OUTdis}$	MAX440 only, EN = 0V			15		pF
Open-Loop Voltage Gain	A$_{VOL}$	R$_L$ = 75Ω, -2V ≤ V$_{OUT}$ ≤ +2V	T$_A$ = +25°C	50	60		dB
			T$_A$ = T$_{MIN}$ to T$_{MAX}$	46			
Common-Mode Rejection Ratio	CMRR	-2V ≤ V$_{IN}$ ≤ +2V	T$_A$ = +25°C	46	50		dB
			T$_A$ = T$_{MIN}$ to T$_{MAX}$	40			
Power-Supply Rejection Ratio	PSRR	±4.75V to ±5.25V	T$_A$ = +25°C	54	80		dB
			T$_A$ = T$_{MIN}$ to T$_{MAX}$	54			
Output Voltage Swing	V$_{OUT}$	T$_A$ = +25°C		±3			V
		T$_A$ = T$_{MIN}$ to T$_{MAX}$		±2			

(c)

12

Video circuits titles and descriptions

ELECTRICAL CHARACTERISTICS (continued)

(V+ = 5V, V- = -5V, V_{NS} = -5V, R_L = 150Ω, T_A = +25°C, unless otherwise noted.)

PARAMETER	SYMBOL	CONDITIONS			MIN	TYP	MAX	UNITS
DYNAMIC PERFORMANCE								
-3dB Bandwidth	BW1	A_V = 0dB, COMP = GND, R_L = 75Ω				160		MHz
	BW2	A_V = 6dB, COMP = OPEN, R_L = 150Ω				110		
Slew Rate	SR1	A_V = 0dB, COMP = GND, R_L = 75Ω				250		V/µs
	SR2	A_V = 6dB, COMP = OPEN, R_L = 150Ω				370		
Differential Phase Error (Note 1)	DP	V_{NS} = -2.5V to -5V, COMP = OPEN, A_V = 6dB, R_L = 150Ω				0.03		deg
Differential Gain Error (Note 1)	DG	V_{NS} = -2.5V to -5V, COMP = OPEN, A_V = 6dB, R_L = 150Ω				0.04		%
Settling Time	t_S	To 0.1% of final value, A_V = 6dB, COMP = OPEN, 1V step input				65		ns
Adjacent Channel Crosstalk (Note 2)	X_{TALK}	f = 10MHz, R_S = 75Ω, A_V = 0dB	MAX440			-66		dB
			MAX441			-70		
Non-Adjacent Channel Crosstalk (Note 2)	X_{TALK}	f = 10MHz, R_S = 75Ω, A_V = 0dB				-77		dB
Feedthrough with Amplifier Disabled (Note 2)	FT	MAX440 only, f = 10MHz, A_V = 0dB	CH0 -CH6 driven			-71		dB
			CH0 -CH7 driven			-63		
Input Noise-Voltage Density	e_n	f = 10kHz				12		nV/√Hz
POWER-SUPPLY REQUIREMENTS								
Operating Supply-Voltage Range	V_S				±4.75		±5.25	V
Positve Supply Current	I_{CC}	V_{IN} = 0V	T_A = +25°C		33	40	50	mA
			0°C to +70°C		30		52	
			-40°C to +85°C		27		54	
			-55°C to +125°C		27		54	
Negative Supply Current	I_{EE}	V_{IN} = 0V	T_A = +25°C		24	30	40	mA
			0°C to +70°C		20		42	
			-40°C to +85°C		17		44	
			-55°C to +125°C		17		44	
SWITCHING CHARACTERISTICS (see Figure 10)								
Logic Low Threshold	V_{IL}	T_A = T_{MIN} to T_{MAX}					0.8	V
Logic High Threshold	V_{IH}	T_A = T_{MIN} to T_{MAX}			2.4			V
Address Setup Time (Note 3)	t_{AS}						10	ns
Address Hold Time (Note 3)	t_{AH}						10	ns
Address Propagation Delay	t_{APD}					20		ns
Latch Propagation Delay	t_{LPD}					20		ns
Channel Switching Time (Note 4)	t_{SW}	V_{NS} = -2.5V				15		ns
		V_{NS} = -5V				25		ns
Enable Propagation Delay	t_{ENPD}	MAX440 only				15		ns
Output Disable Time	t_{DA}	MAX440 only				10		ns
Output Enable Time	t_{EN}	MAX440 only				40		ns
Switching Transient (Note 5)		R_L = 75Ω	V_{NS} = -2.5V			100		mV$_{p-p}$
			V_{NS} = -5V			800		

Note 1: Input test signal: 3.58MHz sine wave of amplitude 40IRE superimposed on a linear ramp (0IRE to 100IRE). IRE is a unit of video signal amplitude developed by the International Radio Engineers. 140IRE = 1.0V.
Note 2: See Figure 9, *Dynamic Test Circuits*.
Note 3: Guaranteed by design.
Note 4: Channel switching time specified for switching between 2 grounded input channels; does not include signal rise/fall times for switching between channels with different input voltages.
Note 5: Measured while switching between 2 grounded channels.

(c) Continued

Continued

PIN		NAME	FUNCTION
MAX440	**MAX441**		
1	1	IN0	Analog Input, Channel 0
2	2	LEVEL/$\overline{\text{EDGE}}$	Digital input that controls the operation of LATCH input as follows: When LEVEL/$\overline{\text{EDGE}}$ = 0V, input data is latched on the rising edge of the LATCH input (edge triggered); when LEVEL/$\overline{\text{EDGE}}$ = 5V, input data is latched when LATCH = 5V (level triggered). Hardwire to +5V or GND for improved crosstalk.
3	4	IN1	Analog Input, Channel 1
4, 10, 12, 14	3, 9	GND	Ground
5	7	IN2	Analog Input, Channel 2
6, 21, 22	5, 16	V+	Positive Power Supply, +5V
7	10	IN3	Analog Input, Channel 3
8, 16, 17	6, 12	V-	Negative Power Supply, -5V
9	–	IN4	Analog Input, Channel 4
11	–	IN5	Analog Input, Channel 5
13	–	IN6	Analog Input, Channel 6
15	–	IN7	Analog Input, Channel 7
18	13	IN-	Amplifier Inverting Input
19	14	V_{OUT}	Amplifier Output
20	15	COMP	Amplifier Compensation Input. Ground for unity-gain application, or use to adjust compensation for higher-gain applications (see text).
23	17	LATCH	Latch control for digital inputs. If LEVEL/$\overline{\text{EDGE}}$ = 0V, data is latched on the rising edge of LATCH. If LEVEL/$\overline{\text{EDGE}}$ = 5V, the input register is transparent when LATCH = 0V and latched when LATCH = 5V.
24	18	A0	Channel Address Input 0, LSB
25	19	A1	Channel Address Input 1, MSB for MAX441
26	–	A2	Channel Address Input 2, MSB
27	–	EN	Amplifier Output Enable control, active high. This is internally latched, along with A0 to A2.
28	20	V_{NS}	Normally -5V, minimize switching time and transients by tying this pin to -2.5V. Analog input voltage must never be more negative than the voltage on this pin.
–	8, 11	N.C.	No Internal Connection

(d)

Video circuits titles and descriptions

Continued

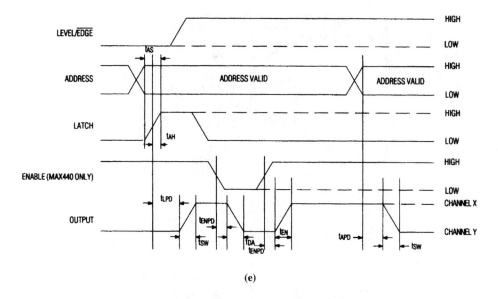

(e)

Closed-Loop Gain		COMP Pin State
V/V	dB	
$1 \leq A_{VCL} \leq 2$	$0 \leq A_{VCL} \leq 6$	GND
$2 \leq A_{VCL} \leq 10$	$6 \leq A_{VCL} \leq 20$	OPEN
$A_{VCL} \geq 10$	$A_{VCL} \geq 20$	V_{OUT}

(f)

High-speed video multiplexer/amplifier

Fig. 12-19 The circuit in the illustration shows a MAX440 multiplexer/amplifier connected to form an eight-channel coaxial-cable driver. Figures 12-19B, C, D, E, and F show the channel-selection, characteristics, pin descriptions, timing, and compensation-pin state, respectively. The test circuits for the characteristics given in Fig. 12-19C are shown in Figs. 12-20 through 12-23. MAXIM NEW RELEASES DATA BOOK, 1993, PP. 8-6, 7, 10, 11, 12, 13, 16.

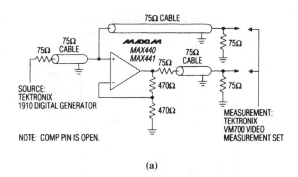

(a)

Vns VOLTAGE (V)	DIFFERENTIAL GAIN ERROR (%)	DIFFERENTIAL PHASE ERROR (°)
-1.0	0.05	0.04
-1.5	0.04	0.04
-2.0	0.04	0.03
-2.5	0.04	0.03
-5.0	0.04	0.03

(b)

Differential-gain and phase-error test circuit

Fig. 12-20 The typical values for this test are given in Fig. 12-20B. MAXIM NEW RELEASES DATA BOOK, 1993, P. 8-12.

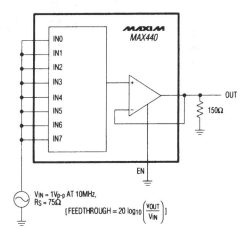

Disabled amplifier feedthrough test circuit

Fig. 12-21 The typical values for this test are –63 dB to –71 dB (Fig. 12-19C). MAXIM NEW RELEASES DATA BOOK, 1993, P. 8-14.

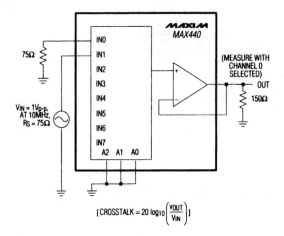

$$[\text{CROSSTALK} = 20 \log_{10} \left(\frac{V_{OUT}}{V_{IN}} \right)]$$

Adjacent-channel crosstalk test circuit (MAX440)

Fig. 12-22 The typical values for this test are –66 dB to –70 dB (Fig. 12-19C). MAXIM NEW RELEASES DATA BOOK, 1993, P. 8-14.

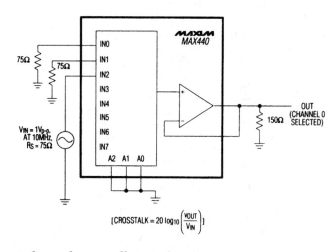

$$[\text{CROSSTALK} = 20 \log_{10} \left(\frac{V_{OUT}}{V_{IN}} \right)]$$

Nonadjacent-channel crosstalk test circuit

Fig. 12-23 The typical value for this test is –77 dB (Fig. 12-19C). MAXIM NEW RELEASES DATA BOOK, 1993, P. 8-14.

Video circuits

(a)

Video circuits titles and descriptions

A3	A2	A1	A0	SELECTED CHANNEL
0	0	0	0	0
0	0	0	1	1
0	0	1	0	2
0	0	1	1	3
0	1	0	0	4
0	1	0	1	5
0	1	1	0	6
0	1	1	1	7
1	0	0	0	8
1	0	0	1	9
1	0	1	0	10
1	0	1	1	11
1	1	0	0	12
1	1	0	1	13
1	1	1	0	14
1	1	1	1	15

(b)

Video multiplexer (1-of-16)

Fig. 12-24 The circuit in the illustration shows two MAX440 multiplexers connected in cascade to form a 1-of-16 video multiplexer. Figure 12-24B shows the channel selection. MAXIM NEW RELEASES DATA BOOK, 1993, P. 8-15.

Video circuits

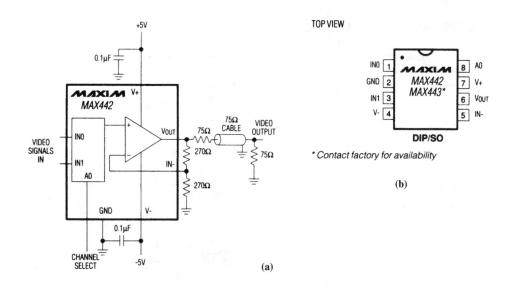

TOP VIEW

* Contact factory for availability

(b)

(a)

ELECTRICAL CHARACTERISTICS

(V+ = 5V, V- = -5V, R_L = 150Ω, T_A = T_{MIN} to T_{MAX}, unless otherwise noted.)

PARAMETER	SYMBOL	CONDITIONS		MIN	TYP	MAX	UNITS
DC PERFORMANCE							
Input Voltage Range	V_{IN}			-2		2	V
Input Offset Voltage (All Channels)	V_{OS}	T_A = +25°C			±1.5	±7.0	mV
		MAX442C				±10	
		MAX442E				±12	
Offset Matching (V_{OS0} - V_{OS1})		T_A = +25°C			±1.0	±2.5	mV
		T_A = T_{MIN} to T_{MAX}				±5.0	
Input Bias Current (Channel On)	I_B	V_{IN} = 0V	T_A = +25°C		±1	±2	μA
			T_A = T_{MIN} to T_{MAX}			±5	
Input Leakage Current (Channel Off)	I_{LKG}	V_{IN} = 0V	T_A = +25°C		±0.5	±50	nA
			T_A = T_{MIN} to T_{MAX}			±1	μA
Input Resistance (Channel On) (Note 1)	R_{IN}	-2V ≤ V_{CM} ≤ 2V	T_A = +25°C	0.5	2.0		MΩ
			T_A = T_{MIN} to T_{MAX}	0.2			
Input Capacitance	C_{IN}	Channel on or off			4		pF
DC Output Resistance	R_{OUT}	A_V = 0dB			25		mΩ
		A_V = 6dB			50		
Open-Loop Voltage Gain	A_{VOL}	R_L = 75Ω, -2V ≤ V_{OUT} ≤ +2V	T_A = +25°C	50	60		dB
			T_A = T_{MIN} to T_{MAX}	46			
Common-Mode Rejection Ratio	CMRR	-2V ≤ V_{IN} ≤ +2V	T_A = +25°C	46	50		dB
			T_A = T_{MIN} to T_{MAX}	46			
Power-Supply Rejection Ratio	PSRR	±4.75V to ±5.25V	T_A = +25°C	54	80		dB
			T_A = T_{MIN} to T_{MAX}	54			
Output Voltage Swing	V_{OUT}	R_L = 75Ω	T_A = +25°C	±2.5	±3.0		V
			T_A = T_{MIN} to T_{MAX}	±2.0			

(c)

Continued

ELECTRICAL CHARACTERISTICS (continued)

(V+ = 5V, V- = -5V, R$_L$ = 150Ω, T$_A$ = T$_{MIN}$ to T$_{MAX}$, unless otherwise noted.)

PARAMETER	SYMBOL	CONDITIONS		MIN	TYP	MAX	UNITS
DYNAMIC PERFORMANCE							
-3dB Bandwidth	BW	A$_V$ = 0dB, R$_L$ = 100Ω			140		MHz
Slew Rate	SR1				250		V/μs
Differential Phase Error	DP				0.09		deg
Differential Gain Error	DG				0.07		%
Settling Time	t$_s$	To 0.1% of final value, A$_V$ = 0dB, R$_L$ = 150Ω, 2V step input			50		ns
Crosstalk	X$_{TALK}$	f = 10MHz, R$_S$ = 75Ω, A$_V$ = 0dB			76		dB
Input Noise-Voltage Density	e$_n$	f = 10kHz			12		nV/√Hz
POWER REQUIREMENTS							
Operating Supply-Voltage Range	V$_S$			±4.75		±5.25	V
Positve Supply Current	I$_{CC}$	V$_{IN}$ = 0V	T$_A$ = +25°C	25	30	35	mA
			MAX442C	22		38	
			MAX442E	19		41	
Negative Supply Current	I$_{EE}$	V$_{IN}$ = 0V	T$_A$ = +25°C	23	28	35	mA
			MAX442C	20		38	
			MAX442E	17		41	
SWITCHING CHARACTERISTICS							
Logic Low Threshold	V$_{IL}$					0.8	V
Logic High Threshold	V$_{IH}$			2.4			V
Address Propagation Delay	t$_{APD}$				24		ns
Channel Switching Time	t$_{SW}$				36		ns

Note 1: Incremental resistance for a common-mode voltage between ±2V.

(c) Continued

PIN	NAME	FUNCTION
1	IN0	Analog Input, channel 0
2	GND	Ground
3	IN1	Analog Input, channel 1
4	V-	Negative Power Supply, -5V
5	IN-	Amplifier Inverting Input
6	V$_{OUT}$	Amplifier Output
7	V+	Positive Power Supply, +5V
8	A0	Channel Address Input: A0 = logic 0 selects channel 0, A0 = logic 1 selects channel 1

(d)

Video multiplexer/amplifier (140-MHz, two-channel)

Fig. 12-25 The circuit in the illustration shows a MAX442 connected as a two-channel video cable multiplexer. Figures 12-25B, C, and D show the pin configuration, characteristics, and pin descriptions, respectively. MAXIM NEW RELEASES DATA BOOK, 1993, PP. 8-17, 18, 19.

Video circuits

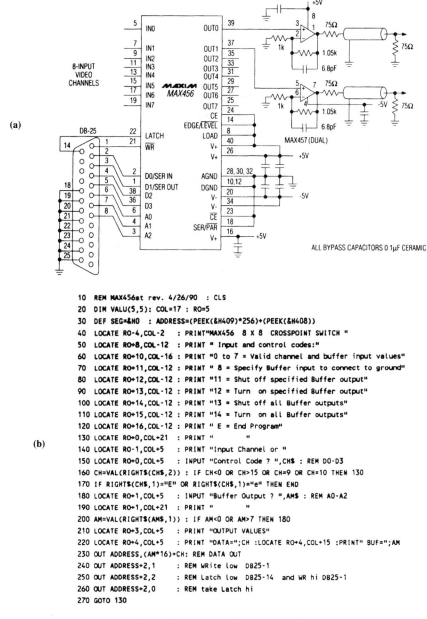

(a)

(b)

```
10  REM MAX456st rev. 4/26/90  : CLS
20  DIM VALU(5,5): COL=17 : RO=5
30  DEF SEG=&H0  : ADDRESS=(PEEK(&H409)*256)+(PEEK(&H408))
40  LOCATE RO-4,COL-2   : PRINT"MAX456 8 X 8  CROSSPOINT SWITCH "
50  LOCATE RO+8,COL-12  : PRINT " Input and control codes:"
60  LOCATE RO+10,COL-16 : PRINT "0 to 7 = Valid channel and buffer input values"
70  LOCATE RO+11,COL-12 : PRINT " 8 = Specify Buffer input to connect to ground"
80  LOCATE RO+12,COL-12 : PRINT "11 = Shut off specified Buffer output"
90  LOCATE RO+13,COL-12 : PRINT "12 = Turn  on specified Buffer output"
100 LOCATE RO+14,COL-12 : PRINT "13 = Shut off all Buffer outputs"
110 LOCATE RO+15,COL-12 : PRINT "14 = Turn  on all Buffer outputs"
120 LOCATE RO+16,COL-12 : PRINT " E = End Program"
130 LOCATE RO+0,COL+21  : PRINT "      "
140 LOCATE RO-1,COL+5   : PRINT "Input Channel or "
150 LOCATE RO+0,COL+5   : INPUT "Control Code ? ",CH$ : REM D0-D3
160 CH=VAL(RIGHT$(CH$,2)) : IF CH<0 OR CH>15 OR CH=9 OR CH=10 THEN 130
170 IF RIGHT$(CH$,1)="E" OR RIGHT$(CH$,1)="e" THEN END
180 LOCATE RO+1,COL+5   : INPUT "Buffer Output ? ",AM$ : REM A0-A2
190 LOCATE RO+1,COL+21  : PRINT "      "
200 AM=VAL(RIGHT$(AM$,1)) : IF AM<0 OR AM>7 THEN 180
210 LOCATE RO+3,COL+5   : PRINT "OUTPUT VALUES"
220 LOCATE RO+4,COL+5   : PRINT "DATA=";CH :LOCATE RO+4,COL+15 :PRINT" BUF=";AM
230 OUT ADDRESS,(AM*16)+CH: REM DATA OUT
240 OUT ADDRESS+2,1   : REM WRite low  DB25-1
250 OUT ADDRESS+2,2   : REM Latch low  DB25-14  and WR hi DB25-1
260 OUT ADDRESS+2,0   : REM take Latch hi
270 GOTO 130
```

Video cross-point switch with PC interface (8 × 8)

Fig. 12-26 The circuit in the illustration shows a MAX456 digital switch connected to interface a PC. The input/output information is presented to the chip at A2-A0 and D3-D0 by a parallel printer port. The data bits are stored in the 1st-rank registers on the rising edge of $\overline{\text{WR}}$. When the LATCH line goes high, the switch configuration is loaded into the second-rank registers, and all eight outputs enter simultaneously. Figure 12-26B shows a BASIC program for programming data into the MAX456 from an IBM PC or compatible. MAXIM NEW RELEASES DATA BOOK, 1993, PP. 8-28, 29.

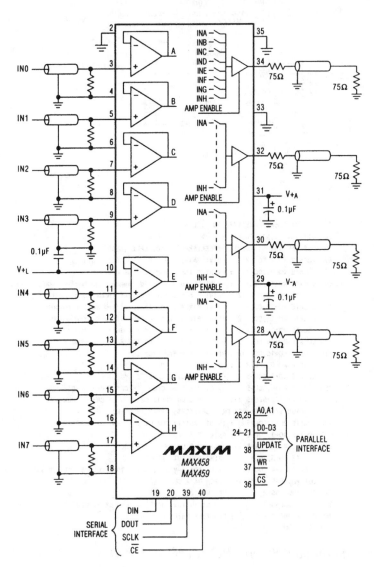

Video cross-point switch with buffers (8 × 4)

Fig. 12-27 The circuit in the illustration shows a MAX458/459 digital switch connected as a video cable (75 Ω) driver. Each channel has a 100-MHz unity-gain

Video circuits

bandwidth, and a 300 V/μs slew rate. Switching time is a typical 40 ns. Both 16-bit serial and 6-bit parallel address modes are available. MAXIM NEW RELEASES DATA BOOK, 1994, P. 8-22.

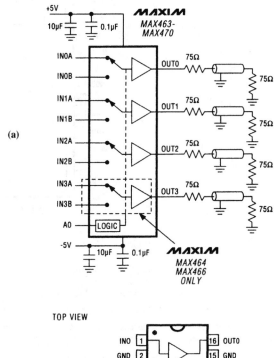

(a)

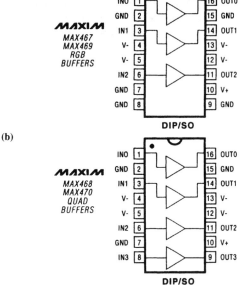

(b)

Pin Configurations and Typical Operating Circuit continued on next pages.

Video circuits titles and descriptions

Continued

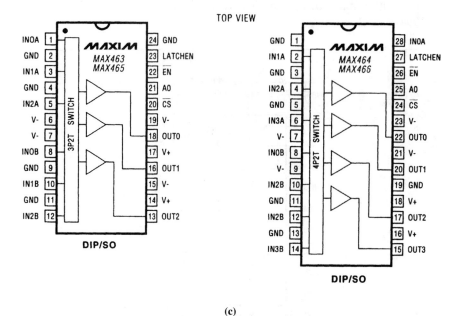

TOP VIEW

(c)

Two-channel, triple, and quad RGB video switches

Fig. 12-28 The circuit in the illustration shows a MAX463-470 digital switch with buffers connected as a four-channel RGB interface and driver. Figures 12-28B and C show the pin configurations. This series of switches has 100-MHz unity-gain bandwidth, 90-MHz bandwidth with a 2 V/V gain, 0.02%/0.06% differential gain/phase error, 300 V/μs slew rate (2 V/V gain), and a typical 20-ns channel switching time. MAXIM NEW RELEASES DATA BOOK, 1994, PP. 8-23, 24, 25.

12

592

Video circuits

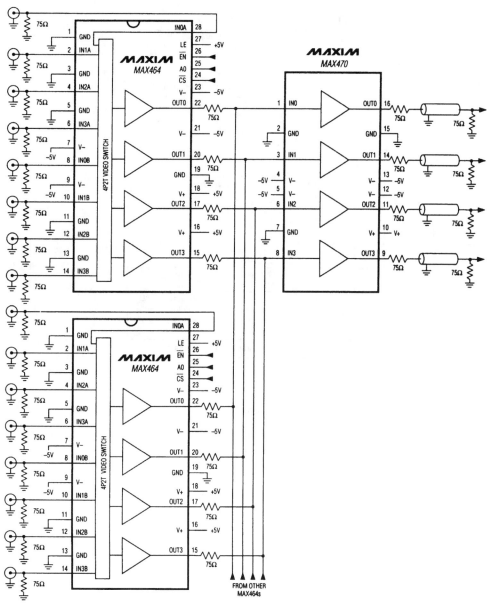

(a)

Video circuits titles and descriptions

Continued

$\overline{CS}$	$\overline{EN}$	A0	FUNCTION
0	0	0	Enables amplifier outputs. Selects channel A.
0	0	1	Enables amplifier outputs. Selects channel B.
0	1	X	Disables amplifiers. Outputs high-Z.
1	X	X	Latches all input registers. Changes nothing.

(b)

$\overline{CS}$	$\overline{EN}$	A0	FUNCTION
0	0	0	Enables amplifier outputs. Selects channel A.
0	0	1	Enables amplifier outputs. Selects channel B.
0	1	0	Disables amplifiers. Outputs high-Z. A0 register = channel A
0	1	1	Disables amplifiers. Outputs high-Z. A0 register = channel B
1	0	X	Enables amplifier outputs, latches A0 register, programs outputs to output A or B, according to the setting of A0 at $\overline{CS}$'s last edge.
1	1	X	Disables amplifiers. Outputs high-Z.

(c)

High-order RGB and sync video multiplexer

Fig. 12-29 The circuit in the illustration shows two MAX464 switches (Fig. 12-28) connected in parallel to form an RGB and sync video multiplexer. Figure 12-29B shows the amplifier and channel-selection truth tables for LE (pin 27) connected to V+. Figure 12-29C shows the same information for LE connected to ground. The test circuits for the switch ICs are given in Figs. 12-30 through 12-33. MAXIM NEW RELEASES DATA BOOK, 1995, PP. 8-33, 35.

12

594

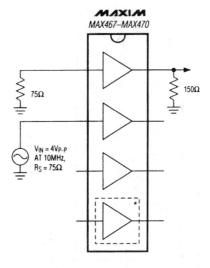

Adjacent-channel crosstalk test circuit (MAX467-470)

Fig. 12-30 The typical value for this test is 60 dB. MAXIM NEW RELEASES DATA BOOK, 1995, P. 8-34.

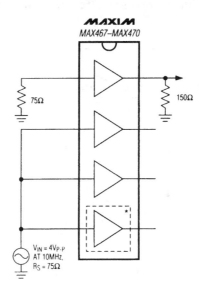

All-hostile crosstalk test circuit (MAX467-470)

Fig. 12-31 The typical value for this test is 50 dB. MAXIM NEW RELEASES DATA BOOK, 1995, P. 8-34.

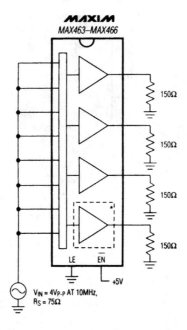

All-hostile OFF isolation test circuit (MAX463-466)

Fig. 12-32   The typical value for this test is 70 dB. MAXIM NEW RELEASES DATA BOOK, 1995, P. 8-34.

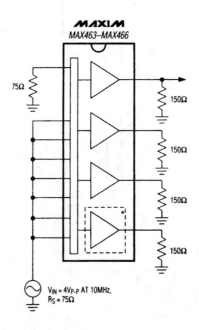

Video circuits

All-hostile crosstalk test circuit (MAX463-466)

Fig. 12-33 The typical value for this test is 50 dB. MAXIM NEW RELEASES DATA BOOK, 1995, P. 8-34.

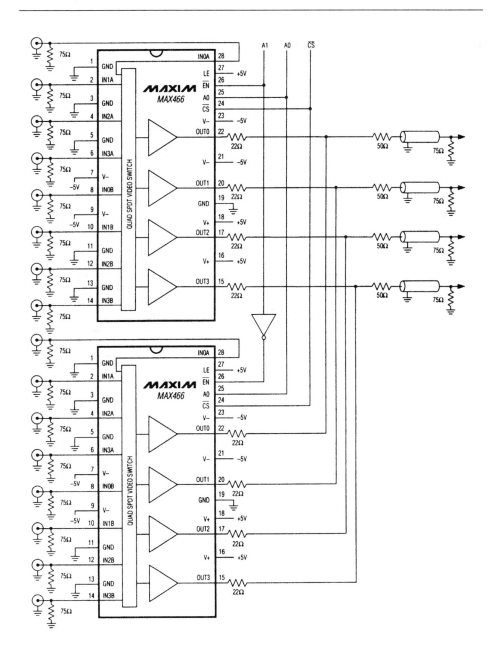

RGB and sync video multiplexer (1-of-4)

Fig. 12-34 The circuit in the illustration shows two MAX466 switches (Fig. 12-29) connected in parallel to form a 1-of-4 RGB and sync video multiplexer. Figure 12-29B shows the amplifier and channel-selection truth tables for LE (pin 27) connected to V+. Figure 12-29C shows the same information for LE connected to ground. The test circuits for the switch ICs are given in Figs. 12-30 through 12-33. MAXIM NEW RELEASES DATA BOOK, 1995, P. 8-36.

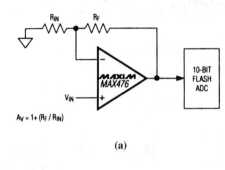

$A_V = 1 + (R_F / R_{IN})$

(a)

TOP VIEW

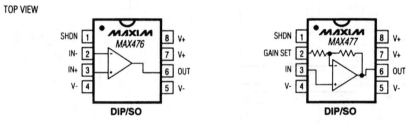

(b)

425-MHz high-speed amplifier and buffer

Fig. 12-35 The circuit in the illustration shows a MAX476 amplifier/buffer connected as an amplifier for a 10-bit flash ADC (Chapter 9). This IC has a 425-MHz small-signal bandwidth ($A_V = +1$), a 175-MHz full-power bandwidth ($A_V = +1$, $V_0 = 4 V_{p-p}$), a 2200 V/μs slew rate, 2-μA input bias current, 9-mA quiescent current, and 300-μA shutdown. Figure 12-35B shows the pin configurations. MAXIM NEW RELEASES DATA BOOK, 1995, PP. 8-39, 40.

=13=

Audio circuits

This chapter is devoted to audio-frequency circuits. Most of the circuits are a form of amplifier or buffer, so the test/troubleshooting procedures described in Chapter 1 apply to the circuits in this chapter. Where there are differences in test/troubleshooting, the recommended test circuits (and procedures where it is not obvious) are given for the ICs involved.

Audio circuit titles and descriptions

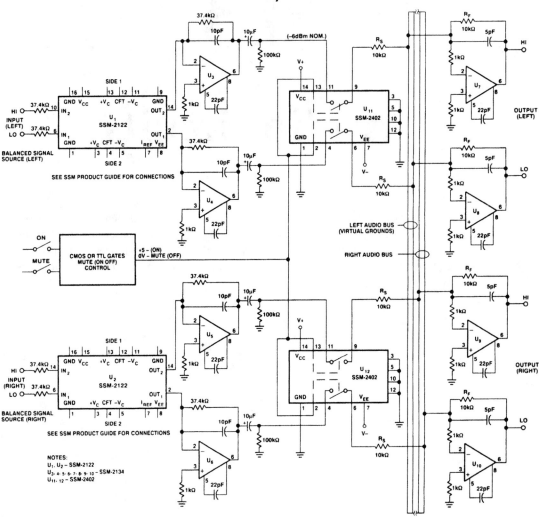

(a)

Continued

Max Input Level	+30dBu
Input Impedence, Balanced	75kΩ
Common-Mode Rejection (20Hz to 20kHz)	>70dB
Common-Mode Voltage Limit	±12V Peak
Max Output Level	+30dBu
Output Voltage Slew Rate	12V/µs
Frequency Response (±0.05dBu)	20Hz to 20kHz
Frequency Response (±0.5dBu)	10Hz to 50kHz
THD + Noise (20Hz to 20kHz, +8dBu)	0.005%
THD + Noise (20Hz to 20kHz, +24dBu)	0.03%
IMD (SMPTE 60Hz & 4kHz, 4:1, +24dBu)	0.02%
ON/MUTE Isolation (20Hz to 20kHz)	>85dB
S/N Ratio @ 0dB Gain	135dB

(b)

Balanced mute circuit for audio-mixing consoles

Fig. 13-1 The circuit in the illustration provides a channel-mute (on/off) function for an audio mixer, with negligible transient noise (as a result of signal switching) and low signal distortion over a wide dynamic range. Figure 13-1B shows the circuit performance specifications. Note that the *IMD* (intermodulation distortion) uses 60 Hz and 4 kHz, instead of the 7 kHz given in Chapter 1. ANALOG DEVICES, APPLICATIONS REFERENCE MANUAL, 1993, PP. 4-19, 20.

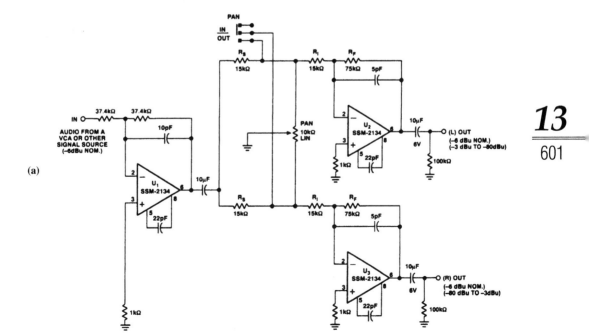

(a)

Audio circuit titles and descriptions

Continued

PAN Range, L ← C → R (L Out)	+3dB ← 0dB →	−80dB
PAN Range, R ← C → L (R Out)	+3dB ← 0dB →	−80dB
Max Input Level		+24dBu
Input Impedance, Balanced		37.4kΩ
Max Output Level (> 600Ω ±18V$_{DC}$ PS)		+24dBu
Headroom		30dB
Output Voltage Slew Rate		<6V/μs
Frequency Response (±0.05dB)	20 Hz to 20kHz	
Frequency Response (±0.5dB)	10 Hz to 50kHz	
THD + Noise (20Hz to 20kHz, +8dBu)		0.005%
THD + Noise (20Hz to 20kHz, +24dBu)		0.03%
IMD (SMPTE 60Hz & 4kHz, 4:1, +24dBu)		0.02%
S/N Ratio		130dB

(b)

Constant-power "Pan" control circuit for microphone audio mixing

Fig. 13-2 The circuit in the illustration provides a constant-power transient-free PAN control for use in audio-mixing consoles. The PAN IN/OUT switches do not introduce transient noise or interruptions in the audio when activated or deactivated. When panning, an accurate constant-power output is maintained between the sum of the two channels. This allows punching-in and punching-out of the PAN circuit while mixing down or on-the-air, without transient clicks or holes in the mix. Figure 13-2B shows the circuit performance specifications. ANALOG DEVICES, APPLICATIONS REFERENCE MANUAL, 1993, PP. 4-21, 22.

13

602

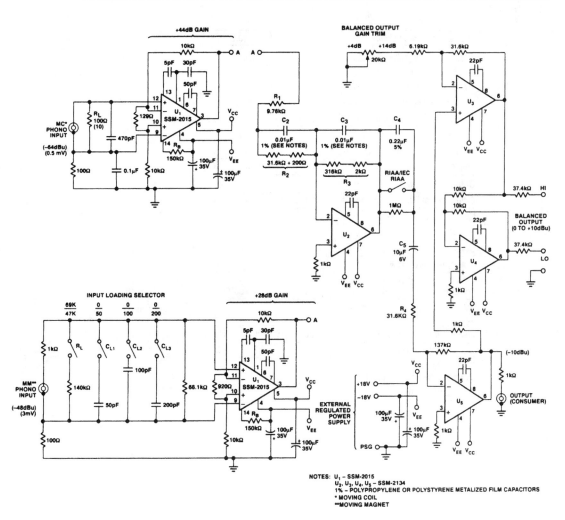

NOTES: U₁ – SSM-2015
U₂, U₃, U₄, U₅ – SSM-2134
1% – POLYPROPYLENE OR POLYSTYRENE METALIZED FILM CAPACITORS
* MOVING COIL
** MOVING MAGNET

(a)

Audio circuit titles and descriptions

Continued

MC Nominal Input Level	−64dBu (0.5mV)
MC Input Impedance	100Ω
MM Nominal Input Level	−48dBu (3.0mV)
MM Input Impedance, Resistive	69kΩ or 47kΩ
MM Input Impedance, Capacitive	50pF to 350pF
Common-Mode Rejection (20Hz to 20kHz)	>50dB
Common-Mode Voltage Limit	±10V Peak
Nominal Output Level, Balanced	+8dBu/dBm
Max Output Level, Balanced	+30dBu/dBm
Output Impedance, Balanced	70Ω
Gain Control Range, Balanced	0.0dBu to 10dBu/dBm
Nominal Output Level, Unbalanced	−10dBu
Max Output Level, Unbalanced	+24dBu
Output Impedance, Unbalanced	1,000Ω
Output Voltage Slew Rate	>6V/ μs
RIAA Reproduction Characteristics (20Hz to 20kHz)	±0.25dB
RIAA/IEC Reproduction Characteristic (2Hz to 20kHz)	±1.0dB
Wideband Frequency Response (±1.0dB)	0.0Hz to 70kHz
Signal-to-Noise Ratio (20Hz to 20kHz)	>90dB
THD + Noise (20Hz to 20kHz +8dBu, Any Output)	0.01%
IMD (SMPTE 60Hz & 4kHz, 4:1)	0.02%

(b)

13

604

Audio circuits

Continued

Frequency (Hz)	RIAA /IEC Relative Level (dB)	RIAA Relative Level (dB)
2.0	−0.2	
2.5	+1.8	
3.15	+3.7	
4.0	+5.7	
5.0	+7.6	
6.3	+9.4	
8.0	+11.2	
10.0	+12.8	
12.5	+14.1	
16.0	+15.4	
20.0	+16.3	+19.3
25.0	+16.8	+19.0
31.5	+17.0	+18.5
40.0	+16.8	+17.8
50.0	+16.3	+16.9
63.0	+15.4	+15.8
80.0	+14.2	+14.5
100	+12.9	+13.1
125	+11.5	+11.6
160	+9.7	+9.8
200	+8.2	+8.2
250	+6.7	+6.7
315	+5.2	+5.2
400	+3.8	+3.8
500	+2.6	+2.6
630	+0.8	+0.8
1,000	0.0	0.0
1,250	−0.8	−0.7
1,600	−1.6	−1.6
2,000	−2.6	−2.6
2,500	−3.7	−3.7
3,150	−5.0	−5.0
4,000	−6.6	−6.6
5,000	−8.2	−8.2
6,300	−10.0	−10.0
8,000	−11.9	−11.9
10,000	−13.7	−13.7
12,500	−15.6	−15.6
16,000	−17.7	−17.7
20,000	−19.6	−19.6

(c)

13

Audio circuit titles and descriptions

RIAA/IEC MC and MM phono preamplifier (high accuracy)

Fig. 13-3 The circuit in the illustration has both *MC* (moving coil) and *MM* (moving magnet, or variable reluctance) inputs, and selectable old *RIAA* (Recording Industries Association of America) equalization or RIAA/IEC (International Electro-Technical Commission) curves. Figure 13-3B shows both RIAA and RIAA/IEC playback characteristics. Figure 13-3C shows circuit performance specifications. ANALOG DEVICES, APPLICATIONS REFERENCE MANUAL, 1993, PP. 4-23, 26.

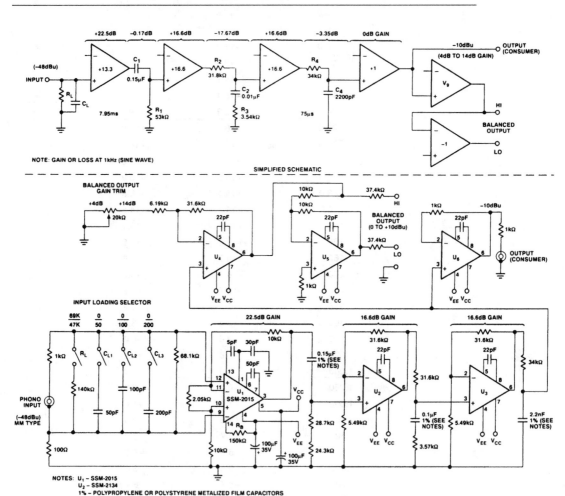

(a)

Continued

MM Nominal Input Level	−48dBu (3.0mV)
MM Input Impedance, Resistive	69kΩ or 47kΩ
MM Input Impedance, Capacitive	50pF to 350pF
Common-Mode Rejection (20Hz to 20kHz)	> 50 dB
Common-Mode Voltage Limit	±10V Peak
Max Output Level, Balanced	+30dBu/dBm
Nominal Output Level, Balanced	+8dBu/dBm
Output Impedance, Balanced	70Ω
Gain Control Range, Balanced	0.0dBu to 10dBu/dBm
Nominal Output Level, Unbalanced	−10dBu
Max Output Level, Unbalanced	+24dBm
Output Impedance, Unbalanced	1,000Ω
Output Voltage Slew Rate	>6V/μs
RIAA Reproduction Characteristic (20Hz to 20kHz)	±0.25dB
RIAA/IEC Reproduction Characteristic (2Hz to 20kHz)	±0.5dB
Wideband Frequency Response (±1.0dB)	0.0Hz to 70kHz
Signal-to-Noise Ratio (20Hz to 20kHz)	>90dB
THD + Noise (20Hz to 20kHz +8dBu, Any Output)	0.01%
IMD (SMPTE 60Hz & 4kHz, 4:1)	0.02%

(b)

RIAA/IEC MM phono preamplifier (passive multifilter)

Fig. 13-4 The circuit in the illustration is similar to that of Fig. 13-3, except the input is intended to be MM only, and passive filters are used. Figure 13-4B shows circuit performance specifications. ANALOG DEVICES, APPLICATIONS REFERENCE MANUAL, 1993, PP. 4-24, 27.

13

607

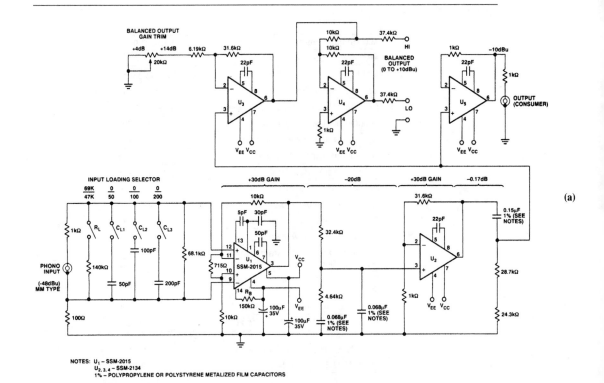

NOTES: U_1 – SSM-2015
$U_{2,3,4}$ – SSM-2134
1% – POLYPROPYLENE OR POLYSTYRENE METALIZED FILM CAPACITORS

(a)

MM Nominal Input Level	−48dBu (3.0mV)
MM Input Impedance, Resistive	69kΩ or 47kΩ
MM Input Impedance, Capacitive	50pF to 350pF
Common-Mode Rejection (20Hz to 20kHz)	> 50dB
Commom-Mode Voltage Limit	±10V Peak
Max Output Level, Balanced	+30dBu/dBm
Nominal Output Level, Balanced	+8dBu/dBm
Output Impedance, Balanced	70Ω
Gain Control Range, Balanced	0.0dBu to 10dBu/dBm
Nominal Output Level, Unbalanced	−10dBu
Max Output Level, Unbalanced	+24dBu
Output Impedance, Unbalanced	1,000Ω
Output Voltage Slew Rate	>6V/μs
RIAA Reproduction Characteristic (20Hz to 20kHz)	±0.5dB
RIAA/IEC Reproduction Characteristic (2Hz to 20kHz)	±1.0dB
Wideband Frequency Response (±1.0dB)	0.0Hz to 70kHz
Signal-to-Noise Ratio (20Hz to 20kHz)	>90dB
THD + Noise (20Hz to 20kHz, +8dBu, Any Output)	0.01%
IMD (SMPTE 60Hz & 4kHz, 4:1)	0.02%

(b)

Audio circuits

RIAA/IEC MM phono preamplifier (low cost)

Fig. 13-5 The circuit in the illustration is similar to that of Fig. 13-3, except that the input is MM only, and passive filters are used. Figure 13-5B shows circuit performance specifications. ANALOG DEVICES, APPLICATIONS REFERENCE MANUAL, 1993, PP. 4-25, 27.

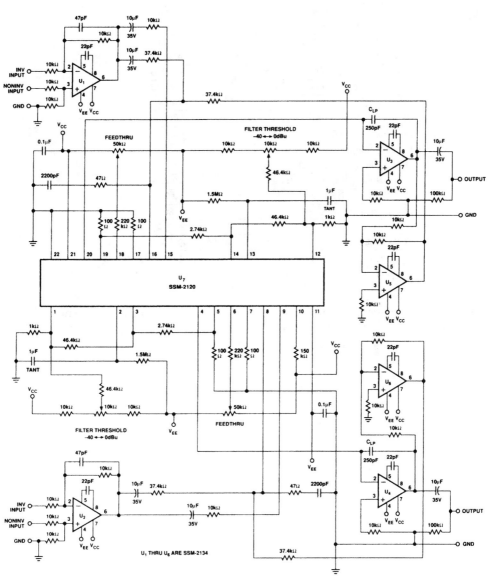

(a)

Continued

Nominal Input Voltage (−10dBu Out)	−10dBu
Headroom (−10dBu Out)	+30dBu
Input Voltage Range	−20dBu to +10dBu
Input Type/Impedance, Balanced	20kΩ
Input Type/Impedance, Unbalanced	10kΩ
Dynamic Noise Reduction Class	Dynamic Low-Pass
Filter Activate Time Constant (6dB)	350ms
Threshold Range (Level)	−40dBu to 0dBu
Filter Deactivate Time Constant	6ms
Signal Rectifier Type	Full Wave Averaging
Modulation Feedthrough, Trimmed	−100dB
Frequency Response (20Hz to 16kHz)	±1dB
Filter Type, Low-Pass	Single Pole, 6dB/Oct
Input 10dB Below Threshold Setting	f_{C1} = 3,800Hz
Input 20dB Below Threshold Setting	f_{C2} = 1,400Hz
Dynamic Range	
@ 0dB Gain (Ref. +22dBu)	106dB
THD + Noise (20Hz to 20kHz)	0.02%
IMD (SMPTE 60Hz & 4kHz, 4:1)	0.05%
Output Voltage (2kΩ Load)	+22dBu
Output Type	Unbalanced
Power Supply	±18V_{DC} Regulated

(b)

Two-channel dynamic-filter noise-reduction system

Fig. 13-6 The circuit in the illustration provides a dynamic noise reduction where the input-signal level and threshold-control determine the corner frequency of a lowpass filter. The SSM-2120 contains two class-A *VCAs* (voltage controlled amplifiers) that are used as the filter-control element and two wide dynamic range full-wave rectifiers with control amplifiers. Figure 13-6B shows the circuit performance specifications. ANALOG DEVICES, APPLICATIONS REFERENCE MANUAL, 1993, PP. 4-29, 30.

13

610

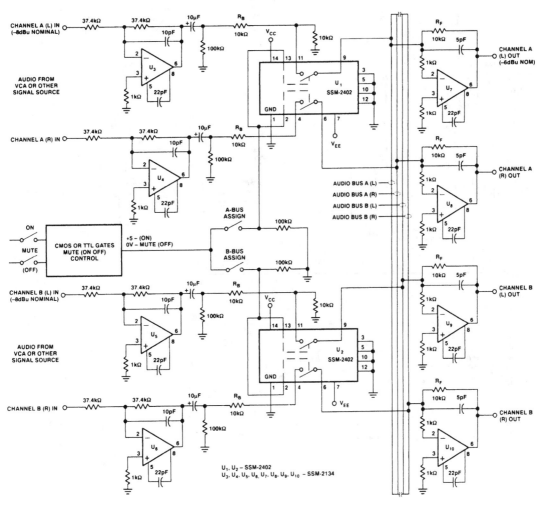

Unbalanced mute circuit for audio-mixing consoles

Fig. 13-7   The circuit in the illustration is similar to that of Fig. 13-1, except that the design is unbalanced, with virtual ground switching. ANALOG DEVICES, APPLICATIONS REFERENCE MANUAL, 1993, P. 4-31.

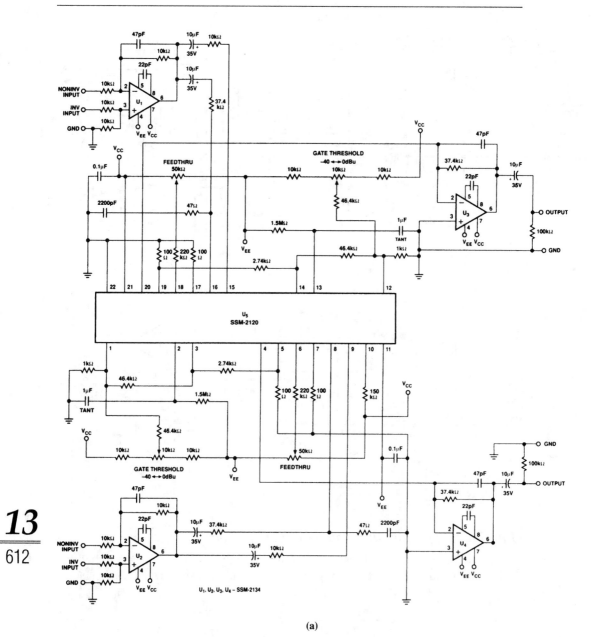

(a)

Audio circuits

Continued

Nominal Input Voltage (–10dBu Out)	–10dBu
Headroom (–10dBu Out)	+30dB
Input Type/Impedance, Balanced	20kΩ
Unbalanced	10kΩ
Downward Expander Class	Feedthrough
Threshold Sense Time Constant (6dB)	350ms
Threshold Range (Level)	–40dBu to 0dBu
Gate Deactivate Time Constant	6ms
Signal Rectifier Type	Full-Wave Averaging
Modulation Feedthrough, Trimmed	< –60dBV
Gain Reduction Ratio, Downward Expansion	1 to 2 (–2dB/dB)
Frequency Response (20Hz to 20kHz)	±0.25dB
Dynamic Range	100dB
THD + Noise (20Hz to 20kHz)	0.02%
IMD (SMPTE 60Hz & 4kHz, 4:1)	0.05%
Output Voltage Slew Rate	6V/µs
Output Voltage (2kΩ Load)	+22dBu
Output Type	Unbalanced
Power Supply	± 18V$_{DC}$ Regulated

(b)

Two-channel noise gate

Fig. 13-8 The circuit in the illustration is a noise gate (a type of noise-reduction system that fully attenuates a voltage-controlled amplifier when no audio signal is present) and also is known as an adjustable-threshold downward expander. Figure 13-8B shows the circuit performance specifications. ANALOG DEVICES, APPLICATIONS REFERENCE MANUAL, 1993, PP. 4-33, 34.

13

613

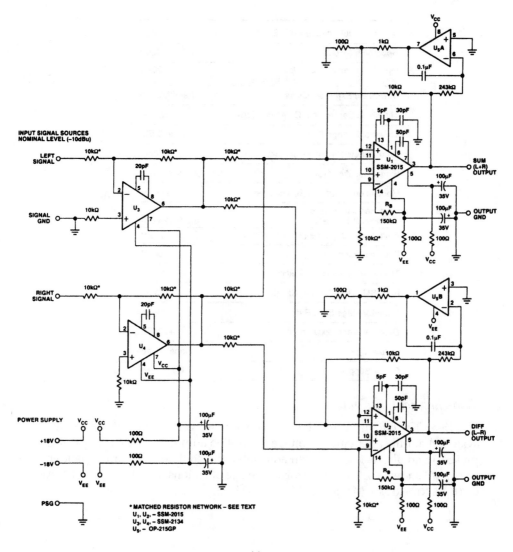

(a)

Audio circuits

Continued

Frequency Response (± 0.02dB)	20Hz to 20kHz
Dynamic Range (20kHz Bandwidth)	104dB
THD + Noise (20Hz to 20kHz, +24dBu)	0.007%
IMD (SMPTE 60Hz & 4kHz, 4:1, +24dBu)	0.015%
Slew Rate	10V/μs
Nominal Signal Level	−10dBu
Maximum Output Voltage (2kΩ Load)	+23.3dBu or 11.3V$_{RMS}$
Amplitude Accuracy	0.05%
Differential Error	<10ns

(b)

Precision sum and difference (audio matrix)

Fig. 13-9 The circuit in the illustration produces accurate sum (L+R) and difference (L–R) signals from stereo left (L) and right (R) inputs. The circuit uses matched (laser trimmed) resistor networks combined with high open-loop-gain differential amplifiers to produce virtually no phase and amplitude error in the sum and difference channels. Figure 13-9B shows the circuit performance specifications. Note that the 10-k resistors must be matched within 0.05% of each other but can be 5% in value tolerance. The 0.1-μF capacitor in the integrator circuit should be metalized polyester film with 10% tolerance. ANALOG DEVICES, APPLICATIONS REFERENCE MANUAL, 1993, PP. 4-35, 36.

13

615

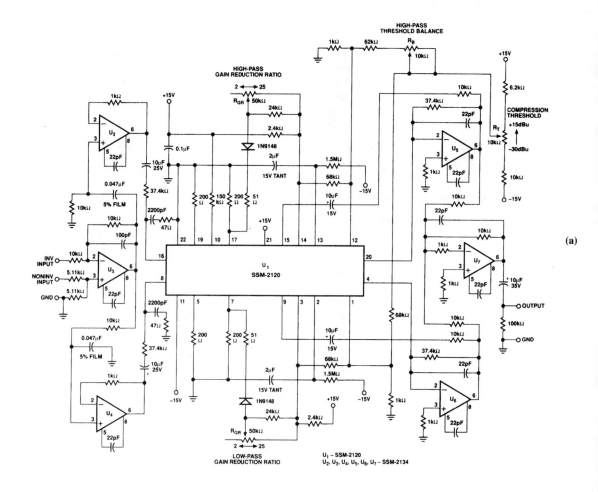

(a)

Input Voltage Range	−10dBu to 0dBu
(Nominal for 0dBu Out)	245mV to 755mV
Rectifier Type	Averaging
Compressor Amplifier Class	Feedback
Attack Time (+10dB or Greater Level Change)	6ms
Recovery Rate	1.67dB/ms
Feedthrough, Trimmed	−100dB
Gain Reduction Range	2 to 25
Frequency Response (20Hz to 20kHz)	0.2dB
Dynamic Range @ 0dB Gain	100dB
THD + Noise (20Hz to 20kHz)	0.02%
IMD (SMPTE 60Hz & 4kHz, 4:1)	0.05%
Output Voltage Slew Rate	6V/μs
Output Voltage (2kΩ Load)	+22dBu or 10V$_{RMS}$

(b)

Audio circuits

Continued

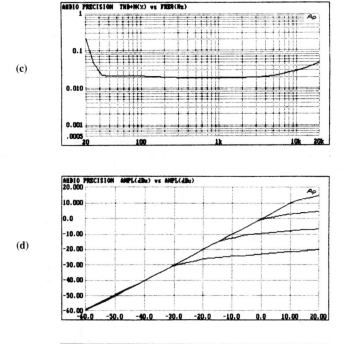

(c)

(d)

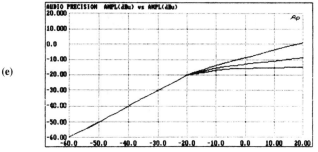

(e)

Two-band audio compressor/limiter

Fig. 13-10 The circuit in the illustration is a two-band audio compressor, featuring separate and adjustable signal-compression ratios and threshold levels. Figures 13-10B, C, D, and E show the circuit performance specifications, *THD+N* versus frequency (80-kHz lowpass filter), compression threshold variance characteristics (–30 dBu to +15 dBu), and ratio characteristics (2:1 to 25:1), respectively. ANALOG DEVICES, APPLICATIONS REFERENCE MANUAL, 1993, PP. 4-37, 38.

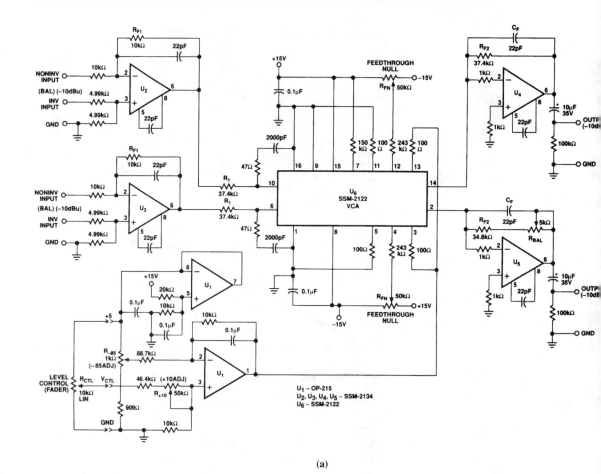

(a)

Input Voltage (Nominal for –10dBu Out)	–10dBu or 245mV$_{RMS}$
Input Impedance, Unbalanced	10kΩ
Imput Impedance, Balanced	20kΩ
Headroom (Nominal for –10dBu In &Out)	32dB
Feedthrough, Trimmed	<750µV
Gain Control Range (Nominal)	+10dB to –85dB
Gain Control Voltage (+10dB to –85dB)	5V$_{DC}$ to 0V$_{DC}$
Frequency Response (20Hz to 20kHz)	±0.1dB
S/N Ratio @ 10dB Gain	100dB
THD + Noise (20Hz to 20kHz, +22dBu)	0.005%
IMD (SMPTE 60Hz & 4kHz, 4:1, +22dBu)	0.02%
Output Voltage Slew Rate	6V/µs
Output Voltage (1kΩ Load)	+22dBu or 10V$_{RMS}$
Output Impedance	<10Ω

(b)

Audio circuits

Two-channel VCA level control

Fig. 13-11 The circuit in the illustration is a dual-channel VCA (voltage-controlled amplifier) level (volume) control. The circuit is most useful when extremely close gain-matching of a stereo audio source is desired, such as ON-AIR and production-audio consoles. Figure 13-11B shows the circuit performance specifications. ANALOG DEVICES, APPLICATIONS REFERENCE MANUAL, 1993, PP. 4-39, 40.

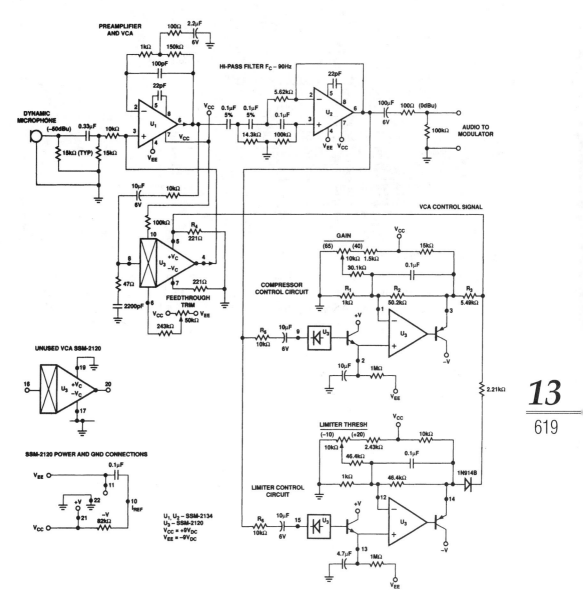

(a)

Audio circuit titles and descriptions

Continued

Input Voltage Range (Nominal for 0dBu Out)	−65dBu to −40dBu
Compressor/Limiter Rectifier Type	Averaging
Compressor Amplifier Class	Feedback
Compressor Attack Time (to 3dB of Final Value)	26ms
Compressor Recovery Rate	3ms/dB
Feedthrough (Trimmed)	−70dBV
Compressor Gain Reduction Ratio	2:1
Limiter Attack Time (to 3dB of Final Value)	13ms
Limiter Recovery Rate	1.5ms/dB
Limiter Gain Reduction Ratio	4.6 to 1
High-Pass Filer, F_c	90Hz
High-Pass Filter Type	3rd Order Butterworth
Frequency Response (90Hz to 20kHz)	±0.5dB
S/N Ratio @ 0dB Gain	100dB
THD + Noise (% 1kHz to 20kHz)	0.05
IMD (% SMPTE 60Hz & 4kHz, 4:1)	0.05
Output Voltage Slew Rate	6V/µs
Power Supply, (Battery)	±9V$_{DC}$
Output Voltage (2kΩ Load, ±9V$_{DC}$	+15dBu

(b)

Compressor-limiter for wireless audio systems

Fig. 13-12 The circuit in the illustration shows the compressor-limiter for the transmitter portion of a wireless audio system. Figure 13-12B shows the circuit performance specifications. ANALOG DEVICES, APPLICATIONS REFERENCE MANUAL, 1993, PP. 4-41, 44.

Audio circuits

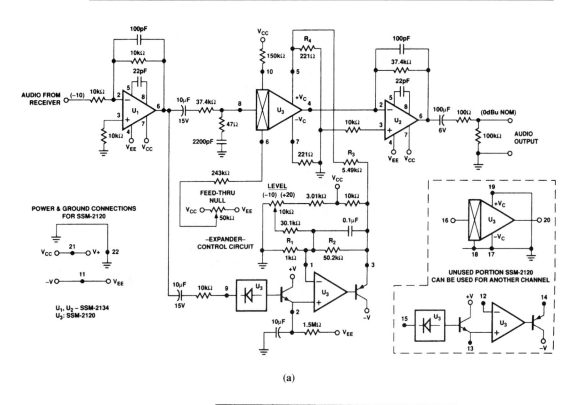

(a)

Input Voltage Range (Nominal for 0dBu Out)	−10dBu
Expander Rectifier Type	Averaging
Expander Amplifier Class	Control Feed-Forward
Expander Attach Time (to 3dB of Final Value)	26ms
Expander Recovery Rate	3ms/dB
Feedthrough (Trimmed)	−70dBV
Gain Expander Ratio	1:2
Frequency Response (20Hz to 20kHz)	±1.0dB
S/N Ratio @ 0dB Gain (Dynamic Range)	100dB
THD + Noise (% 1kHz to 20kHz)	0.20%
IMD (% SMPTE 60Hz % 4kHz, 4:1)	0.25%
Output Voltage Slew Rate	6V/μs
Power Supply	±15V$_{DC}$
Output Voltage (2kΩ Load)	+21dBu

(b)

13

621

Expander for wireless audio systems

Fig. 13-13 The circuit in the illustration shows the expander for the receiver portion of a wireless audio system. Figure 13-13B shows the circuit performance specifications. ANALOG DEVICES, APPLICATIONS REFERENCE MANUAL, 1993, PP. 4- 43, 44.

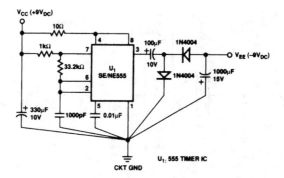

dc-to-dc converter for wireless audio systems

Fig. 13-14 The circuit in the illustration provides the –9 V required for operation of the circuits in Figs. 13-12 and 13-13. The supply is not necessary when two 9-V batteries can be used in the wireless audio system. However, for smaller hand-held wireless microphones, a single 9-V battery is required. ANALOG DEVICES, APPLICATIONS REFERENCE MANUAL, 1993, PP. 4-42.

Audio circuits

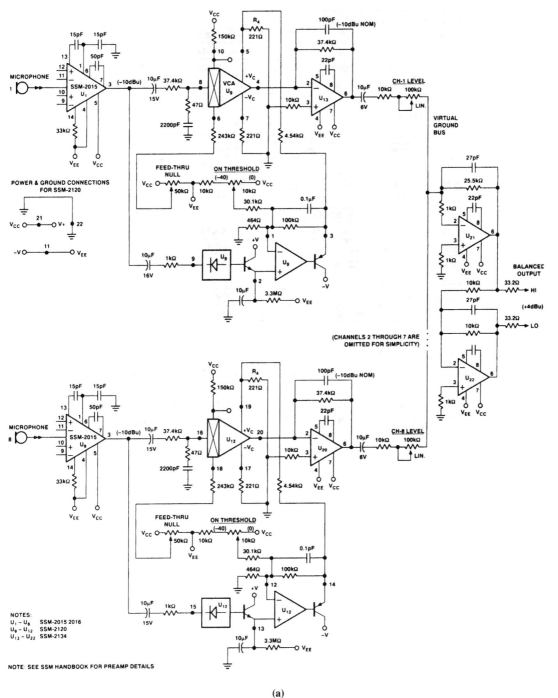

(a)

Audio circuit titles and descriptions

Continued

Input Voltage, without Preamplifier, (for +4dBu Out)	−10dB
Input Impedance, Unbalanced	~1kΩ
Headroom (Nominal for −10dBu In and Out)	32dB
Turn ON Time (to 3dB of Final Value)	30ms
Turn OFF Time (No Signal)	~3sec
Turn OFF Ramp Time	100ms
Feedthrough (Trimmed)	>1mV
ON/OFF Threshold Range (Nominal)	0dBu to −40dBu
ON/OFF Gain Extent	0dB to −90dB
Frequency Response for ±0.1dB	20Hz to 20kHz
S/N Ratio @ 0dB Gain	110dB
THD + Noise (from 20Hz to 20kHz)	0.005%
IMD (SMPTE 60Hz and 4kHz, 4:1)	0.02%
Output Voltage Slew Rate	12V/μs
Rated Output Level (600Ω Load)	+24dBu
Output Impedance	68Ω
Output Type	Balanced
Power Supply Requirements	±15V$_{DC}$ Regulated

(b)

Automatic microphone mixer

Fig. 13-15 The circuit in the illustration shows an audio-activated microphone mixer designed to accommodate eight input channels. The circuit provides for un-attended microphone-mixing functions, such as in a conference room that requires sound reinforcement or conversation recording. The circuit accommodates automatic and transparent channel ON/OFF operation. The audio output automatically turns ON in less than 1 ms, and back OFF after two to four seconds of no audio. Each channel incorporates independent and automatic operation, with ON threshold-sensitivity and level-adjustment (trim) controls. Figure 13-15B shows circuit performance specifications. ANALOG DEVICES, APPLICATIONS REFERENCE MANUAL, 1993, PP. 4-45, 46.

13

624

Audio circuits

+15V

C1
+
100μF/25V

C2
0.1μF

U1
AD744JN

U2
AD811AN
(NOTE 3)

R6
49.9Ω

OUTPUT HIGH

R_{S1}
500k
(SEE TEXT)

V_{IN}

R_{S2}

R5
750Ω

OUTPUT LO

R2
1.82k

R1
7.5k

R3
7.5k (NOTE 2)

C3
0.1μF

(NOTE 4)

Q1

2N3906

R4
1.82k (NOTE 2)

C4
100μF/25V

NOTES:
1. G = (R1/R2) + 1.
2. R3/R4 EQUAL TO OR > R1/R2.
3. HEAT SINK RECOMMENDED, SEE TEXT.
4. OPEN JUMPER FOR BOOTSTRAP ON.

-15V

Bootstrap circuit to lower distortion in op amps

Fig. 13-16 The circuit in the illustration shows a means of bootstrapping the substrate of JFET op amps (Chapter 1) to minimize distortion (THD) caused by nonlinear input capacitance. The bootstrap circuit consists of Q1 and associated parts. Resistors R_{S1} and R_{S2} provide input-impedance compensation. The 500-kΩ R_{S1} is for test purposes. In a practical application, R_{S1} can be part of the total source impedance. Resistor R_{S2} is made equal to this total source impedance. Figure 3-17 shows the output signal (top) and distortion component (bottom, 0.03% of full scale) of the circuit without bootstrap (jumper across Q1). Figure 13-18 shows the same characteristics with bootstrap (jumper removed from Q1). Both tests were made with ±15-V supplies and both show 3-V_{rms} output. When supply voltages are above ±12 V, a heatsink (Aavid 5801) is recommended for the AD811AN (U2). ANALOG DEVICES, APPLICATIONS REFERENCE MANUAL, 1993, PP. 4-74, 4-75.

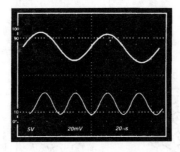

Test results of circuit without bootstrap

Fig. 13-17 Output signal (top) and distortion components (bottom, 0.3% full scale). Conditions: $V_S \pm 15$ V, Figure 1 circuit, without bootstrapping, 3 V rms out.

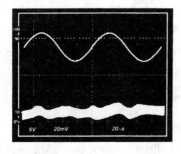

Test results of circuit with bootstrap

Fig. 13-18 Output signal (top) and distortion/noise components (bottom, 0.03% full scale). Conditions: $V_S \pm 15$ V, Figure 1 circuit, with bootstrapping, 3 V rms out.

Audio circuits

14

Special applications for digital/analog converters

This chapter is devoted to special applications for the digital/analog (DAC or D/A) converter ICs covered in Chapter 6. Because such ICs are essentially digital, all the test and troubleshooting information of Chapter 6 applies. Also, since many of the special applications in this chapter also include amplifiers, the test and troubleshooting of Chapter 1 apply.

DAC special-application circuit titles and descriptions

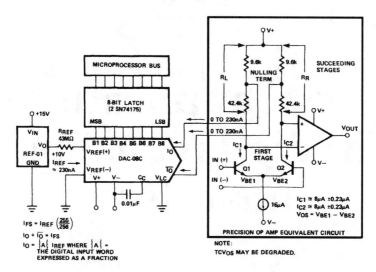

Precision op amp with digital nulling

Fig. 14-1 The circuit in the illustration shows a DAC-08 connected to provide digitally controlled offset nulling of a precision op amp (OP-5, OP-7, OP-77, etc.). ANALOG DEVICES, APPLICATIONS REFERENCE MANUAL, 1993, P. 8-49.

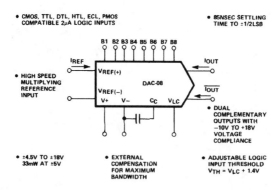

Basic DAC-08 characteristics

Fig. 14-2 The circuit in the illustration shows the basic DAC-08 characteristics. ANALOG DEVICES, APPLICATIONS REFERENCE MANUAL, 1993, P. 8-51.

Special applications for digital/analog converters

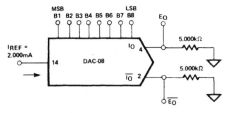

(a)

	B1	B2	B3	B4	B5	B6	B7	B8	I_O mA	$\overline{I_O}$ mA	E_O	$\overline{E_O}$
FULL SCALE −1LSB	1	1	1	1	1	1	1	1	1.992	0.000	−9.960	0.000
FULL SCALE −2LSB	1	1	1	1	1	1	1	0	1.984	0.008	−9.920	− 0.40
HALF SCALE +LSB	1	0	0	0	0	0	0	1	1.008	0.984	−5.040	−4.920
HALF SCALE	1	0	0	0	0	0	0	0	1.000	0.992	−5.000	−4.960
HALF SCALE −LSB	0	1	1	1	1	1	1	1	0.992	1.000	−4.960	−5.000
ZERO SCALE +LSB	0	0	0	0	0	0	0	1	0.008	1.984	−0.040	−9.920
ZERO SCALE	0	0	0	0	0	0	0	0	0.000	1.992	0.000	−9.960

(b)

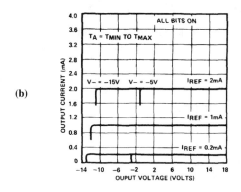

Basic unipolar negative operation of the DAC-08

Fig. 14-3   The circuit in the illustration shows the DAC-08 connected for unipolar negative operation. Figure 14-3B shows the output current versus output voltages (output voltage compliance). ANALOG DEVICES, APPLICATIONS REFERENCE MANUAL, 1993, PP. 8-51, 52.

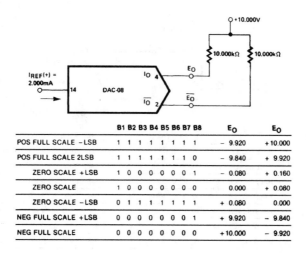

	B1 B2 B3 B4 B5 B6 B7 B8	E_O	$\overline{E_O}$
POS FULL SCALE −LSB	1 1 1 1 1 1 1 1	− 9.920	+ 10.000
POS FULL SCALE 2LSB	1 1 1 1 1 1 1 0	− 9.840	+ 9.920
ZERO SCALE +LSB	1 0 0 0 0 0 0 1	− 0.080	+ 0.160
ZERO SCALE	1 0 0 0 0 0 0 0	0.000	+ 0.080
ZERO SCALE −LSB	0 1 1 1 1 1 1 1	+ 0.080	0.000
NEG FULL SCALE +LSB	0 0 0 0 0 0 0 1	+ 9.920	− 9.840
NEG FULL SCALE	0 0 0 0 0 0 0 0	+10.000	− 9.920

Basic bipolar operation of the DAC-08

Fig. 14-4 The circuit in the illustration shows the DAC-08 connected for bipolar operation. ANALOG DEVICES, APPLICATIONS REFERENCE MANUAL, 1993, P. 8-52.

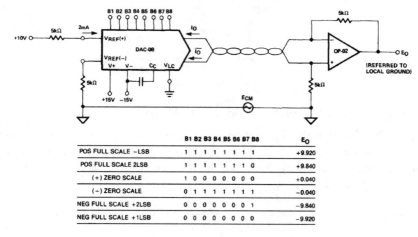

	B1 B2 B3 B4 B5 B6 B7 B8	E_O
POS FULL SCALE −LSB	1 1 1 1 1 1 1 1	+9.920
POS FULL SCALE 2LSB	1 1 1 1 1 1 1 0	+9.840
(+) ZERO SCALE	1 0 0 0 0 0 0 0	+0.040
(−) ZERO SCALE	0 1 1 1 1 1 1 1	−0.040
NEG FULL SCALE +2LSB	0 0 0 0 0 0 0 1	−9.840
NEG FULL SCALE +1LSB	0 0 0 0 0 0 0 0	−9.920

Current-to-voltage conversion with high noise immunity

Fig. 14-5 The circuit in the illustration shows the DAC-08 combined with an OP-02 to provide voltage-current conversion with high noise immunity. ANALOG DEVICES, APPLICATIONS REFERENCE MANUAL, 1993, P. 8-52.

Special applications for digital/analog converters

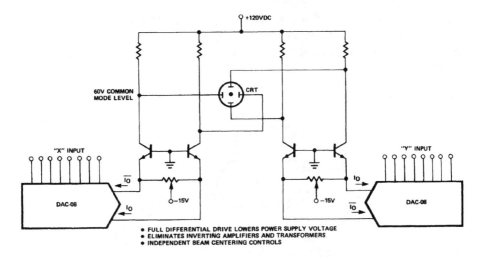

- FULL DIFFERENTIAL DRIVE LOWERS POWER SUPPLY VOLTAGE
- ELIMINATES INVERTING AMPLIFIERS AND TRANSFORMERS
- INDEPENDENT BEAM CENTERING CONTROLS

CRT display driver

Fig. 14-6 The circuit in the illustration shows two DAC-08s connected to drive a CRT without transformers. ANALOG DEVICES, APPLICATIONS REFERENCE MANUAL, 1993, P. 8-53.

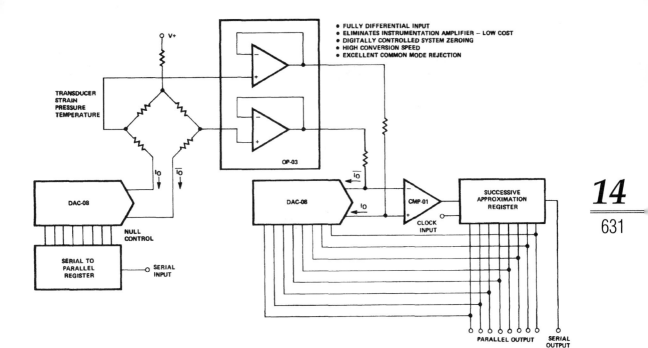

- FULLY DIFFERENTIAL INPUT
- ELIMINATES INSTRUMENTATION AMPLIFIER – LOW COST
- DIGITALLY CONTROLLED SYSTEM ZEROING
- HIGH CONVERSION SPEED
- EXCELLENT COMMON MODE REJECTION

DAC special-application circuit titles and descriptions

Bridge-transducer control system

Fig. 14-7 The circuit in the illustration shows two DAC-08s combined with an amplifier, SAR, and serial-to-parallel register to provide a control system for a bridge transducer (with full differential input). ANALOG DEVICES, APPLICATIONS REFERENCE MANUAL, 1993, P. 8-53.

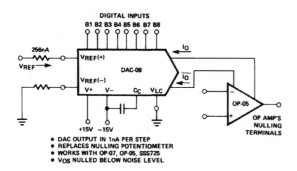

DIGITAL INPUTS
B1 B2 B3 B4 B5 B6 B7 B8
256nA
VREF
VREF(+)
DAC-08
VREF(−)
V+ V− CC VLC
IO
IO
OP-05
OP AMP'S NULLING TERMINALS
+15V −15V
• DAC OUTPUT IN 1nA PER STEP
• REPLACES NULLING POTENTIOMETER
• WORKS WITH OP-07, OP-05, SSS725
• VOS NULLED BELOW NOISE LEVEL

Digitally controlled offset nulling

Fig. 14-8 The circuit in the illustration shows the DAC-08 connected to provide offset nulling of an op amp. ANALOG DEVICES, APPLICATIONS REFERENCE MANUAL, 1993, P. 8-54.

DIGITAL INPUT
VREF
B1 B2 B3 B4 B5 B6 B7 B8
VREF(+)
DAC-08
VREF(−)
V+ V− CC VLC
EIN
IO
IO
RL
• BIPOLAR OUTPUT WITH OFFSET BINARY CODING
• PROVIDES DC ISOLATION BETWEEN SOURCE AND LOAD
• HIGH VOLTAGE OUTPUT CAPABILITY
• USEFUL WITH PULSE OR SINE WAVE REFERENCE INPUT
• USEFUL WITH PULSE OR SINE FUNCTION DIGITAL INPUT

Providing isolation between digital source and analog load

Fig. 14-9 The circuit in the illustration shows the DAC-08 analog output applied to the load through a balanced transformer. ANALOG DEVICES, APPLICATIONS REFERENCE MANUAL, 1993, P. 8-54.

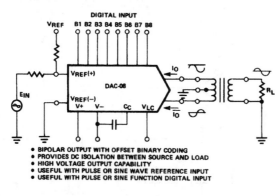

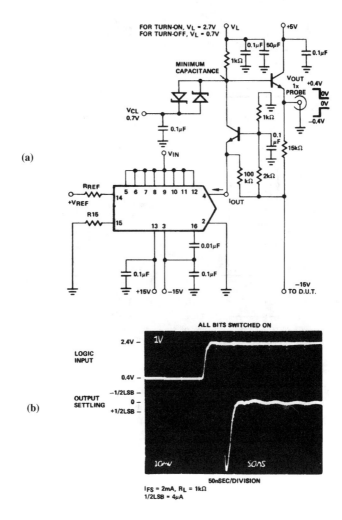

Settling-time measurement

Fig. 14-10 The circuit in the illustration shows test connections for measurement of settling time. Figure 14-10B shows the scope display for full-scale (all bits on) settling time. ANALOG DEVICES, APPLICATIONS REFERENCE MANUAL, 1993, P. 8-54.

DAC special-application circuit titles and descriptions

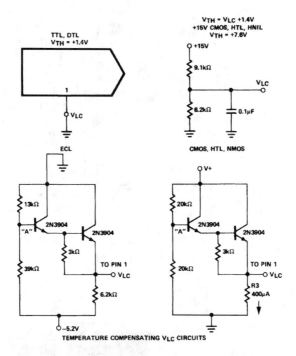

Interfacing with various logic families

Fig. 14-11 These circuits provide interfacing between the DAC-08 and various logic families. ANALOG DEVICES, APPLICATIONS REFERENCE MANUAL, 1993, P. 8-55.

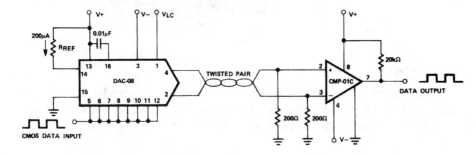

CMOS differential line driver/receiver

Fig. 14-12 The circuit in the illustration shows a DAC-08 output applied to an op amp through a twisted pair line. ANALOG DEVICES, APPLICATIONS REFERENCE MANUAL, 1993, P. 8-55.

Special applications for digital/analog converters

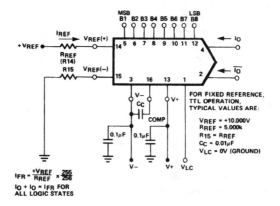

Basic positive reference operation of the DAC-08

Fig. 14-13 The circuit in the illustration shows the DAC-08 connected for positive reference operation. ANALOG DEVICES, APPLICATIONS REFERENCE MANUAL, 1993, P. 8-56.

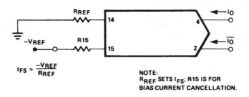

Basic negative reference operation of the DAC-08

Fig. 14-14 The circuit in the illustration shows the DAC-08 connected for positive reference operation. ANALOG DEVICES, APPLICATIONS REFERENCE MANUAL, 1993, P. 8-57.

DAC special-application circuit titles and descriptions

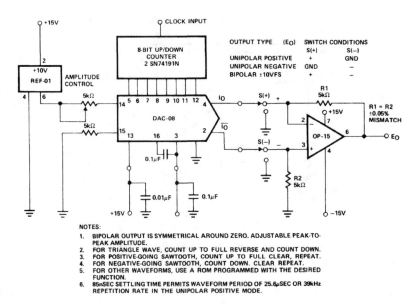

OUTPUT TYPE	(E$_O$)	SWITCH CONDITIONS	
		S(+)	S(−)
UNIPOLAR POSITIVE		+	GND
UNIPOLAR NEGATIVE	GND		−
BIPOLAR ±10VFS		+	−

NOTES:
1. BIPOLAR OUTPUT IS SYMMETRICAL AROUND ZERO. ADJUSTABLE PEAK-TO-PEAK AMPLITUDE.
2. FOR TRIANGLE WAVE, COUNT UP TO FULL REVERSE AND COUNT DOWN.
3. FOR POSITIVE-GOING SAWTOOTH, COUNT UP TO FULL CLEAR, REPEAT.
4. FOR NEGATIVE-GOING SAWTOOTH, COUNT DOWN. CLEAR REPEAT.
5. FOR OTHER WAVEFORMS, USE A ROM PROGRAMMED WITH THE DESIRED FUNCTION.
6. 85nSEC SETTLING TIME PERMITS WAVEFORM PERIOD OF 25.6μSEC OR 39kHz REPETITION RATE IN THE UNIPOLAR POSITIVE MODE.

High-speed waveform generator

Fig. 14-15 The circuit in the illustration shows the DAC-08 connected as a high-speed, digitally controlled waveform generator. ANALOG DEVICES, APPLICATIONS REFERENCE MANUAL, 1993, P. 8-58.

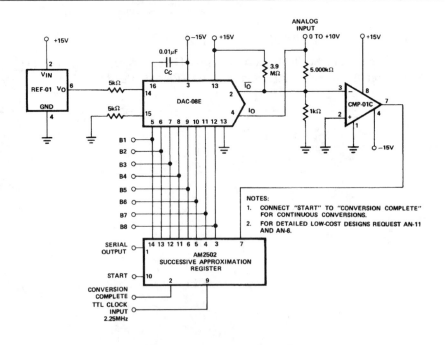

NOTES:
1. CONNECT "START" TO "CONVERSION COMPLETE" FOR CONTINUOUS CONVERSIONS.
2. FOR DETAILED LOW-COST DESIGNS REQUEST AN-11 AND AN-6.

Special applications for digital/analog converters

Low-cost A/D converter

Fig. 14-16 The circuit in the illustration shows the DAC-08 connected as a low-cost A/D converter (Chapter 9). ANALOG DEVICES, APPLICATIONS REFERENCE MANUAL, 1993, P. 8-58.

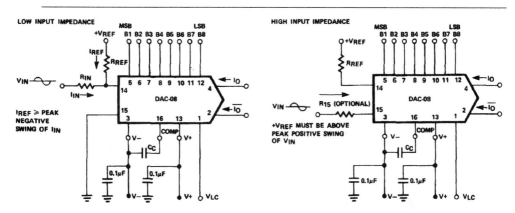

Connections for bipolar reference operation of the DAC-08

Fig. 14-17 The circuit in the illustration shows the connections for bipolar reference operation of the DAC-08. ANALOG DEVICES, APPLICATIONS REFERENCE MANUAL, 1993, P. 8-64.

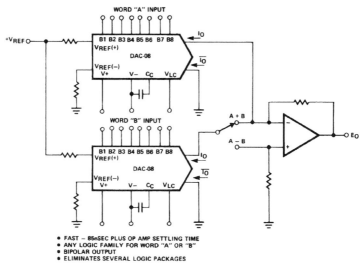

Digital addition or subtraction with an analog output

Fig. 14-18 The circuit in the illustration shows two DAC-08s connected to provide digital addition or subtraction, with an analog output. ANALOG DEVICES, APPLICATIONS REFERENCE MANUAL, 1993, P. 8-59.

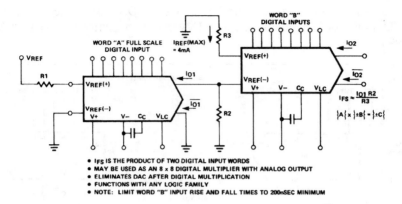

- IFS IS THE PRODUCT OF TWO DIGITAL INPUT WORDS
- MAY BE USED AS AN 8 x 8 DIGITAL MULTIPLIER WITH ANALOG OUTPUT
- ELIMINATES DAC AFTER DIGITAL MULTIPLICATION
- FUNCTIONS WITH ANY LOGIC FAMILY
- NOTE: LIMIT WORD "B" INPUT RISE AND FALL TIMES TO 200nSEC MINIMUM

Digital multiplier (8 × 8) with analog output

Fig. 14-19 The circuit in the illustration shows two DAC-08s connected to provide 8 × 8 digital multiplication with analog output. ANALOG DEVICES, APPLICATIONS REFERENCE MANUAL, 1993, P. 8-59.

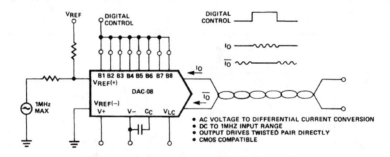

- AC VOLTAGE TO DIFFERENTIAL CURRENT CONVERSION
- DC TO 1MHZ INPUT RANGE
- OUTPUT DRIVES TWISTED PAIR DIRECTLY
- CMOS COMPATIBLE

Modem transmitter

Fig. 14-20 The circuit in the illustration shows the DAC-08 connected as a modem transmitter. ANALOG DEVICES, APPLICATIONS REFERENCE MANUAL, 1993, P. 8-60.

14

638

Special applications for digital/analog converters

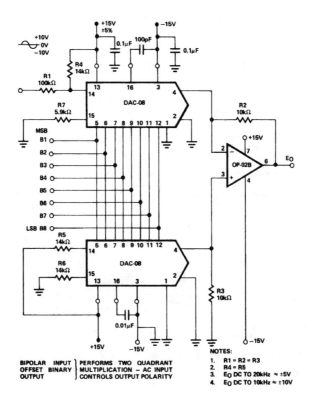

Programmable-gain amplifier with dc-coupled digital attenuation

Fig. 14-21   The circuit in the illustration shows two DAC-08s connected to provide digitally controlled attenuation for a programmable-gain amplifier. ANALOG DEVICES, APPLICATIONS REFERENCE MANUAL, 1993, P. 8-60.

DAC special-application circuit titles and descriptions

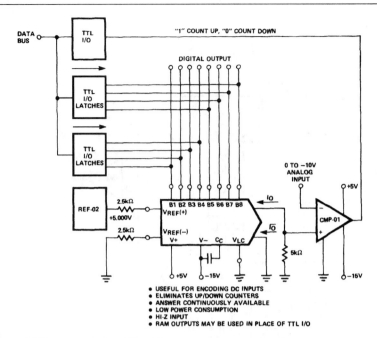

- USEFUL FOR ENCODING DC INPUTS
- ELIMINATES UP/DOWN COUNTERS
- ANSWER CONTINUOUSLY AVAILABLE
- LOW POWER CONSUMPTION
- HI-Z INPUT
- RAM OUTPUTS MAY BE USED IN PLACE OF TTL I/O

Tracking A/D converter with microprocessor control

Fig. 14-22 The circuit in the illustration shows a DAC-08 connected to produce an A/D converter with microprocessor control. ANALOG DEVICES, APPLICATIONS REFERENCE MANUAL, 1993, P. 8-61.

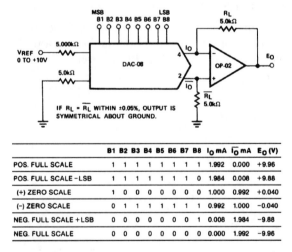

	B1	B2	B3	B4	B5	B6	B7	B8	I_O mA	$\overline{I_O}$ mA	E_O (V)
POS. FULL SCALE	1	1	1	1	1	1	1	1	1.992	0.000	+9.96
POS. FULL SCALE - LSB	1	1	1	1	1	1	1	0	1.984	0.008	+9.88
(+) ZERO SCALE	1	0	0	0	0	0	0	0	1.000	0.992	+0.040
(−) ZERO SCALE	0	1	1	1	1	1	1	1	0.992	1.000	−0.040
NEG. FULL SCALE + LSB	0	0	0	0	0	0	0	1	0.008	1.984	−9.88
NEG. FULL SCALE	0	0	0	0	0	0	0	0	0.000	1.992	−9.96

Bipolar digital two-quadrant multiplication

Fig. 14-23 The circuit in the illustration shows the DAC-08 connected to provide digital two-quadrant multiplication, where the digital input word controls output polarity. ANALOG DEVICES, APPLICATIONS REFERENCE MANUAL, 1993, P. 8-65.

Special applications for digital/analog converters

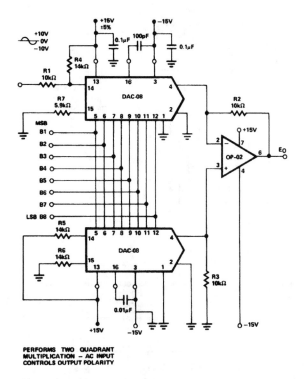

PERFORMS TWO QUADRANT
MULTIPLICATION – AC INPUT
CONTROLS OUTPUT POLARITY

Bipolar analog two-quadrant multiplication

Fig. 14-24 The circuit in the illustration shows the DAC-08 connected to provide analog two-quadrant multiplications, where the analog reference input controls output polarity. ANALOG DEVICES, APPLICATIONS REFERENCE MANUAL, 1993, P. 8-65.

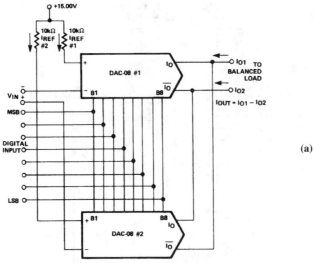

(a)

- HIGH SPEED MONOTONIC OPERATION OVER ENTIRE RANGE
- HIGH IMPEDANCE DIFFERENTIAL INPUTS
- ±10V DIFFERENTIAL INPUT RANGE
- 2 PACKAGES VS 3 FOR VOLTAGE SWITCHED DAC'S
- TRUE CURRENT OUTPUT – WIDE COMPLIANCE
- ADJUSTABLE LOGIC THRESHOLD
- WIDE POWER SUPPLY RANGE

DIGITAL INPUT	$V_{IN}(+)$	$V_{IN}(-)$	V_{IN} DIFF.	I_{REF} #1 (mA)	I_{REF} #2 (mA)	I_O#1 (mA)	I_O#2 (mA)	I_{O1} (mA)	I_O#2 (mA)	I_O#1 (mA)	I_{O2} (mA)	I_{OUT} DIFF.
1111 1111	+5V	−5V	+10V	2.000	1.000	1.992	0	1.992	0.996	0	0.996	0.996mA
1000 0000	+5V	−5V	+10V	2.000	1.000	1.000	0.496	1.496	0.500	0.992	1.492	0.004mA
0111 1111	+5V	−5V	+10V	2.000	1.000	0.992	0.500	1.492	0.496	1.000	1.496	−0.004mA
0000 0000	+5V	−5V	+10V	2.000	1.000	0	0.996	0.996	0	1.992	1.992	−0.996mA
1111 1111	0V	0V	0V	1.500	1.500	1.494	0	1.494	1.494	0	1.494	0.000mA
1000 0000	−10V	−10V	0V	2.500	2.500	1.250	1.240	2.490	1.250	1.240	2.490	0.000mA
0111 1111	+10V	+10V	0V	0.500	0.500	0.248	0.250	0.498	0.248	0.250	0.498	0.000mA
0000 0000	0V	0V	0V	1.500	1.500	0	1.494	1.494	0	1.494	1.494	0.000mA
1111 1111	−5V	+5V	−10V	1.000	2.000	0.996	0	0.996	1.992	0	1.992	−0.996mA
1000 0000	−5V	+5V	−10V	1.000	2.000	0.500	0.992	1.492	1.000	0.496	1.496	−0.004mA
0111 1111	−5V	+5V	−10V	1.000	2.000	0.496	1.000	1.496	0.992	0.500	1.492	0.004mA
0000 0000	−5V	+5V	−10V	1.000	2.000	0	1.992	1.992	0	0.996	0.996	0.996mA

(b)

Special applications for digital/analog converters

Continued

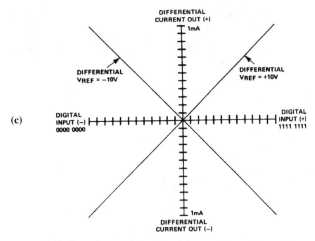

(c)

Four-quadrant multiplication with fixed-impedance input

Fig. 14-25 The circuit in the illustration shows two DAC-08s connected to provide four-quadrant multiplication, with a fixed-impedance input. Figures 14-25B and C show the multiplying current values and transfer functions, respectively. ANALOG DEVICES, APPLICATIONS REFERENCE MANUAL, 1993, PP. 8-65, 66.

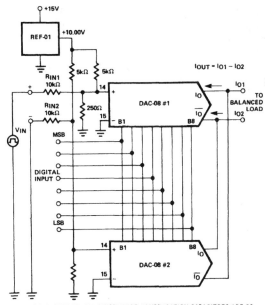

- 250Ω RESISTORS AND OMISSION OF COMPENSATION CAPACITORS ARE OP-TIONAL FOR FAST PULSED REFERENCE OPERATION.
- INPUT DIFFERENTIAL AND COMMON MODE RANGES ARE EXTENDABLE BY INCREASING 10kΩ RESISTORS. EXAMPLE: 100kΩ FOR ±100V.
- HIGH SPEED MULTIPLYING CONNECTION WITH MONOTONIC OPERATION OVER ENTIRE RANGE.
- NOT NECESSARY TO MODULATE BOTH DAC'S WITH THE REFERENCE INPUT.

Four-quadrant multiplication with extendable input range (high speed)

Fig. 14-26 The circuit in the illustration is similar to that of Fig. 14-25 except that the input ranges (differential and common-mode) are extendable, and the reference slew rate (circuit speed) is about 16 mA/μs. ANALOG DEVICES, APPLICATIONS REFERENCE MANUAL, 1993, P. 8-66.

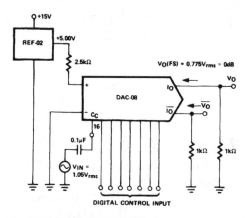

High input-impedance ac-coupled multiplication

Fig. 14-27 The circuit in the illustration shows the DAC-08 connected for ac-coupled multiplication with high input-impedance. (The circuit is essentially an audio-frequency digital attenuator.) ANALOG DEVICES, APPLICATIONS REFERENCE MANUAL, 1993, P. 8-67.

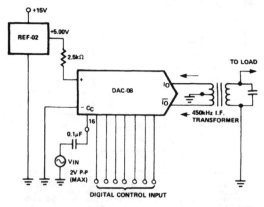

Special applications for digital/analog converters

High input-impedance ac-coupled multiplication (transformer coupled)

Fig. 14-28 The circuit in the illustration shows the DAC-08 connected for ac-coupled multiplication with high input impedance. (The circuit is essentially an intermediate-frequency digital attenuator.) The highest recommended operating frequency for the circuit is 455 kHz. ANALOG DEVICES, APPLICATIONS REFERENCE MANUAL, 1993, P. 8-67.

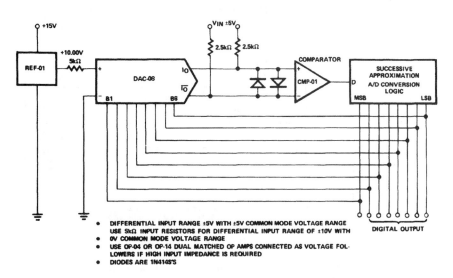

- DIFFERENTIAL INPUT RANGE ±5V WITH ±5V COMMON MODE VOLTAGE RANGE
  USE 5kΩ INPUT RESISTORS FOR DIFFERENTIAL INPUT RANGE OF ±10V WITH
- 0V COMMON MODE VOLTAGE RANGE
- USE OP-04 OR OP-14 DUAL MATCHED OP AMPS CONNECTED AS VOLTAGE FOL-
  LOWERS IF HIGH INPUT IMPEDANCE IS REQUIRED
- DIODES ARE 1N414S'S

Basic A/D converter with differential input

Fig. 14-29 The circuit in the illustration shows the basic connections for an A/D converter with differential input. Conversion is made within 2.0 µs. ANALOG DEVICES, APPLICATIONS REFERENCE MANUAL, 1993, P. 8-67.

<u>**14**</u>

645

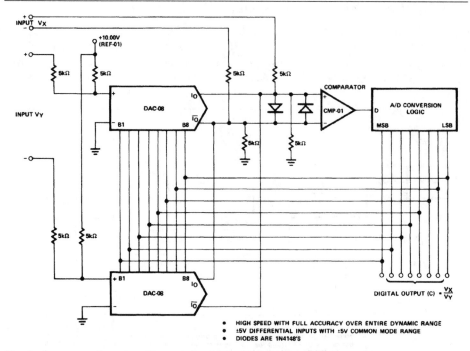

Basic four-quadrant ratiometric A/D conversion

Fig. 14-30 The circuit in the illustration shows the basic connections for a four-quadrant ratiometric A/D converter. ANALOG DEVICES, APPLICATIONS REFERENCE MANUAL, 1993, P. 8-68.

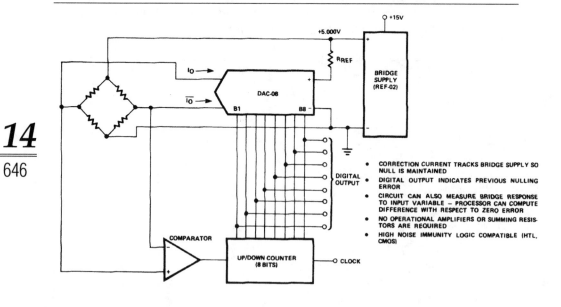

Special applications for digital/analog converters

Null and indicator circuit for bridge transducer

Fig. 14-31 The circuit in the illustration uses a DAC-08 to provide automatic nulling for a transducer bridge, as well as a digital representation of the bridge error. ANALOG DEVICES, APPLICATIONS REFERENCE MANUAL, 1993, P. 8-69.

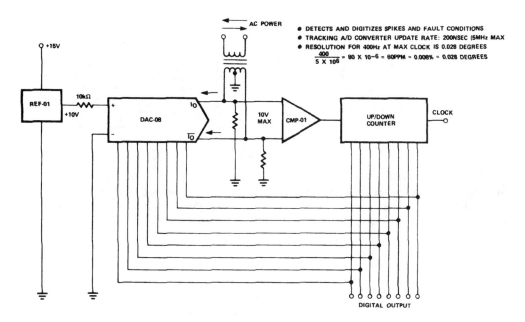

Power-fault monitor and detector

Fig. 14-32 The circuit in the illustration monitors a power line to detect (and digitize) spikes or other fault conditions. Common-mode voltage at the comparator input must not exceed ±10 V, and differential voltage must not exceed 11 V. Voltage-limiting resistors at the comparator inputs are recommended. ANALOG DEVICES, APPLICATIONS REFERENCE MANUAL, 1993, P. 8-69.

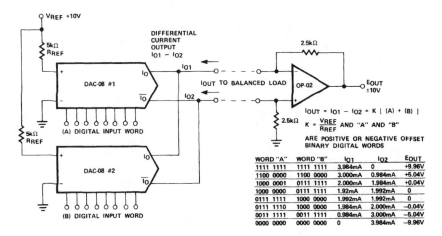

WORD "A"	WORD "B"	IO1	IO2	EOUT
1111 1111	1111 1111	3.984mA	0	+9.96V
1100 0000	1100 0000	3.000mA	0.984mA	+5.04V
1000 0001	0111 1111	2.000mA	1.984mA	+0.04V
1000 0000	0111 1111	1.92mA	1.992mA	0
0111 1111	1000 0000	1.992mA	1.992mA	0
0111 1110	1000 0000	1.984mA	2.000mA	−0.04V
0011 1111	0011 1111	0.984mA	3.000mA	−5.04V
0000 0000	0000 0000	0	3.984mA	−9.96V

$$I_{OUT} = I_{O1} - I_{O2} = K \; [\; (A) + (B) \;]$$

$$K = \frac{V_{REF}}{R_{REF}} \text{ AND "A" AND "B"}$$

ARE POSITIVE OR NEGATIVE OFFSET BINARY DIGITAL WORDS

Power-quadrant algebraic digital computation

Fig. 14-33   The circuit in the illustration uses two DAC-08s connected to provide a fast algebraic sum with a direct analog readout. (The output is the algebraic sum of word A and word B in all four quadrants.) ANALOG DEVICES, APPLICATIONS REFERENCE MANUAL, 1993, P. 8-70.

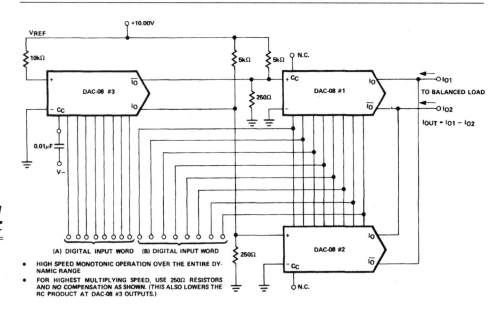

- HIGH SPEED MONOTONIC OPERATION OVER THE ENTIRE DYNAMIC RANGE
- FOR HIGHEST MULTIPLYING SPEED, USE 250Ω RESISTORS AND NO COMPENSATION AS SHOWN. (THIS ALSO LOWERS THE RC PRODUCT AT DAC-08 #3 OUTPUTS.)

Four-quadrant 8-bit × 8-bit digital multiplier

Fig. 14-34   The circuit in the illustration shows three DAC-08s connected to provide high-speed multiplication of two 8-bit digital words with an analog output. ANALOG DEVICES, APPLICATIONS REFERENCE MANUAL, 1993, P. 8-70.

Special applications for digital/analog converters

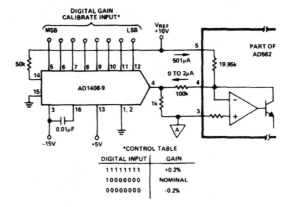

Programmable-gain trimming or calibration

Fig. 14-35 The circuit in the illustration shows an AD1408-9 connected to provide an incremental adjustment range (in 256 adjustment steps) for the scale factor of an AD562 12-bit DAC. ANALOG DEVICES, APPLICATIONS REFERENCE MANUAL, 1993, P. 8-73.

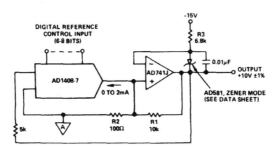

Direct reference-voltage calibration

Fig. 14-36 The circuit in the illustration shows an AD1408-7 connected to calibrate the output of a buffered voltage-reference circuit. ANALOG DEVICES, APPLICATIONS REFERENCE MANUAL, 1993, P. 8-73.

14

649

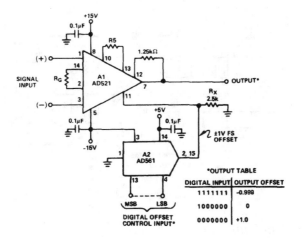

OUTPUT TABLE

DIGITAL INPUT	OUTPUT OFFSET
1111111	-0.998
1000000	0
0000000	+1.0

Instrumentation amplifier with programmable offset

Fig. 14-37 The circuit in the illustration uses an AD561 to provide a programmed constant-offset (or offset-zeroing voltage) introduced at the reference input of an instrumentation amplifier. This voltage produces an output offset that is independent of gain. ANALOG DEVICES, APPLICATIONS REFERENCE MANUAL, 1993, P. 8-73.

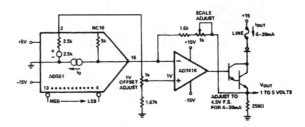

Simple 4- to 20-mA current controller

Fig. 14-38 The circuit in the illustration shows an AD561 used to transmit data in the form of a 4- to 20-mA current (4 mA corresponds to zero, and 20 mA corresponds to full scale). Such a circuit minimizes the effects of ground-potential differences, series resistance and voltage-noise pickup. To calibrate, apply all-zeros at the input, adjust the 1-kΩ offset pot for 4-mA output current. Then apply all-ones and set the scale adjust pot for 20 mA (or 19.98 mA, to be more precise!) of output current. ANALOG DEVICES, APPLICATIONS REFERENCE MANUAL, 1993, P. 8-74.

Special applications for digital/analog converters

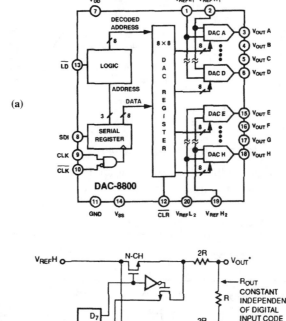

(a)

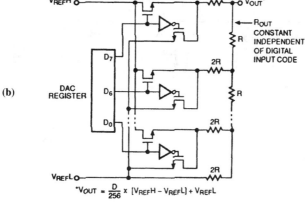

(b)

$$V_{OUT}^* = \frac{D}{256} \times [V_{REF}H - V_{REF}L] + V_{REF}L$$

Electronic trimming potentiometer

Fig. 14-39 The circuit in the illustration shows the DAC-8800 octal 8-bit TrimDAC® used for voltage-adjustment applications. The DAC-8800 is a digitally controlled voltage adjustment device used primarily to replace mechanical potentiometers. As shown in Fig. 14-39B, the IC has eight individual DACs divided into two groups of four, each group having its own high and low reference inputs. Each DAC output is independently controlled by a serial interface through which the 8-bit data word and 3-bit address are loaded. Each DAC contains an R-2R ladder connected between the high and low reference inputs as shown. The output voltage is set by the position of the switches according to the equation shown in Fig. 14-39B (where *D* is the digital code). ANALOG DEVICES, APPLICATIONS REFERENCE MANUAL, 1993, PP. 8-85, 8-99.

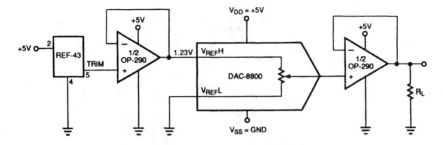

DAC-8800 connected for single-supply operation

Fig. 14-40 The circuit in the illustration shows the DAC-8800 (Fig. 14-39) connected +5-V operation. ANALOG DEVICES, APPLICATIONS REFERENCE MANUAL, 1993, P. 8-88.

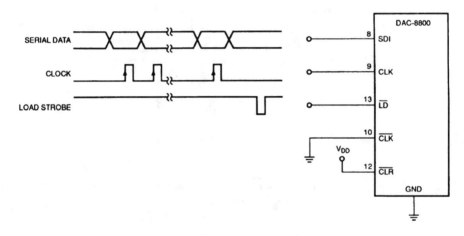

Serial interfacing for the DAC-8800

Fig. 14-41 The circuit in the illustration shows serial interfacing for the DAC-8800 (Fig. 14-39). Note that when loading data, three address bits are loaded with *MSB* (most significant bit) first, followed by eight bits of data (again MSB first). Thus 11 bits in all are loaded through the SDI pin to control each DAC. The clock can be 6.6 MHz, making it possible to load all eight DACs in as little as 14 μs. ANALOG DEVICES, APPLICATIONS REFERENCE MANUAL, 1993, P. 8-88.

Special applications for digital/analog converters

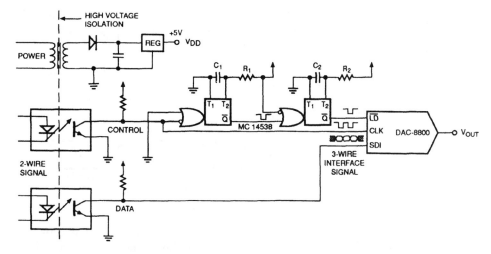

Two-wire interface for process environments

Fig. 14-42 The circuit in the illustration shows a basic, isolated, two-wire serial interface for process environments. The first one-shot's timeout must be longer than the clock period. Each succeeding clock pulse will retrigger the one-shot until all 11 bits are loaded into the DAC. Then the clock must pause long enough to allow the one-shot to time out. When the first one-shot's output goes low, the second one-shot is triggered, producing the LOAD pulse and completing the load cycle. ANALOG DEVICES, APPLICATIONS REFERENCE MANUAL, 1993, P. 8-89.

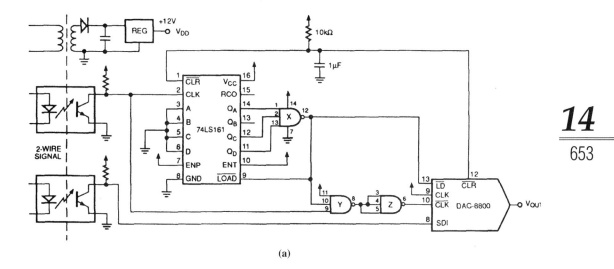

(a)

DAC special-application circuit titles and descriptions

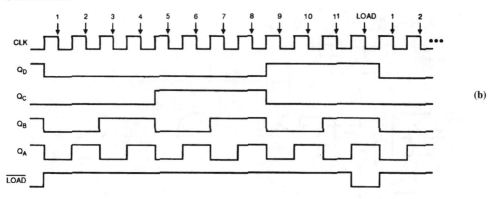

(b)

Isolated two-wire serial interface with a counter

Fig. 14-43 The circuit in the illustration is similar to that of Fig. 14-42, except that the counter keeps track of the number of clock cycles and, when all data bits have been input to the DAC, the external logic creates the $\overline{\text{LOAD}}$ pulse. Figure 14-43B shows the timing diagram. ANALOG DEVICES, APPLICATIONS REFERENCE MANUAL, 1993, P. 8-89.

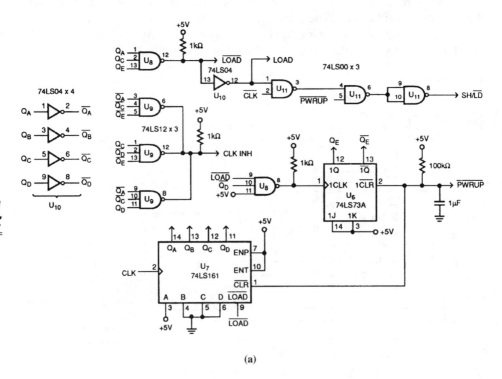

(a)

Special applications for digital/analog converters

Continued

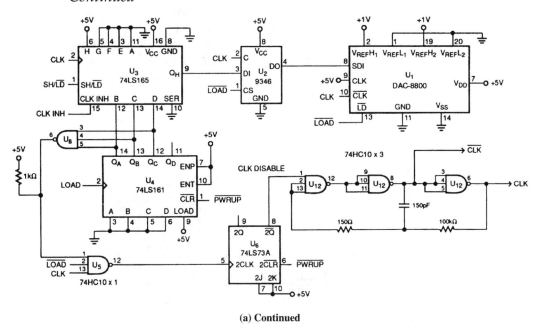

(a) Continued

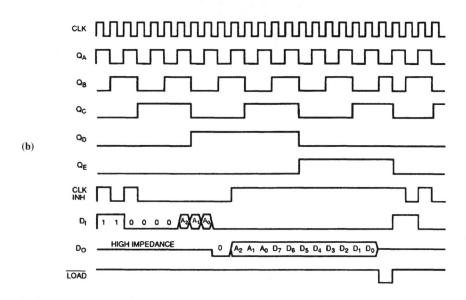

(b)

Stand-alone operation (nonmicroprocessor)

Fig. 14-44 The circuit in the illustration automatically loads the DAC on power-up. The DAC-8800 must have all eight data words loaded to set the proper dc output voltages. (This requirement is not a problem where a microprocessor is used, but it can be with stand-alone operation.) The core of the circuit is the serial input/output *EEPROM* (electrically erasable programmable read-only memory) device U2, preprogrammed with the appropriate data for the DAC. The timing diagram of Fig. 14-44B shows the loading of one address. When the system is powered up, R1 and C1 create a $\overline{PWRUP}$ pulse to asynchronously clear all of the counters and the DAC. After the $\overline{PWRUP}$ pulse goes high, the free-running clock begins to load all eight addresses. After the eighth address is loaded, the clock is disabled to remove any digital switching noise in the analog circuits. ANALOG DEVICES, APPLICATIONS REFERENCE MANUAL, 1993, PP. 8-91, 92.

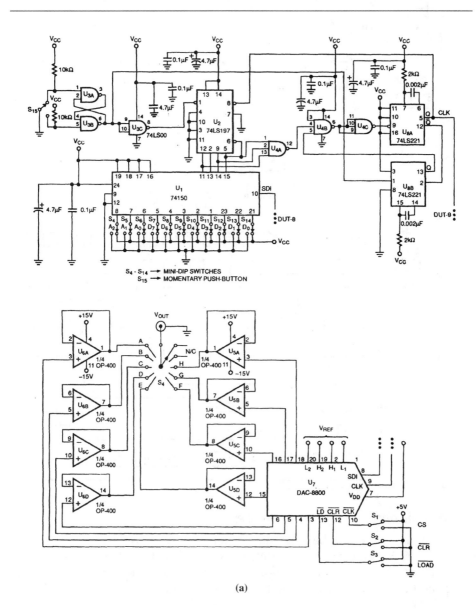

(a)

Special applications for digital/analog converters

Continued

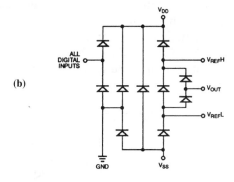

(b)

DAC-8800 test fixture

Fig. 14-45 The circuit in the illustration shows a test fixture for the DAC-8800. Figure 14-45B shows diodes within the IC designed for *ESD* (electrostatic discharge) and latch-up protection. The test circuit includes switches to set the address and data so each DAC can be loaded with any digital word. After the switches are set, pressing the push-button activates the monostable multivibrator, which generates the clock signal to load the DAC register. The counter selects the successive MUX channels, which switch the bits in proper loading order. After all eleven bits are loaded, the $\overline{\text{LOAD}}$ switch is manually toggled to generate a $\overline{\text{LOAD}}$ pulse. The $\overline{\text{CLR}}$ switch should be high at all times, except to clear all eight DACs, in which case $\overline{\text{CLR}}$ needs to be switched low and then back to high. The CS switch should always be set low to keep the DAC selected at all times. The DAC-8800 outputs are buffered by OP-400 op amps. However, the outputs can be configured many different ways for the actual tests required. ANALOG DEVICES, APPLICATIONS REFERENCE MANUAL, 1993, PP. 8-93, 94.

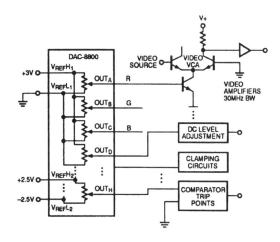

DAC special-application circuit titles and descriptions

Trimming/adjustment operations

Fig. 14-46 The circuit in the illustration suggests various basic trimming and adjustment applications for which the DAC-8800 can be used. For example, a comparator trip point can be digitally altered for different signal conditions. The gain of a VCA can be controlled by altering the collector current through the differential pair, thus changing transconductance and gain. ANALOG DEVICES, APPLICATIONS REFERENCE MANUAL, 1993, P. 8-94.

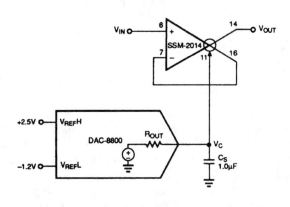

$$20 \text{ LOG} \left(\frac{V_{OUT}}{V_{IN}} \right) = \left(\frac{-1dB}{30mV} \right) V_C$$

$$\text{WHERE } V_C = \frac{D}{256} (V_{REFH} - V_{REFL}) + V_{REFL} = \frac{D}{256} (3.7V) - 1.2V$$

Digitally controlled VCA

Fig. 14-47 The circuit in the illustration shows a DAC-8800 used as a digital gain control for a VCA (the audio-range SSM-2014 in this circuit). The SSM2014 has more than 100 dB of dynamic range, and the gain is logarithmically proportional to the control voltage (V_C). Using the reference voltages shown provides a gain range of –80 dB to +40 dB at frequencies well above 20 kHz. The circuit has a typical control feedthrough of 1.3 mV/V at 100 Hz, so capacitor C_S is used to slow down the DAC transitions. Using 1.0 μF for C_S produces a pole at 13 Hz to filter out any glitch or high-frequency noise. ANALOG DEVICES, APPLICATIONS REFERENCE MANUAL, 1993, P. 8-94.

14

658

Special applications for digital/analog converters

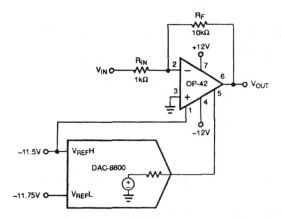

Trimming op-amp offset voltage (null pins)

Fig. 14-48 The circuit in the illustration shows a DAC-8800 used to trim the offset voltage of an op amp in digital steps. Note that the op amp is trimmed (through the null pins, 1 and 5) to the negative supply, over a ±40-mV range and that the negative supply cannot exceed −12 V. ANALOG DEVICES, APPLICATIONS REFERENCE MANUAL, 1993, P. 8-95.

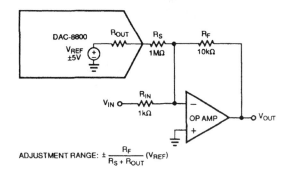

Trimming op-amp offset voltage (summing node)

Fig. 14-49 The circuit in the illustration shows a DAC-8800 used to trim the offset voltage of an op amp in digital steps. The circuit in the illustration is similar to that of Fig. 14-48, except that the circuit is trimmed to the summing node of the op amp, thus eliminating the voltage limitation. The adjustment range for the Fig. 14-49 circuit is ±50 mV. ANALOG DEVICES, APPLICATIONS REFERENCE MANUAL, 1993, P. 8-95.

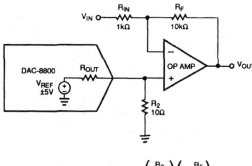

ADJUSTMENT RANGE: $\pm \left(\dfrac{R_2}{R_{OUT}}\right)\left(1+\dfrac{R_F}{R_{IN}}\right) V_{REF}$

Trimming op-amp offset voltage (noninverting node)

Fig. 14-50 The circuit in the illustration shows a DAC-8800 used to trim the offset voltage of an op amp in digital steps. The circuit in the illustration is similar to that of Fig. 14-48, except that the circuit is trimmed to the noninverting node of the op amp, thus eliminating the voltage limitation. The adjustment range for the Fig. 14-50 circuit is ±42 mV. Resistor R2 is added to reduce any noise in the DAC-8800 to a point where the noise is insignificant compared to the op-amp noise. With an R_2 of 10 Ω, the DAC-8800 noise is 15 pV/√Hz. ANALOG DEVICES, APPLICATIONS REFERENCE MANUAL, 1993, P. 8-95.

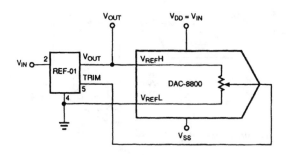

V_{OUT} ADJUSTMENT RANGE: ±300mV

Trimming voltage references

Fig. 14-51 The circuit in the illustration shows a DAC-8800 used to trim a voltage reference (an RF-01 in this circuit) in digital steps. The circuit can be used with virtually any voltage reference, provided that there is at least 4-V of headroom between V_{DD} and V_{OUT}. ANALOG DEVICES, APPLICATIONS REFERENCE MANUAL, 1993, P. 8-96.

Special applications for digital/analog converters

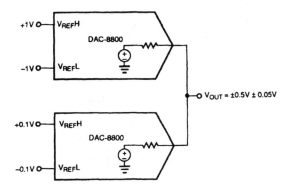

Coarse-fine control (separate reference)

Fig. 14-52 The circuit in the illustration shows two DAC-8800s connected in parallel to provide coarse-fine control by averaging the outputs in digital steps. The circuit in the illustration is limited by the voltage difference between references (typically to less than 1 V), and the output can never go above the high reference. ANALOG DEVICES, APPLICATIONS REFERENCE MANUAL, 1993, P. 8-96.

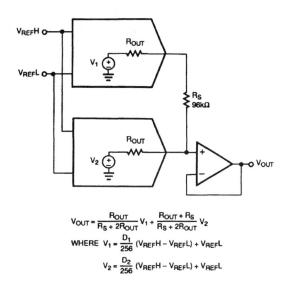

$$V_{OUT} = \frac{R_{OUT}}{R_S + 2R_{OUT}} V_1 + \frac{R_{OUT} + R_S}{R_S + 2R_{OUT}} V_2$$

$$\text{WHERE } V_1 = \frac{D_1}{256} (V_{REFH} - V_{REFL}) + V_{REFL}$$

$$V_2 = \frac{D_2}{256} (V_{REFH} - V_{REFL}) + V_{REFL}$$

Coarse-fine control (same reference)

Fig. 14-53 The circuit in the illustration is similar to that of Fig. 14-52, except that the same reference voltage is used for both DACs, making the maximum output voltage equal to the DAC reference, as shown by the equations. ANALOG DEVICES, APPLICATIONS REFERENCE MANUAL, 1993, P. 8-97.

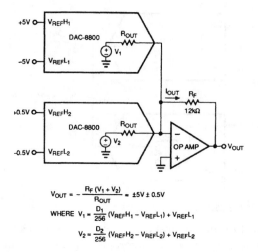

$$V_{OUT} = -\frac{R_F\,(V_1 + V_2)}{R_{OUT}} = \pm 5V \pm 0.5V$$

$$\text{WHERE } V_1 = \frac{D_1}{256}\,(V_{REF}H_1 - V_{REF}L_1) + V_{REF}L_1$$

$$V_2 = \frac{D_2}{256}\,(V_{REF}H_2 - V_{REF}L_2) + V_{REF}L_2$$

Coarse-fine control (current summing)

Fig. 14-54 The circuit in the illustration is similar to that of Fig. 14-52, except that the DAC outputs are applied to the summing node of an op amp (the virtual-ground point). This arrangement makes it possible to operate the circuit with differences in reference voltage well above 1.6 V. (In the circuit of Fig. 14-52, if the difference in reference voltages is greater than 1.6 V, it is possible that the switches in the DACs will turn on, without regard to the actual digital code.) ANALOG DEVICES, APPLICATIONS REFERENCE MANUAL, 1993, P. 8-97.

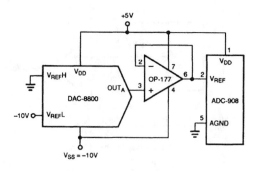

Adjustable reference for A/D converters

Fig. 14-55 The circuit in the illustration shows a DAC-8800 and an OP-177 buffer connected to provide a digitally adjustable reference for an ADC-908. The buffer is required because of the typical ADC low-reference-input impedance. ANALOG DEVICES, APPLICATIONS REFERENCE MANUAL, 1993, P. 8-97.

Special applications for digital/analog converters

=15=

3- and 3.3-V circuits (portable computer)

This chapter is devoted to circuits that operate with (or produce) 3-V or 3.3-V power. Such circuits are especially useful for portable computers (laptops, palmtops, notebooks, etc.). Because most of the circuits are power supplies, all the test/troubleshooting information of Chapters 2, 3, and 4 apply here. The interfaces (Fig. 15-15) and transceivers (Fig. 15-14) also include amplifiers, so the test/troubleshooting of Chapter 1 apply.

3-V and 3.3-V circuit titles and descriptions

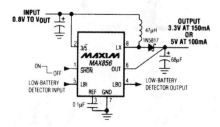

3.3 V from two cells

Fig. 15-1 The circuit in the illustration shows a MAX856 connected to provide 3.3 V at 150 mA from a two-cell input. MAXIM ANALOG DESIGN SOLUTIONS, 4TH EDITION.

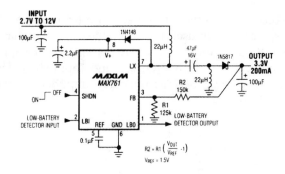

3.3 V from three or four cells

Fig. 15-2 The circuit in the illustration shows a MAX761 connected to provide 3.3 V at 200 mA from a three- to four-cell input. MAXIM ANALOG DESIGN SOLUTIONS, 4TH EDITION.

3- and 3.3-V circuits (portable computer)

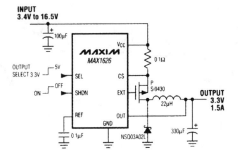

3.3 V from four or eight cells

Fig. 15-3 The circuit in the illustration shows a MAX761 connected to provide 3.3 V at 1.5 A from a four- to eight-cell input. MAXIM ANALOG DESIGN SOLUTIONS, 4TH EDITION.

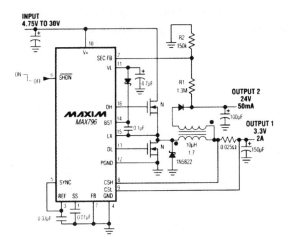

3.3 V and 24 V from five cells

Fig. 15-4 The circuit in the illustration shows a MAX796 connected to provide 3.3 V at 2 A, and 24 V at 50 mA from a five-cell input. MAXIM ANALOG DESIGN SOLUTIONS, 4TH EDITION.

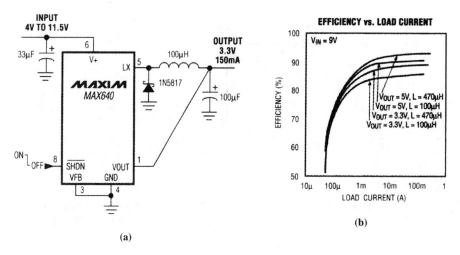

(a)

(b)

3.3 V from 4 V to 11.5 V

Fig. 15-5 The circuit in the illustration shows a MAX640 connected to provide 3.3 V at 150 mA, using only 10-μA supply current. Figure 15-5B shows the characteristics. MAXIM ANALOG DESIGN GUIDE, 6TH EDITION.

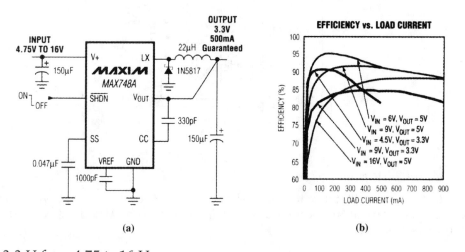

(a)

(b)

3.3 V from 4.75 to 16 V

Fig. 15-6 The circuit in the illustration shows a MAX748A connected to provide 3.3 V at 500 mA guaranteed (750 mA possible), using PWM with no subharmonic switching noise. Figure 15-6B shows the characteristics. MAXIM ANALOG DESIGN GUIDE, 6TH EDITION.

3- and 3.3-V circuits (portable computer)

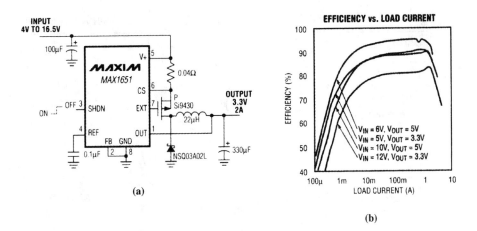

(a)

(b)

3.3 V from 4 V to 15.5 V

Fig. 15-7 The circuit in the illustration shows a MAX1651 connected to provide 3.3 V at 2 A, using only 100-µA supply current. Figure 15-7B shows the characteristics. MAXIM ANALOG DESIGN GUIDE, 6TH EDITION.

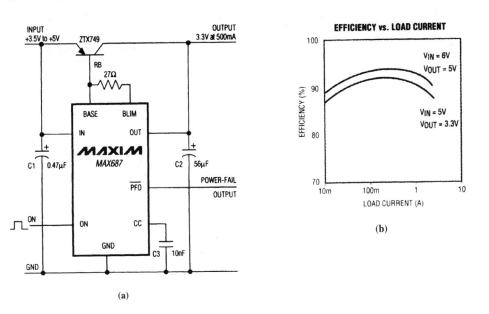

(a)

(b)

3.3 V from 4.5 V to 15 V

Fig. 15-8 The circuit in the illustration shows a MAX747 connected to provide 3.3 V at 2.5 A, using only 860-µA supply current. Figure 15-8B shows the characteristics. MAXIM ANALOG DESIGN GUIDE, 6TH EDITION.

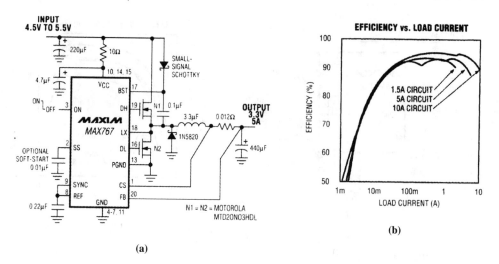

(a)

3.3 V from 4.5 V to 5.5 V

Fig. 15-9 The circuit in the illustration shows a MAX767 connected to provide 3.3 V at 5 A, using only 700 μA supply current. Figure 15-9B shows the characteristics. Maxim Analog Design Guide, 6th Edition.

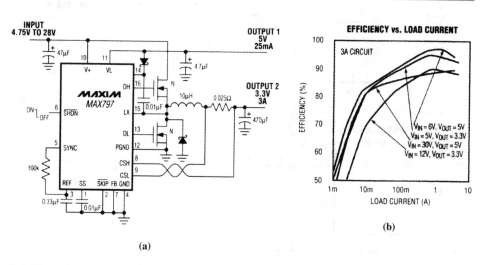

(a)

3.3 V and 5 V from 4.75 V to 28 V

Fig. 15-10 The circuit in the illustration shows a MAX797 connected to provide 3.3 V at 3 A and 5 V at 25 mA, using 375 μA supply current. Figure 15-10B shows the characteristics. Maxim Analog Design Guide, 6th Edition.

3- and 3.3-V circuits (portable computer)

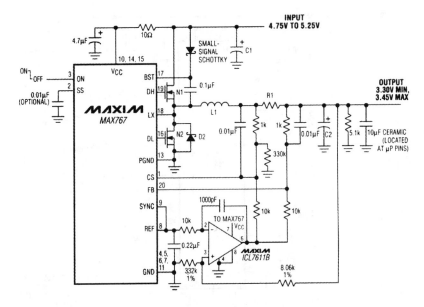

3.30 V to 3.45 V from 4.75 V to 5.25 V

Fig. 15-11 The circuit in the illustration shows a MAX767 connected to provide 3.3 V ±2.5% at 5 A or 10 A. MAXIM ANALOG DESIGN GUIDE, 6TH EDITION.

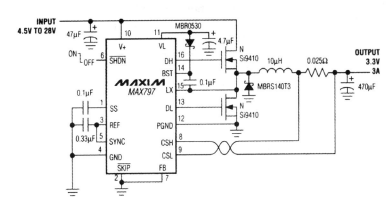

3.3 V from 4.5 V to 28 V

Fig. 15-12 The circuit in the illustration shows a MAX797 connected to provide 3.3 V at 3 A. MAXIM ANALOG DESIGN GUIDE, 6TH EDITION.

15

3-V and 3.3-V circuit titles and descriptions

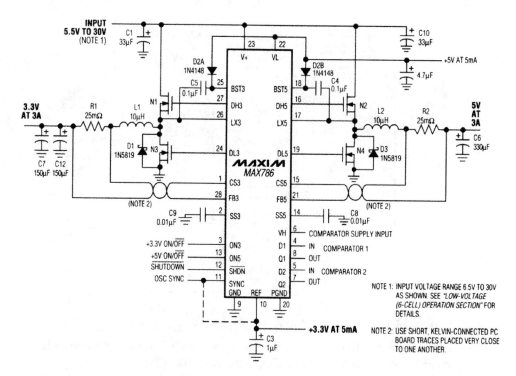

3.3 V and 5 V from 5.5 V to 30 V

Fig. 15-13 The circuit in the illustration shows a MAX786 connected to provide 3.3 V at 3 A and 5 V at 3 A. MAXIM ANALOG DESIGN GUIDE, 6TH EDITION.

3- and 3.3-V circuits (portable computer)

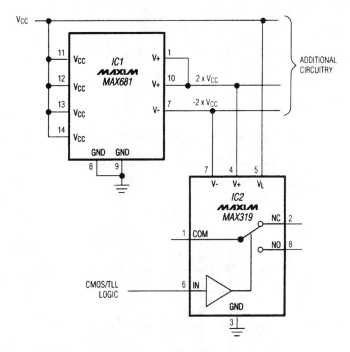

Operating CMOS/TTL switches on a 3-V board

Fig. 15-14 The circuit in the illustration shows a MAX631 charge-pump converter (IC1) providing a $+2 \times V_{CC}$ and $-2 \times V_{CC}$ to power a MAX319 CMOS/TTL analog switch (IC2). This makes it possible to use fast-switching CMOS/TTL switches (with low ON-resistance) on a board designed for 3-V operation only. For example, a V_{CC} of 3 V produces ±6-V rails for IC2, resulting in on-resistance of less than 30 Ω, switching times below 200 ns, leakage of less than 0.1 nA, and an N_{CC} current of 0.5 mA. MAXIM ENGINEERING JOURNAL, VOLUME NINETEEN, P. 14.

15

671

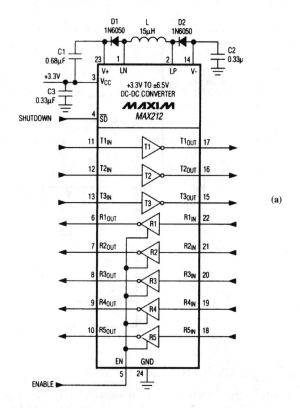

(a)

MANUFACTURER	PART NUMBER	PHONE NUMBER	FAX NUMBER
Allegro	TMPD6050LT	USA (508) 853-5000	USA (508) 853-5049
Motorola	MMBD6050LT1	USA (408) 749-0510	USA (408) 991-7420
Murata	LQH4N150K-TA	USA (404) 831-9172 Japan (075) 951-9111	USA (404) 436-3030 Japan (075) 955-6526
Philips	PMBD6050	USA (401) 762-3800	USA (401) 767-4493
Sumida	CD43150	USA (708) 956-0666 Japan (03) 3607-5111	USA (708) 956-0702 Japan (03) 3607-5428
TDK	NLC453232T-150K	USA (708) 803-6100 Japan (03) 3278-5111	USA (708) 803-6296 Japan (03) 3278-5358

(b)

15

True RS-232 transceiver with 3-V power

Fig. 15-15 The circuit in the illustration shows a MAX212 transceiver intended for 3-V to 3.6-V operation, with EIA/TIA-232E, -562, and V.28N.24 communication interfaces, where three drivers and five receivers are needed. Shutdown current is 1 µA, with 3 mA maximum (unloaded) supply current. The guaranteed data rate is 120 kbps, making the IC LapLink™ compatible. Figure 15-15B shows suggested component suppliers. MAXIM NEW RELEASES DATA BOOK, 1995, P. 2-61.

3- and 3.3-V circuits (portable computer)

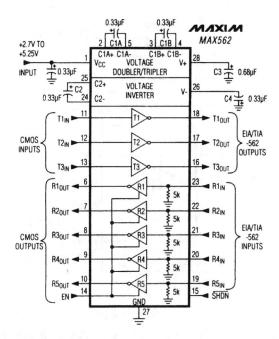

Serial interface for notebook computers

Fig. 15-16   The circuit in the illustration shows a MAX562 serial interface intended for 2.7-V to 5.25-V operation, with EIA/TIA-562 compatibility. This guarantees compatibility with RS-232 interfaces. The data rate is 230 kbps, and the IC is LapLink™ compatible. The slew rate is 4 V/µs. MAXIM NEW RELEASES DATA BOOK, 1995, P. 2-175.

3-V and 3.3-V circuit titles and descriptions

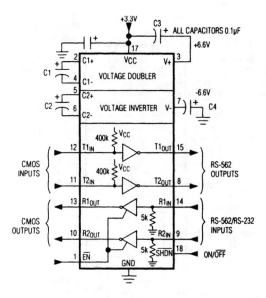

Dual EIA/TIA-562 transceiver with 3.3-V power

Fig. 15-17 The circuit in the illustration shows a MAX563 transceiver intended for 3-V to 3.6-V operation, with EIA/TIA-562 interfaces. The IC communicates with RS-232 transceivers yet consumes far less power (10 μA maximum in shutdown). The guaranteed data rate is 116 kbps while maintaining ±3.7-V signal levels, making the IC LapLink™ compatible. Maxim New Releases Data Book, 1995, P. 2-183.

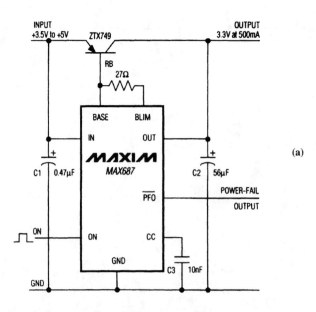

(a)

3- and 3.3-V circuits (portable computer)

Continued

TOP VIEW

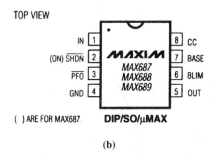

( ) ARE FOR MAX687

(b)

•Fixed Outputs:
  3.3V (MAX687/MAX688)
  3.0V (MAX689)

•Directly drives external PNP transistor

•10mA min base-current drive for >1A output

•Low dropout voltage:
  <200mV dropout at 500mA output (ZTX749)
  <200mV dropout at 200mA output (2n2907A)
  <100mV dropout at 150mA output (ZTX749)

•±2% accurate power-fail monitior

•Automatic, latched shutdown when output falls
  out of regulation (MAX687)

•Precision threshold shutdown control
  (MAX688/MAX689)

•Low supply current:
  <250μA operating
  <1μA shutdown

•2.7 to 11.0V supply range

•8-Pin DIP/SO/μMAX packages

(c)

Linear regulator for 3-V and 3.3-V circuits

Fig. 15-18 The circuit in the illustration shows a MAX687 linear regulator connected to provide 3.3 V at 500 mA. Figures 15-18B and C show the pin configurations, and features, respectively, for the IC. MAXIM NEW RELEASES DATA BOOK, 1995, P. 4-83.

15

675

3-V and 3.3-V circuit titles and descriptions

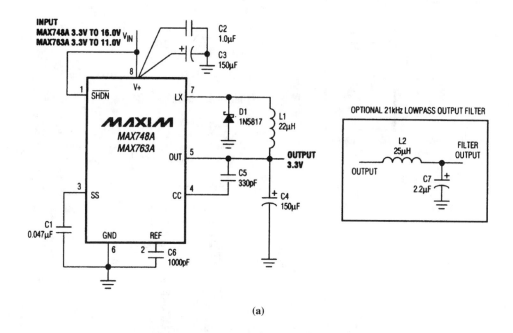

(a)

	C1(μF)	C2(μF)	C3(μF)	C4(μF)	C5(pF)	C6(pF)	L1(μH)
Through-Hole	0.047	1.0	150*	220*	330	1000	22
SO	0.047	1.0	68**	100***	330	1000	22

(b)

* Sanyo OS-CON Series (very low ESR)

** 16V or greater maximum voltage rating.

*** 6.3V or greater maximum voltage rating.

Circuit Cond.		Soft-Start Time (ms) vs. C1 (μF)			
V+ (V)	I_OUT (mA)	C1 = 0.01	C1 = 0.047	C1 = 0.1	C1 = 0.47
8	0	1	4	7	12
12*	0	1	2	3	6
8	200	10	33	50	200
12*	200	7	17	20	80
8	300	13	44	65	325
12*	300	8	25	35	140

(c)

* MAX748A only

3- and 3.3-V circuits (portable computer)

Continued

Production Method	Inductors	Capacitors
Surface Mount	Sumida CD105 series Coiltronics CTX series Coilcraft DT series	Matsuo 267 series Sprague 595D/293D series
High Performance/ Miniature Through-Hole	Sumida RCH895 series	Sanyo OS-CON series (very low ESR)
Through-Hole	Renco RL1284 series	Nichicon PL series (low ESR)

Phone and FAX Numbers:

Coilcraft	USA:	(708) 639-6400, FAX: (708) 639-1469	Renco	USA:	(516) 586-5566, FAX: (516) 586-5562
Coiltronics	USA:	(305) 781-8900, FAX: (305) 782-4163	Sanyo	USA:	(0720) 70-1005, FAX: (0720) 70-1174
Matsuo	USA:	(714) 969-2491, FAX: (714) 960-6492	Sprague Elec. Co.	USA:	(603) 224-1961, FAX: (603) 224-1430
	Japan:	(06) 332-0871	Sumida	USA:	(708) 956-0666, FAX: (708) 956-0702
Nichicon	USA:	(708) 843-7500, FAX: (708) 843-2798			
	Japan:	(03) 3607-5111, FAX: (03) 3607-5428			

(d)

Current-mode PWM (pulse-width modulation) converter (commercial temperature range)

Fig. 15-19 The circuit in the illustration shows a MAX748A or MAX763A connected to provide a 3.3-V output, using through-hole components, over commercial temperature ranges. Figure 15-19B shows component values for wide-temperature applications. The MAX748A delivers a guaranteed 300 mA for input voltages of 4 V to 16 V and a guaranteed 500 mA for inputs of 4.75 V to 16 V, with 800 mA typical output currents. The MAX763A delivers a guaranteed 300 mA for inputs of 4 V to 11 V, a guaranteed 500 mA for inputs of 4.75 V to 11 V, and has 700-mA typical output currents. Both ICs operate with an input down to 3 V. Figures 15-19C and D show the soft-start times and component suppliers, respectively. MAXIM NEW RELEASES DATA BOOK, 1995, PP. 4-93, 94, 95.

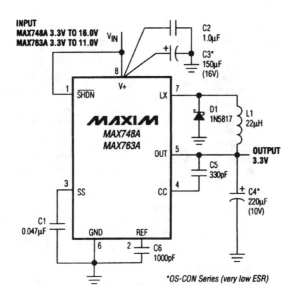

Current-mode PWM converter (all temperature ranges)

Fig. 15-20 The circuit in the illustration is similar to that of Fig. 15-19, except with components selected for all temperature ranges (see Fig. 15-19B). MAXIM NEW RELEASES DATA BOOK, 1995, P. 4-96.

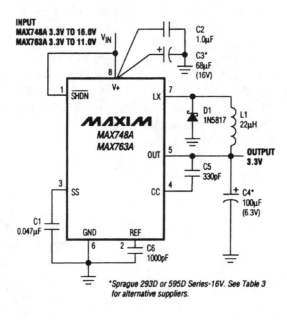

3- and 3.3-V circuits (portable computer)

Current-mode PWM converter (surface mount)

Fig. 15-21   The circuit in the illustration is similar to that of Fig. 15-19, except that surface-mount components (Fig. 15-19D) are used. The components are selected for commercial and extended-industrial temperature ranges (see Fig. 15-19B). MAXIM NEW RELEASES DATA BOOK, 1995, P. 4-96.

3-V and 3.3-V circuit titles and descriptions

Index

analog-to-digital converters, *continued*
  three-wire interface and timer, III-250 ksps, III-480
  three-wire parallel interface, III-250 ksps, III-479-480
  tracking A/D converter with microprocessor control, III-640
  unipolar operation, III-474-475
  unipolar operation, gain adjustment, III-475
  video converter, III-485-486, III-486-487, III-488-490, III-491-494
  voltage/frequency (V/F) circuits, III-399
  Z80 interface, II-352, II-353
AND gate, I-331, I-560
anemometer, thermal, II-636
antennas, VHF booster, I-108
AppleTalk interface transceiver, III-364
astable multivibrator, II-278, III-315, III-316
attenuation
  gain/attenuation amp, digitally programmable, III-66-68
audio amplifiers, I-23-71, II-538
  boosted-gain, I-57-58
  bridge, I-55, I-56, I-57
  compressor/expander, I-24
  constant-voltage crossover, I-62-63
  discrete-component, I-37, I-39,-41, I-41-43, I-43-45
  discrete output/IC input (12 W), I-46
  for piezoelectric transducers, I-48
  intercom, I-57
  line-operated IC, I-62
  low noise/drift dc, I-28, I-29
  low-cost high-level preamp and tone control, I-50-51
  magnetic cartridge preamp, I-26
  microphone preamp, I-24, I-26-27, I-63, I-64, I-69
  minimum-component, I-70
  mixer, I-53
  mixer/selector, I-71
  NAB tape preamp with fast turn-on, I-52
  noise reduction system, I-67
  phono, I-49, I-53, I-54
  power, I-59, I-60, I-61
  power output/gain, I-8-9
  ratiometric bridge with A/D conversion, I-32

  RIAA preamp, I-23, I-25, I-55, I-69
  short-circuit protection, I-32
  single-chip, II-44, II-45, II-46, II-47, II-48, II-49
  strain-gauge, I-30, I-31
  switchable, I-27
  tape playback, I-51
  tape-head preamp, I-67, I-68
  tone control, I-25, I-47, I-48
  ultra-high-gain, I-50
  ultrasonic receiver, I-31
  volume control, I-533
audio circuits (*see also* RF circuits), I-1-71, III-599-626
  amplifiers (*see* audio amplifiers)
  bootstrap circuit to lower distortion in op amps, III-625
  compressor/limiter for wireless audio, III-619-620
  compressor/limiter, two band, III-616-617
  dc, I-23-71
  dc-to-dc converter for wireless audio systems, III-622
  expander for wireless audio systems, III-621
  microphone mixer, III-623-624
  mute circuit, balanced 600-601
  mute circuit, unbalanced, III-611
  noise gate, two channel, III-612-613
  noise-reduction system, dynamic filter, two channel, III-609-610
  pan control circuit, microphone mixing, III-601-602
  preamp, RIAA/IEC MC and MM phono, III-603-606
  preamp, RIAA/IEC MM phono, III-606-607
  preamp, RIAA/IEC MM phono, III-608-609
  sum and difference amplifier, audio matrix, III-614-615
  testing and troubleshooting, I-16-22
  VCA level control, two channel, III-618-619
audio oscillator
  variable, III-313
audio voltmeters, I-2-5
  characteristics, I-2, I-4
  measurements/signal tracing, I-4-5
automatic gain control (AGC)
  amplifier, III-86
  amplifier with adjustable attack/release, III-62-64

B
background noise, II-12-14
  IC amplifier circuits, III-19-20, III-20
backup power source from capacitor, III-356, III-357
band-elimination filters, III-541
band-rejection filters, III-541
band-stop filters, III-541
band-suppression filters, III-541
bandpass amplifier, WTA, III-52
bandpass filters, III-541
  Chebyshev, fourth order, III-549, III-550
  high Q, III-545
bandwidth, II-6
  IC amplifier circuits, III-11, III-12
bar-graph display, II-187, II-188
bar-graph driver, LED, II-677
battery-powered circuits, II-437-471, III-245-296
  +5 V from 3 V with no inductors, I-258
  +/-5 V from 9-V battery, III-269
  3.3-V supply for notebook/laptop computers, III-288
  5 V at 100 mA from 3-V lithium battery, III-249
  5 V from 3-or 9-V battery, III-273
  5 V from 3-V lithium battery, III-272
  5 V power-distribution system, logic controlled, III-254-255
  5 V supply for notebook/laptop computers, III-287
  3-state indicator, I-271
  backup capacitor source, III-356, 333-357
  backup monitor, III-349
  backup switch, II-474
  backup regulator, II-453
  cascading for increased output voltage, III-271
  charger, I-257, I-457, I-466, I-527, II-407
    constant-current, III-277
    dual rate, III-278
    lead-acid, III-281
    NiCad, III-280
    NiCad and NiMH, step-down regulator, III-276
    programmable, III-279
    switch mode, III-275
  converter
    1.5 to 5 V, III-284, III-285
    2.5 W 3 V to 5 V, III-260
    3 to 5 V, low-battery frequency shift, III-260

IC amplifiers, *continued*
  notch filter, III-90
  sample and hold, III-91
  short-circuit protected, III-100
  simulated inductor, III-88
  single-supply, III-93
  tuned circuits, III-89
  video, III-99
cascade, II-537
cascode, I-159
chopper-stabilized op amps, III-37-39
clipping amplifier, III-87
CMOS-inverter output stage, II-30
common-emitter output stage, II-31
composite, II-37
current monitor, III-81
current pump, III-71
current source, bilateral, III-71
current-feedback amplifiers (CFA), III-32-37
dc-stabilized, II-40, II-41, II-42
decibel addition, III-4
decibel measurement basics, III-2-4
decibels and reference levels, III-5
decibels to compare voltage and current, III-4-5
difference, II-511
distortion, III-44
driver, large-capacitive loads, III-99
dynamic output impedance or resistance, III-12-14
equipment for testing amplifiers, III-1-2
error band, III-21
feedback measurement, III-25, III-26
FET, II-39
frequency response, III-5-10
fundamental-suppression principle, III-16
gain-control, II-546
gain
  closed-loop gain, III-6, III-7
  low-gain, III-43-44
  measurement in discrete stages, III-42-43
  open loop gain, III-5, III-6, III-10
  power output and power gain, III-11
  programmable-gain amplifier, III-71
  voltage gain, III-10

gain/attenuation amp, digitally programmable, III-66-68
harmonic distortion, III-16, III-17, III-18
headphone amplifier, III-77
high-current, II-32
high-level amplifier, balanced-input, III-57-58
highpass filter, buffer amplifier, III-90
IC, II-28
impedance, II-461
impedance transforming amplifier, III-85
impedance, III-12-13, III-13-14
increasing output current, II-165, II-166
input-bias current, III-25-26, III-27
input-offset voltage and current, III-27, III-28
instrumentation, II-25, II-167, II-168, II-458, II-470, II-527, II-533, II-652, III-84, III-96
intermodulation distortion, III-18-19
inverting, I-507, II-510, II-511, II-512, II-513, II-526, II-551
isolation, II-653, III-74
leakage, III-44, III-45
light detector, op amp, III-73
linear, I-138, I-141, I-142, II-144, II-151, III-79-80, III-80-81
lock-in, II-26
log (*see* log amplifiers)
logarithmic/limiting, I-162, II-79
lowpass filter, buffer amp, III-89
meter, II-456, II-462
microphone, II-457, III-59-61
MOS, II-134, II-136, II-138, II-140
multiplexer/amplifier, III-102
  large capacitive loads, III-103
  minimal phase distortion, III-103
  unity gain, III-104
multiplier/divider, III-82
narrowband, II-115
noninverting, I-508, II-510, II-512, II-513, II-525
Norton (*see* Norton amplifiers)
notch filter, III-75, III-90
operational (*see* operational amplifiers)
operational transconductance amplifiers (OTA), III-31-32
output voltage vs. frequency, III-9
overshoot, III-21, III-23-24

phase shift, III-24-25
phase/frequency plots, III-6-7, III-8
parallel, II-43
power operational amplifier (*see* power op amp)
power ratios, doubling, III-4
power supply sensitivity (PSS), III-28, III-29
photodiode, I-496
preamps (*see* preamplifiers)
pulse, I-480
push-pull, II-127, II-129
reference levels and decibels, III-5
resistance, III-12-14
RF/IF (*see* RF/IF amplifiers)
rise time, III-21, III-22, III-23-24
rolloff point, III-5
sensitivity, input, III-11
sensitivity, load, III-11-12
sensor, I-603
servo, II-538, II-539
servomotor amplifier, III-69-70
settling time, III-21, III-23-24
signal to noise ratio (S/N), III-16, III-17, III-20-21
signal tracing, III-41-42
sine-wave analysis, III-14-15
slew rate, III-21, III-22
square-wave analysis, III-15-16
squelch amplifier, III-87-88
SSB driver, II-147, II-149
summing amplifier, virtual-ground, unbalanced, III-58-59
switch, stereo routing switcher, III-65
testing, II-1-14, II-1-29
thermocouple, II-524, III-83
threshold detector/amplifier, op-amp, III-72-73
total harmonic distortion (THD), III-16, III-17
transducer amplifier, piezoelectric, III-76
transient response, III-21, III-22
troubleshooting, II-14-22, III-45-48
UHF, I-139-141
ultra-fast current-boosted, II-29
variable-gain, II-164
video (*see* video amplifiers)
voltage-controlled amplifier (VCA), I-532, III-94-95
weigh-scale amplifier, III-97-98
wideband (*see* wideband amplifiers)
wideband transconductance amplifiers (WTA), III-39-41, III-48-50

About the author

John D. Lenk is a technical author specializing in practical electronic service and troubleshooting guides for more than 40 years. A long-time writer of international bestsellers in the electronics field, he is the author of more than 87 books on electronics, which together have sold well over 2 million copies in nine languages.

Mr. Lenk's guides regularly become classics in their fields, such as *Lenk's Video Handbook, Lenk's Audio Handbook, Lenk's Laser Handbook, Lenk's RF Handbook, Lenk's Digital Handbook, Lenk's Television Handbook, McGraw-Hill Electronic Testing Handbook, McGraw-Hill Electronic Troubleshooting Handbook, Simplified Design of Linear Power Supplies, Simplified Design of Switching Power Supplies, Simplified Design of Micropower/Battery Circuits, Simplified Design of IC Amplifiers,* and *McGraw-Hill Circuit Encyclopedia & Troubleshooting Guide, Volumes 1 and 2.*